PERSPECTIVES IN BIOMEDICAL ENGINEERING

PERSPECTIVES IN BIOMEDICAL ENGINEERING

Proceedings of a Symposium organised in association with the
Biological Engineering Society and held in the University
of Strathclyde, Glasgow, June 1972

Edited by

R. M. KENEDI

Bioengineering Unit, Wolfson Centre, University of Strathclyde

PALGRAVE MACMILLAN

ISBN 978-1-349-01606-8   ISBN 978-1-349-01604-4 (eBook)
DOI 10.1007/978-1-349-01604-4

First published 1973

THE MACMILLAN PRESS LTD

London and Basingstoke
Associated companies in New York, Melbourne,
Dublin, Johannesburg and Madras

SBN 333 13843 0

PREFACE

Work in Bioengineering commenced at the University of Strathclyde in the Department of Mechanical Engineering some seventeen years ago in close cooperation with surgeons and physicians in Glasgow's teaching hospitals. Multidisciplinary medical/engineering project teams drawn from several of the relevant specialties worked in the University engineering laboratories, the hospital wards, the operating theatres and the medical research departments. The activity received organisational recognition in 1962 by the formation of a Medical Research Council Biomechanics Group. In the past ten years this has grown into an independent research and post-graduate teaching department—the Bioengineering Unit. The Unit is now in its new building, the Wolfson Centre which was funded by the Wolfson Trust and formally opened by Lady Wolfson on June 21st, 1972.

To commemorate the occasion a Symposium was organised in association with the Biological Engineering Society during which recent advances and perspectives of developments world wide were discussed in the main areas of the Unit's activities. Participation was gratifying: 256 attended, consisting of representatives from Belgium, Canada, Denmark, East Germany, Eire, France, Israel, Italy, Netherlands, Norway, Sweden, Switzerland, Turkey, West Germany, the United Kingdom, the United States and Yugoslavia.

The details of the Scientific and Technical Sessions are given in the contents list. There was also an Exhibition on Medical Engineering in which 20 firms and organisations participated (see list on page xvii). A visit to the Wolfson Centre was part of the programme during which a variety of displays showing research projects in progress in the Bioengineering Unit were on show.

All the papers, presented together with additional contributions and the discussions, are printed complete in this volume grouped in the four sections shown in the list of contents.

The organising committee acknowledge with gratitude their sense of indebtedness to the Western Regional Hospital Board for their support of the Symposium and for the most enjoyable reception given by them, in association with the University of Strathclyde, for participants and guests. The committee also wish to record their grateful thanks to the Biological Engineering Society of the United Kingdom for their support and cooperation in assembling the programme and publicising the Symposium.

All the work connected with the Symposium as regards advance information, registration, accommodation arrangements, correspondence and the day-to-day running of the meetings was carried out by the staff of the Bioengineering Unit with the unstinted cooperation of the University Administration.

By courtesy of the Bank of Scotland and Messrs Arbuckle Smith & Co., banking and travel agent facilities were made available to participants at the Symposium.

The organising committee gratefully acknowledges the cooperation and whole-hearted assistance of all concerned.

R. M. KENEDI
University of Strathclyde, Glasgow, Scotland
August 1972

LIST OF PARTICIPANTS AND AUTHOR INDEX

(*a*, participating author; *a**, non-attending author; *c*, Chairman and Chairman-Associate)

Aas-Aune, Dr G., Central Sykehouset 1, p.t. Nevrologisk Avdeling, 7000 Trondheim, Norway.

a (page 187) **Abrahams, M.,** B.Sc., Ph.D. Research Unit, G.K.N. Group Technological Centre, Birmingham New Road, Wolverhampton WV4 6BW, England.

Agbo, Dr D. C., Orthopaedic Department, Doncaster Royal Infirmary, Doncaster, England.

Andersson, Dr Gunnar, Department of Orthopaedic Surgery 1, University of Göteborg, Sahlgren Hospital, 413 45 Göteborg, Sweden.

Ashley, Dr C., University of Birmingham, P.O. Box 363, Birmingham 15 2TT, England.

a (pages 241, 245) **Bain, W. H.,** M.D., F.R.C.S.(Edin. and Glas.), Cardiovascular Surgical Unit, Royal Infirmary, Glasgow G4 0SF, Scotland.

a (page 165) **Barbenel, J. C.,** B.D.S., M.Sc., L.D.S., R.C.S.(Eng.), University of Strathclyde, Bioengineering Unit, Wolfson Centre, 106 Rottenrow, Glasgow G4 0NW, Scotland.

Barnes, Prof. Roland, C.B.E., B.Sc., M.B., Ch.B., F.R.C.S.(Eng. Ed., Glasg.), 35 Boclair Road, Bearsden, Glasgow G61 2AF, Scotland.

Bazin, Dr Raphael, 909 President Street, Near Prospect Park West, Brooklyn, New York 11215, U.S.A.

a (page 233) **Bekkering, Prof. D. H.,** Institute of Medical Physics TNO, Da Kostakade 45, Utrecht, Netherlands.

Bell, F., B.Sc., Ph.D., University of Strathclyde, Bioengineering Unit, Wolfson Centre, 106 Rottenrow, Glasgow G4 0NW, Scotland.

a (page 281) **Bengi, Prof. Halil,** Chairman, Department of Electrical Engineering, Middle East Technical University, Ankara, Turkey.

Bennett, Mr Leon, 6 Rivercrest Road, New York City, N.Y., U.S.A.

a (page 21) **Black, M. M.,** B.Sc., M.Sc., Ph.D., C.Eng., A.F.R.Ae.S., Biomedical Engineering Research Group, School of Applied Sciences, University of Sussex, Falmer, Brighton BM1 9QT, Sussex, England.

Blietz, Dr Rudolf, 28 Bremen, An der Weide 41, West Germany.

Boenick, Prof. Dr U., Technische Universität Berlin, 1 Berlin 12, Strasse des 17. Juni 135, East Germany.

Boitano, Miss Marilyn, Birmingham Accident Hospital, Bath Row, Birmingham 15, England.

Bolton, M. P., B.Sc., University of Strathclyde, Bioengineering Unit, Wolfson Centre, 106 Rottenrow, Glasgow G4 0NW, Scotland.

a* (page 173) **Bornstein, Mr P.,** Department of Biochemistry, University of Washington, Seattle, Washington 98105, U.S.A.

Boyle, Mr K. H. M., 2321 Great Western Road, Glasgow G15 6RT, Scotland.

Brettle, Mr J., Building A9.1B, AWRE, Aldermaston, Reading RG7 4PR, Berks., England.

Briggs, J. D., M.B., Ch.B., M.R.C.P., Consultant Physician in Renal Diseases, Renal Unit, Western Infirmary, Glasgow G11 6NT, Scotland.

Brinckmann, Dr P., Lutfridstrasse 1, 53 Bonn, West Germany.

Brown, Dr M. C., Bio-Engineering and Medical Physics Unit, University of Liverpool, P.O. Box 147, Liverpool L69 3BX, England.

Bultitude, Dr F. W., Atomic Weapons Research Establishment, U.K.A.E.A., Building B8A, Aldermaston, Berks, England.

a (page 131) **Burstein, A. H.,** Ph.D., Biomechanics Laboratory, Case Western Reserve University, Bingham Engineering Building, 10900 Euclid Avenue, Cleveland, Ohio 44016, U.S.A.

Carter, Paul, B.A., National Heart Hospital, Westmoreland Street, London W1M 8BA, England.

a (page 35)   **Cattell, W. R.,** M.D., F.R.C.P.E., F.R.C.P., Consultant Nephrologist, Department of Nephrology, St Bartholomew's Hospital, West Smithfield. London EC1 A7BE, England.

a (page 39)   **Chang, T. M. S.,** M.D., C.M., Ph.D., Associate Professor of Physiology, Department of Physiology, Faculty of Medicine, McGill University, P.O. Box 6070, Montreal 101 Qué., Canada.

Chmiel, Dr Horst, Helmholz-Institut für Biomed. Technik, 51 Aachen, Goethestr. 27129, West Germany.

a (page 81)   **Chodera, J. D.,** C.Fc., Department of Health and Social Security, Biomechanical Research and Development Unit, Roehampton, London SW15 5PR, England.

a (page 147)   **Clarke, I. C.,** B.Sc., Ph.D., University of Strathclyde, Bioengineering Unit, Wolfson Centre, 106 Rottenrow, Glasgow G4 0NW, Scotland.

Cleland, R., B.Sc., University of Strathclyde, Bioengineering Unit, Wolfson Centre, 106 Rottenrow, Glasgow G4 0NW, Scotland.

Clelland, Mr R., 11 Blairbeth Drive, Glasgow G44, Scotland.

Clifford, Mr J., Unilever Ltd., Port Sunlight Laboratory, Port Sunlight, Wirral, Cheshire L62 4XN, England.

Cole, Mr Stephen, Department of Ergonomics and Cybernetics, University of Technology, Loughborough, Leics. LE11 3TU, England.

Combée, Mr B., Philips Medical Systems Division, Eindhoven, Netherlands.

Condie, D. N., B.Sc., Dundee Limb Fitting Centre, 133 Queen Street, Broughty Ferry, Dundee DD5 1AG, Scotland.

Conner, Dr A. N., Orthopaedic Department, Royal Hospital for Sick Children, Yorkhill, Glasgow, Scotland

Cooke, Mr A. F., Department of Mechanical Engineering, University of Leeds, Leeds LS2 9JT, England.

a (page 45)   **Courtney, J. M.,** B.Sc., Ph.D., A.R.C.S.T., A.R.I.C., University of Strathclyde, Bioengineering Unit, Wolfson Centre, 106 Rottenrow, Glasgow G4 0NW, Scotland.

Cox, Mr Peter, National Heart Hospital, Westmoreland Street, London W1M 8BA, England.

Creasey, Mr G. H., 151 Bruntsfield Place, Edinburgh 10, Scotland.

Cruickshank, Mr J. S., Triadynamics (M & P) Ltd., Glenearn Road, Perth, Scotland.

Curran, Dr J., 44 Campsie Drive, Bearsden, Glasgow, Scotland.

a   **Curran, Sir Samuel,** D.L., F.R.S., Principal and Vice-Chancellor, University of Strathclyde Royal College, 204 George Street, Glasgow C1 1XW, Scotland

Dabbous, Dr O., Orthopaedic Department, Hairmyres Hospital, East Kilbride, Glasgow, Scotland.

Dacquino, Mr G., Instituto di Elettrotecnica ed Elettronica, Politecnico di Milano, Italy.

a (pages 173, 181)   **Daly, C. H.,** Ph.D., Department of Mechanical Engineering, University of Washington, Seattle, Washington 98105, U.S.A.

Davidson, Mr J. M. F., Department of Engineering Science, Oxford University, Oxford, England.

Davies, Dr R. M., Department of Mechanical Engineering, University College, Gower Street, London WC1E 6AS, England.

Davis, Mr J. K., Bio Medical Engineering Unit, The North Staffordshire Polytechnic, College Road, Stoke-on-Trent, England.

Diskin, Prof. M. H., Head, Department of Bio-Medical Engineering Technion, Haifa, Israel.

Doig, Dr W. B., Royal Hospital for Sick Children, Yorkhill, Glasgow, Scotland.

Donegan, Miss Norah, 91 St Vincent Street, Broughty Ferry, Dundee, Scotland.

a (page 103)   **Dowson, Prof. D.,** B.Sc., Ph.D., C.Eng., M.I.Mech.E., A.M.A.S.M.E., Director, Institute of Tribology, Department of Mechanical Engineering, University of Leeds, Leeds LS2 9JT, England.

Edwards, Dr J., Department of Mechanical Engineering, University of Surrey, Guildford, Surrey, England.

Edwards, R. O., B.Sc., University of Strathclyde, Bioengineering Unit, Wolfson Centre, 106 Rottenrow, Glasgow G4 0NW, Scotland.

Jobbins, Mr Brian, Department of Mechanical Engineering, University of Leeds, Leeds LS2 9JT, England.

Johnson, Mr Garth, Rheumatism Research Unit, 44 Clarendon Road, Leeds 2, England.

a (page 273) **Jordan, Monica M.,** M.Sc., University of Strathclyde, Bioengineering Unit, Wolfson Centre, 106 Rottenrow, Glasgow G4 0NW, Scotland.

Jørgensen, Mr Torbet E., Ortopaedisk Hospital, Randersvej 1, 8200 Århus N., Denmark.

Juhasz, L., Dipl. Ing., University of Strathclyde, Bioengineering Unit, Wolfson Centre, 106 Rottenrow, Glasgow G4 0NW, Scotland.

Kalen, Mr Ragnar, Karolinska Institute, Norrbackainstitutet, Box 6403, 113 82 Stockholm, Sweden.

Kazarian, Dr Leon, Department of Orthopaedic Surgery, Karolinska Institute, Norrbacka-Institutet, Box 6403, 113 82 Stockholm, Sweden.

Kelly, Mr A. H., Lucas Medical Equipment, Electral House, Neasden Lane, London N.W.10, England.

a (page 157) **Kempson, G. E.,** B.Sc., Ph.D., D.I.C., Biomechanics Unit, Imperial College of Science and Technology, Exhibition Road, London SW7 2BX, England.

Kenedi, Prof. R. M., B.Sc., Ph.D., A.R.C.S.T., C.Eng., F.I.Mech.E., A.F.R.Ae.S., F.R.S.E. Head, Bioengineering Unit, University of Strathclyde, Wolfson Centre, 106 Rottenrow, Glasgow G4 0NW, Scotland.

Kirkpatrick, Dr G., Department of Rehabilitation Medicine, University of Washington, Seattle, Washington 98105, U.S.A.

a (page 49) **Klinkmann, Prof. H.,** M.D., D.Sc., Professor of Internal Medicine, Medizinische Universitäts-Poliklinik, Department of Nephrology, DDR 25 Rostock, Rembrandtstr. 18, East Germany.

Knudsen, Mr Palle Th., 103 Dronningensgade, 7000 Fredericia, Denmark.

Kölbel, Dr Reinhard, 1000 Berlin 33, Rheinbabenalee 19, East Germany.

a* (page 15) **Kolff, Prof. W. J.,** M.D., Ph.D., University of Utah, College of Medicine, Department of Surgery, Bldg. 512, Salt Lake City, Utah 84112, U.S.A.

Kralj, Dr Alojz, Faculty for Electrical Engineering, Trzaska 25, 61000 Ljubljana, Yugoslavia.

Krenz, Dr Joachim, Orthopaedic Hospital, D-44 Münster, Hüfferstrasse 27, West Germany.

Kumar, Dr Shrawan, Engineering School, Trinity College, 21 Lincoln Place, Dublin 2, Eire.

a* (page 181) **Kydd, W. L.,** D.D.S., Department of Oral Biology, University of Washington, Seattle, Washington 98195, U.S.A.

Ladbrook, Mr Peter, Disa Electronics (UK), 116 College Road, Harrow, Middx. HA1 1HQ, England.

Lambert, Prof. T. H., Department of Mechanical Engineering, University College, Gower Street, London WC1E 6AS, England.

Lanyon, Mr L. E., Department of Veterinary Anatomy, University of Bristol, Park Row, Bristol BS1 52S, England.

Larke, Dr John, Soft Lens Research, 17 Highfield Road, Edgbaston, Birmingham B13 3DU, England.

Laumann, Dr Udo, Orthopädische Universitätsklinik, D-44 Münster, Hüfferstrasse 27, West Germany.

a (page 177) **LaVigne, Dr Angela B.,** Department of Orthopaedics, SA-10, University of Washington, Seattle, Washington 98195, U.S.A.

Lawrence, R. B., C.Eng., M.I.E.E., Department of Health and Social Security, Biomechanical Research and Development Unit, Roehampton, London SW15 5PR, England.

a (page 267) **Lawrie, Prof. T. D. V.,** M.D., B.Sc., F.R.C.P.(Glasg. and Edin.), Department of Medical Cardiology, University of Glasgow, Royal Infirmary, Glasgow G4 0SF, Scotland.

Lees, Mr D. W., Department of Health and Social Security, Alexander Fleming House, Elephant and Castle, London SE1, England.

c **Lenihan, J. M. A.,** O.B.E., M.Sc., Ph.D., C.Eng., F.I.E.E., F.Inst.P., F.R.S.E., Regional Physicist, Western Regional Hospital Board, Department of Clinical Physics and Bio-engineering, 11 West Graham Street, Glasgow G4 9LF, Scotland.

a (page 81) **Levell, R. W.,** B.Sc.(Eng.), Department of Health and Social Security, Biomechanical Research and Development Unit, Roehampton, London SW15 5PR, England.

Polster, Prof. Jürgen, Orthopaedic Hospital, D-44 Münster, Hüfferstrasse 27, West Germany.

a* (page 215)   **Portnoy, H. D.,** M.D., Oakland Neurological Clinic, P.C., 445 West Huron Street, Pontiac, Michigan 48053, U.S.A.

Rabischong, Prof. Pierre, Unité de Recherche Biomechanique, Avenue des Moulins, Route de Ganges, 34-Montpellier, France.

a (page 95)   **Radin, E. L.,** M.D., Harvard Medical School, Orthopaedic Research Laboratories, The Children's Hospital Medical Center, 300 Longwood Avenue, Boston, Massachusetts 02115, U.S.A.

Reberšek, Dr Stanislav, Faculty for Electrical Engineering, Trzaska 25, 61000 Ljubljana, Yugoslavia.

Redhead, R. G., M.B., B.S., M.R.C.S., L.R.C.P., Department of Health and Social Security, Biomechanical Research and Development Unit, Roehampton, London SW15 5PR, England.

a* (page 131)   **Reilly, D. T.,** M.S., Case Western Reserve University, Biomechanics Laboratory, Bingham Engineering Building, 10900 Euclid Avenue, Cleveland, Ohio 44016, U.S.A.

Reul, Dr Helmut, Helmholtz-Institut für Biomed. Technik, 51 Aachen, Goethestrasse 27/29, West Germany.

Richardson, Elspeth, B.A., University of Strathclyde, Bioengineering Unit, Wolfson Centre, 106 Rottenrow, Glasgow G4 0NW, Scotland.

a (page 215)   **Roberts, V. L.,** Ph.D., Head, Biosciences Division, Highway Safety Research Institute, University of Michigan, Huron Parkway and Baxter Road, Ann Arbor, Michigan 48105, U.S.A.

Robson, Dr Martha, Medical Research Council, 20 Park Crescent, London W1N 4AL, England.

Rodseth, C. P., B.Sc.(Physiotherapy), M.C.S.P., Dip.T.P., University of Strathclyde, Bioengineering Unit, Wolfson Centre, 106 Rottenrow, Glasgow G4 0NW, Scotland.

Rogers, Mr Thomas, 46 Roselea Drive, Milngavie, Glasgow G62 8HF, Scotland.

Romanus, Dr Bertil, Department of Orthopaedic Surgery II, Sahlgren Hospital, 413 45 Gothenburg, Sweden.

Rosa, Prof. Pierre, 204 Chausée de Charleroi, 1060 Brussels, Belgium.

a* (page 95)   **Rose, R. M.,** Sc.D., Department of Metallurgy and Materials Sciences, Massachusetts Institute of Technology, Cambridge, Massachusetts 02139, U.S.A.

a* (page 173)   **Ross, Mr R.,** Department of Biochemistry and Medicine, University of Washington, Seattle, Washington 98105, U.S.A.

Sandover, Mr J., Department of Ergonomics and Cybernetics, University of Technology, Loughborough, Leics. LE11 3TU, England.

Schyvens, Mr Ton, Eindhoven University of Technology, HG 4-13, Eindhoven, Netherlands.

Scott, Dr Peter, 206 Peckham Rye, London SE22 0LU, England.

Seliktar, R., M.Sc., Ph.D., University of Strathclyde, Bioengineering Unit, Wolfson Centre, 106 Rottenrow, Glasgow G4 0NW, Scotland.

a (page 251)   **Semple, T.,** M.D., B.Sc., F.R.C.P., Department of Cardiology, Victoria Infirmary, Langside, Glasgow G42 9TY, Scotland.

Seroo, Mr Thys, Eindhoven University of Technology, HG 4-13, Eindhoven, Netherlands.

Shannon, Dr F. T., 42 Katrine Avenue, Bishopbriggs, Glasgow G64 1HA, Scotland.

Shrive, Mr Nigel G., Department of Engineering Science, Parks Road, Oxford, England.

c   **Simpson, D. C.,** M.B.E., B.Sc., Ph.D., F.R.S.E., Director, Orthopaedic Bioengineering Unit, Princess Margaret Rose Orthopaedic Hospital, Fairmilehead, Edinburgh EH10 7ED, Scotland.

Smith, Mr R. N., Department of Veterinary Anatomy, University of Bristol, Park Row, Bristol BS1 52S, England.

Snijder, Mr J. G. N., Technische Hogeschool Te Eindhoven, Postbus 513, Eindhoven, Netherlands.

Snowden, Dr John, Biomechanics Unit, Mechanical Engineering Department, Imperial College of Science and Technology, Exhibition Road, London SW7 2BX, England.

Wachtel, Mr Thos. L., U.S. Army Aeromedical Research Laboratory, P.O. Box 577, Fort Rucker, Alabama 36360, U.S.A.

Wagstaff, Mr Malcolm, National Research Development Corporation, Kingsgate House, 66–74 Victoria Street, London SW1 E6SL, England.

a* (page 161) **Walker, Dr T. W.,** Institute of Biomedical Electronics and Engineering, University of Toronto, Toronto 5, Ontario, Canada.

Wall, Mr J. C., Crystallography Department, Birkbeck College, Malet Street, London WC1E 7HX, England.

Wall, Mr J. G., Road Research Laboratory, Crowthorne, Berkshire, England.

a* (page 177) **Watkins, Mr R. P.,** Department of Orthopaedics, School of Medicine, University of Washington, Seattle, Washington 98105, U.S.A.

Watson, Mr P. A., Smith and Nephew Research Ltd., Gilston Park, Harlow, Essex, England.

Watts, Mr N. H., Department of Mechanical Engineering, Sheffield University, Mappin Street, Sheffield S1 3JD, England.

Weber, Dr Henrik, Hoslegt 19, 1340 Bekkestua, Norway.

Webster, Mr Martyn, Regional Plastic Surgery Unit, Canniesburn Hospital, Bearsden, Glasgow G61 1QL, Scotland.

Weightman, Mr B., Room 603, Department of Mechanical Engineering, Imperial College of Science and Technology, Exhibition Road, London SW7 2BX, England.

a* (page 29) **Welkowitz, Prof. W.,** Department of Electrical Engineering, Rutgers University, New Brunswick, New Jersey 08903, U.S.A.

Whittaker, Mr G. E., Regional Medical Physics Department, Weston Park Hospital, Sheffield, England.

a* (page 203) **Wickstrom, J. K.,** M.D., Biomechanics Laboratory, Department of Surgery, Division of Orthopaedics, Tulane University, School of Medicine, 1430 Tulane Avenue, New Orleans, LA. 70112, U.S.A.

Wilkinson, Rosemary, B.A., University of Strathclyde, Bioengineering Unit, Wolfson Centre, 106 Rottenrow, Glasgow G4 0NW, Scotland.

Wilson, Dr A. B. K., M.R.C. Powered Limbs Unit, West Hendon Hospital, Goldsmith Avenue, The Hyde, London NW9 7HR, England.

Wright, Dr J. T. M., Bio-Engineering and Medical Physics Unit, University of Liverpool, P.O. Box 147, Liverpool L69 3BX, England.

Wright, Mr K. W. J., Brunel University, Department of Mechanical Engineering, Kingston Lane, Uxbridge, Middlesex, England.

a (page 103) **Wright, Prof. V.,** M.D., F.R.C.P., Rheumatism Research Unit, University of Leeds, School of Medicine, 44 Clarendon Road, Leeds LS2 9PJ, England.

a (page 305) **Wolff, H. S.,** B.Sc., Head, Bioengineering Division, Medical Research Council, Clinical Research Centre, Watford Road, Harrow, Middx., HA1 3UJ, England.

Wood, Miss Majorie, Hillview, Inverbay, Invergowrie, Dundee, Scotland.

Wyatt, Mr Miles, 1 Ravens Cross, Long Ashton, Bristol, England.

Zamenhof, R. G., Dipl.E.T., University of Strathclyde, Bioengineering Unit, Wolfson Centre, 106 Rottenrow, Glasgow G4 0NW, Scotland.

EXHIBITORS

Becton Dickinson U.K. Ltd., York House, Empire Way, Wembley, Middlesex, HA9 0PS, England.

Biological Engineering Society, Secretary, Mr K. Copeland, C.Eng., M.I.E.E., M.I.E.R.E., Biophysics Department, Faculty of Medical Sciences, University College London, Gower Street, London, England.

BioMedical Engineering, 42/43 Gerrard Street, London W1V 7LP, England.

G. & E. Bradley Ltd., Electral House, Neasden Lane, London NW10 1RR, England.

British Olivetti Ltd., 69 West Nile Street, Glasgow G1 2SG, Scotland.

Cardiac Recorders Ltd., 375–7 City Road, London EC1V 1NB, England.

Centre for Industrial Innovation, University of Strathclyde, Glasgow G4 0LZ, Scotland.

Data Laboratories Ltd., Wates Way, Mitcham, Surrey, CR4 4HR, England.

Dansk Industri Syndikat A/s, 116 College Road, Harrow, Middlesex HA1 1HQ, England.

Griffin & George Ltd., Braeview Place, Nerston, East Kilbride, Glasgow, Scotland.

Hewlett-Packard Ltd., 224 Bath Road, Slough, Bucks, SL1 4DS, England.

Instron Ltd., Coronation Road, High Wycombe, Bucks, England.

Mercury Electronics (Scot) Ltd., Pollok Castle Estate, Newton Mearns, Glasgow G77 6NU, Scotland.

P. K. Morgan Ltd., 28 Alderney Street, London S.W.1., England.

Morison & Miller Engineering Ltd., 180 West Regent Street, Glasgow, Scotland.

Princess Margaret Rose Orthopaedic Hospital, Orthopaedic Bioengineering Unit, Fairmilehead, Edinburgh, Scotland.

Queen's University, Belfast, Northern Ireland.

Scottish Reactor Centre, East Kilbride, Glasgow, Scotland.

Sierex Ltd., 17 West Lenziemill, Cumbernauld G67 2XU, Scotland.

Telecare Ltd., 8 Dixon Place, College Milton, East Kilbride, Scotland.

CONTENTS

OPENING REMARKS AND WELCOME TO PARTICIPANTS

SIR SAMUEL CURRAN DL FRS[1].

There never seems to me to be a great deal of point in having Opening Remarks at the beginning of an international Symposium such as the one starting today. Experts have come together to discuss specialist topics. However that may be in a general way, it is probably true to say that there is more point to Opening Remarks now than in the past. For example, we have just recently had a rather huge conference on the environment taking place in Stockholm primarily under the auspices of the United Nations. None of us can escape, even if we would, the pressures that are mounted in all forms of communication on the themes of ecology, the environment and pollution. Newspapers, radio and television remind us constantly of these topical themes. It has, for most scientists and engineers, a possibly serious repercussion. Mixed up in much of the argument and propaganda is the implication that technology is driving man along routes which might lead to his doom. In fact, there is now a real danger that a substantial number of the most able young people who would have entered science and technology will be dissuaded from doing so.

Two things surely have to be said. One is that science and technology as such have not produced environmental problems. No scientist or engineer worth his salt has failed to outline both the good and the bad in his discoveries, inventions or industrial innovations. Society, and in particular industry, makes use of the innovations stemming from science and it is perfectly true that, in the urge to accomplish this or that project, hazards and dangers have been to some degree overlooked. The next thing to say, of course, is that the answer surely is not to turn the clock back, with the very substantial decline in living standards that would thereby arise. It is rather to engage scientists and engineers in close collaboration with forward-looking industrialists in efforts to improve existing practice and to ensure that due attention is paid to all new industrial procedures and practices in the future. We need scientists and engineers to do this because, strangely enough, the serious effects of

environmental pollution are generally long-term in their nature and measurement of them is extremely difficult. New methods of measuring may have to be devised and I know this will engage the attention of some of those in our own Department of Applied Physics for several years.

I have embarked however briefly on this topic because I wanted to draw attention to the very important fact that since 1962 this University has had a full Department (our Department of Bio-engineering) engaged in work which is dedicated to the improvement of the way of life of all for whom their research can prove effective. In fact, we started work in the subject fifteen years ago. I have no doubt that many of you in the company have had similar experience and that as scientists and engineers engaged in this relatively modern multidisciplinary study, labelled somewhat inadequately Bioengineering or Medical Bioengineering, you have been yourselves dedicating your efforts to the improvement of the way of life of those suffering from a great variety of disabilities and diseases—and this was indeed the case long before the recent excitement about the environment.

It is very simple to point to defects and inadequacies in the present-day world but from time to time it is vital that the tremendous improvement in the standard of living of most people and in the care of the disabled and diseased are stressed. I would hope that during the coming days when you exchange information on the progress of your researches and sketch your ideas for future work it will be obvious that in your field, and of course in many others, very notable things are being achieved —none of them questionable in their final effect; that is, in their contribution to the welfare of man.

This University, under its Charter, states clearly the essential message that I have been trying to convey today. It started as a university-type institution in 1796 and followed within two years the founding of the École Polytechnic (1794), the first institution of its kind in the world. Our Founder was a scientist, John Anderson, who became aware two hundred years ago of the great challenge that was presented by the first Industrial Revolution. He was particularly concerned with the effects of this Revolution on the lives of artisans and was

[1] Principal and Vice-Chancellor, University of Strathclyde, Glasgow, Scotland.

1

most anxious to ensure that within our universities men of learning devoted much of their skill, knowledge and thought to the improvement of the minds and living conditions of all of the people. In this he was in some senses a real pioneer and it was because he found so many obstacles in the way within universities of his time that he asked a number of Trustees in the City to start a new kind of university. We hope that we have pursued over nearly two hundred years the same philosophy and that we continue to do it today. We are sure that in subjects like Medical Bioengineering, and especially in this new Wolfson Centre of Bioengineering, John Anderson would find reassurance that the spirit in which he wrote his philosophy is very much alive.

SESSION A

Biochemical Engineering

Chairman: Professor E. M. McGirr, B.Sc., M.D., F.R.C.P. (Lon. Edin. and Glasg.), F.R.S.E., F.F.C.M. (Lon., Edin. and Glasg.).

Associate: Dr J. M. A. Lenihan, O.B.E., M.Sc., Ph.D., C.Eng., F.I.E.E., F.Inst.P., F.R.S.E.

HEART REPLACEMENT AND CIRCULATORY SUPPORT

D. B. LONGMORE[1]

INTRODUCTION

A short contribution to a Symposium may not be meaningful if it is not considered in context. Thus I welcome the opportunity to outline the background to areas of research undertaken by myself and my teams. These fit into a policy of studying fields which have been overlooked by current research programmes, but which are immediately relevant by merit of being possible. It is important to realise that we are not committing scientific treason if instead of following surgical and bioengineering fashion we look elsewhere. The question I have tried to ask is not where would we like to be in the field of heart replacement in x years' time, but where is it possible for us to be. In my view, unless bioengineering is, as its name implies, a real life science and not engineering divorced from biological reality, our potential advancement will depend on chance; a proven but exceedingly slow evolutionary process.

It is strange that in spite of the long history of bioengineering the whole discipline, with its multitude of bioengineering departments, seems at present to be still in its infancy. Its beginnings date from such early ancestors as Galileo (1564–1642), Young (of Young's Modulus) and the wave theory of light (1773–1829), and the first real bioengineer, Willem Kolff, a contributor to this Symposium, who pioneered practical extra corporeal circulation over a quarter of a century ago in difficult wartime conditions. Yet, inadequacies and failures of communication between doctors, biologists and engineers are still the rule. Nowhere are the failures more obvious than in the area of heart replacement.

On the one hand, physicians have failed to appreciate the enormity of the clinical problems which make heart replacement or circulatory assistance mandatory for the patient's survival. We have no clear indication whether in the United Kingdom, we could save 10 000, 20 000, or as some figures suggest, 90 000 patients each year with circulatory support or heart replacement. More importantly, we have no accurate figure for the number of young patients who die daily. What is most disturbing is that in the forseeable future we are not likely to achieve

more than an informed guess. Even if we did there is very little scientific evidence that any of the existing or projected devices will be able to maintain life if there is insufficient myocardium to support the circulation.

Such is the anarchic structureless form of medicine that we do not have the elements of information retrieval to answer these questions. The basis of this communication, which examines possible methods for supporting, replacing, supplementing or repairing an inadequate heart in a patient who is otherwise viable, may in fact be based entirely upon wishful thinking and false premise.

There are four possible methods for supplementing or replacing a failing heart, as follows:

Mechanical Devices. Replacement of the heart with a mechanical pump; or support of the circulation with a mechanical device for a short time, to maintain the patient whilst one of the three possible biological forms of heart replacement is undertaken.

Animal to Man Transplantation. The use of animal to man transplantation with all its attendant biological complications.

Man to Man Transplantation. The use of man to man transportation, a proven technique with all its attendant social problems.

Biological Organ Repair and Regeneration. The possible use of foetal material or extracts, and in the distant future the use of other biological methods to cause replication of cells, repair, and revascularisation in the area surrounding a myocardial infarction. A step doubtless bristling with new and unforeseen difficulties.

I propose to review the first of these in detailed, critical terms and refer briefly to the remaining three in summary terms under the generic title of Biological Heart Replacement.

MECHANICAL DEVICES

I propose to use the various attempts at producing the mechanical heart, to illustrate the obstacles which block progress in our joint discipline of bioengineering. Never before in medicine has there been a bigger expenditure of public money, was there more ballyhoo, publicity and fund raising, based on 'pie-in-the-sky' promises than in the area of the artificial heart.

[1] National Heart Hospital, London, England.

The result of the publicity surrounding the fund raising and exaggerated forecasts about mechanical hearts was that the general public and the information media, had for the first time an insight into the mechanisms of the growing edge of medicine, and its allied disciplines. Thus it became apparent that any news editor could make his name as a hard hitting incisive journalist by turning his camera lens onto any aspect of the medical scene. Here he would find so many plums ripe for picking that he would be spoiled for choice. Nowhere were the plums bigger and riper than in the life support industry.

Even the slowest-witted journalist could see that the social euthanasia, which passes for treatment of the thousands of patients with renal failure every year, is unjustifiable whilst we have artificial kidney machines and renal transplantation. So also with heart failure. There can be very few families which have not experienced sudden death in one of their members due to myocardial infarction—currently not even sporadically treated by the type of organ replacement hinted at by the artificial heart programme. Rightly, the murmur of discontent which arises following occasional publicised cases will eventually become a popular clamour.

Before this happens, and bioengineers are accused of failing to do their share, we would do well to see why there are no signs of real progress in this field. The huge effort by talented engineers, material technologists, and doctors, has provided for the first time *some* sound fundamental knowledge and principles in the field of biocompatability. No doubt we all have benefited from this fall-out of fundamental knowledge. The sad feature is that although the bioengineers have striven towards the distant goal of an implantable, internally powered and reliable artificial heart, the lack of a clear medical lead and of achievable specifications has meant that most of their time and effort has been wasted. Engineers have worked diligently, frequently with apparently inspired ideas. Biologists and doctors should have worked alongside them to avert the production of ideas and hardware incompatible with human physiology and biochemistry. We must have a workable research programme. If we now examine in detail the current status of the artificial heart programme and the attempts at circulatory support, we can salvage from the enormous amount of work already done some valuable findings, which though mostly negative are nevertheless useful.

BIOCOMPATABILITY

This title brings to mind some of the well-known factors associated with placing foreign materials in the human body. These are highlighted when these foreign materials come into contact with circulating blood. This contains a continuously replaceable supply of corrosive chemicals capable of attacking most metals and plastics and even such inert materials as silicone rubber.

Blood clots rapidly when in contact with foreign surfaces—there are two approaches to this problem: either we use the surface properties of materials to deliberately promote clotting, or we opt for the use of materials designed to prevent clotting. Both apparently contradictory methods come within sight of success. If a material is used which will rapidly accumulate and retain a thin stable platelet thrombus, endothelial cells will migrate from adjacent vessels and be deposited on the surface of the thrombus. This will soon line the implant with a smooth, streamlined and stable natural surface from which emboli will not be liberated. Various velours have these properties, and have worked reasonably well in a few clinical trials.

The opposite philosophy is to try to incorporate either the heparin molecule, or other negatively charged molecules, on to the surface of the plastic or silicone rubber from which the blood containing wall is made. Some extracorporeal circulations, such as in left heart by-pass of limited duration for aortic arch surgery, have been successful without heparinisation, due to the use of tubing in which the benzylconium molecule is incorporated in the surface; but it is not yet known how long such a surface will retain its platelet repelling properties. Some of the newer dialysing membranes using this principle will soon provide the answer. Some forms of carbon do not readily initiate thrombus and may be useful in prosthetic heart valves.

INCORPORATION

Biocompatibility should mean more to us than the presence of a thrombogenic surface in the blood stream, setting in motion the cascade of clotting reactions. We know little about the incorporation and fixation of foreign tissues and devices in the body. Since the beginnings of surgery we have relied on needle and thread and the ever present healing process to reassemble human tissues. This is simply not adequate when the object we are trying to implant in the subject is a mechanical pump.

A pump will change its shape and its volume and, because of its function in accelerating blood, it will have a considerable recoil. Fixation of a pump in the orthotopic position presents a special set of difficulties—the pericardium is not suitable for attachment by suture because it is fibrous and has a poor blood supply, and the stitches will pull out. Attachment to the adjacent vertebral column and sternum is of doubtful value, for it is well known that bone will reabsorb and disappear in the face of even modest non-physiological pressures (a fact that every orthodontist is aware of when he applies a dental brace to straighten crooked teeth). In addition, the sternum is very susceptible to infection and reabsorbtion. The vertebral column is separated from the pericardium by the great vessels and oesophagus. Some alternative methods of attachment seem to offer greater possibilities.

Adhesives are frequently unsuitable for attaching implants because of the presence of abundant free fat and lipids which act as parting agents; however, the relatively fat-free inner surface of the pericardium, whilst unsuitable for suturing might offer a good site for attachment for an adhesive in the form of a mesh. This would allow tissue ingrowth and incorporate soundly. Development of new ceramic materials offers a chance to improve the strength of junctional areas since they offer the possibility of tissue ingrowth on a large proportion of their surface. The problem of obtaining sound union of the junctional areas of a prosthesis is complicated because no materials, including those from which the pump itself is made, can resist infection, since they cannot have a blood supply. Infection is still a feature of surgery which causes significant mortality and considerable morbidity. The ecology of the surgical patient is continually changing with the development of antibiotics and antifungal drugs. Nevertheless, the overall risk of infection remains significant and is compounded by increasing amounts of nonviable implant.

MECHANICAL RELIABILITY

Even if reasonable reliable incorporation can be achieved, there still remains the achievement of mechanical reliability with these pumps. Very few of the tissues of the body have much permanence. The very bones with which we move today will be rebuilt within a few months; repair and replacement of most tissues continues unabated within us until age and degeneration overtake us. We have yet to contemplate the manufacture of a self-repairing material. Even in such financially sound and quality-controlled ventures as space exploration, component failure is not unknown. The mass production methods which would be required if mechanical hearts were to ever become a routine surgical appliance seem to increase the risk of mechanical failure. The time gap between the teething troubles on receipt of a new appliance and its wearing out is so narrow with mass production engineering that a new industry even more rigidly controlled than the aircraft industry is required—but at much lower cost. Miniaturisation and increasing the performance to size ratio can stress components more and increase the risk of failure. Few of todays efficient forms of transport will have a longevity comparable with their clumsy unstressed stream multi-million-mile predecessors. To make a mechanical pump which will beat reliably forty million times a year, pumping several tons of corrosive fluid each day without requiring service for decades, may be beyond the technical skills of today's mass production engineers.

DESIGN

At first sight it might seem that a well-trained fluid dynamicist or hydraulics engineer could design a satisfactory pump to perform the relatively modest functions of a heart. If the output of the mechanical pump were fixed it might be possible to achieve a pump design which did not introduce problems of blood cell destruction or protein denaturation by putting shearing forces on the delicately coiled protein molecules. However, a design which achieves this fixed output objective is not adequate.

CONTROL SYSTEMS

Control of the output of an artificial heart is an area which seems to have received less attention than it merits. The physiological control systems influence the normally innervated heart by changing its stroke rate, velocity of contraction, force of contraction and stroke volume. In addition, the nervous system and circulating hormones also influence the peripheral circulation. Combined, these ensure that blood flow requirements of all essential organs are met whatever the demands made on the circulation. The feedback mechanisms and the afferent controls are at present only vaguely understood. Those of us who have heard a patient with a fixed rate pacemaker which takes no account of the inbuilt circadian clock, complain of insomnia, cannot help asking what will become of the circulation of a patient with only a limited range of cardiac outputs, controlled by arbitrary methods. Even so, it is likely that the apparently insuperable complications associated with a relatively fixed heart pump output will be compensated for by the remarkable resilience of the body's physiological mechanisms.

FLUID DYNAMICS

We live under an atmosphere some fifteen miles thick, which consists of approximately 80% nitrogen, consequently the water in our tissues and blood contains this gas in solution at a pressure equal to its ambient atmospheric pressure. Any pump which has restricted cross sectional areas increasing the velocity of flow, sharp bends, or which arrests rapidly flowing columns of blood with the closure of rigid valves or, worst of all, has leaky valves producing venturi jets, will cause unacceptable pressure drops. Although blood is not an ideal fluid the effects of Bernouilli's equation still apply. Thus, any local increase in velocity is accompanied by a corresponding decrease in pressure (some bioengineers would still do well to remember this), and the effect of even a modest increase in velocity is to bring dissolved gas out of solution. Because of the gas-carrying properties of haemoglobin the potential volume of bubbles is great.

The formation of bubbles is exacerbated by turbulence when even lower localised pressure drops occur; imperfect walls in the pump increase turbulence. Red blood cells act as nuclei for the formation of bubbles of water vapour containing nitrogen. There are two potential fates for these bubbles. Rarely, they may persist and act as a continuous stream of emboli blocking small vessels, but usually they will collapse when they move into regions of higher pressure. Enormous energy, for their size, is released when these small cavitation bubbles collapse. This destructive effect is believed to be associated with very small 'Munro' jets arising from the concave surfaces of asymmetrically collapsing bubbles. Very high pressures are often produced when such jets strike a solid object. An artificial heart producing continuous streams of cavitation bubbles will be capable of both blood and self destruction.

Natural hearts do not produce cavitation bubbles despite variations of cardiac output which can range from 2 litres/minute to over 50 litres/minute. This is due to two factors which enable laminar flow to be maintained: smooth vessel walls and the non-Newtonian character of blood related to the structure of the plasma proteins, which are able to change their shape under the influence of shearing forces. The coiled molecules lengthen under stress and can return to their original shape if not damaged.

The main reasons for the lack of turbulence energy dissipation in the heart and vessels are, however, more fundamental. The heart with its pumping action, acceleration of blood and changes in direction flow, might seem to share the hazards of low pressure zones with mechanical hearts. That it does not do so is due to the fact that the heart and great vessels in the embryo develop from the same mass of tissue. The heart tube and brachial arches begin to form in the embryo from a mass of mesodermal tissue; this becomes denser and assumes the morphological configuration of these structures. Soon waves of peristaltic contraction begin in the area destined to become the heart, and the central core of the heart and vessels breaks down to form the foetal blood. The ebbing and flowing of this blood percursor helps to mould the very vessels through which it is to flow, thus making them truly streamline. Where the blood flow is through a spongy network of vessels, it soon concentrates into a few main streams, rather as in a flip-flop pneumatic valve where a small biasing stream will divert the total flow.

The relative roles of evolutionary genetic transference and random development of the heart and vessels are not clear. The outcome, though, is a highly efficient non-turbulent cardiovascular system. It remains an academic discussion whether there is a detailed genetic blueprint for each tiny vessel or whether there is a genetic sketch plan leaving the detailed architecture to change. Either way, the result is both more adaptable to flow changes, and safer, than any prosthesis design which has yet been suggested.

POWER SOURCES

It is when we come to the problems of power sources that the inadequacies of our present day technology manifest themselves most.

Obviously, electrical power appears as a locally clean source, for all its major pollution is concentrated at the generating source, although heat remains as a local pollutant when it is used to power a mechanical heart. The shortcomings of electric power in this context are mainly related to the possible methods of obtaining the 12–200 watts of power required to provide an adequate cardiac output without wires piercing the skin. Implantable batteries with sufficient capacity cannot be made small enough. Rechargeable batteries are cumbersome and may not be sufficiently reliable.

The use of an induction coil system to provide power directly, instead of batteries requiring to be recharged, is clumsy but has been proved to be feasible by its use as a power supply for cardiac pacemakers. It requires the implantation of a large subcutaneous secondary coil and needs fairly accurate alignment of the external primary coil which in turn needs a large power source.

The use of the body's own electrolytes and dissimilar metals is attractive; in theory it offers a method of tapping the almost inexhaustible supply of chemical energy produced by metabolism. Polarisation and the liberation of toxic products at the cathode make it impracticable.

Radio-frequency coupling could theoretically power an artificial heart, but no attempt has yet been made to make it do so, because of the potential disruption of communication.

The use of atomic power packs would seem to be of extreme interest to the bioengineer, but in spite of the recent publicity from the United States, a clean source of sufficient power without excessive heat production has yet to be made. The heat burden which can be borne by the human body is limited; it must be remembered that not only is there the production of heat from the power source, but there is also the inefficiency of the heart mechanism to consider. Any turbine- or piston-driven heart will be unlikely to approach the extraordinary efficiency of the myocardium in which very high power outputs are accompanied by a relatively modest heat production.

Other methods of providing sufficient power have been considered and found wanting. One is the generation of electrical energy by the use of a battery of piezo electric crystals within the jaws of a nut-cracker shaped device. The jaws would then be actuated by the movements of, say the aorta, or the respiratory muscles. If sufficient energy is to be produced in this way, however, the loading on the attachment points exceeds the tolerance limits of any tissues by many orders of magnitude.

CIRCULATORY SUPPORT

Every aspect of total heart replacement we look at appears to require a major research effort in order to solve the problems. If total heart replacement by mechanical pumps is 'pie-in-the-sky', temporary support of the circulation might be 'crust-on-earth'. The question which has to be answered about the various forms of circulatory support is whether the potential benefit to the patient outweighs the risk, discomfort, and complications of the obligatory invasive procedure.

Reducing the burden of the circulation

Theoretically an inadequate circulation might be helped in three ways: either by boosting it, by reducing the demands made upon it, or by both helping it and reducing the demands made on it. Various methods of cutting down the metabolic needs of a patient are available and have been tried. Normal metabolic processes depend upon millions of enzyme- and catalyst-aided chemical reactions; a reduction in body temperature of 5° centigrade should, if normal chemical laws are obeyed, reduce metabolic rate by 25%. It would do so if all the enzymes involved had a linear response to temperature changes. Some enzymes work more efficiently at 25° centigrade than at 37° centigrade, others are inhibited almost completely at this temperature; consequently abnormal metabolic pathways and intermediate products result. Hypothermia is always accompanied by a certain amount of cumulative biochemical disturbance, increasing with duration and depth of hypothermia. A logical extension of hypothermia techniques was to attempt to mimic hibernation. Hibernating animals which are normally homiothermic can progressively lower their body temperatures in conditions of reduced external temperature and shortened daylight hours. This they do first by allowing the normal diurnal swings of temperatures to increase, then by failing to achieve a normal daytime temperature, and finally by settling down to a low temperature requiring a minimal circulation to support life. The mechanism by which this progressive metabolic slowing down is achieved is not understood at all nor is the absence of clotting in the stagnant blood vessels. Although there have been attempts since 1952 to produce hibernation-like states with 'lytic cocktails', we are still unable to discover a drug, or combination of drugs, to induce hibernation hypothermia. It is doubtful whether most patients with circulatory collapse would survive the prolonged induction required by this type of therapy.

If oxygenation were the only function of the circulation, hyperbaric oxygen should succeed in providing a suitably enriched environment for an infarct to recover. Hyperbaric chambers have been in use for over a decade without any conclusive demonstration of their value to coronary patients. Bed-sized chambers carry risk of fire and explosion and produce a physical barrier between patient and attendant. Chambers big enough to accommodate the whole staff expose them to the risks of oxygen or nitrogen poisoning.

Relieving the patient from the burden of breathing by using a mechanical ventilator indirectly reduces cardiac work. There are two types of ventilator: patient triggered-pressure cycled, and automatic-volume cycled. Pressure cycled ventilators have the theoretical advantage that they use the patient's own physiological mechanisms to regulate the rate and depth of breathing, and allow the periodic sigh essential to avoid inadequate expansion of segments of the lung. Nevertheless, volume cycled ventilators are favoured by most anaesthetists. In the conscious patient these rely on excessive ventilation to eliminate enough carbon dioxide to remove the patient's desire to breathe, supplemented by partial paralysis of the respiratory muscles to remove the patient's ability to fight the machine, thus negating any beneficial effects from removing the burden of breathing.

None of the currently available methods of resting the circulation has much advantage over the time honoured medical tradition of giving morphia, raising the foot of the bed, and keeping the patient peaceful and quiet.

Many methods of boosting the heart have been tried. Some are simply engineered devices requiring considerable surgical interference to connect them to the patient. Others are more complex but require less surgery. All interfere with the physiological state of the patient, already rendered abnormal by circulatory inadequacies. That no single circulatory support technique has any clear advantage over the others suggests that they may all have failed to turn promise into performance. If we are to analyse why this is so, assessment is required of the possible benefit assessed against:—the trauma of surgical intervention; the price to be paid in physiological disturbance, and the electro-mechanical complexity and reliability of the apparatus and, importantly, its control mechanism.

Having outlined some of the more important methods of circulatory assistance, I will use an arbitrary scoring system for each area in each method (table 1, below). Because of my surgical background and respect for the hazards of operating, I will start with the procedures requiring least surgical intervention, and end with the procedures requiring thoracotomy and connections to the heart and great vessels.

Body acceleration synchronised with heartbeat (BASH)

Pumping fluid along a pipe can be achieved by accelerating the fluid forwards along the pipe, or by accelerating the pipe rapidly in the opposite direction. The achievement of this second method of pumping in a human necessitates fixing the patient to a table and moving him with an acceleration of over 1 g caudally and slightly to the left, just as the heart is ejecting. This is achieved using an ECG trigger. Recovery is at a smaller acceleration, during the longer diastolic period. This method has great potential benefit for, as well as relieving the volume load of both ventricles, it offers an increased coronary perfusion pressure during the diastolic recovery phase. It requires no surgery and the engineering is straight forward, though formidably

massive. There is no extracorporeal circulation or blood/plastic interface. This method shares with most other circulatory support methods the disadvantage of relying on the ECG trigger, but unlike some of them, a failure to trigger, or asynchrony with the heart due to an artefactual triggering, is not lethal. The physiological disturbance is hard to assess, for the effects of this violent motion are not known, nor is the amount of potential trauma. We do not even know whether the liver will remain attached to the inferior vena cava or if brain damage will ensue. Body acceleration has great potential and the experimental results seem to corroborate this optimism.

Peripheral cuffs

Various possibilities are available using cuffs or special clothing which share with body acceleration the advantage that surgery is avoided. The simplest technique is to achieve a sequestration of blood by a cuff which is above venous but below arterial pressure. This provides an uncomfortable, but harmless, temporary venesection and lowering of venous return pressure and is a modern way of achieving the results of applying leeches. Pulsatile leggings (similar to those of a g-suit, and capable of ECG triggering, so that during diastole they can exceed arterial pressure and during systole they can be released to accommodate a proportion of the left ventricular output) offer a simple and safe form of diastolic counterpulsation. As with the body acceleration, failure of the ECG trigger is non-lethal.

Method	Potential benefit	Surgical intervention	Physiological disturbance	Bioengineering complexity
Body acceleration	Good	None	Unknown	Simple but massive engineering Requires ECG trigger
Non-invasive counterpulsation	Poor	None	Upsets balance between right and left sides of the heart	Requires ECG trigger
Invasive counterpulsation	Fair	Varies—can involve thoracotomy	More efficient than non-invasive counterpulsation, more disturbance	Requires ECG trigger Potentially fatal if mistimed
Left ventricular by-pass	Fair	Can require thoracotomy	Very severe imbalance between right and left sides of the heart	All the problems of the artificial heart except power source Requires ECG trigger Potentially fatal if mistimed
Cardiopulmonary by-pass	Can attain total circulatory support	If cannulation of great vessels is in the groin, risk of infection	There is a lower limit of usefulness: in by-pass of less than about 30% total flow does not reduce cardiac oxygen consumption	Blood damage limits duration of by-pass Long perfusions only possible with membrane oxygenators—still in development stage

TABLE 1: Comparative assessment of methods of circulatory assistance.

Invasive diastolic counterpulsation

Various effective forms of diastolic counter-pulsation have been tried. All use an ECG trigger and aim to empty as much of the arterial system as possible just before and during systole, and then to replace the blood during diastole. Whatever method is used, the ECG trigger is critical. If femoral or subclavian arterial cannulae are used to withdraw blood and to replace it during diastole, trigger failure would mean that the left ventricle would be attempting to eject and the aortic valve would be open whilst the extracorporeal pump was delivering. This would produce a rise in the aortic pressure. Acute failure of an ailing left ventricle would be certain under these conditions. The amount of surgery involved varies with technique from simple peripheral cannulation or the insertion of intra-aortic balloons, to thoracotomy for the insertion of para-aortic apparatus. The engineering can vary from simple extracorporeal piston pumps to complex intrathoracic apparatus.

The most important defect of diastolic counterpulsation whatever method is used, and which it shares with left heart by-pass, is that it can only help the left ventricle. Although the left ventricle is responsible for a major part of the work load, as shown by its using the lion's share of the oxygen consumption of the heart, it must be remembered that the right ventricle has to pump the same volume load as the left, although up a smaller pressure gradient. Furthermore, nine hundred million years of evolution have given us, in health, an infallible control mechanism for synchronisation of the two sides of the heart. Over any given period of time the output of the two sides of the heart, when working steadily, is perfectly balanced and appropriate to the given conditions. The control mechanism is delicate enough to cope with evaporative losses from the lesser circulation, via the lungs, and with the whole body water balance of absorption from the gut, and loss through excretion and secretion. It is not surprising therefore that support of the left ventricle alone, which disregards the sophisticated intrinsic cardiac control mechanisms, is of little clinical value. This is because as the volume through-put of the left side of the heart is reduced, so the right side reduces its output correspondingly.

Left ventricular by-pass

This is the best known and currently the most developed circulatory assist device, and is misplaced effort if it is assessed using the same criteria we have applied to all the other assist devices. It requires major invasion of the patient with thorocotomy and cardiotomy if the flow rates are to be high, although peripheral cannulation is possible if low flows are acceptable. The engineering is difficult, involving blood interfaces with tubes and pumps.

Other circulatory support techniques

Other methods which have been considered, but have not warranted clinical trial, show a different approach to the problem. One such method is to pump blood from a peripheral artery and to inject it through a catheter inserted in the descending aorta in such a way that a jet is formed accelerating the blood during systole. This venturi pump is a simple form of diastolic counterpulsation, but the pressures required are high and potentially destructive. Another method is more ingenious, but is aimed only at improving the blood supply to the myocardium. It uses the femoral artery pulse to actuate a hydraulic multiplier, which in turn actuates a venturi to increase coronary perfusion pressure. The losses in the system are so great that it is unlikely to produce any measurable boost to the coronary circulation.

The simplest way of reducing cardiac work would be to reduce the pressure load, although this is only readily applicable to the left ventricle. The pressure load cannot usefully be reduced by simply vasodilating the patient's peripheral circulation because the coronary circulation would also be reduced, as it requires a physiological diastolic aortic pressure. Coronary vasodilator drugs to reduce the resistance of the coronary vessels correspondingly do not compensate for a reduced perfusion pressure as such drugs cannot dilate diseased vessels. Any increase in blood flow they cause may be due to the opening up of shunts which actually by-pass the areas which require oxygenation. Even if the coronary circulation could be maintained, reduction of aortic pressure will still do little to aid the right ventricle.

Cardiopulmonary by-pass

Reduction of the total volume load with a cardiopulmonary by-pass has been the clinician's goal since the advent of the practical heart-lung machine. It is simple to support the circulation for a limited period at the end of an open heart operation, when the patient is already cannulated, connected to a machine and fully monitored. Those of us who have been in the unfortunate position so many times of needing to do this know that even under these circumstances how difficult it can be to help a failing heart. If too little volume is by-passed there seems to be no reduction in cardiac size and no clinical improvement of the heart, although at first the remainder of the patient improves. Soon, the protein denaturation arising from the raw gas-to-blood interface in the oxygenator, and the haemolysis and cell damage arising in the pump and tubing, begin to overwhelm the patient. If a greater proportion of the circulation is by-passed, the cardiac assistance is more obvious but the blood destruction is more rapid. The real troubles develop when the left ventricle cannot eject against the artificially elevated pressure in the aorta. The left atrium then fills at arterial pressure from the bronchial arteries and causes a back pressure on the lung capillaries. If the pulmonary valve is competent lung destruction soon follows. In spite of this knowledge, the production of membrane oxygenators which can be used for very long periods with comparatively low blood damage has rekindled the interest in circulatory support. Certainly the difficulties I have already outlined with relation to by-passing only the left side of the heart do not apply.

COST EFFECTIVENESS

Before going on to comment on biological methods of replacing the heart, *viz.* heterograft, homograft and foetal implant, it would be instructive to consider the very important factor of cost effectiveness.

Mechanical devices can be costed more readily and accurately than biological methods of replacing the heart. In human terms, there is no price limit to be set against any form of live-saving treatment. There are in fact a handful of severely paralysed patients who have been kept alive with biomechanical aids for many years without any hope of recovery and with doubtful quality of life. Apart from these few and regrettable cases, such forms of patient care rightly do not become the subject for public debate.

Life enhancement is somewhat different; if there is an overall shortage of facilities, there may in this field be a case for trying to equate the value of an operation to its cost. For example, cosmetic surgery might be abandoned in favour of operations for painful piles. The practical difficulties arise when a new and expensive or potentially and numerically important form of treatment is translated from the laboratory to the clinical area. In countries with private medicine, the provident and the insured may be treated—if, by geographical fortune, they meet a doctor able to advise them about the new form of therapy. In countries with centrally organised and funded medical services the patient is rather more likely to be referred for treatment, but little more likely to receive it, as the potentially open-ended demands of medicine are dutifully restrained by administrators and committees. We have an example of this in renal disease, where only a fraction of the patients requiring dialysis or transplantation receive treatment. Physicians, whose duty it is to try to treat all patients sent to them, find to their shame that they have to sit in judgment over whom will be neglected. I am in no doubt that the same restrictions will beset the medical bioengineering teams when heart replacement becomes a real threat to the medical accountants.

The cost of the heart, the cost of its insertion, and the maintenance of the patient will all be paraded. All failures will be publicised and equated with an ideal 100% success rate, rather than compared with a realistic small salvage rate appropriate to a lethal disease. Such is the $2\frac{1}{2}\%$ one-year survival rate for gastrectomy for stomach cancer. If there is a balance sheet, the cost of treating the patient will be set against the patient's theoretical contribution to the Gross National Product. No account will be taken of the cost of allowing a patient to die a chronic death in hospital, of his dependents' costs, his insurance payments, or the less visible loss to the economy due to the loss of payment of insurance premiums and taxes, consumer spending, etc.

There is little which can be done to combat this attitudinal climate which compares the new with the theoretical and not with the established. What we can do is to make certain that all funds will be utilised optimally. Using our example of renal disease, we must not parallel the costly and wasteful 19th century concepts, building and staffing small special units susceptible to infection. We must use 20th century techniques with centrally organised mobile services to bring expertise rapidly, cheaply and efficiently to the patient. We must avoid the situation where, by having cost ineffective forms of treatment for a few, there is no money for efficient treatment for all the sufferers.

BIOLOGICAL HEART REPLACEMENT

Some of the high cost of the mechanical heart replacement might be avoided by using biological tissues. Every day, thousands of sound organs are cremated or interred, and tons of human foetal tissues are incinerated annually. The real arguments in favour of using living tissue to replace or repair the heart are much more powerful than the mere economic factors. Biological tissues repair themselves, resist infection, and are powered by highly efficient metabolic energy sources.

Animal-to-man transplants are not a contemporary concept. The antiquity and multiplicity of examples of heterografts are only limited by the diligence of the historian. They have, however, been overtaken by man-to-man transplantation.

The operation performed by Barnard on Louis Washanski was no surprise for the group of workers who had for years been preparing for this venture. Alexis Carell and Lindburgh in the United States had started experimental work in this area in the 1930s. John Hunter of the Royal College of Surgeons of England had anticipated them by over a century. Vladimir Demikov in the 1940s had devised his method of transplanting the heart and both lungs without cardiopulmonary by-pass and had short-term dog survivors. Norman Shunway at Stanford University Hospital had by 1950 devised his method of transplanting the heart alone, using heart-lung by-pass, and producing longer survivals. Lord Brock of Guy's Hospital reported the loss of two transplanted dogs' hearts, from bleeding from the suture lines in the mid 1950s. In the first part of the 1960s Richard Lower and I also started, along different lines; Lower following Shunway's lead, whilst I was interested in storage of the heart-lung preparation, and its complete implantation, unknowingly following Demikov's techniques.

The coverage, by the information media, of medical and bioengineering activities in the artificial heart programme, has rendered any comment about man-to-man heart transplantation, other than reviewing the results, superfluous. By any medical standards the surgical figures achieved by Barnard and Shunway show remarkably good one- and two-years results. The quality of life for most of their patients, dying uncomfortably of a lethal disease, was improved by more than could be attributed to hope. I name these two surgeons because they have continued to provide a data base in this area.

In my view it is not rash to forecast that heart transplantation will find its proper place in medicine's fight against cardiac insufficiency, decades before artificial hearts.

The use of adult cadaver material, however, whether animal or human, implies that part of the life-span, dictated by the biological time clock, is used up. Against this, human *foetal cells* have a potential three score and ten years before them and could be used by organ culture. If the findings of the Peel Commission are implemented, foetal heart tissue will potentially be available in quantity. Thus, should tissue matching be important, its implantation will still be practical, even if eventually we find that the likelihood of an acceptable match is worse than the three hundred plus to one we expect with our limited knownledge of antigen histocompatibility.

CONCLUSIONS

Of the mechanical and biological methods of helping the patient with lethal heart disease that I have outlined there are several with potential. Circulatory support using a membrane oxygenator for partial cardiopulmonary by-pass shows slight promise. Animal-to-man transplantation is worthy of further research effort. Man-to-man transplantation has been shown to work and we are fortunate that a handful of workers continue to study this simple practical approach, despite a rather strange antipathetic social climate. The use of foetal tissues is still in its early stages, but like the embryo, it may have great potential in the correct environment.

When we examine heart replacement in critical terms, the apparent *embarrasse de richesse* disappears; we are left with only a few practical possibilities and a number of impracticable mechanical alternatives.

It might seem strange that in a bioengineering symposium I have chosen to criticise heavily the fashionable bioengineering lines of mechanical heart replacement. My reasons are simple; I believe that the future of our discipline is at risk. Our continued existence lies in a corporate medical, biological and engineering effort. We must not contract the serious disease of science, caused by inbreeding within a discipline. This arises when super specialists in a narrow field talk only to other super experts doing the same work, often unrelated to real needs. We must keep the breed healthy with cross pollination between fertile teams. If bioengineers are left to pursue impossible unworkable goals, governments, grant giving bodies and the man in the street will all soon notice the lack of progress.

For those of us who care about the quality of life the carrot is obvious; for those of us who care about the progress of science in our discipline, the stick is clearly imminent. There will be no funds for bioengineers if we do not have realistic and realisable goals. Charity and industry will soon tire of financing failure.

We should be fearless in our reappraisal of the structure and objectives of bioengineering. Whereas Galileo was imprisoned by the Inquisition for misbeliefs, we only have to put up with the criticisms of an uninformed establishment. The reluctance to abandon established but useless lines of research has hitherto stemmed from a lack of credible alternatives. We now have these alternatives. When the medical historians look back on the last third of the 20th century, I hope they will see not the extinction of bioengineering due to inflexibility but its turning promise into performance. This could be done by contributing realistically to heart replacement. Bioengineers will then help to relieve suffering, in the era before preventive medicine leaves us to die an ideal death as expressed three centuries ago in Dryden's Oedipus:

Of no distemper, of no blast he died,
But fell like autumn-fruit that mellowed long,
Ev'n wondered at because he dropp'd no sooner.
Fate seemed to wind him up for fourscore years
Yet freely ran he on ten winters more;
'Till, like a clock worn out with eating time,
The weary wheels of life at last stood still.

Acknowledgments
Where major research efforts are involved, the list of helpers is so great as to occupy too much space. I would like to take this opportunity of thanking so many of my colleagues, administrators, nurses, Ph.D. students, and surgical post-graduate students,

anaesthetists, and many others, who have helped to make the work that I am privileged to present today possible. Perhaps my greatest thanks should go to Charlie Gilmour and the group of Hospital technicians who have unstintingly given their spare time and set up and cleared up the hundreds of experiments we have done in various fields of research I have mentioned in this communication.

I would also like to thank the National Heart Hospital, The British Heart Foundation, and many industrial organisations and drug firms for their financial and practical support.

REFERENCES

BERGLUND, E., MUNROE, R. G., and SCHREINER, G. L. 1957: *Acta Physiol. Scan.* **41**: 261–268.

BRAUNWALD, E. 1971: *Am. J. Cardiol.* **27**: 416–432.

HUGHES, D. M., and LONGMORE, D. B. 1972: *Nature* **235**: 334–336.

LONGMORE, D. B. 1968: 'Spare Part Surgery'. Aldus Books, London.

LONGMORE, D. B. 1969: 'Machines in Medicine'. Aldus Books, London.

LONGMORE, D. B. 1971: 'The Heart'. Weidenfeld and Nicolson, London.

LONGMORE, D. B., and HUGHES, D. M. 1972: *Nature* **238**: 40–41.

RODBARD, S., WILLIAMS, C. B., and RODBARD, D. 1964: *Cir. Res.* **14**: 139–149.

SARNOFF, S. J., BRAUNWALD, E., WELCH, G. H. JR., CASE, R. B., STAINSBY, W. N., and MAGRUZ, R. 1958: *Am. J. Physiol.* **192**: 148–156.

SONNENBLICK, E. H., ROSS, J. JR., and COVELL, J. W. 1965: *Am. J. Physiol.* **209**: 919–927.

CARDIAC ASSIST DEVICES, TOTAL ARTIFICIAL HEART AND UNCONVENTIONAL DIALYSIS

YUKIHIKO NOSÉ[1] AND W. J. KOLFF[2]

TOTAL ARTIFICIAL HEART

It was 1957 when one of the authors (W.J.K.) began research on total mechanical artificial heart implantation. From that time on it was difficult, until only very recently, to keep an experimental animal alive with an artificial heart for longer than two or three days. However, in 1971 the long-standing question of whether the artificial heart could sustain the animal's entire circulation beyond the time of recovery from surgery and maintain the animal in a reasonable physiological state with a mechanical pump was answered. To date at least three groups in the country have been able to sustain a calf with an implanted mechanical heart for over three days. These long-term survivors with total mechanical hearts have brought forth a few unanswered questions.

In 1971 in Utah four calves implanted with artificial hearts survived over 100 hours. One of them lived for almost 11 days. The average was 179 hours; the range was 102–260 hours. Two of the four hearts were implanted with the aid of cardiopulmonary bypass; the other two under deep hypothermia with circulatory arrest for less than 60 minutes. The cardiac prostheses used for these studies were one heart with a smooth Silastic inner surface, three with Dacron fabric inner surfaces. After surgery the general condition of all four calves was good during most of the pumping time. This was indicated by self support, eating, micturition, defaecation and response to surroundings. Although emboli formation was minimised by the inner surface of the heart, the long-term effect of the artificial heart revealed a few problems:

1. Pulmonary insufficiency requiring eventual respiratory support (with spontaneous respiration of room air the pO_2 was approximately 60 mm Hg).
2. Infection.
3. Symptoms of right heart failure with high central venous pressure (over 10 mm Hg) which led to ascites formation, liver damage with central lobular necrosis, bile retention and renal failure,

marked tubular necrosis, infarction, with haemoglobin cast and increased blood volume, hydrothorax and oedema of the extremities and abdominal organs.

These findings are the same as those found in the animal after only a few days' survival with an implanted artificial heart. However, these long-term survivors have certainly changed the scope of the artificial heart programme and have given more confidence about the possibilities of a mechanical cardiac prosthesis and its reality.

Surgery on an animal such as a calf is more difficult than on a human. Of course many failures can be attributed to minor errors in surgical technique. The accumulation of these various problems certainly limits the survival time of the animal. In our own experience we know that many of the problems were not created by the mechanical heart but by poor handling of the animal, poor surgery, air emboli after surgery, excess bleeding, overtransfusion of the animal, etc. It is also true that in the early stages of implantation, improper design and fabrication of the cardiac prosthesis, poor quality control, and other mechanical faults were apparent. We can now say that at present these obvious problems have been eliminated. However, after overcoming the initial problems and maintaining the experimental animal over weeks, three major problems became gradually obvious: 1, poor haemodynamic performance of the cardiac prosthesis; 2, limited durability of the mechanical heart; 3, blood coagulation problems (see figure 1).

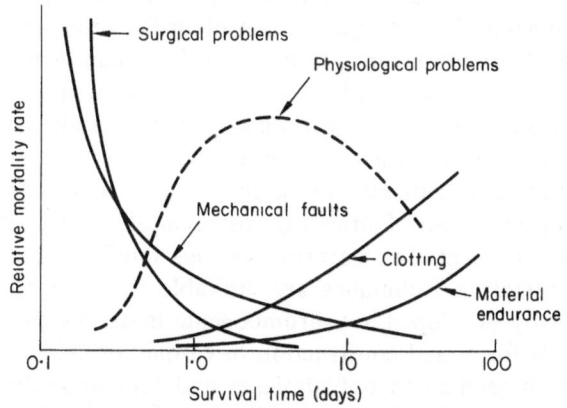

Figure 1. Artificial total heart recipient mortality.

[1] Department of Artificial Organs, Cleveland Clinic, Cleveland, Ohio, U.S.A.
[2] Department of Surgery, University of Utah, Salt Lake City, Utah, U.S.A.

It is surprising that groups having long-term survivals recently, have rarely seen clot formation to the heart, major infarcts, or intravascular disseminated coagulopathy. The group at the University of Mississippi uses a silicone heart with an inside surface of Silastic. Prior to use they coat the inside surface with Silastic adhesive diluted with Xylene thus providing a fresh coating of silicone rubber all over it. This eliminates connection protuberances and unevenness. The group at the University of Utah uses a Silastic surface with Dacron fabric. The Cleveland Clinic group uses aldehyde-treated biological tissue such as aorta intima or pericardium. To date the longest survival has only been 11 days, but once this time is extended, clotting will probably be a problem again. It has been reported that a Dacron fabric-inlaid cardiac prosthesis developed by the Boston group can remain patent in the animal for 180 days and the Cleveland Clinic group showed that three calves with cardiac prostheses having inflow and outflow valves (the entire inside surface being made of biological tissue) implanted in the aorta survived over five months. This so-called 'biolised' artificial heart showed increased neo-intima formation up to three weeks with a decrease in neo-intima formation thereafter. This is somewhat different from what we have seen with the velour- or fabric-coated artificial heart. With those hearts it was difficult to control the thickness of the neo-intima formation, and again the anticoagulation treatment was recommended.

Many attempts have been made to develop thromboresistant materials. So far any newly created material has presented some problems in one way or the other. Silicone rubber is still considered to be the most reliable material for cardiac prosthesis. While some materials have shown better thromboresistant properties than silicone rubber, their mechanical strength when used in constructing a cardiac prosthesis is rather disappointing. The problems of mechanical durability have also become more noticeable lately. The production of a more durable mechanical heart is more important now than before because animals can be maintained for longer periods of time. To date all cardiac prostheses used for total implantation have been handmade. They were fabricated in the laboratory from moulds by dipping or layering. Because they are handmade it is difficult to achieve reproducible and flawless cardiac prostheses. Certainly a more industrial approach for fabricating mechanical hearts is necessary at this moment. Various reinforcing or laminating techniques are desirable to make the newly developed antithrombogenic materials suitable for actual construction of cardiac prostheses. Both segmented polyurethane and biocompatible natural rubber have a rather short flex life which make them unsuitable for cardiac prosthesis construction as a singular component. More emphasis should be placed on this problem which has almost been overlooked in the past.

The final problem, and probably the most important, is just how much of the function of the natural heart should a cardiac prosthesis simulate. The cause of death in the long-term surviving animal is still right heart failure or left heart failure, particularly the former. Again this comes back to the poor performance of the inflow mechanical artificial heart valves. To augment reasonable drainage of the blood into the artificial heart it is necessary to have a low resistance, low regurgitating inflow valve. So far none of the commercially available heart valves have been able to meet these requirements. Various investigators have been using handmade leaflet heart valves, but the durability and reproducibility of these leaflet valves is still uncertain.

At the Cleveland Clinic the aldehyde-treated homologous heart valve was used as the inflow and outflow valve for the cardiac prosthesis. The entire surface of the prosthesis was constructed with biological tissue which was treated with aldehyde. Compared with any mechanical leaflet heart valve, this natural valve showed excellent haemodynamic performance. For the inflow valve the large aortic valve from an adult bovine can be used. The heart valves and aorta can be utilised as a unit, thus preserving the natural configuration of the aorta; this also produces reasonable haemodynamics and eliminates blood coagulation around the valve area.

For driving the artificial heart, there is a tendency toward using simpler driving schemes and various investigators have decided not to have any closed-loop feedback mechanism. In all the long-term survivals the driving system used was quite simple and all the cardiac output regulation was expected to be performed by the heart device itself. The Starling's Law curve is becoming more and more important.

The use of atomic energy for the driving scheme has also proved applicable. Recently the Artificial Heart Program has used an atomic energy driving scheme to drive a Harmison–Thermo Electron pump showing the feasibility of this approach. Recently the Atomic Energy Commission (AEC) has decided to develop an atomic energy total heart driving and control scheme. With the interest of the AEC and NIH (National Institutes of Health), this totally implantable atomic energy driven artificial heart is realistic at this moment. This type of total implantable driving scheme seems to be gaining more and more interest. The experiences of the Michigan group when Dr A. Kantrowitz implanted a patch graft to the aorta and used it as an implantable booster heart were unfortunate.

The inside surface was Dacron fabric, electro-conductive polyurethane ('electrolour') and was sutured into the aorta of the patient. After several months of survival this booster heart improved the cardiac condition of the patient, but in the end the patient died of infection. Certainly this is good evidence supporting the use of a totally implantable mechanical device. At present most investigators are using conventional, large driving systems outside the body and whether it is driven by air or liquid, the driving line has to cross the chest wall. That this type of physiological access method can be feasible was demonstrated by the Scribner–Quinton shunt for dialysis; however, the driving lines leading into the chest cavity are of a reasonably large diameter and the multiple tubes need reasonable physiological access devices. So far, the development of a physiological access plug has been disappointing. Probably, the solution will be a totally implantable mechanical device five years from now. However, the future development of a durable, antithrombogenic and haemodynamically reasonable and compact cardiac prosthesis should not be underestimated. Many physiological phenomena, physiological parameters, and animal-pump interface problems have not yet been solved.

BLOOD COMPATIBLE MATERIALS

To date the development of necessary biocompatible materials has been under-emphasised. This is due to the rather long survivals and less incidence of blood coagulation, the reasonable success with the fabric inlaid cardiac prosthesis and the frustration generated by spending large amounts of research funds in attempts to produce better antithrombogenic materials. About 10 years ago the key to understanding thromboresistant properties seemed to be rather simple, and it was thought that these phenomena were understood. However, after more investigators became involved, introducing new theories and materials, the factors thought to be essential for keeping the prosthesis antithrombogenic were seen not always to be true. Even a simple question like whether a smooth or rough surface is better remains unanswered. Long-term patency has been seen for both types of surfaces. Additionally, even though to the naked eye a surface looks smooth, it may still be a rough surface to the 'eye' of the blood and *vice versa*. Probably the electroconductivity, electrical charge, surface finish, inertness and heparin coating which were thought to be contributory factors to antithrombogenicity, were not always reproducible. An entirely new approach is needed to the understanding of antithrombogenic problems.

One of the authors (Y.N.) has proposed a process, biolisation, to generate biocompatible materials.

It has been well documented that any artificial material whether or not it is hydrophilic, hydrophobic, inert, electroconductive, or heparinised is covered with protein almost instantaneously after implantation in the vascular system. Regardless of this protein coating some of the materials still showed better thromboresistance than the others. While many theories have been proposed to explain the thromboresistant properties of such materials, it is difficult at this moment to be specific regarding thromboresistant properties. Coating the surface of a cardiac prosthesis with a monomolecular layer of denatured albumin has been proposed, but its effects were rather difficult to reproduce. It was also found that while not every protein surface showed thromboresistance, when it is treated by aldehyde or heat it did show reasonable and reproducible thromboresistance. Such protein can be mixed in a synthetic material as in aldehyde-treated albumin, natural rubber coated over plastic or it can be in natural tissue. The treatment described probably makes the protein insoluble, cross-linked or denatured. What kind of process constitutes the main role is not completely understood at present, consequently it is difficult to select an appropriate term to describe it. Thus the term 'biolisation' came into use. Together with this term, a possible process for production of biocompatible (particularly blood compatible) material was also proposed. It is suspected, incidentally, that not only proteins but other biological components such as polysaccharides have a similar effect. Thus the material to be used for this process can be either of natural or artificial origin, and should contain protein, polysaccharides, or other biological components which are not specifically known yet.

Artificial synthetic materials can be appropriately conditioned, either by the admixture of, or coating by biological components (biological activation). When the surface is then treated with aldehyde or heat it becomes biologically inactivated. This probably means insolubilisation, denaturation, or cross-linkage of the biological components and the others. Natural tissue of a different or remote tissue typing group can also be treated and inactivated in the same way so that such tissue can then be used for implantation without triggering the intensive immunorejection process.

Heterologous or homologous preserved aortic valves have already been successfully used clinically. A low incidence of thromboembolic complications was unanimously reported. In the past attempts have been made to implant biological tissue as in transplantation to keep the biological activity as normal as possible. However, when the experimental species is different, this biological activity stimulates the defence mechanism of the body. By making the surface protein non-specific heterologous tissue can be implanted. Some of the tissue used is actually serving, not as a biologically functioning organ but as part of the mechanical structure (vascular graft or heart valves). For this type of application this non-viable graft will serve the purpose of implantation. If nonviable tissue material can be used as a part of the cardiovascular prosthesis combined with synthetic material, it will open up a new field for the construction of thromboresistant and haemodynamically ideal devices.

In summary, for biolisation to occur there must be a protein or biological component on the surface; this protein or biological component must be rendered non-specific, cross-linked, or insoluble by aldehyde, heat, or some other method; the surface must be wet. From the viewpoint of the recipient of the device it is probable that the differentiation between transplanted natural tissue and implanted prosthesis is not possible. Thus the reaction of the recipient system to the foreign system is probably the same whether the foreign system is of natural or synthetic origin. Or more simply stated: the biolisation process converts material in such a way that the body recognises it as biological tissue, but most importantly, the tissue is viewed as being inactive, of no threat, and therefore compatible with the body and blood. It should be noted that this theory is only an hypothesis proposed by one of the authors, and is not yet generally recognised.

CARDIAC ASSIST DEVICES

The use of circulatory assistance can support selected patients through periods of failure of the natural heart, thereby providing time for accurate diagnosis and possible medical, surgical, or spontaneous correction of basic cardiac defects. The timely use of circulatory assistance in some patients may be able to arrest progression of the basic disease, or even prevent it in critical periods. A permanent circulatory assist device could extend the life of a patient with a functionally inadequate heart.

It is certainly helpful if cardiac assist devices can be used for patients with myocardial infarction. The need for this is more stressed recently because of the current expansion of the surgical indication for coronal revascularisation procedures.

In principle, a circulatory assist device must be capable of taking over from the diseased heart both the volume and pressure work necessary for the maintenance of adequate cardiac output.

Since the majority of the work necessary for the heart is in the left side, left heart support is generally of major interest. The object is to reduce myocardial work and thus diminish oxygen demand. The

following methods were generally used for this purpose.

1 Reducing the heart's volume work by pumping blood from the venous system to the arterial system. As a venous system a large vein, atrium, or ventricle itself can be used. The left heart bypass pump developed by the Boston group (Thermo Electron–Harmison pump) is one of the foremost of this type.

The double-valved pneumatic bladder pump accepts blood directly from the left ventricular chamber during systole at low pressure and pumps it at arterial pressure into the aorta. This device is capable of reducing left ventricular work to a minimum. Implantation of such pumps in calves has been studied for periods of up to four months. This pump was coupled with a nuclear powered driving system so it can be implanted permanently for probably up to 10 years. A fraction of the radio-isotope heat from a Pu-238 fuel capsule is converted into hydraulic power for driving a blood pump via a miniature thermal engine. Certainly after further development of this driving system it can be coupled with any cardiac assist pump and made available for a totally implantable cardiac assist system.

2 Reducing the end-diastolic pressure in the aorta. Because about 75% of total myocardial work is expended in opening the aortic valve (pressure work), a circulatory assist device must in one way or another reduce the end-diastolic pressure gradient across the aortic valve.

Several modes of circulatory assistance were developed and investigated. For this purpose simplicity and minimum surgical or related procedures are more advantageous because of the need in emergency or ambulatory applications. External body compression by a pressure suit in systole has shown limited effects in clinical cases.

Several types of counterpulsation via external catheters and pumps were studied. Blood is withdrawn from the body during cardiac systole and reinjected during systole. A primary arterial-arterial counterpulsation system can be expanded to a veno-arterial or aortic-aortic system.

Intraortic balloon pumping is one of the simple methods for providing counterpulsation. For this reason, this procedure has been investigated probably to an extreme producing at the same time much controversy and discussion. Especially immediately after the onset of coronary thrombosis and before the time for surgery, this intraaortic balloon pumping will certainly support the patient in shock or at least supply more blood to the marginal area of the infarct. It is also true that after a coronary revascularisation procedure is performed, some of the patients obtain great benefit from intraaortic balloon pumping. The mortality after this procedure is reduced and the surgical procedure is applicable to patients who at present die before they could get the benefit of more advanced coronary surgery. The procedure was developed by Moulopoulos and Kolff in 1962, but it took 5–10 years to get the recognition of the medical community.

3 Combination of approaches 1 and 2. For assisting the heart the indication for intraaortic balloon pumping and counterpulsation is limited, and as an additional possibility the Zwart–Kolff catheter approach for the left heart bypass (figure 2) is worthy of consideration. This certainly will help patients who would benefit from counterpulsation

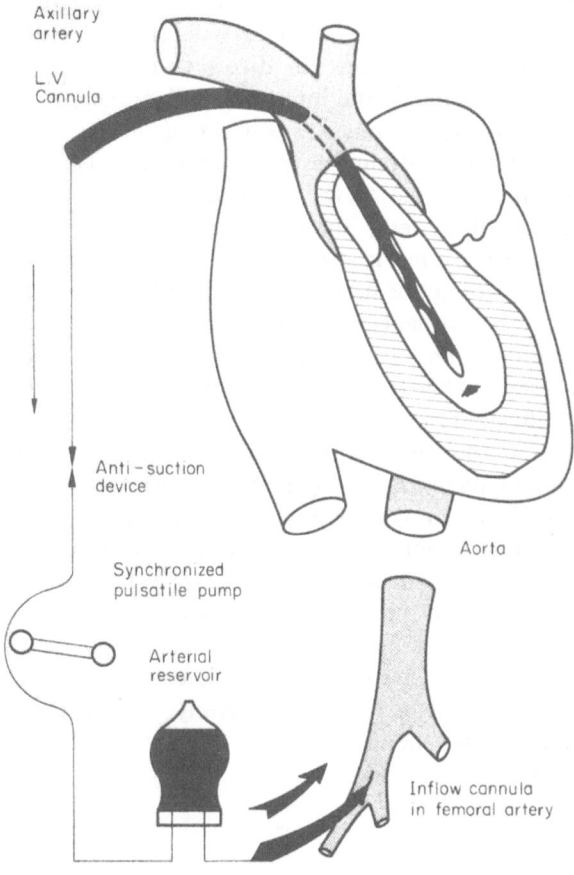

Figure 2. Zwart-Kolff catheter approach for the left heart bypass.

and in addition other patients who have developed fibrillation or myocardial pump failure. In such cases this left heart bypass approach is more favourable than just simple counterpulsation. It is felt that in the future the development of this type of procedure should be greatly emphasised.

UNCONVENTIONAL DIALYSIS

Haemodialysis is now well accepted and at present probably 10 000 patients are on haemodialysis throughout the world. Still this haemodialysis population represents barely one-third to one-half of those who need the treatment. Other

pessimistic thoughts on haemodialysis are that it is a rather complicated procedure, expensive, it is dangerous because blood must be handled, and there is always the danger of viral hepatitis which could be fatal to physicians, nurses, and technicians treating the haemodialysis patient. It is felt that such thoughts relating to the undesirable effects of haemodialysis are probably caused by insufficient degree of haemodialysis. If the number of haemodialysis hours is increased utilising better membranes then possibly these 'effect' problems encountered may disappear. Certainly it would be ideal to develop a more permeable membrane, particularly for larger molecular weight substances. With frequent dialysis the complications of chronic uraemia tend to disappear. However, the progress in making a more permeable membrane has been disappointingly slow and it is certainly hoped that the 1970s will see it improve. Enzymes could probably be grafted to polymers to promote transportation. Peritoneal lavage offers a very permeable membrane, but it has fallen into discredit in the treatment of chronic renal failure. Dr Tenckhoff has however convincingly shown that it should be revived in the 1970s. He has children on peritoneal lavage who continue to grow. He has emphasised the necessity of doing away with the multiple batch dialysis system and has produced a method of sterilising batches of dialysate with heat. The University of Utah has developed a machine that sterilises the fluid by reverse osmosis. This will be further developed in the 1970s and peritoneal lavage could become an alternate treatment for patients in whom access to blood vessels is difficult or impossible, also for diabetics and for young children. In East Germany, where this practice is employed, the deaths from glomerulonephritis in small children is 0.75 per million per year. In the United States (population over 200 000 000) that means that over 150 children per year could be helped with peritoneal lavage.

The efficiency of chemical absorbents in the dialysate solution was demonstrated by Gordon. Utilising these chemical absorbents it is now possible to dialyse with 1–2 litres of dialysate instead of the normal 300 litres needed for a period of 10 hours. It seems to be particularly appealing to run the blood directly over a chemical absorbent such as charcoal as was done years ago by Yatzidis. Presently Dr Andrade at the University of Utah can coat the charcoal with either Hydron or albumin and thereby eliminate the unfavourable effect of thrombocytes while maintaining the adsorbing effect. Since the charcoal absorbs particularly well what cellophane based dialysis does least, that is the removal of large molecular substances, the combination of the two seems to be hopeful for the future.

For direct contact to the blood or for the digestion

method (eatable adsorbents) the adsorbent must be encapsulated. This is necessary to prevent interaction with the blood constituents, with food, and with gastrointestinal fluid. Chang carried out two-hour haemoperfusion using microencapsulated activated charcoal and found it comparable or superior to standard haemodialysis for the removal of creatinine and uric acid. Gordon has reported maintenance dialysis with 1–2 litres of dialysate containing chemical adsorbents (activated charcoal, zirconium phosphate, zirconium oxide and the enzyme urease). His two patients were maintained for 9–10 months with dialysis twice weekly for 7–8 hours per dialysis. In this system encapsulation of the adsorbent is not necessary since the adsorbents are not in contact with blood constituents and macromolecules which would limit the removal capabilities.

If this chemical adsorbent concept is feasible it is probably also possible to revise intestinal dialysis. A two-year followup on a patient dialysed by isolated intestinal dialysis was reported by Clark in 1965. This has shown potential and definite advantages. A non-sterile dialysate can be used and no blood access is required. Thus there is no danger of air emboli, bleeding, or blood coagulation; no risk of infection, such as occurs in the shunt site for haemodialysis, or peritonitis occurring with peritoneal lavage and no sophisticated or complicated equipment is required. The procedure is simple and easy to perform and requires no specialised training as required for haemodialysis and peritoneal dialysis. However, there are also disadvantages and problems associated with intestinal dialysis: laparotomy is required; the rate of dialysis is less than haemodialysis or probably peritoneal dialysis particularly for uric acid; and it is rather difficult to control the water, electrolyte and phosphate balance.

By combining the two approaches, isolated intestinal dialysis and chemical adsorbents, 24-hour intestinal dialysis without any connection to external tubes or a dialysate supply line is feasible. This system could function 24 hours per day just like the normal kidney. The addition of chemical adsorbents would serve to maximise the transport capabilities of the perfused isolated intestinal loop for any specific metabolites or electrolytes to be removed. Because of the low volume of dialysate good control of the electrolytes and water balance would be possible and by having the dialysate mixture in direct contact with the dialysing area of the intestinal wall, transport will not depend on dialysate flow rate as is the case in the present dialysing system. Further as adsorbents are not exposed to the functioning of the gastro-intestinal system, these dialysis procedures will not affect normal gastrointestinal function. Encapsulation will not be necessary to prevent adsorption of food constituents, organic macromolecules, food products and gastrointestinal juices as none will be present. This can be said to be an implantable artificial kidney.

For conventional haemodialysis one thing we should mention is single needle dialysis. In 1964 Dr Twiss in the Netherlands applied the single cannula principle in a Shaldon cannula for dialysis. Few people paid attention. In the University of Utah Dr Kopp invented a single cannula method to be used with the usual fistula. It was proved that a single Teflon needle with side holes can effectively maintain haemodialysis. Hopefully, this procedure will expand the use of the A–V fistula for haemodialysis.

DEVELOPMENT AND TESTING OF PROSTHETIC HEART VALVES: CARDIOVASCULAR SIMULATION AND LIFE SUPPORT SYSTEMS

M. M. BLACK[1]

INTRODUCTION

The major part of this paper deals with the design, development and testing of flexible leaflet prosthetic heart valves. Other applications of the valves are indicated so as to give as complete a picture as possible of the likely biomedical uses of this type of valve. Various *in vitro* test rigs have been developed to evaluate the performance of these valves including a simulator for the left side action of the heart. Details of this equipment are given as they may be of interest to others working in this field.

It should be noted at the outset that it is not intended to review in detail the very wide range of research in this field which has been carried out in many other centres. Nevertheless, where appropriate, reference will be made to other forms of valve development, if only to indicate the diversity of designs already tried and why there is still a need for further work on this topic.

PROSTHETIC HEART VALVES

The earliest recorded clinical use of a prosthetic valve was made by Hufnagel (1954) although Denton (1951) had previously described the successful use of prosthetic mitral valves in animal experiments. According to McGoon (1971), there is now a 'vast number of types and models' of such valves. This latter fact is in itself not surprising considering the number of researchers in different countries working on this topic over a period of twenty years.

The unusual result of all this effort is that even today the most successful valve so far, that is the Starr–Edwards ball and cage type, bears no resemblance to any of its biological counterparts. In a recent paper by Wright (1972) prosthetic heart valves are classified into five major types: (i) unicusp or flap valves; (ii) tricuspid flexible leaflet valves; (iii) ball and cage valves; (iv) low profile disc valves; and (v) pivoting disc valves. Examples of valves corresponding to each of these categories are also indicated by Wright. Even this classification is not sufficiently comprehensive since for example the hinged-leaflet valve developed by Gott (1964) is not readily included in any of the above categories.

In the range of valves developed so far the only form which is in anyway similar to the human valve is the flexible leaflet type. The history of prosthetic valves of this format is almost as long as that of prosthetic valves themselves. Roe (1958) produced a tricuspid flexible leaflet valve which unfortunately was subject to fatigue failure of the leaflets. Indeed the major problems common to the majority of valves of this type have been leaflet fatigue and thromboembolism. Thus although the development of such valves occurred very early in this field of prosthetic devices they have as yet never proved satisfactory, especially when compared with the results obtained by the use of the Starr-Edwards valves. Most thoracic surgeons do not appear to have much faith in the idea of a flexible leaflet valve being produced which will yield successful clinical application.

However, advances in polymer science are continuing and it seems likely that a suitable material will be produced. Such a material may be either a specially developed polymer or a bio-polymer, possibly using tropocollagen as a base. Work on this latter type of material in relation to heart valves has been reported by Carpentier (1971).

The flexible leaflet valve offers several advantages over all other forms since it is conceptually similar to its biological counterpart. It is centrally opening and will create very little turbulence. The closing action of the leaflets combined with the low level of turbulence will greatly reduce mechanical damage of the blood (a low level of haemolysis). In addition, it is likely that the flexibility of the leaflets will assist in valve closure when placed in the low pressure right side of the heart. On the other hand, very thin leaflets of a suitable material could react to the haemodynamic conditions of the aortic and mitral positions where the development of vortices has been shown by Bellhouse (1972) to assist in valve closure.

For all these reasons it is justifiable to continue the search for a satisfactory flexible leaflet prosthetic valve. In the section which follows, details are given

[1] School of Applied Sciences, University of Sussex, Brighton, England.

21

of the particular forms and methods of manufacture and testing which have been adopted in the development of such valves at Sussex.

DESIGN AND TESTING OF FLEXIBLE LEAFLET VALVES

As already noted, the tendency for fatigue failure of leaflet valves is very pronounced. To an engineer this phenomenon is a function of material behaviour and stress distribution arising from deformations. Flaws in the material are a likely source of fatigue failure. If such flaws coincide with positions of 'stress concentration' the probability of failure is obviously very much increased. A good design must therefore aim at minimising these effects.

The majority of leaflet valves developed so far have been made from man-made materials usually in the form of a woven mesh suitably coated to produce the required impermeability. The possibility of inherent flaws in such a material is likely to be high. In addition the use of such material usually implies a manufacturing process which involves several steps and inevitably requires the common attachment of several pieces to form the finished product. Once again the introduction of flaws becomes probable.

To overcome these problems the major part of the Sussex valves is a single homogeneous polymeric material which is hot moulded in a single stage operation. The dacron sewing cuff is incorporated into the valve in the moulding process such that it is impregnated in part by the polymeric material of the valve. In designing the moulds particular attention has been paid to reducing sharp re-entrant corners so as to reduce stress concentration levels.

Another important feature of a leaflet valve is to produce a geometrical format which permits the maximum amount of orifice area to be available for flow. In this sense the ideal would be a leaflet which folds back completely to the base diameter of the orifice. No leaflet valve form of the earlier designs had this specific property.

The range of flexible leaflet prosthetic heart valves produced so far are shown in figure 1.

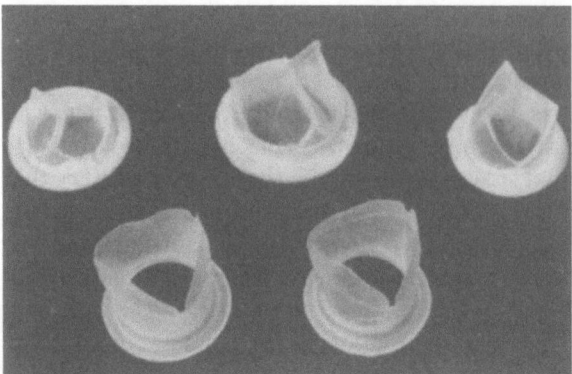

Figure 1. Flexible leaflet valves produced at Sussex.

The top three versions are either curved or plane tri-leaflet in format and examples of each valve with their corresponding moulds are shown in figures 2, 3 and 4.

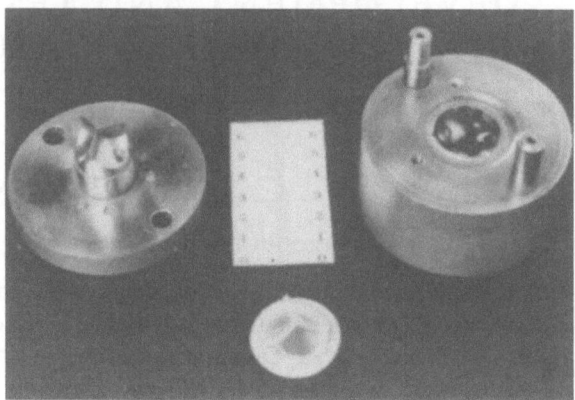

Figure 2. Curved tri-leaflet valve and mould.

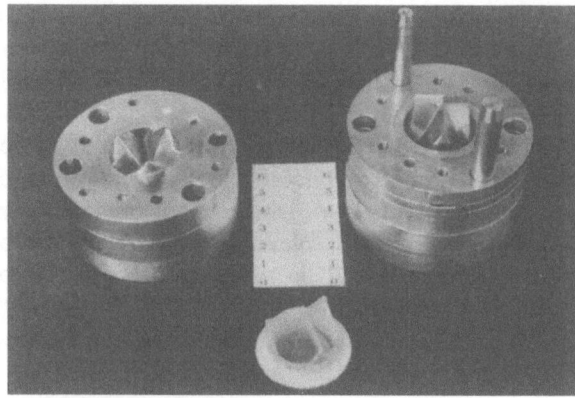

Figure 3. Plane tri-leaflet valve and mould.

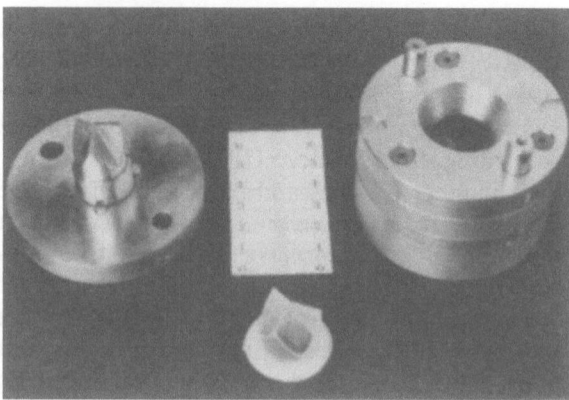

Figure 4. Plane tri-leaflet valve with reduced wall height and mould.

The curved tri-leaflet valve shown in figure 2 was not considered ideal as the maximum amount of leaflet deformation possible still did not provide a sufficiently large flow compared with the available base area of the orifice. The plane tri-leaflet valves shown in figures 3 and 4 are so designed as to allow the leaflet to fold back much more closely to the base. diameter thus providing a greater proportion of the orifice for flow. These two valves have been the subject of considerable *in vitro* testing which will be discussed later.

The two valves shown in the bottom of figure 1 and again with the corresponding mould in figure 5

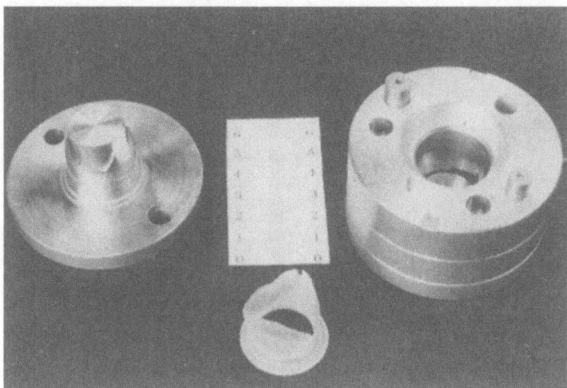

Figure 5. Truncated conical leaflet valve and mould.

are the most recent in concept which have been produced. Initially this valve is taken from the truncated conical mould without any apparent leaflet format present. However, the mould is so designed as to produce thickening of the wall at appropriate positions such that when the valve, held in a special jig, is cured, the final result is the triangulated shape of the downstream end as shown in figure 1. This valve provides the maximum possible flow area and is designed for the aortic and mitral positions where the vortices already referred to assist in closing the valve.

In all the valves described so far the downstream

leaflet edges are so arranged as to form a surface to surface, rather than edge to edge, contact on closure. This surface contact in the closed position can be seen clearly in the top row of valves in figure 1.

Each of the valve types produced so far has been subjected to substantial *in vitro* test programmes.

Figure 6. General view of simulator for the left side action of the heart.

These programmes involve two quite distinct lines of investigation. In the first case the valves are incorporated in a simulator for the left-side action of the heart. This equipment is shown in figures 6 and 7. The system functions by subjecting a latex 'ventricle' to pressure-time variations corresponding to those obtaining in the human heart. The pressure-time waveform is produced by means of a specially designed piston with air ports, a cylinder with

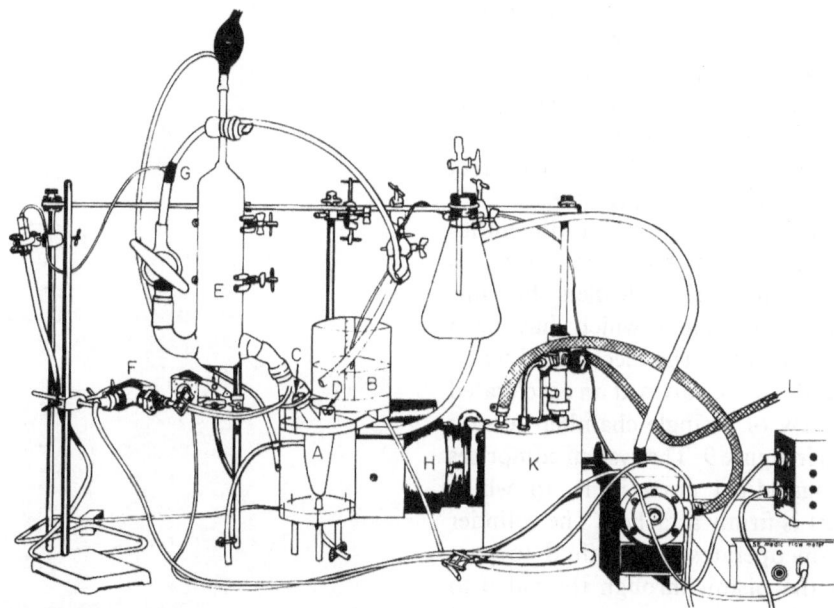

Figure 7. Diagrammatic view of simulator for the left side action of the heart. A latex ventricle, B atrial chamber, C 'aortic' valve, D 'mitral' valve, E aortic compliance chamber, F pressure transducer, G blood flow cannula, H variable speed drive, J piston and cylinder producing pressure conditions, K constant pressure air reservoir, L air supply to reservoir.

appropriately positioned pressure release valves and a constant pressure air reservoir.

This arrangement can produce both normal and abnormal ventricular pressure-time waveforms and can operate at pulse rates varying from 50 to 350 cycles per minute. A typical ventricular pressure-time trace from this equipment, recorded on a storage oscilloscope is shown in figure 8. The system is so designed as to enable the 'atrial'

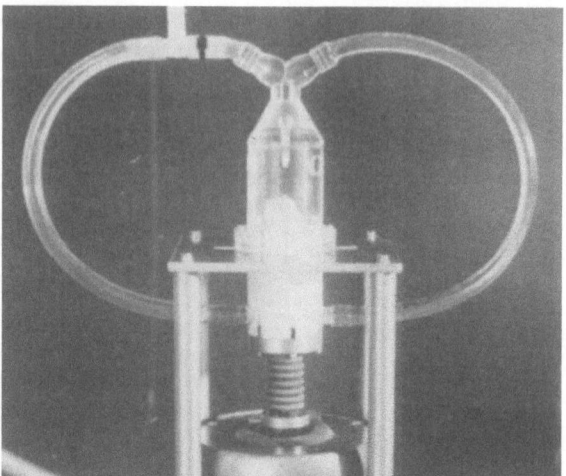

Figure 9. Single chamber valve fatigue test rig.

by the opening and closing action of the valve. The vibrator driving the piston is easily controlled for both frequency and amplitude. By this means the complete opening and closing cycle of the valve can be produced at varying rates at present up to a

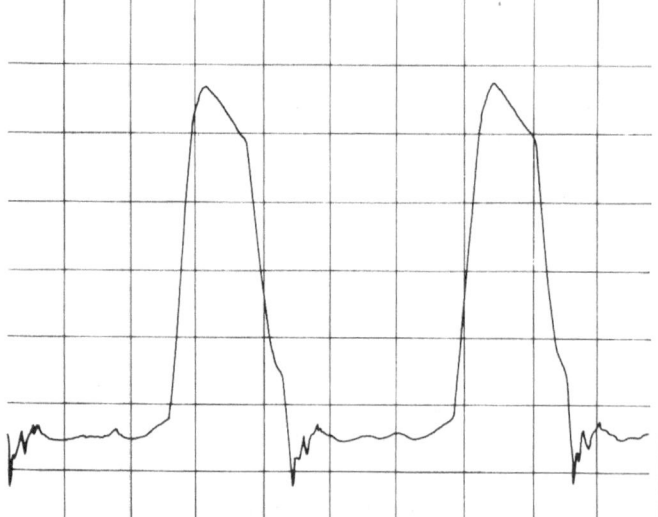

Figure 8. Typical ventricular pressure-time waveform reproduced by the simulator.

pressure and 'aortic' compliance to be varied. Flow rates are metered by incorporating a cannula of a blood flow measuring device in the 'aorta'. Pressure readings at different positions are recorded either from tapping points in the vessel walls or by standard cardiac pressure catheters.

Using this test rig it is possible to obtain *in vitro* measurements of the effects of various parameters, for example pulse rate and aortic compliance, on the behaviour of valves. Apart from evaluating the competence of the valve this rig also establishes whether or not a particular valve allows the correct volume of flow at the appropriate pressure.

As noted earlier, one of the major difficulties experienced in the use of flexible leaflet valves so far has been fatigue failure of the leaflets. For this reason the second test system which has been developed is designed to test the mechanical wearability and fatigue life of the valves at an accelerated rate. A close-up view of a single chamber version of this rig is shown in figure 9. The system comprises a piston with longitudinal ports in it, to which the valve under test is firmly attached. The cylinder in which the piston operates allows fluid to pass from one side of the piston through the valve to the other. The complete perspex cylinder and ancillary flow pipes are filled with fluid and the valve is made to move up and down through the fluid. In this way fluid is pumped round the system

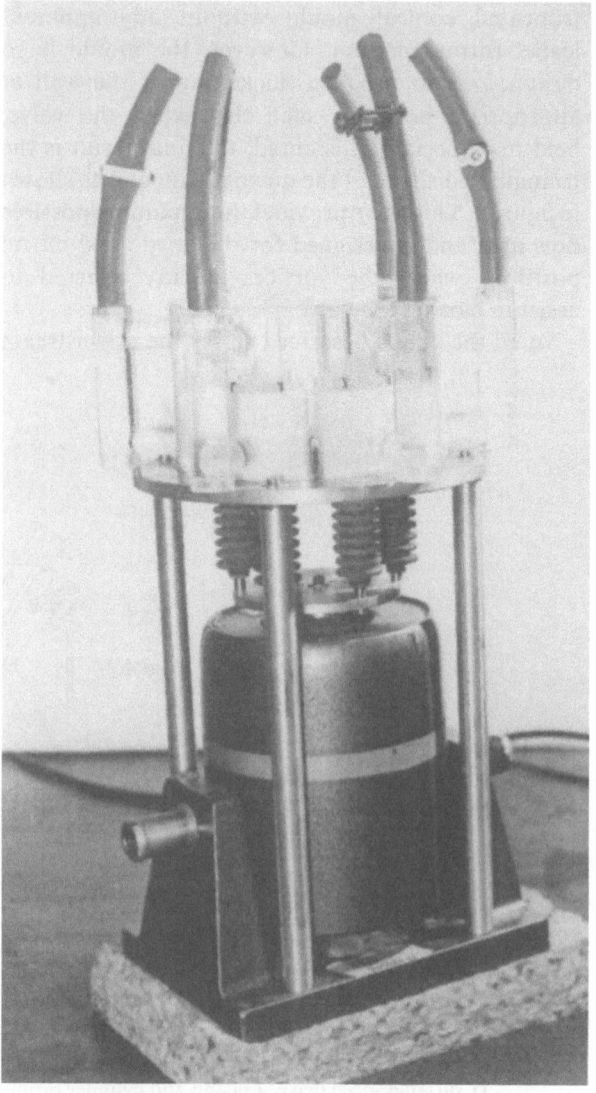

Figure 10. Six-chamber valve fatigue test rig.

maximum of 7.5×10^3 cycles per minute. This means that with continuous running, for which the equipment is designed, one year's normal heart valve operation can be reproduced in less than two days. Results from this test rig have shown the valves to be capable of more than ten years normal life wear when operating in water. Unfortunately the effect of the biological environment on the valve material is likely to reduce this life.

In order to test simultaneously a number of valves in different fluids, a second rig was produced which works on the same principle and comprises six test chambers. This equipment is shown in figure 10. Valves are being tested in this rig in water, lipid solution and blood plasma and the results so far, indicate considerable promise for the life of the leaflets even in the lipid and plasma environment.

These tests cannot however truly establish the effects of the biological environment on the material and a series of test specimens of the particular silicone rubber being used for the valves are being implanted in rabbits to study in more details the degradation of the material. These specimens and their moulds are shown in figures 11 and 12. This particular part of the research programme is not yet complete although initial results do indicate a drop in the ultimate tensile strength of the material after four weeks of implantation in the animals.

A programme of animal experiments involving goats has now been started for the actual implanting of the valves. Results of one test, where a valve of the type shown in figure 4 was put in the pulmonary

position, are shown in figures 13 and 14. The pressure-time waveforms describe the right ventricular and pulmonary artery pressures before and

Figure 11. Flat test specimen of material and mould.

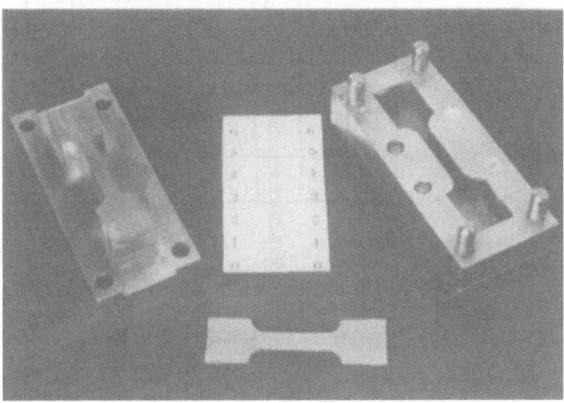

Figure 12. Circular test specimen of material and mould.

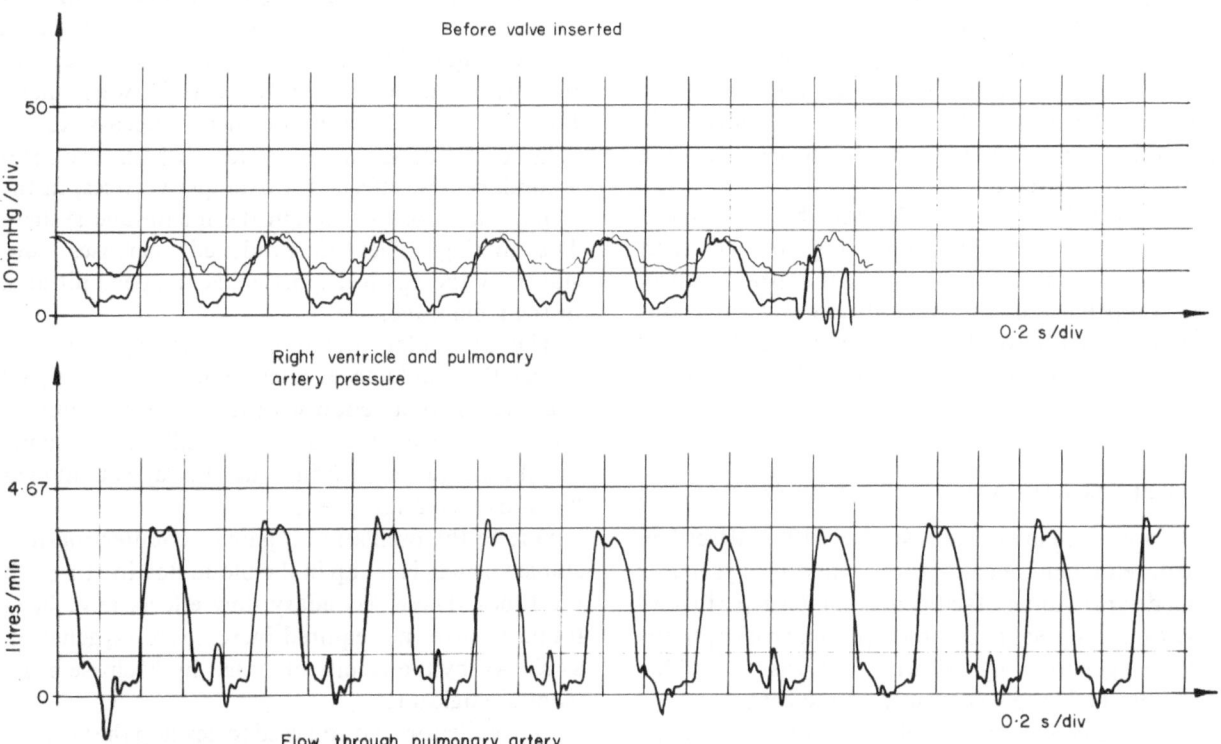

Figure 13. Recordings of goat's right ventricle and pulmonary artery pressure and pulmonary artery flow before valve implant.

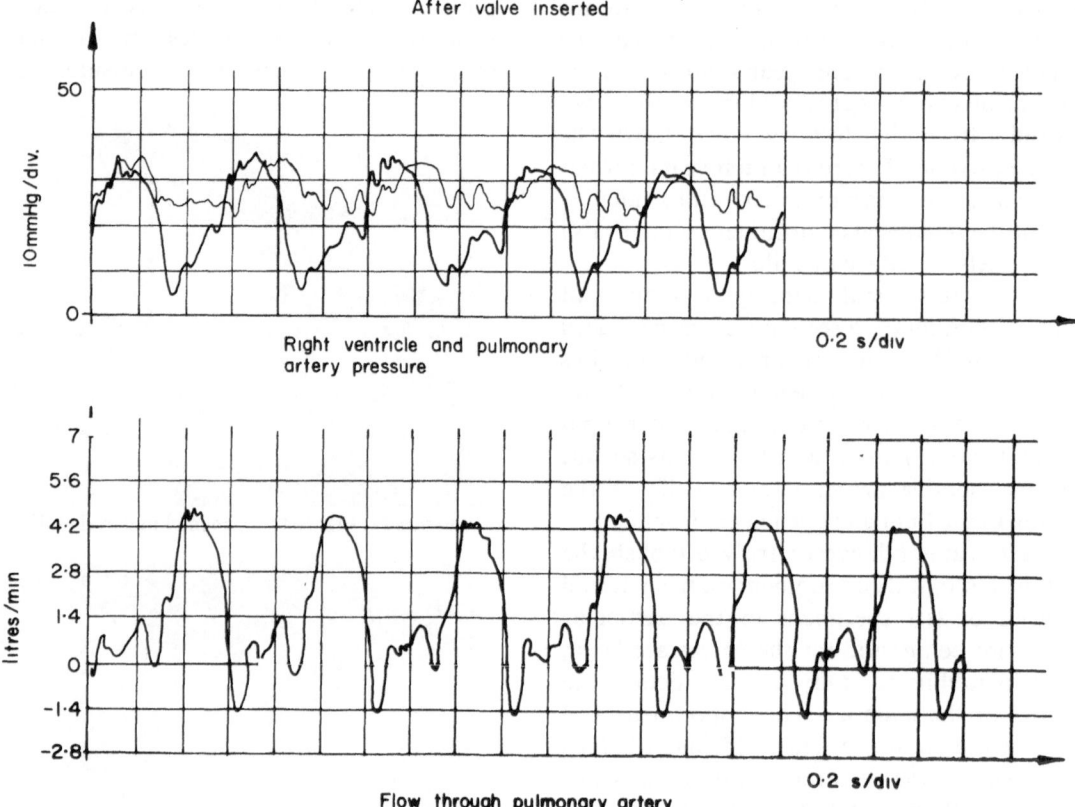

Figure 14. Recordings of goat's right ventricle and pulmonary artery pressure and pulmonary artery flow after valve implant.

after valve implant. In addition, the lower trace shows the blood flow through the pulmonary artery, again before and after valve implant. The satisfactory performance of the valve *in vivo*, is clearly illustrated by these results which show a slight rise in pressure and corresponding flow after implant. The traces shown in figure 13 were taken after the goat had been anaesthetised. It is also noticeable that there is minimal pressure difference across the valve at peak systole.

All the valves tested so far have been produced without any metal stiffening rings in the base. The lack of such stiffening does not appear to affect the competence of the valve in any way. Previous flexible leaflet valves have usually included a metal former in order to maintain both the valve form and its competence.

FLEXIBLE LEAFLET VALVES IN A PUMP-OXYGENATOR

In a recent paper, Melrose *et al.* (1972) described the development of a new form of pump-oxygenator aimed at providing a facility for long-term perfusion. Existing equipment using conventional roller pumps and oxygenating by allowing the gas to bubble through the blood has been shown to cause damage which limits the period of time for which such equipment can be used. Considerable research effort is now being expended in a number of centres

in attempts to produce a satisfactory oxygenator, based on the use of newly developed silicone membranes, in which the blood does not come into direct contact with the oxygen. The major difficulty appears to be the very large area of membrane surface required to provide sufficient oxygen uptake if uniform blood flow only exists between membranes. However, this problem appears to be soluble by designing a system which produces secondary flow effects, for example vortices, in the blood, as it passes through the membrane system. The mixing effect so produced increases substantially the gas transport efficiency per unit area of the membrane.

The particular design being developed at the Royal Postgraduate Medical School, London, by Melrose in conjunction with the research group at Sussex involves the use of oscillating silicone membrane tubes carrying the blood, surrounded by an oxygen atmosphere.

Due to the oscillation a pair of counter-rotating stable vortices is set up in the blood flowing through the tubes. These secondary flow effects have been found to give the required increase in oxygenation as shown by the results obtained by Melrose *et al.* from biological tests.

The oscillatory motion used to develop the vortices in the blood flow induces a proportion of the blood contained to move in synchrony with the motion.

Thus, if the system is equipped with suitable one-way valves, the reversing flow can be converted into a single flow direction and hence an effective pumping action is produced. The work now being undertaken at Sussex on this project involves the design of a suitable manifold and valving arrangement as well as the development of a simple low-cost oscillating drive system.

A view from above of the prototype version of the oxygenator is shown in figure 15. The membrane tubes carrying the blood are contained in the

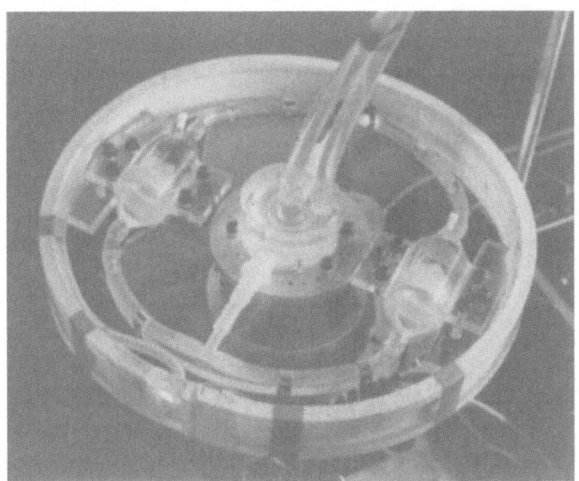

Figure 15. Prototype version of oscillating pump-oxygenator.

cylindrical section. These tubes are fed by a manifold system at one point in the cylinder which is in turn fed by tubes coming from the central support position. The blood having passed through the tubes is returned to the central position by way of two chambers containing flexible leaflet one-way valves. Having the two diametrically opposed valve chambers ensures that pumping takes place during both directions of oscillation of the tube cylinder. Preliminary tests have shown this arrangement to provide a highly efficient pumping system.

Flexible leaflet valves are particularly suitable for this application as their closing action produces less mechanical damage to the blood compared with say, a ball-type valve. The valve chambers can be removed to enable easy replacement of the valves which, due to their low cost, can be considered dispensable.

The oscillatory drive system which is being developed involves a normal electric motor which is adapted electrically such that the rotor oscillates rather than rotates in one direction. The oscillation amplitude is at present controlled by a torsion spring which has a natural frequency of torsional oscillation corresponding to that desired for operating the system. In this way the energy input required to drive the motor is kept to a minimum and in

effect is such as is required to overcome the various friction losses in the system. A new design is now being developed in which the torsion spring will be replaced by a torsion tube.

CONCLUSIONS

The main aim of this paper was to illustrate the need for further work on the development of flexible leaflet prosthetic heart valves. Although previous criticism of the premature fatigue failure of the leaflets of such valves is acknowledged, it must be remembered that polymer science is a rapidly advancing subject. This cannot and must not be ignored as it is almost certain that a new and biologically satisfactory material will emerge, quite possibly in the form of a 'bio-polymer'. Such a material might overcome the fatigue and thromboembolic problems experienced so far in the clinical use of earlier flexible leaflet valves.

It has also been shown that such valves can have considerable applications in the extra-corporeal circulation of body fluids, particularly blood as illustrated by the new designs for pump-oxygenators.

To summarise briefly: this form of valve must be made of a suitable material and produced, if possible, by a single-stage manufacturing process. The design should take into account the problems of stress concentration and the aim should always be to produce a format which allows leaflet deformation such as will provide maximum opening compared with the available base orifice area. The valves need not necessarily be produced with the leaflets in a closed form. If the leaflets are sufficiently flexible, then the haemodynamic conditions which obtain for example in the aortic and mitral valve positions can provide the required closing action. Such valves offer many advantages compared with those at present in use, not least of which will be the comparative low cost of the finished product.

Acknowledgments

The author wishes to thank the South East Metropolitan Regional Hospital Board for their generous and continuing financial support for this research programme. Thanks are also due to colleagues in the University and the Brighton and Lewes Hospital Group for their continued assistance and advice, and to King's College Hospital Medical School for their cooperation in performing some of the animal experiments.

REFERENCES

BELLHOUSE, B. J. 1972: The fluid mechanics of aortic and mitral valves. *In* 'Developments in Biomedical Engineering' (Ed. M. M. Black). Chatto and Windus for Sussex University Press.

CARPENTIER, A. *et al.* 1971: Collagen-derived heart valves. *J. thorac. cardiovasc. Surg.* **62**: 5: 707.

DENTON, G. R. 1951: A plastic prosthesis without moving parts for the A.V. valves. *Proc. Surg. Forum.* Clinical Congress, College of Surgeons, 1950: 239. Saunders, Philadelphia.

GOTT, V. L. *et al.* 1964: A hinged-leaflet valve for total replacement of the human aortic valve. *J. thorac. cardiovasc. Surg.* 48: 713.

HUFNAGEL, C. A. *et al.* 1954: Surgical correction of aortic insufficiency. *Surgery* 35: 673.

McGOON, D. C. 1971: Choice of grafts or prostheses for valvular replacement. *Br. Heart J.* 33 (Suppl.): 35.

MELROSE, D. G. *et al.* 1972: Oscillating silicone membrane tubes: a new principle of extracorporeal respiration. *Bio-Medical Engineering,* 7: 2: 60.

ROE, B. B., and MOORE, D. 1958: Design and fabrication of prosthetic valves. *Experimental Medicine and Surgery* 16: 177.

WRIGHT, J. T. M. 1972: The heart, its valves and their replacement. *Bio-Medical Engineering* 7: 1: 26.

PERSPECTIVES IN CARDIOVASCULAR ANALYSIS AND ASSISTANCE

S. FICH[1] AND W. WELKOWITZ[1]

INTRODUCTION

The basic objective of the research on the cardio-vascular system conducted by the Biomedical Engineering Group in the Department of Electrical Engineering at Rutgers University is to obtain a model of the aorta and the left ventricle that can be used to determine abnormalities in the cardio-vascular system and also to aid in the optimisation of the action of cardiac assist devices.

The minimum requirement to obtain the necessary perspectives of the cardiovascular system and assisted circulation is a quantitative analysis of the haemodynamic system including both the left ventricle and the aorta. In order to accomplish this result, the following theoretical and experimental projects were undertaken: (a) construction of a model of the aorta that is approximately consistent with the principal physiological properties of the system, and is also mathematically tractable; (b) verification of the model by comparing calculated and observed pressure and flow waveforms along the aorta; (c) use of the model to determine aorta parameters from pressure transfer coefficients; (d) modelling of the left ventricle as a pressure source with an internal impedance, both of which can be determined by dynamic *in vivo* haemodynamic measurements; (e) use of both the source and aorta models in the analysis of assisted circulation; and (f) relating the pressure and impedance of the model to certain mechanical properties of the left ventricle.

MODEL OF THE AORTA

The model postulated by the Rutgers group is shown in figure 1 where an exponentially tapered radius and a non-reflective termination at the femoral bifurcation is assumed (Fich *et al.*, 1966). In addition, since the elastic modulus of the vessel increases with pressure and distance along the aorta (Bergel, 1964; Apter, 1965), it is assumed that the modulus variation is approximately given by

$$E(x) = \frac{E_2}{r(x)} \qquad (1)$$

---

[1] Department of Electrical Engineering, Rutgers University, New Jersey, U.S.A.

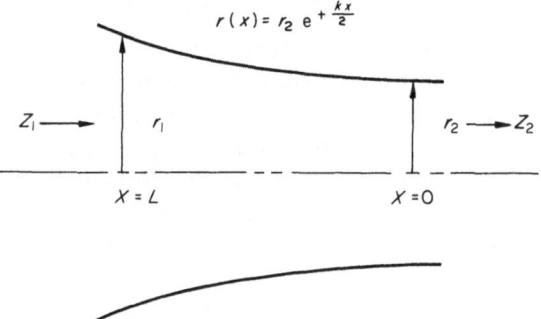

Figure 1. Model of tapered aorta.

E_2 is the modulus at the bifurcation, and the tapered radius is given by

$$r(x) = r_2 e^{kx/2} \qquad (2)$$

where r_2 is the radius at the bifurcation, and k is the taper coefficient. It can be shown that under these conditions the effective capacitance per unit length is given (Fich *et al.*, 1966) as

$$\bar{C}(x) = K_1[r(x)]^4 \qquad (3)$$

Assuming Poiseuille flow, the resistance per unit length is

$$\bar{R}(x) = \frac{K_2}{[r(x)]^4} \qquad (4)$$

and

$$R(x)\bar{C}(x) = K_1 K_2 = \text{constant} \qquad (5)$$

Equation 5 makes it possible to obtain traceable solutions for pressure and flow in the tapered sections if the blood mass is temporarily neglected. These solutions have the following form (Welkowitz and Fich, 1967):

$$P_1(s) = A(s)P_2(s) + B(s)Q_2(s) \qquad (6)$$

$$Q_1(s) = C(s)P_2(s) + D(s)Q_2(s) \qquad (7)$$

where $A(s)$, $B(s)$, $C(s)$, and $D(s)$ are transmission coefficients that are functions only of the parameters of the vessel and the complex frequency, s. The blood mass is introduced as a lumped inertance, $L_0 = \rho l/A$, where ρ, l, and A are respectively the density, length, and average area of the section. The assumed configuration is shown in figure 2.

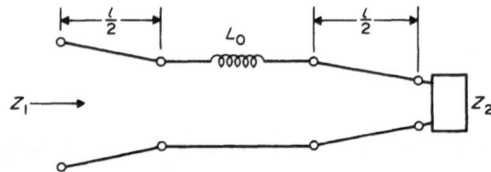

Figure 2. Model of tapered aorta with lumped blood Mass.

The resultant transmission matrix for this configuration is given by

$$\begin{bmatrix} A & B \\ C & D \end{bmatrix} = \begin{bmatrix} A_1 & B_1 \\ C_1 & D_1 \end{bmatrix} \times \begin{bmatrix} A_0 & B_0 \\ C_0 & D_0 \end{bmatrix} \times \begin{bmatrix} A_2 & B_2 \\ C_2 & D_2 \end{bmatrix} \quad (8)$$

Equations 6, 7, and 8 can be used to find pressure and flow transfer coefficients for any known length of the aorta. If only the pulsatile steady state is of interest, $j\omega$ is substituted for s in Equations 6, 7, and 8. The input at the root of the aorta is resolved into its Fourier components. Application of the transfer function to these components yields the Fourier components of pressure and flow at any point along the aorta. The time domain waveforms of pressure and flow can then be constructed from the significant Fourier components. Figures 3 and 4 show the pressure and flow waveform in the aorta of a dog (Welkowitz and Fich, 1967). Reference to equation 6 shows that, using similar methods, the flow at any point can be determined from two pressure measurements.

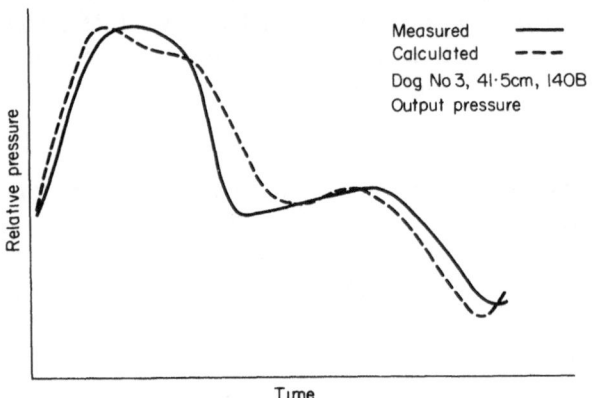

Figure 3. Calculated and measured pressure.

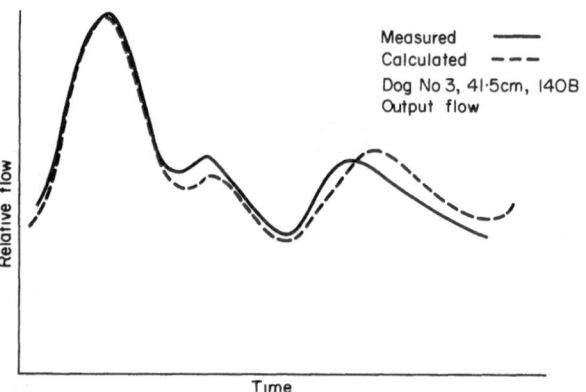

Figure 4. Calculated and measured flow.

DETERMINATION OF AORTA PARAMETERS

Development of a reliable method for finding aorta parameters could be of considerable importance in the determination of abnormalities in that part of the cardiovascular system, and in the evaluation and adjustment of cardiac assist devices. If the model shown in figure 2 is accepted, and one measures a pulsatile pressure transfer function in the aorta and relates it to the pulsatile transfer function of the model, it is possible to adjust the values of the physical parameters that appear in the model until the transfer functions approximately coincide. The crux of the problem is to devise an algorithm by means of which the values of the physical parameters are perturbed in a systematic fashion to find the optimum fit of the magnitude and phase of the transfer functions. The optimum fit in this case is defined as one which minimises a weighted error cost function derived from the transfer functions. The procedure followed for perturbing the parameters involves both the use of a search technique and the method of steepest descent. This method has been rather successfully applied to find the geometric taper coefficient, input area, and hoop elasticity at the input for the aorta of both a dog and a chicken as shown in table 1 (Strano *et al.*, 1972).

Parameter	Units	Dog aorta		Chicken aorta	
		Determined by combined optimisation method	Reported in the literature	Determined by combined optimisation method	Reported in the literature
Geometric taper coefficient, k	m^{-1}	4·2	4·72[a]	6·6	7·9
Input area, A_1	m^2	$3·52 \times 10^{-4}$	$3·81 \times 10^{-4}$[a]	$0·217 \times 10^{-4}$	$0·2 \times 10^{-4}$
Hoop elasticity at the input, β_{e_1}	N/m^2	$9·0 \times 10^4$	$7·84 \times 10^4$[a]	$1·29 \times 10^4$	0.506×10^4[c] Turkey aorta
Effective blood viscosity, v	Ns/m^2	0·243	$0·04 - 0·12$[b]	0·225	Not available

[a] PATEL, D. J. *et al.* 1963: Relationship of radius to pressure along the aorta in living dogs. *J. Appl. Physiol.* **18**: 6, 1111.

[b] FRY, D. L., GRIGGS, D. M., and GREENFIELD, J. C. 1964: *In vivo* studies of pulsatile blood flow: the relationship of the pressure gradient to the blood velocity. *In* 'Pulsatile Blood Flow' (Ed. E. O. Attinger). McGraw-Hill, New York.

[c] SPECKMANN, E. W., and RINGER, R. K. 1964: Static elastic modulus of the turkey aorta. *Canadian Journal of Physiology and Pharmacology*, **42**.

TABLE 1: Comparison of results

MODEL OF THE LEFT VENTRICLE

A model of the left ventricle is shown in figure 5 where P_g, P_{1v}, and P_a are respectively source, left

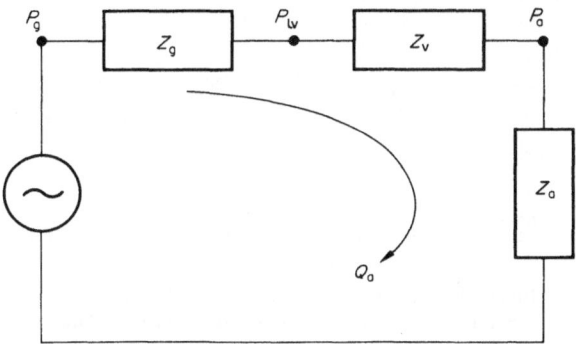

Figure 5. Model of left ventricle.

ventricular, and aortic pressure. Z_g, Z_v, and Z_a are source, valve, and aortic input impedance, and Q_a is the aortic flow. Considering any Fourier component of pulsatile flow, it follows that:

$$P_{1v} = P_g - Z_g Q_a \qquad (9)$$

It is seen from equation 9 that P_g and Z_g can be determined from two different values of P_{1v} and Q_a.

The parameters of the equivalent pressure source are obtained by measuring left ventricular pressure, aortic pressure, and aortic flow under two conditions. The first condition is the normal state. In the second condition, an intra-aortic assist device is inserted to change the dynamic input impedance at the root of the aorta in order to obtain a significant change in pressure and flow as shown in figure 6. In order to eliminate the effects of the physiological controls upon the muscle parameters,

AP

before pumping   during pumping

LVP

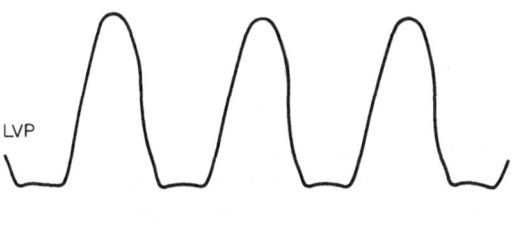

AF

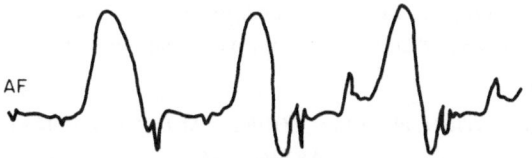

Figure 6. Dynamic changes in pressure and flow caused by pumping. AP, aortic pressure; AF, aortic flow; LVP, left ventricular pressure.

measurements of pressure and flow are recorded after one or two cycles of pumping in each state. This procedure is used because the time constants of the cardiovascular control paths have been estimated to be much greater than the period corresponding to one or two cycles of pumping in a dog with a heart rate of about 150 beats per minute (Puri and Gido, 1968). The experimental method is to record time domain variations of pressure and flow and to obtain Fourier components from these variations. Although the essential parameters are time varying, Fourier analysis is valid because all parameters have the same fundamental repetition rate (Min, 1972).

Using prime and double prime notations to indicate the unassisted and assisted states, the pressure and internal impedance, phasors are given by the following equations for each Fourier component:

$$P_g = \frac{P'_{1v}Q''_a - P''_{1v}Q'_a}{Q''_a - Q'_a} \qquad (10)$$

$$Z_g = \frac{P_g - P'_{1v}}{Q'_a} = \frac{P_g - P''_{1v}}{Q''_a} \qquad (11)$$

Figure 7 shows the variation of the magnitude of the fundamental components of P_g and Z_g with and without pharmacological nerve blockage.

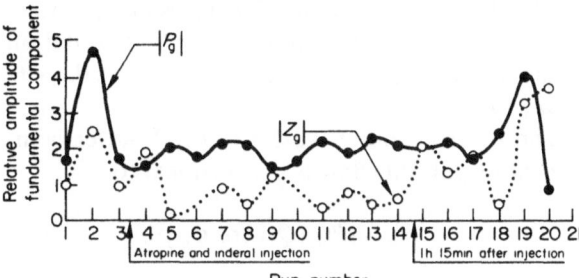

Figure 7. Effect of denervation on source pressure and impedance.

When Inderal (0·5 mg/kg) and Atrophine (3 mg/kg) were intravenously injected slowly over periods of 3 and 5 minutes, the fluctuations are seen to be greatly reduced. This indicates that P_g and Z_g are related to physiological parameters that respond to the neural control systems.

ANALYSIS OF ASSISTED CIRCULATION

As the field of assisted circulation develops, the determination of objective criteria for satisfactory therapeutic performance becomes increasingly important. The incorporation of such criteria into automatic control systems for assist devices is highly desirable. As a first step in this direction, an attempt was made to analyze the operation of an intraaortic balloon pump, considered as a flow source. The pump is shown in figure 8 in conjunction with the models of the aorta and left ventricle shown in figures 2 and 5.

Assuming an approximately linear system, the principle of superposition can be applied to calculate the combined effect of the action of the left ventricle and the pump. Figure 8a applies for the case when only the ventricle operates, and figure 8b applies

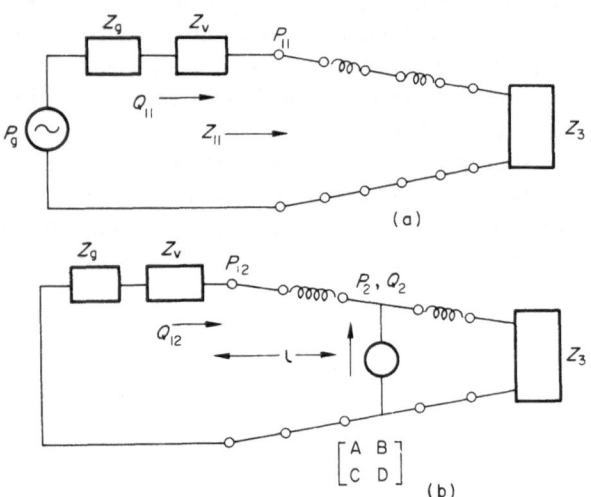

Figure 8. Model of left ventricle, aorta and flow source.

when only the pump is in operation. Using relations similar to equations 6 and 7, expressions for total aortic pressure and flow for the combined excitation of the ventricle and pump can be derived (Welkowitz *et al.*, 1972).

The average source work for any Fourier component of pulsatile flow is proportional to

$$W_g = |P_g|\,|Q_1|\cos\Phi_{g,1} \qquad (12)$$

where $\Phi_{g,1}$ is the angle between the source pressure and aortic flow phasers. The average aortic work is proportional to

$$W_a = |P_a|\,|Q_1|\cos\Phi_{g,1} \qquad (13)$$

where Φ_{a1} is the angle between aortic pressure and flow phasers. Figure 9 shows the calculated fundamental component of source work as a function of the aortic impedance phase angle, and figure 10 shows the corresponding variation of aortic work. Figure 11 shows the experimentally determined fundamental aortic work reduction as a function of the aortic input impedance phase angle. The fundamental component was used because most of the pulsatile work is done at the fundamental frequency. The three curves have one important feature in common. The source and aortic work are minimum when the impedance phase angle is about π radians. Our analysis of experimental data obtained during

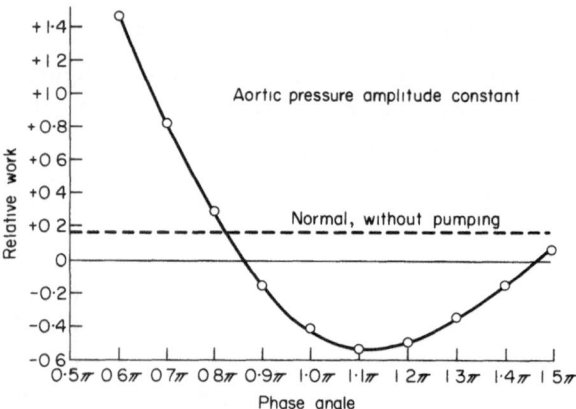

Figure 9. Calculated variation in pulsatile fundamental source work as a function of aortic impedance phase angle.

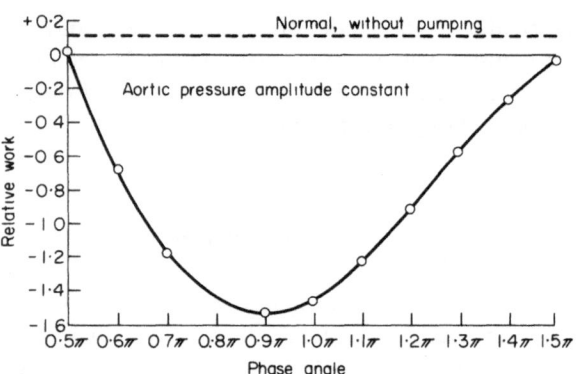

Figure 10. Calculated variation in pulsatile fundamental aortic work as a function of aortic impedance phase angle.

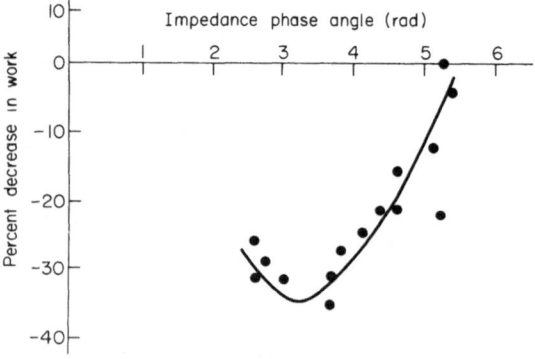

Figure 11. Measured pulsatile fundamental work reduction as a function of aortic impedance phase angle.

therapeutic pumping has also shown that the aortic work reduction during pumping is a maximum when the aortic input impedance phase angle is about π radians (Coffen and Bay, 1956). These results are being used in the design of an experimental driving system for an intraaortic assist device.

RELATION BETWEEN SOURCE AND MECHANICAL PARAMETERS

Continued research has shown that a rather simple relationship exists between the maximum isovolumic stress and the dynamic length variations of the left ventricle and the fundamental Fourier components of source pressure and impedance (Min, 1972). Curves showing the calculated stress-length relation for four dogs agree reasonably well with those found by Cotten and Bay (1956) (Fich *et al.*, 1971). Calculations, based upon haemodynamic measurements, have also shown a linear relationship between stroke volume and end-diastolic pressure for a dog with pharmacological nerve blockage.

SUMMARY

A model of the aorta has been postulated and verified with respect to both pressure and flow waveforms along the aorta. This model has also been used to determine essential aorta parameters from pressure transfer functions. The left ventricle has been modelled as a pressure source with an internal impedance. The pressure and source phasers have been calculated from Fourier analysis of time domain pressure and flow waveforms under normal and dynamic loading conditions. Dynamic loading has been achieved by use of an intra-aortic assist device. A model consisting of the postulated aorta, source, and cardiac assist device has been used to determine a critical criterion for the optimisation of assisted circulation. This criterion is in good agreement with experimental results. The parameters of the pressure source have been related to the stress-length variation of the left ventricle. Calculated values agree reasonably well with experimental results found by other investigators. The combined model is being used in the preliminary design of an optimum control system for driving an intra-aortic assist device.

Acknowledgment

This research has been supported in part by a grant from the Rutgers Research Council and USPHS Grant HE-11173.

REFERENCES

APTER, J. T. 1965: Analysis of some effects of aortic smooth muscle on *in vivo* aortic pressure curves. *Bull. Math. Biophys.* **27**: 119.

BERGEL, D. H. 1964: Arterial viscoelasticity. *In* 'Pulsatile Blood Flow' (Ed. E. O. Attinger). McGraw-Hill, New York.

COTTEN, M. D., and BAY, F. 1956: Direct measurements of changes in cardiac contractile force. *Amer. J. Physiol.* **187**: 122.

FICH, S., WELKOWITZ, W., and HILTON, R. 1966: Pulsatile blood flow in the aorta. *Biomedical Fluid Mechanics Symposium*. Fluids Engineering Conference, ASME, **34**.

FICH, S., WELKOWITZ, W., MIN, B., SHASTRI, S., KANTROWITZ, A., and JARON, D. 1971: *In vivo* determination of stress-mean length relation in left ventricle. *Proceedings 24th ACEMB* **32**.

JARON, D. *et al.* 1970: Measurements of ventricular load phase angle as an operating criterion for in series assist devices. *Trans. Amer. Soc. Artif. Int. Organs.* **XVI**: 466.

MIN, B. 1972: Analysis and optimization of balloon pumping. *Ph.D. Thesis*, Department of Electrical Engineering, Rutgers University, New Brunswick, N.J.

PURI, N. N. and GIDO, J. F. 1968: Control systems for circulating assist devices. *University City Science Center*, Philadelphia, Pa., 3.

STRANO, J. J., WELKOWITZ, W., and FICH, S. 1972: Measurement and utilization of *in vivo* blood transfer functions of dog and chicken aortas. *I.E.E.E. Transactions on Biomedical Engineering*, BME-19, No. 4.

WELKOWITZ, W., and FICH, S. 1967: A nonuniform hybrid model of the aorta. *Trans. N. Y. Acad. Sci.* Series II, **29**: 316.

MAINTENANCE HAEMODIALYSIS AND PREVENTIVE MEASURES IN RENAL DISEASE

W. R. CATTELL[1]

INTRODUCTION

Recent years have seen striking advances in the field of biomedical engineering. As this Symposium indicates, many areas of medical practice are involved, from the provision of limb prosthesis to artificial hearts. Among the striking success stories is the development of systems which provide renal replacement for patients with terminal kidney disease. International statistics (Parsons *et al.*, 1971) show five year survivals of between 60 and 90% for patients whose lives are maintained by artificial kidneys—figures which compare very favourably with the results of treatment of other killing conditions such as coronary artery disease and lung cancer.

Gratifying as these survival statistics may be, they are however achieved only at a considerable cost—both financial and social. The patient is alive but the quality of life is far from ideal (Gordon and Cattell, 1972). Moreover, the very existence of such treatment has posed considerable social, financial, political and ethical questions. How much can we afford to pay for the treatment?; who should receive it?; and to what extent should society at large support artificial life? Perhaps most relevant of all is the observation that this is last ditch treatment. Surely money could be much better spent on the prevention of progressive renal disease rather than on the treatment of terminal kidney failure? This use-of-resources question is a cause of recurring doubts among those biomedical engineers who are actively engaged in the further development of haemodialysis systems. It is for this reason that this paper re-examines the theme 'Prevention is Better than a Cure' as it applies to terminal renal failure.

RENAL DISEASE, TREATMENT AND PREVENTION

What is the position of haemodialysis versus preventive medicine? Sadly, it must be admitted that doctors are much better at treating end-stage renal failure than preventing its development. There are, of course, many causes of progressive renal failure; the renal disease may be a condition which primarily affects the kidney or it may be part and parcel of

some more generalised systemic disease. Of the primary renal diseases, those most commonly leading to terminal renal failure (table 1) are

Chronic glomerulonephritis
Chronic pyelonephritis
Calculus (stone) disease
Analgesic abuse
Urinary tract obstruction
Polycystic renal disease

TABLE 1: Common causes of progressive renal failure

chronic glomerulonephritis, chronic pyelonephritis, calculus disease, analgesic abuse, obstruction to the outflow of urine and hereditary polycystic disease of the kidney.

Taking these in turn, chronic glomerulonephritis (table 2) is a disease of disturbed immunity mechanisms—a sort of allergic disorder. Antigen is some

A disease of disturbed immunity mechanisms initiated by: Streptococcal infection
? – – – – – –
? – – – – – –
Progressive deposition of antigen–antibody complex in kidney
Progressive loss of kidney function 1–30 years

TABLE 2: Chronic glomerulonephritis

foreign material or 'foreign body'—often a germ—and antibody is produced by the individual in an attempt to neutralise this antigen, sometimes complexing or binding with it to do so. Renal damage results from the deposition of the antigen-antibody complex on the glomerulus, or filtering mechanism of the kidney. Blood in the kidney passes through a delicate capillary system. Water and waste products are removed from the blood by filtration across the capillary wall to form urine. Endothelial cells line the inside of the capillary, then there is the basement membrane and, on the outside, epithelial cells. Antibody-antigen complexes are deposited on the basement membrane and in so doing 'silt up' the system. An actual kidney section is shown in figure 1. The black areas represent deposited complex. The deposition of this material, or 'silting up' leads to acute or subacute damage to

---

[1] St Bartholomew's Hospital, London, England.

35

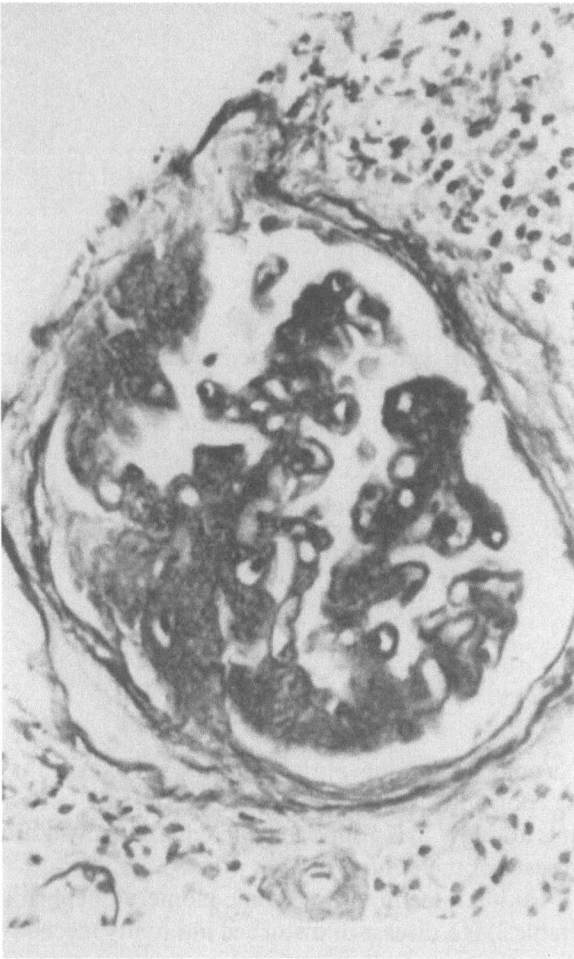

Figure 1. Kidney biopsy showing heavy deposition of antigen-antibody complex on glomerular capillaries.

1	Prevent initial disease
	Requires elimination of cause
	(a) Streptococcal infections
	(b) Mostly unknown
2	Prevent progression of disease
	Requires paralysis of *all* immune mechanisms
	Not possible or indeed desirable

TABLE 3: Chronic glomerulonephritis—preventive measures

the kidney which can slowly progress over months or many years to severe kidney failure (table 2). The initiating factor is commonly infection with a particular germ, called a streptococcus, such as causes a sore throat. However, not all sore throats predispose to glomerulonephritis, nor indeed do all streptococcal infections lead to this. Rather special streptococci are required. Infection with these particular streptococci may occur randomly in the community or in small epidemics. Thus, without detailed bacteriological surveillance it is not possible to identify which sore throats will lead to nephritis. Quite apart from streptococcal infections there are almost certainly many other as yet ill-understood factors leading to glomerulonephritis. Once the condition is established it may heal spontaneously. On the other hand it may gradually progress to serious kidney damage.

How then can the problem of preventing the development of chronic end-stage glomerulonephritis be tackled? This might be done either by preventing the initial disease (table 3) or by preventing progression of the disease once it is established. In terms of the former this would obviously require elimination of the cause. The only clearly understood cause is streptococcal infection, and to prevent nephritis developing we would have to eliminate from society streptococcal infections in general; and this is not a practical proposition. As many cases of glomerulonephritis are unknown, measures to eliminate them are obviously impossible. Turning to the prevention of progression of the disease, there is unfortunately no known effective treatment. The problem is one of preventing antigen-antibody complexing and this can probably only be achieved by total paralysis of all immunity mechanisms. However, many immune mechanisms are vital to normal health, acting as a defence against infection. Thus, were total immuno-paresis possible, which it is not, this is almost certainly not desirable —the treatment would probably accelerate rather than defer death.

Turning to chronic pyelonephritis, this results from the direct invasion of the kidney by infecting organisms. It is characterised by a scarred, contracted and poorly functioning kidney as shown in figure 2. Formerly it was thought that such invasion of the kidney with progressive damage was a constant threat throughout life but it is now generally believed (table 4) that this condition, more popularly called atrophic pyelonephritis, is established in early childhood, perhaps before the age of 5 years. While the initiating infection may be associated with urinary symptoms, all too often the child experiences no symptoms whatsoever. Hence the disease may not be diagnosed until adult life or until renal failure has developed. Once the condition is fully established, and particularly when the diagnosis is first made in adult life, there is no method of preventing progressive kidney scarring. Any effective treatment requires detection in infancy and aggressive therapy. Thus, considering possible preventive measures, because the condition is commonly asymptomatic the disease could only be diagnosed by massive, nation-wide medical screening of *all* infants and children. This would require both medical examination and culture of the urine, a formidable undertaking which would give a very small return in terms of the total population. Without such massive screening many cases would remain undiagnosed until the damage was far advanced. Once recognised, affected children would require regular medical surveillance and, where necessary, treatment. Most physicians experienced

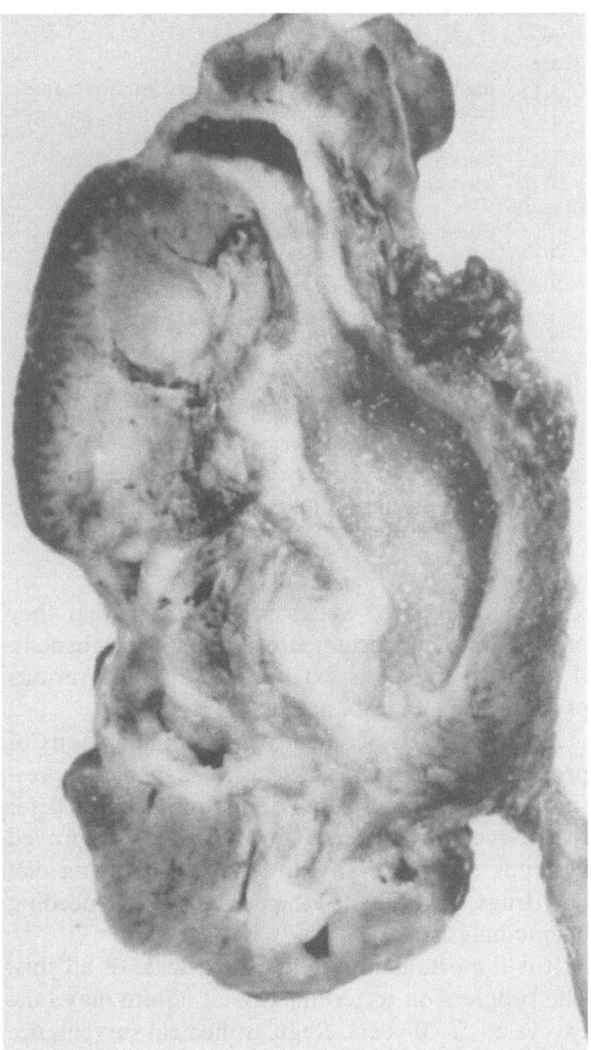

Figure 2. Section of scarred and deformed pyelonephritic kidney.

Stone disease	Life-long surveillance
	Life-long dietary modification
Analgesic abuse	Publicise hazard
	Withdrawal of dangerous drugs
	Analgesics on prescription only
Stones ⎫	Life-long surveillance
Analgesics ⎬ + Infection	Life-long antibiotics
	Miscellaneous

TABLE 5

Description
  A disease of childhood
  Commonly symptom-free
  Requires detection and treatment in infancy
Preventive measures
  Massive country-wide medical screening of *all*
    infants and children
  Regular medical supervision
  ? Compulsory treatment

TABLE 4: Atrophic pyelonephritis

in such work recognise that a great problem in this is failure of children to attend for follow up examination and treatment. If, therefore, society demanded that this cause of terminal renal failure be eradicated, compulsory treatment might have to be considered.

With respect to the miscellaneous group of stone disease, analgesic abuse, etc. (table 5), it is undoubtedly true that a great deal of preventive medicine should be possible in these conditions. Thus, any patient diagnosed as having stone disease must have life-long surveillance as the presence of stones may occasionally result in symptomless obstruction to the kidney and progressive damage. Should there be any underlying cause such as excessive absorption of calcium, the patient might have to envisage life-long dietary modification with all the inconvenience this involves.

Analgesic abuse is a condition only relatively recently recognised. By abuse is meant the habitual excessive ingestion of certain, but not all, headache medicines. Prevention of such induced kidney disease requires publicising the hazards and discouraging people in general from becoming habituated to analgesics. Unfortunately, medicine-taking can become a habit and there are very large numbers of people habituated to headache powders. To protect individuals from themselves it may then be argued that potentially harmful analgesic drugs should be forbidden. Alternatively, it may be argued that analgesics should be obtainable only by prescription from a doctor. This, however, would scarcely be sensible as the vast majority such as aspirin, etc., taken on occasion, are entirely safe. To demand that these should only be prescribed by doctors would add yet one more major burden to the over-taxed health services.

The combination of stones, or analgesic abuse, and infection in adult life is a cause of serious kidney damage. This is potentially preventable provided there is life-long follow up and possibly the administration of antibiotics over many years. This demands cooperation both by the medical profession and the patient.

Polycystic renal disease (table 6) is an hereditary condition associated with multiple small developmental cysts in the kidney (figure 3). Over the course of years these cysts gradually enlarge, distending the kidney and destroying the normal intervening tissue. Polycystic disease is inherited as a Mendelian dominant (table 6). Thus, if one parent has the

Description
  Hereditary
  Mendelian dominant
    50–50 chance with each child
  Inevitable progression
Prevention
  Genetic counselling
  ? Eugenic breeding

TABLE 6: Polycystic renal disease

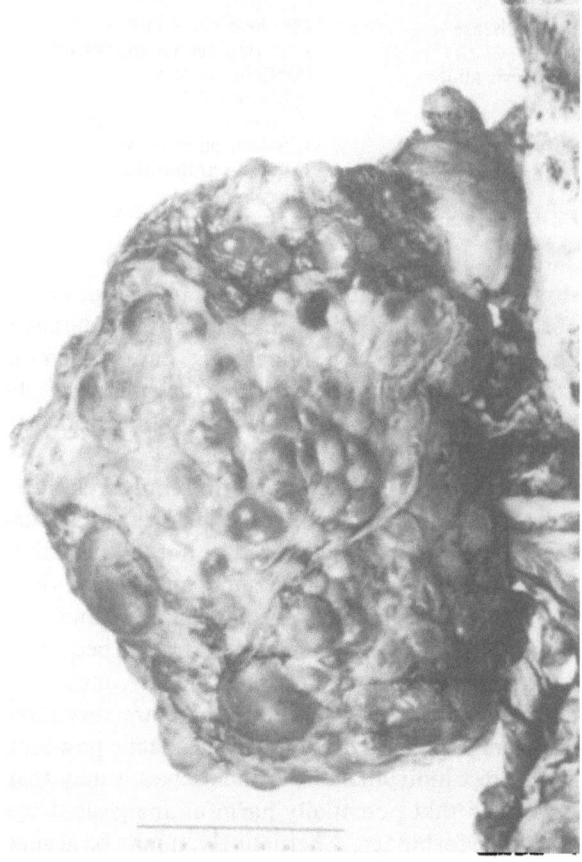

Figure 3. Polycystic kidney.

condition there is a 50–50 chance that any offspring of such a marriage would have this disease. There is no known method of arresting progression of this condition. Prevention can only be achieved by genetic counselling and it must be clearly understood that the physician can only advise parents, not direct them. It can, of course, be argued that this condition should be eliminated by eugenic breeding. Few, however, would be prepared to accept controlled mating in a free society.

There are many other causes of progressive kidney damage but the same sorry tale applies to these. It must therefore be admitted that at this time treatment is singularly ineffective in preventing the progression towards terminal renal failure once a renal lesion is established. Furthermore, it should be emphasised that such measures as can be applied may be unacceptable to society. Thus, table 7 sets down the steps which would be required to implement such measures as can be taken at this time. First of all, the disease must be identified and, as already indicated, this means nation-wide screening examinations, followed up by regular, long-term surveillance. Both such measures raise immense financial problems. Furthermore, they introduce the problem of the psychological side-effects of compulsory surveillance. Experience from regular, compulsory medical check-ups on business execu-

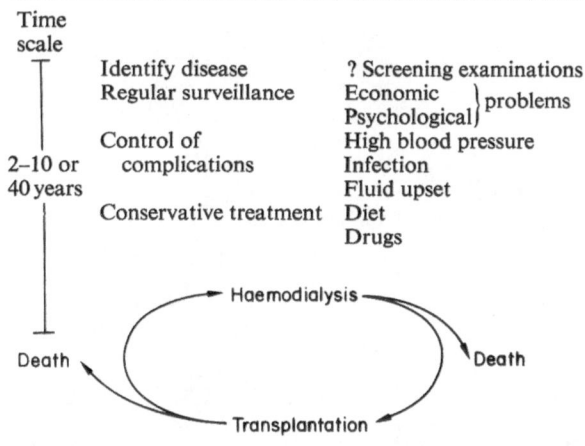

TABLE 7: Summary of measures required for the prevention of renal failure

tives in the United States has indicated that they produce more medical neuroses than health benefit. If such experience is extended to children serious psychological disturbances could result.

More aggressive control of the complications of progressive renal disease such as high blood pressure, infection and fluid upset undoubtedly can defer the onset of serious renal failure. Similarly, improved methods of conservative treatment, involving diet and drugs, has extended the period prior to needing haemodialysis.

It is important to note the time-scale of all this. The progression to terminal renal failure may take two years or 40 years. Regular medical surveillance over such very long periods is immensely difficult and often fails because the patient feels quite well.

SUMMARY

Overall, the outlook for any major advances in the prevention of terminal renal failure is at present poor. It must be accepted that haemodialysis complemented by transplantation is, and will remain for the foreseeable future, a necessary and desirable area for medical research and development. It has been shown to be effective, now its practice must be refined. Much remains to be done by clinicians and biomedical engineers working in concert to improve available equipment so that life on a kidney machine is not merely life saving but more pleasantly endurable. Time, money and effort spent on improving haemodialysis techniques and equipment is money very well spent.

REFERENCES

PARSONS, F. M., BRUNNER, F. P., GURLAND, H. J., and HARLEN, H. 1971: *Proceedings of the European Dialysis and Transplant Assoc.* 8: 3.
GORDON, P. M., and CATTELL, W. R. 1972: *Proceedings of the European Dialysis and Transplant Assoc.* 9: 35.

RECENT TRENDS IN BIOCOMPATIBLE MATERIALS AND MICROCAPSULAR ABSORBENTS FOR KIDNEY FUNCTION REPLACEMENT

T. M. S. CHANG [1]

Semipermeable microcapsules for kidney function replacement (Chang, 1964, 1966, 1969, 1972; Chang et al., 1966) are semipermeable ultrathin polymer membrane envelopes each enclosing an aqueous interior or coating a solid particle. Direct coating is more convenient for use in haemoperfusion, since this results in rigid particles which give less resistance to flow and, in addition, the rigid particles themselves offer a rigid support for the ultrathin membrane.

Ten millilitres of 20 micron diameter micro-

membrane area (about 9 000 cm^2) will still be at least 40 times faster than in the standard artificial kidney (table 1). A compact microcapsule artificial kidney which has biocompatible materials for the microcapsule membranes and adsorbents inside these microcapsules to trap the uraemic metabolites, was proposed (Chang, 1966) and developed (Chang, 1966, 1969, 1972; Chang et al., 1967, 1968, 1970). It is now in the third year of clinical trial (Chang et al., 1970, 1971, 1972). This approach has been supported and extended by a number of other

Type of artificial kidney	Membrane area A (cm^2)	Membrane thickness Δs (Å)	Rate of solute transport $A/\Delta s \alpha ds/dt$ (cm^2/Å)
Standard hemodialysers	less than 20 000	50 000 to 500 000	0·04 to 0·4
Microcapsules			
10 ml 20 μ diameter	25 000	500	50
33 ml 100 μ diameter	20 000	500	40
300 ml 2-mm diameter	9 000	500	18

TABLE 1: Efficiency of types of artificial kidney

capsules or 33 ml of 100 micron diameter microcapsules have a total surface area (about 20 000 cm^2) which is larger than that available in a conventional artificial kidney (less than 20 000 cm^2). In addition, the membrane thickness of the microcapsules (less than 500 Å) is much less than that used in standard artificial kidneys (usually 10 000 cm^2). If we pack 33 ml of 100 micron diameter microcapsules into a shunt, then the total membrane area available for diffusion will be slightly greater than that of an artificial kidney. Since the membrane thickness is at least one-hundredth that in the standard artificial kidney, metabolites from blood flowing past these microcapsules can cross the membrane into the microcapsules at least 100 times faster than in a standard artificial kidney. The ultrathin membrane (less than 500 Å) is such that even in 300 ml of much larger microcapsules (2 mm diameter) the initial rate of movement of metabolites across the total

laboratories, especially Levine and La Course (1967), Andrade et al. (1971), Sparks et al. (1969, 1971) and Gardner et al. (1971). This paper discusses the present trends in research in two of the important components of this microcapsule artificial kidney: biocompatible materials and microcapsular absorbents.

BIOCOMPATIBLE MATERIALS

We shall confine our discussions on biocompatible materials to the recent trends in research on blood-compatible materials. The many excellent detailed reviews on this subject include a symposium volume (Conference on Mechanical Surface and Gas Layer Effects on Moving Blood, 1971). For the convenience of discussion let us divide the recent trends of research in blood-compatible materials into the following categories: (1) surfaces with anionic radicals or negative electrical charges; (2) heparinised surfaces; (3) surfaces with inert materials; (4) albumin surfaces.

Sawyer and Pate (1953) proposed that the blood

[1] Faculty of Medicine, McGill University, Montreal, Quebec, Canada.

compatibility of intima is due to the negative charge of the endothelium. Further investigations showed that this may not necessarily be the only factor. Despite the lack of any absolute correlation between surface charge and the degree of thromboresistance of polymers (Milligan *et al.*, 1968), a number of thromboresistant materials have been obtained as a result of this proposal. Thus, some recently developed charged membranes have been found to be both thromboresistant and highly selective in their permeability characteristics (Muir *et al.*, 1972). Also demonstrated have been the relationship between thrombogenicity and the position of metals in the electromotive series (Sawyer and Srinivasan, 1967), and the increased thromboresistance of some polymers with anionic radicals or negative electrical charges added to their surfaces (Sharp *et al.*, 1968; Lyman *et al.*, 1969; Nemchin *et al.*, 1969). Similarly, semipermeable microcapsules when specially prepared with negatively charged surfaces (Chang, 1964, 1972; Chang *et al.*, 1966) are found to be more biocompatible.

The original studies by Gott's group (Gott *et al.*, 1963) on graphite-benzalkonium-heparin (GBH) demonstrated the thromboresistance of heparinised surfaces. Here, heparin is attached by ionic bonding to the cationic benzalkonium previously attached to graphite. Other approaches include the ionic bonding of heparin to quaternary amines or other cationic groups previously attached to the polymer surfaces by a number of methods (Merrill *et al.*, 1966; Leininger *et al.*, 1966; Eriksson *et al.*, 1967; Grode *et al.*, 1969); the covalent linkage of heparin to polymers (Merrill *et al.*, 1970); and the addition of bulk heparin to the polymers (Hufnagel *et al.*, 1967). The exact reason for the thromboresistance of heparinised surfaces is not known, but recent evidence shows that heparinised surfaces do not activate the Hageman factor (factor XII) (Vroman and Adams, 1969; Merrill *et al.*, 1970). The platelets adhere to certain types of heparinised surfaces (Salzman *et al.*, 1969) so that recirculation of blood across a large heparinised surface may result in thrombocytopenia. Coating the heparinised surface with albumin prevents platelet adhesions to the heparinised surfaces (Salzman *et al.*, 1969). Semipermeable microcapsules, including those prepared by the coating of charcoal granules, have been prepared with heparinised surfaces and found to be thromboresistant (Chang *et al.*, 1967, 1968; Chang, 1969).

Different approaches have been used in the preparation of surfaces which have inert materials. Lyman's group (Lyman *et al.*, 1965, 1970) proposed that the thromboresistance of polymer surfaces is inversely related to the surface-free energy. On the basis of this type of proposal a number of hydrophobic polymers have been prepared and have been found to retard coagulation (Lyman *et al.*, 1965, 1970; Boretos and Pierce, 1967). Polyurethane (Boretos and Pierce, 1967; Lyman *et al.*, 1969), fluorinated silicone rubber (Musolf *et al.*, 1969), and graphite in the form of pyrolytic carbon (Gott *et al.*, 1961; Bokros *et al.*, 1969) have been tested with some success. Another approach involves the use of very hydrophilic hydrogel, for example, an acrylic hydrogel, Hydron. It has been proposed that although hydrophilic surfaces have higher surface-free energy than hydrophobic surfaces, their interfacial free energies in aqueous solutions are much lower. Preliminary results show that Hydron is thromboresistant (Levowitz *et al.*, 1968). Hydron had been used for coating activated charcoal in the preparation of semipermeable microcapsules (Andrade *et al.*, 1971).

The adsorption of albumin to surfaces inhibits the adhesion to them of platelets (Chang, 1969; Packham *et al.*, 1969; Salzman *et al.*, 1969; Merrill *et al.*, 1970; Lyman *et al.*, 1970); whereas the adsorption of fibrinogen to foreign surfaces promotes the adhesion of platelets (Packham *et al.*, 1969). The platelet-protecting albumin coating probably acts by preventing the deposition of fibrinogen from plasma (Vroman *et al.*, 1971). Specially prepared albuminated polystyrene has been found to be thromboresistant (Lyman *et al.*, in press). Albumin-coated microencapsulated, activated charcoal does not adversely affect blood platelets either in animal studies (Chang, 1969; Chang and Malave, 1970; Andrade *et al.*, 1971) or in clinical trials (Chang, 1972; Chang *et al.*, 1970, 1971, 1972).

MICROCAPSULAR ABSORBENTS

Blood-compatible membranes have been used to microencapsulate absorbents for use in the microcapsule artificial kidney (Chang, 1966, 1969, 1972; Chang *et al.*, 1968, 1970, 1971; Andrade *et al.*, 1971). Microencapsulated activated charcoal is one of the absorbents first tested experimentally (Chang, 1966, 1972; Chang *et al.*, 1967, 1968, 1970; Andrade *et al.*, 1971) and clinically (Chang *et al.*, 1970, 1971, 1972). Charcoal is used in the microencapsulated form because the efficiency of activated charcoal granules for the absorption of creatinine, uric acid, indican, phenols, guanidine bases, organic acid urochrome, barbiturates, salicylates, and glutethimide has been well established both experimentally and clinically (Yatzidis, 1964; Dunea and Kolff, 1965; Hagstam *et al.*, 1966; De Myttenaere *et al.*, 1967; Dedrick *et al.*, 1967). Unfortunately, free, activated charcoal cannot be used routinely for haemoperfusion because it releases particles which cause embolisms (Hagstam, 1966) and it lowers the systemic platelet level (Dunea and Kolff, 1965; Hagstam *et al.*, 1966; Dutton *et al.*, 1969). Its selective absorption of uraemic metabolites and poisons and its undesirable side-effects of embolisms and platelet depletion make it an ideal material for use as microcapsular absorbent in the microcapsule artificial kidney, since in the microencapsulated form the activated charcoal cannot give off embolising particles, and the enclosing blood-compatible membrane does not affect the formed elements of blood. At the same time, permeant uraemic metabolites can diffuse across the enclosing membrane to be absorbed by the activated charcoal (Chang, 1966, 1969, 1972; Chang *et al.*, 1970, 1971, 1972).

Initial studies involved the use of nylon or cellulose nitrate as the enclosing membranes (Chang, 1966). Later, a heparin-complex membrane was used both as a spherical membrane (Chang *et al.*, 1967, 1968) and as membrane coating (Chang *et al.*, 1968). Still later, an albumin-coated cellulose membrane was used to form microcapsules by direct coating of the activated charcoal (Chang, 1969). *In vitro* clearance studies with creatinine solution show that at a flow rate of 200 ml per minute the clearance was 175 ml per minute in the first 15 minutes, 160 ml per minute after 30 minutes and 135 ml per minute after 60 minutes for 300 g of ACAC-type microencapsulated activated charcoal. In animal testings, it was found that haemoperfusion over 300 g of albumin-cellulose coated microencapsulated charcoal (ACAC) significantly lowered the systemic arterial blood creatinine levels. If this rate is used as an index, in two hours, 300 g of encapsulated activated charcoal would remove all the creatinine produced by a 70-kg man during a $1\frac{1}{2}$-day period. No embolising charcoal powder was detected in the efferent from the shunt or in histological sections of lung, liver, spleen and kidneys of the animals studied. In addition, the albumin-cellulose coated surface did not have any adverse effects on the formed elements of blood. These results led to clinical trial in patients with chronic renal failures (Chang *et al.*, 1970, 1971, 1972). They showed that the internal resistance, measured as pressure drop across the microcapsule artificial kidney, is very low. At a blood-flow rate of 200 ml per minute the internal resistance of the microcapsule artificial kidney is less than that of the Kiil-type haemodialyser, and much lower than the coiled-type haemodialyser. In patients with Scribner A-V shunts, a blood-flow rate of 200 to 250 ml per minute can be obtained with the microcapsule artificial kidneys and without the use of a blood pump. Up to May 1972, a total of 60 clinical procedures had been carried out in 12 patients, including 3 patients who have received treatments for 8 months, 5 months and 4 months respectively. Within a 2-hour haemoperfusion period, the creatinine clearance of 300 g of albumin-coated microencapsulated activated charcoal is comparable to the most efficient type of coil artificial kidney and is much more efficient than the plate type of coil artificial kidney and the plate type of capillary-type artificial kidneys. Uric acid clearance is even more efficient than with the standard haemodialysers. In a patient who excreted a sufficient amount of water, two hours' microcapsule artificial kidney haemoperfusion twice a week removed her previous symptoms of nausea, vomiting, diarrhoea and hiccup, and maintained her symptom-free for two months. Supplementary treatment with the standard artificial kidney for the removal of water and electrolytes is required for those with low urine output. The change in levels of blood platelets was negligible after haemoperfusion over albumin-coated microencapsulated activated charcoal, giving $91 \cdot 8\% \pm 11 \cdot 8\%$ of control platelet levels. Smears of shunt effluent blood did not show any embolic particles. These observations in clinical trials confirm our earlier findings that haemoperfusion over albumin-coated microencapsulated charcoal did not give rise to embolising particles or changes in platelet levels.

For all the patients in general, as far as the alleviation of uraemic symptoms and general feeling of well-being is concerned, 2 hours haemoperfusion with 300 g ACAC-type microcapsules is as effective as 6 hours of haemodialysis with the standard EX01 haemodialyser (Chang *et al.*, 1972). Recent studies (Chang, 1972) have suggested that this is most likely due to the much higher efficiency of the microcapsule artificial kidney for the removal of the medium molecular weight range molecules, expecially those larger than 600 daltons.

In addition to albumin-coated cellulose, other blood-compatible membranes for the microencapsulated activated charcoal are also being investigated. Microcapsules with heparinised surfaces have been studied in this laboratory (Chang *et al.*, 1967, 1968; Chang, 1969). Hydron, a blood-compatible membrane, has been used in another laboratory (Andrade *et al.*, 1971) to form a microcapsular absorbent by coating on activated charcoal and animal studies on this system appear promising. Albumin has also been used to directly coat activated charcoal (Andrade *et al.*, 1971). Other extensions of the proposed idea of microcapsular absorbent (Chang, 1966, 1972) include exploration into the use of microcapsular charcoal for ingestion for the removal of uraemic metabolites (Sparks *et al.*, 1969, 1971; Gardner *et al.*, 1971). The determining factors here would be the rate of diffusion of creatinine, uric acid, and other medium molecular weight uraemic metabolites into the gastrointestinal tract.

One of the uraemic metabolites not removed by microcapsular charcoal is urea. It was originally demonstrated that it may be feasible to remove it by using urease in combination with an ammonia adsorbent to convert urea to ammonia, which would then be removed by an ammonia adsorbent (Chang, 1966, 1972). This approach has been investigated and analysed further (Levine and LaCourse, 1967; Sparks et al., 1969, 1971; Gardner, 1971) and even extended to use in the low-volume dialysate system (Gordon et al., 1971). One interesting extension is that since the rate of movement of urea across the Gl tract is extremely fast, one might be able to use the ingestion approach for the removal of urea (Sparks et al., 1969, 1971; Chang and Loa, 1970; Gardner et al., 1971). In fact, experiments have demonstrated that a micro-encapsulated combination of urease and ammonia absorbent do significantly lower blood urea levels (Chang and Loa, 1970; Gardner et al., 1971). In recent studies, oxidised starch has been micro-encapsulated to remove urea and ammonia from the gastrointestinal tracts (Sparks et al., 1971).

DISCUSSION

The results of clinical trials carried out at McGill indicate that two hours haemoperfusion using 300 g of albumin-cellulose-coated microencapsulated activated charcoal is clinically as effective as 6 hours of haemodialysis with the EX01 standard haemodialyser. This has been related to their much higher efficiency in the removal of medium molecular weight molecules of higher than 600 daltons, especially the removal of those in the 600 to 1200 range (Chang, 1972). The albumin-coated microencapsulated activated charcoal did not have any adverse effects on the formed elements of blood. The feasibility of chronic, intermittent, maintenance haemoperfusion using this system has been demonstrated. The compactness, low cost, and efficiency of this system for the removal of medium-size molecular weight metabolites should contribute to the increased availability of artificial kidneys. The microencapsulated activated charcoal in its present form is only a first step towards making use of the full potential of the principle of micro-capsule artificial kidney. For instance, the present ACAC microencapsulated activated charcoal system, like the standard artificial kidney, requires systemic or regional heparinisation. With further progress in the field of biocompatible material, systemic or regional heparinisation may not be required. The present system using 300 g of micro-capsular activated charcoal is much more compact than the standard artificial kidney system. However, it is still much larger than it could be. The combination of large surface and ultrathin membranes means that a much more compact artificial kidney, perhaps 1/40 the size of the 300 g system would be possible if absorbents with much higher absorbing capacity were available for microencapsulation.

Acknowledgment

This work has been supported by the Medical Research Council of Canada.

REFERENCES

ANDRADE, J. D., KUNITOMO, K., VAN WAGENEN, V., KASTIGIR, B., GOUGH, D., and KOLFF, W. J. 1971: Coated adsorbents for direct blood perfusion: HEMA/activated carbon. Trans. Amer. Soc. Artif. Int. Organs 17: 222.

BOKROS, J. C., GOTT, V. L., LAGRANGE, L. D., FADALL, A. M., VOS, K. D., and RAMOS, M. D. 1969: Correlations between blood compatibility and heparin absorptivity for an impermeable isotropic pyrolytic carbon. J. Biomed. Mat. Res. 3: 497.

BORETOS, J. W., and PIERCE, W. S. 1967: Segmented polyurethane: a new elastomer for biomedical applications. Science 158: 1481.

CHANG, T. M. S. 1964: Semipermeable aqueous microcapsules. Science 146: 524.

CHANG, T. M. S. 1966: Semipermeable aqueous microcapsules ('artificial cells') with emphasis on experiments in an extracorporeal shunt system. Trans. Amer. Soc. Artif. Int. Organs 12: 13.

CHANG, T. M. S. 1969: Removal of endogenous and exogenous toxins by a microencapsulated absorbent. Canad. J. Physiol. Pharmacol. 47: 1043.

CHANG, T. M. S. 1972: 'Artificial Cells'. Monograph. Charles C. Thomas, Springfield, Ill.

CHANG, T. M. S., MacINTOSH, F. C., and MASON, S. G. 1966: Semipermeable aqueous microcapsules. I. Preparations and properties. Canad. J. Physiol. Pharmacol. 44: 115.

CHANG, T. M. S., PONT, A., JOHNSON, L. J., AND MALAVE, N. 1968: Response to intermittent extracorporeal perfusion through shunts containing semipermeable microcapsules. Trans. Amer. Soc. Artif. Int. Organs 13: 163.

CHANG, T. M. S., and MALAVE, N. 1970: The development and first clinical use of semipermeable microcapsules ('artificial cells') as a compact artificial kidney. Trans. Amer. Soc. Artif. Int. Organs 16: 141.

CHANG, T. M. S., GONDA, A. DIRKS, J. H., and MALAVE, N. 1971: Clinical evaluation of chronic intermittent and short term haemoperfusion in patients with chronic renal failure using semipermeable microcapsules ('artificial cells') formed from membrane-coated activated charcoal. Trans. Amer. Soc. Art. Int. Organs 17: 246.

CHANG, T. M. S., GONDA, A., DIRKS, J. H. A., COFFEY, J., and BURNS, T. 1972: Short-term and long-term maintenance of eleven uremic patients using the microcapsule artificial kidney. Trans. Amer. Soc. Art. Int. Organs 18.

CHANG, T. M. S., JOHNSON, L. J., and RANSOME, O. J. 1967: Semipermeable aqueous microcapsules. IV. Nonthrombogenic microcapsules with heparin complexed membranes. Canad. J. Physiol. Pharmacol. 45: 705.

CHANG, T. M. S., and LOA, S. K. 1970: Urea removal by urease and ammonia absorbent in the intestine. Physiologist 13: 70.

Conference on Mechanical Surface and Gas Layer Effects on Moving Blood. 1971. Fed. Proc. 30.

DEDRICK, R. L., VANTOCH, P., GOMBOS, E. A., and MOORE, R. 1967: Kinetics of activated carbon kidney. Trans. Amer. Soc. Artif. Int. Organs 13: 236.

DE MYTTENAERE, M. H., MAHER, J. F., and SCHREINER, G. E. 1967: Haemoperfusion through a charcoal column for glutethimide poisoning. Trans. Amer. Soc. Artif. Int. Organs 13: 190.

DUNEA, G., and KOLFF, W. J. 1965: Clinical experience with the Yatzidis charcoal artificial kidney. Trans. Amer. Soc. Artif. Int. Organs 11: 178.

DUTTON, R. C., DEDRICK, R. L., and BULL, B. S. 1969: A simple technique for the experimental production of acute platelet deficiency. Throm. Diath. Haemorrh. 21: 367.

ERIKSSON, J. C., GILLBERG, G., and LAGERFREN, J. 1967: A new method for preparing nonthrombogenic plastic surfaces. J. Biomed. Mater. Res. 1: 301.

GARDNER, D. L., FALB, R. C., KIM, B. C., and EMMERLING, D. C. 1971: Possible uraemic detoxication via oral-ingested microcapsules. *Trans. Amer. Soc. Artif. Int. Organs* 17: 239.

GORDON, A. BETTER, O. S., GREENBAUM, M. A., MARANTZ, L. B., GRAL, T., and MAXWELL, M. H. 1971: Clinical maintenance haemodialysis with a sorbent-based low-volume dialysate regeneration system. *Trans. Amer. Soc. Artif. Int. Organs* 17: 253.

GOTT, V. L. WHIFFEN, J. D., and DUTTON, R. C. 1963: Heparin bonding on colloidal graphite surfaces. *Science* 142: 1297.

GOTT, V. L., LOEPKE, D. E., DAGGETT, R. L., ZARNSTORFF, W., and YOUNG, W. P. 1961: Coating of intravascular plastic prostheses with colloidal graphite. *Surgery* 50: 382.

GRODE, G. A., ANDERSON, S. J., GROTTA, H. M., and FALB, R. D. 1969: *Trans. Am. Soc. Artif. Int. Organs* 15: 1.

HAGSTAM, K. E., LARSSON, L. E., and THYSELL, H. 1966: Experimental studies on charcoal haemoperfusion in phenobarbital intoxication and uraemia, including histological findings. *Acta. Med. Scand.* 180: 593.

HUFNAGEL, C. A., CONRAD, P. W., GILLESPIE, J. F., PIFARRE, R., ILLANO, A., and YOKOYAMA, T. 1967: Comparative studies of cardiac and vascular implants in relation to thrombosis. *Surgery* 61: 11.

LEININGER, R. I., COOPER, C. W., FALB, R. D., and GRODE, C. A. 1966: Non-thrombogenic plastic surfaces. *Science* 152: 1625.

LEVINE, S. N., and LACROUSE, W. C. 1967: Materials and design consideration for a compact artificial kidney. *J. Biomed. Materials Res.* 1: 275.

LEVOWITZ, B. S., LaGUERRE, J. N., GOULD, F. E., SCHEERER, J., and SCHOENFELD, H. 1968: Biologic compatibility and applications of hydron. *Trans. Amer. Soc. Artif. Intern. Organs* 14: 82.

LYMAN, D. J., ANDRADE, J. D., KLEIN, K. G., BRASH, J. J., FRITZINGER, B. K., and BONOMO, F. S. (in press): Interaction of platelets with protein surface. *Thromb. Diath. Hemorrh. Suppl.*

LYMAN, D. J., BRASH, J. L., and KLEIN, K. G. 1969: p. 113 *in* 'Proceedings Artificial Heart Program Conference' (Ed R. J. Hegyeli). U.S. Govt. Printing Office, Washington, D.C.

LYMAN, D. J., KLEIN, K. G., BRASH, J. L., and FRITZINGER, B. K. 1970: The interaction of platelets with polymer surfaces. 1. Uncharged hydrophobic polymer surfaces. *Thromb. Diath. Haemorrh.* 23: 120.

LYMAN, D. J., MUIR, W. M., and LEE, I. J. 1965: Effect of chemical structure and surface properties of polymers on the coagulation of blood. 1. Surface free energy effects. *Trans. Amer. Soc. Artif. Int. Organs* 11: 301.

MERRILL, E. A., SALZMAN, E. W., LIPPS, B. J., GILLILAND, E. R., AUSTEN, W. G., and JOISON, J. 1966: Antithrombo-

genic cellulose membranes for blood dialysis. *Trans. Amer. Soc. Artif. Intern. Organs* 12: 139.

MERRILL, E. W., SALZMAN, E. W., LONG, P. S. L., and ASHFORD, T. P. 1970: Polyvinyl alcohol-heparin hydrogel 'G'. *J. Appl. Physiol.* 29: 723.

MILLIGAN, H. L., DAVIS, J., and EDMARK, K. W. 1968: The search for correlation between electrokinetic phenomena and blood thrombus formation on implant materials. *J. Biomed. Mat. Res.* 21: 51.

MUIR, W. M., MARTIN, A. M., GILCHRIST, T., and COURTNEY J. M. 1973: The risk in estimating biocompatibility in 'Perspectives in Biomedical Engineering' (Ed. R. M. Kenedi). Macmillan, London.

MUSOLF, M. C., HULCE, V. D., BENNETT, D. R., and RANO, M. 1969: *Trans. Amer. Soc. Artif. Intern. Organs* 15: 18.

NEMCHIN, R. G., PATEL, A. R., ABEL, H. I., SIMS, L., and SPEAKER, M. 1969: *Proceedings Artificial Heart Program Conference.*

PACKHAM, M. A., EVANS, G., GLYNN, M. F., and MUSTARD, J. F. 1969: Effect of plasma proteins on interaction of platelets with glass surfaces. *J. Lab. Clin. Med.* 73: 686.

SALZMAN, E. W., MERRILL, E. W., BINDER, A., WOLF, C. F. W., ASHFORD, T. P., and AUSTEN, W. G. 1969: Protein-platelet interaction on heparinized surfaces. *J. Biomed. Mat. Res.* 3: 69.

SAWYER, P. N., and PATE, J. W. 1953: Bioelectric phenomena as an etiologic factor in intravascular thrombosis. *Amer. J. Physiol.* 175: 103.

SAWYER, P. N., and SRINIVASAN, S. 1967: Metallic and plastic prosthetic devices as vascular wall substitutes: biophysical criteria and methods for evaluation. *J. Biomed. Mat. Res.* 1: 83.

SHARP, W. V., GARDNER, D. L., ANDRESEN, G. J., and WRIGHT, J. 1968: Electrolour: a new vascular interface. *Trans. Amer. Soc. Artif. Intern. Organs* 14: 73.

SPARKS, R. E., MASON, N. S., MEIER, P. M., LITT, M. H., and LINDAN, O. 1971: Removal of uraemic waste metabolites from the intestinal tract by encapsulated carbon and oxidised starch. *Trans. Amer. Soc. Artif. Int. Organs* 17: 229.

SPARKS, R. E., SALEMME, R. M., MEIER, P. M., LITT, M. H., and LINDAN, O. 1969: Removal of waste metabolites in uraemia by microencapsulated reactants. *Trans. Amer. Soc. Artif. Int. Organs* 15: 353.

VROMAN, L., and ADAMS, A. L. 1969: Effect of heparin on reactions at aminated polymer-blood interfaces. *J. Colloid Interface Sci.* 31, 188.

VROMAN, L. ADAMS, A. L., and LINGS, M. 1971: Interactions among human blood proteins at interfaces. *Fed. Proc.* 30: 1494.

YATZIDIS, H. 1964: A convenient haemoperfusion micro-apparatus over charcoal for the treatment of endogenous and exogenous intoxications. *Proc. Europ. Dialysis Transplant. Ass.* 1: 83.

THE RISK IN ESTIMATING BIOCOMPATIBILITY

W. M. MUIR, A. M. MARTIN, T. GILCHRIST and J. M. COURTNEY[1]

For the past ten years, concurrent with the growth of artificial organ engineering for body function replacement, there has been an extremely haphazard development of equipment for lung function replacement, bypass and heart pumps, artificial kidneys, and artificial liver and perfusion equipment. Allied to these design studies by clinicians and bioengineers has been the ever-present and growing awareness that the problem of biocompatibility of devices to which flowing blood, tissues, or bone are directly exposed has been attacked in the most haphazard way.

The carefully deduced current understanding of blood-clotting cascading mechanisms, the availability of bovine mucopolysaccharide for thrombus inhibition, and the realisation that thrombus can be triggered by a variety of mechanisms have all pointed to the complexity of the biocompatibility problem.

Through the 'sixties, with the growth development of the American Space Programme, a number of polymeric materials developed for structure, environment fabrication, and electrical installation applications have found gradual acceptance in the artificial-organ field. With the development of several new products of the 'sixties we have had a material, a chemical, or a fabricated structure which is in effect 'an answer looking for a problem'. Unfortunately we have had throughout the development of implantable devices the opposite situation, namely an unrecognised 'problem looking for an answer'. Despite all the highly optimistic research proposals and programmes which have been funded in the United Kingdom, Europe, and the United States, the only person who appreciated the magnitude of the problem was the late Dr Frank Hastings.

Dr Hastings master-minded the artificial heart programme of the National Institute of Arthritic and Metabolic Diseases at the (British) National Institutes of Health until his recent death. He was responsible for the awareness, if not the implementation by funding, of programmes which would place the biocompatibility problem in perspective with artificial organ development. Due to the lack of proper emphasis on the importance of the use of biocompatible materials and the lack of appreciation of the insufficiency of biocompatible surfaces, we have seen medical engineering projects fail to reach their goals.

Thus, at the end of a decade of expenditure of untold millions, while we in no way intend to belittle the efforts of those workers involved, we have not yet achieved an implantable heart pump with a life expectancy *in vivo* in animals, of more than 1 week (Kwan-Gett *et al.*, 1971). The reason is perfectly clear—the lack of appreciation of what biocompatibility means, and it is a sad comment on the fact that after all the effort expended, the artificial heart syndrome means disseminated intravascular thrombosis arising from faulty heart-pump designs and death from circulatory failure. Such comment can be extended to a lesser extent to almost every extracorporeal circuit or implantable device.

Let us now examine the meaning of biocompatibility.

BIOCOMPATIBILITY

Biocompatibility covers not only the obvious reaction of blood and tissues to the foreign body implant, but also the following dynamic factors inherent in an implantable system:

1 Device manifold cross-sections cause changes in blood flow which may lead to cavitation or emergence from solution of dissolved blood gases, or may lead to phase separation of plasma from red cells when the plasma velocity profile is very different from that of the red cells.

2 Apart from phase separation, extremely high or low plasma velocities lead to the emergence and deposition from solution of plasma proteins, fibrin, and globulins. Deposition of globulins on anionic surfaces may be beneficial in inhibiting fibrin deposition, and pre-coated albumin plastic surfaces may tend to delay thrombus onset, but generally such effects are most certainly not beneficial (Martin *et al.*, 1971).

The static effects of biocompatibility have been well documented, but the techniques of estimation are still in their infancy and it is worth while noting that the logical evaluation now being proposed consists of, in the order given, tests of the following:

1 Muscle-tissue reactions of materials implantable in animals.

---

[1] University of Strathclyde, Glasgow, Scotland

2 Surface-induced coagulation and thrombus effects in short-term exposure of materials with large surface areas, e.g. in the device illustrated in figure 1, A and B, which takes blood *in vivo* on partial bypass.

3 Biochemical and haematological reactions of animal metabolism, following long-term exposure, using long-term-survival dogs over a period of months.

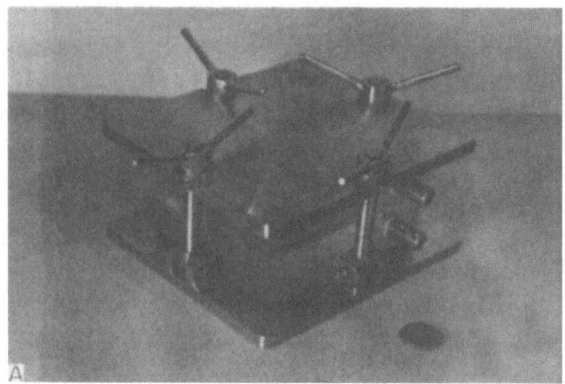

Figure 1. (A) Assessment of blood compatibility; test cell for extracorporeal circuits. Test cell used by authors to hold plastic films in sheets between which blood flows.

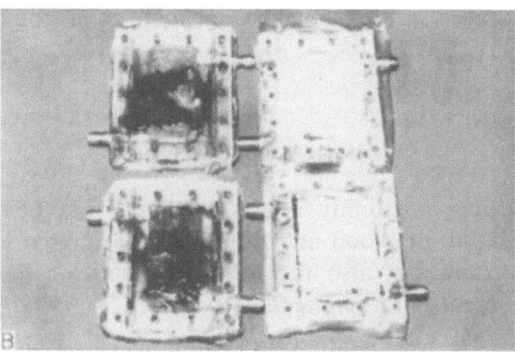

Figure 1. (B) Pair of test cells, after exposure of cuprophan cellulose film, showing dog blood thrombus (left), and new clot-free vinyl plastic given identical dog blood exposure (right).

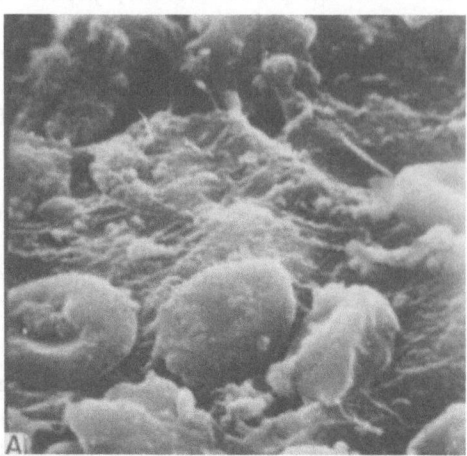

Figure 2. (A) Formed blood element on artificial kidney: cuprophan cellulose film showing fibrin network and enmeshed red cells of full thrombus formation.

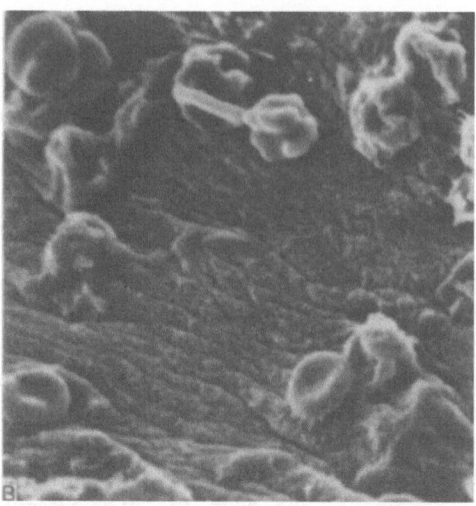

Figure 2. (B) Permane vinyl membrane showing red cells deposited without thrombus.

4 Short-term thrombus effects following exposure of the material in a partial bypass to human blood *in vivo* (figure 2, A).

5 Serum toxicity, using embryonic human lung cells of the heteroploid or diploid types. Choice of new candidate cells would depend on the particular degree of exposure and type of material being evaluated.

6 Biochemical and haematological reactions of formed blood elements following exposure of materials in partial bypass in short-term human tests.

7 Assuming satisfactory outcome of all prior stages and qualified acceptance of these, prototype fabrication should begin after each sequence is repeated in animals and then humans. The effect of foreign materials on the biological system and, in particular, those of the materials of plasma proteins is one of the most important determinants of short-term, leading to long-term, biocompatibility. These interactions are, in fact, the determinants of acceptability of prosthetic materials for artificial organs and prostheses exposed to blood.

Theories of blood compatibility of implantable materials

The most important aspect of material biocompatibility for implantation or extracorporeal circuit is the avoidance of early blood clot formation or thrombus initiation. Clotting may be due to incorrect haemodynamic flow of blood resulting in phase separation of red cells and plasma which subsequently results in thrombus initiation in a manifold, in an artery, or on the foreign material surface. Avoidance of surface initiated clotting has been the subject of much research by workers in haematology and artificial organs. Mechanisms of surface-initiated clotting are thought to be as follows:

1 Zeta potential or flow effects in removing electrons as blood flows over plastic or other materials lead to the establishment of positive electrical charges on implant surface. The result is deposition of blood-clotting materials and thrombus growth.

2 Where the surface energy of the implant materials differs greatly from that of the blood flowing across the surface, surface free-energy discrepancy or gradient causes surface deposition of clotting materials and subsequent thrombus onset (figure 2).

3 Mechanical damage by the implant to surrounding cells results in ADP release and onset of the cascade clotting mechanism.

Generally, thrombus initiation is minimised with varying degrees of success on implant plastic surfaces by the following methods:

1 Injection mould fabrication of prostheses in an electrical field which produces internal charge imbalance with electron surplus at the prosthesis outer surface. Such a device is known as an 'electret'.

2 Use of plastic materials for prosthesis fabrication which do not readily trigger thrombus initiation in certain high blood-velocity situations. The only partially acceptable material used at present is silicone rubber which has serious inherent mechanical deficiencies and cannot be used in load-bearing situations.

3 Intentional coating of prosthesis plastic surfaces by blood proteins or by a neo-intima of precultured host-compatible cells immediately prior to implant. Sometimes there is a fortuitous growth around some implants, especially of silicones, which indicates a good degree of body tolerance.

4 Coating the plastic surface with a nylon velour or woven material to intentionally trap cells on which a stable pseudo-intima grows.

5 Fabrication of prostheses with plasticisers containing quantities of anticoagulants in the formulation. During the life of such devices, the anticoagulant is leached out of the prosthesis and minimises surface-initiated thrombus growth.

6 Coating of prosthesis surface with anticoagulants chemically bound to the plastic surface by quaternary ammonium halides, resulting in reasonably long-life anticoagulated surfaces.

None of the methods outlined is truly satisfactory and there is a constant search under way for new or commercially available materials with good biological tolerance. There is even a good case for screening every new plastic which is marketed for basic biological tolerance.

BIOCOMPATIBILITY AND IMPLANTABLE BLOOD PUMP DEVICES

Ever since the idea of blood pumps was conceived, researchers have endeavoured to achieve a pump which works in as atraumatic a manner as possible. Research over the years has shown that the following are the most important factors in achieving this:

1 Shear forces.
2 Cavitation.
3 The smoothness of the tubing.
4 Pressure changes in the circuit.

Blackshear *et al.* (1965) have shown that shear forces cause significant damage and Domingo (1963) has shown that the use of driven roller pumps significantly reduces blood damage. Longmore (1965) produced a blood pump that had a driven single roller and also gave a pulsatile output. Non-occlusive pumps were also in vogue at one time as they prevented the opposing walls at the pump tubing impinging on each other and hence reduced shear stress. Longmore (1969) has shown that non-occlusive pumps produce cavitation which in turn produces micro-emboli in the shape of gases and polymerised sugar molecules.

As the technology of plastics has advanced, so the smoothness of the pump tubing has approached the ideal, but as Stewart and Sturridge (1959) have shown, the size of the erythrocyte is infinitesimal compared with the roughness of the tubings. Large changes in tubing diameter are to be avoided although there is little evidence (Blackshear, 1965) to show that pressure changes are significant in terms of blood component damage, provided negative pressures are avoided. From the above, it is clear that the ideal pump has yet to be produced but that, within the limitations of present-day technology, clinical blood pumping is possible for quite extended periods.

Despite enormous efforts to develop implantable cardiac booster-pumps and total artificial heart-pumps, the goal of a functioning device with several years' expectancy remains elusive as ever (Lyman, 1971). The immediate problems of massive thrombus formation on foreign surfaces following implantation may be solved or delayed by anticoagulants or proteinaceous surface coatings, but the dynamic problems remain:

1 The material or device is a constant irritant to the natural rejection mechanisms.

2 Dynamic design deficiencies produce secondary lethal consequences.

The net effect is a further syndrome, the 'artificial heart syndrome', which manifests itself in implantable blood pumps, causing death. The syndrome includes the manifestation of emboli initiated by plastic or foreign surface, but not necessarily deposited on it; blood shear; blood component phase separation; gas emboli; or mechanical collisions of blood-formed elements.

These effects manifest themselves as:

1 Emboli formation.
2 Disseminated intravascular thrombosis.

For future prediction, if the artificial heart syndrome is suppressed, the next lethal effect which will manifest itself to researchers in this field will be syndromes associated with:

(*a*) the biological interface where prosthesis meets tissue;
(*b*) progressive loss of function of vital organs supplied with blood from the mechanical pump.

CONCLUSIONS

As has been outlined in this review, a wide variety of materials has been tested for device fabrication in the artificial-organ field. The major obstacle to the satisfactory development and utilisation of body replacement devices is the lack of a range of biocompatible materials with the correct combination of other specifications. In addition, methods for evaluating material compatibility remain unsatisfactory, since material-induced changes are sought in known enzymatic and haematological factors.

For true biocompatibility estimation, a method is required for assessing deleterious effects of material or device on the complete organism.

Preliminary work by the authors suggests that the use of serum toxicity estimation by taking sera from test animals and testing its effect on specific selected cell-colony growth rates might provide a satisfactory answer to this problem.

REFERENCES

BLACKSHEAR, P. L. Jr., DORMAN, F. D., and STEINBACK, J. H. 1965: Some mechanical effects that influence haemolysis. *Trans. Amer. Soc. Artif. Int. Organs*, **11**: 12.

DOMINGO, R. T. 1963: Gear take-up heart pumps—avoidance with internal shear forces. *Trans. Amer. Soc. Artif. Int. Organs* **9**: 275.

KWAN-GETT, C., BADEMANN, D. K., DONOVAN, F. M., EASTWOOD, N., FOOTE, J. L., KAWAI, J., KESSLE, T. R., KRALIOS, A. C., PETERS, J. L., VAN KAMPEN, K. R., WONG, H. K., ZWART, H. H. J., and KOLFF, W. J. 1971: *Trans. Amer. Soc. Artif. Int. Organs* **17**: 474.

LONGMORE, D. B. 1965: Pulsatile flow pump, *Lancet*, May 15, p. 1048.

LONGMORE, D. B. 1969: Cavitation in blood pumps. *Nature* **222**: 5188.

LYMAN, D. J., KWANN-GETT, C., ZWART, H. H. J., BLAND, A., EASTWOOD, N., KWAI, J., and KOLFF, W. J. 1971: *Trans. Amer. Soc. Artif. Int. Organs* **17**: 456.

MARTIN, A. M., MUIR, W. M., BALLANTINE, F., and COURTNEY, J. M. 1971: 'Plastics in Medicine and Surgery'. Symposium, University of Newcastle Medical School, Newcastle, September 15th. The Plastics Institute, London.

STEWART, J. W., and STURRIDGE, M. F. 1959: Haemolysis caused by tubing in extracorporeal circulation. *Lancet*, **1**: 340.

COMMENTS ON HAEMODIALYSIS

HORST KLINKMANN AND MANFRED HOLTZ[1]

INTRODUCTION

It is hard to compete in the modern discussion of unconventional haemodialysis with comments on just conventional dialysis. But knowing that in the last 20 years the success of the conventional artificial kidney in preserving the lives of thousands of people has attracted world-wide attention (Parsons et al., 1971; NIAMD registry, 1971), and that presently 98 % of our dialysis treatment throughout the world is done by conventional dialysis technique, encourage us to make a few comments on this method, which in principle has not changed since the time of Rowntree and Turner (1912).

The two major problems that confront most dialysis programmes throughout the world are excessive cost and secondary medical complications. The problem of economy and our changing view on the pathophysiology of uraemia and its treatment, our belief that, in uraemia, medium-sized and larger molecules may be of great importance for the development of secondary uraemic complications has forced us to reconsider our methods and the frequency of dialysis treatment. Most physicians now want to dialyse their patients more often, according to Shaldon (1966), who termed dialysis as the 'insulin' for the 'chronic nephritic'. To popularise dialysis to this extent artificial kidneys have to be disposable (Klinkmann and Kolff, 1970), but it seems extremely difficult to bring down the cost to levels which would allow for disposable kidneys. Home dialysis offers, to some degree, a solution to the staff shortage in hospitals. Industry, however, tends to make the price of new equipment higher rather than lower. We are therefore seeing a return either to reuse of the dialysers or a return to kidneys that can be reassembled at low cost (Nosé et al., 1970; Acchiardo and Walker, 1972; Gutch et al., 1972). In both attempts the permeability and durability of the dialysis membrane is a central problem. But despite many efforts to bring better membranes into clinical use, most currently used artificial kidneys still have regenerated cellulose membranes, commercially known as Cellophane, Cuprophane, Nephrophane or Neflex. The Strathclyde Bioengineering Unit in Glasgow has taught us how to test membranes (Muir and Ross, 1967; Klinkmann et al., 1971; Muir, 1971) and even with cellulose membranes in use for more than 20 years in routine clinical work, we are still gaining knowledge regarding their specific structure and their mechanism of effectiveness obtained by different processes of manufacture.

PERMEATION OF ESSENTIAL AMINO ACIDS THROUGH CELLULOSE MEMBRANES

It is well known that due to their specific structure the permeability of cellulose membranes depends only on the molecular size and steric configuration of the diffusing substances. This means that there is no selective diffusion through cellulose membranes for either waste uraemic metabolites or life-essential compounds during the dialysis treatment. The enthusiasm over the life-saving results of dialysis of acute and chronic renal failure has led, for quite a time, to an underestimation of this problem.

In a modified dialysis test cell according to Muir and Ross (1967) we determined the mass transfer rate of eight essential amino acids in four different cellulose membranes (Visking cellophane PT 300, Cuprophane PT 150, Nephrophane and Neflex DVF 30 B). Creatinine was used as the reference value (table 1). The graphic representation of these results (figure 1) demonstrates that differently manufactured cellulose membranes show a remarkable difference in their mass transfer for the essential amino acids. The actual dialysance of each individual amino acid is almost identical with that of creatinine in each membrane investigated.

After adjusting the concentration of the test solutions in the test cell to that of the normal plasma levels of the essential amino acids we converted the data from units of $mmol\ m^{-2}\ h^{-1}$ into $mg\ m^{-2}\ h^{-1}$ to get a better comparison with the absolute loss in clinical dialysis. The calculated loss of essential amino acids through an unsupported dialysis membrane, per square metre per hour of dialysis, ranges from 1·3 g in Nephrophane, 1·0 g in Cuprophane, 0·7 g in Visking Cellophane to 0·5 g in Neflex DVF 30 B (table 2).

The highest loss of wanted and unwanted compounds occurs in Nephrophane, the cellulose membrane with the thinnest wall and the largest pores. Cellulose membranes are able to remove

[1] Division of Nephrology. Medizinische Universitäts-poliklinik, Rostock, German Democratic Republic.

Amino-acid	Nephrophane	Cuprophane	Visking-Cellophane	DVF-30 B	mol. wt.
Valine	14·28	13·81	10·85	7·30	117·1
Threonine	17·32	16·00	11·59	3·70	119·1
Leucine	13·91	7·85	5·35	2·48	131·2
Isoleucine	16·95	12·16	10·57	3·72	131·2
Lysine	19·08	14·26	10·43	8·75	146·2
Methionine	10·46	12·45	8·44	6·97	149·2
Phenylalanine	10·08	8·11	4·68	0	165·2
Tryptophan	11·04	10·56	5·72	6·31	204·2
Creatinine	14·20	10·32	8·01	3·46	113·12

TABLE 1: Mass transfer rates (mmol m^{-2} h^{-1}) through four cellulose membranes for essential amino acids and creatinine.

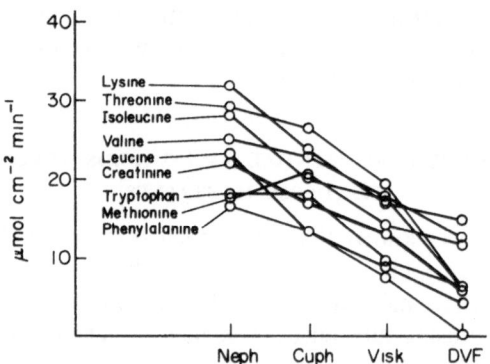

Figure 1. Decrease in mass transfer rates of essential amino acids in four cellulose membranes.

the unwanted uraemic toxins but if their specific diffusion characteristics are not taken into consideration they can also cause a persisting deficiency in life essential compounds. This in turn can lead to the development of muscle wasting and impairments in other organ functions during regular dialysis treatment. A better conventional artificial kidney might incorporate an even 'leakier' cellulose membrane with a selective-block copolymer membrane to mimic the overall action of the natural kidney.

RE-USE OF DIALYSERS—THE PROBLEM OF DECREASING MEMBRANE EFFECTIVENESS

Re-use of dialysers relieves hospitals and patients of quite a sizeable financial burden. But this practice should bring no reduction in the 'quality' of the treatment of the patient. Reports about the number of possible reuses of dialysers have been very contradictory (Hammil, 1969; Acciardo and Walker, 1972; Easterling, 1972; Lavender, 1970). Since in every experiment cellulose membranes have been used, the differences in decrease of efficiency might depend on the hardware construction, especially that of the blood channelling compartment. The re-use of coils with comparably thick blood films did not show a reduction in the performance of the dialyser after five to six successive re-uses (Hammil, 1969). The tube dialyser AUE II with a blood channel height of ~1·5–2 mm already shows a decrease in efficiency after the first performance. To study whether this was due either to the decreasing osmotic gradient between blood and dialysis fluid, or to a deterioration in membrane permeability, we determined the dialysance for urea in the same membrane before and after dialysis. Before the determination of the dialysance after 10 hour dialysis, the dialyser was flushed with tap water until no macroscopically visible blood was left.

There is a significant reduction in membrane effectiveness after using the dialyser in a 10 hour dialysis treatment (figure 2) which corresponds almost ideally with the gradual loss of efficiency during the 10 hour treatment (figure 3). Electron-microscopic examination of the used Nephrophane membrane showed a protein layer on its surface (figure 4) and blocked pores (figure 5).

The same phenomenon has been reported to an

Amino-acid	Nephrophane	Cuprophane	Visking-Cellophane	DVF-30 B	mol. wt.
Valine	267·5	259·0	203·0	137·0	117·1
Threonine	114·0	105·0	71·2	24·2	119·1
Leucine	227·5	88·0	38·6	41·3	131·2
Isoleucine	77·5	56·0	48·6	17·1	131·2
Lysine	264·5	197·5	144·8	121·5	146·2
Methionine	62·3	74·5	50·5	41·4	149·2
Phenylalanine	125·0	94·3	54·2	0	165·2
Tryptophan	169·5	161·0	87·8	96·9	204·2
Σ	1307·8	1035·3	698·7	479·4	

TABLE 2: Calculated loss mg m^{-2} h^{-1} of essential amino acids through four cellulose membranes assuming normal plasma levels.

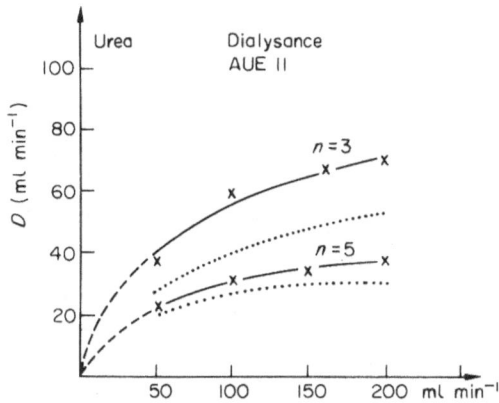

Figure 2. Urea dialysance for the AUE II dialyser before ($n = 3$) and after ($n = 5$) use in a 10 hours dialysis.

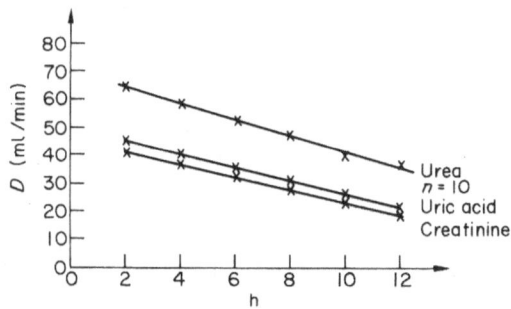

Figure 3. Gradual decrease of the dialysances of urea, creatinine and uric acid in a AUE II dialyser during a 12 hours dialysis.

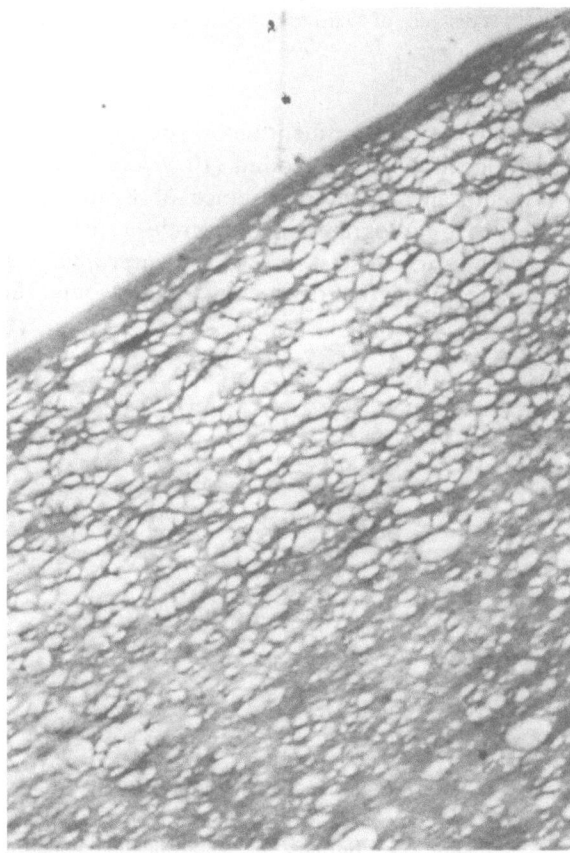

Figure 4. Nephrophane membrane before use (magnification ×10 000).

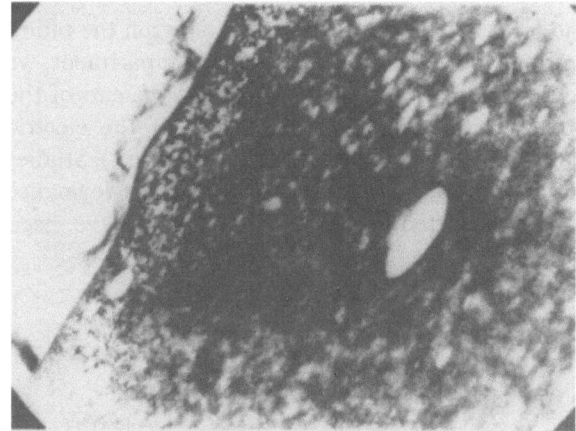

Figure 5. Nephrophane membrane after use in a AUE II dialyser for 12 hours (magnification ×10000).

even greater extent for other small, low-blood-volume dialysers with a 'tight' blood channel, e.g. the Argonne–Lavender Adhesive Bonded Dialyser (Lavender, 1970). We believe that the protein deposit is a consequence of flow velocity and high filtration pressure gradient. Therefore, re-use of these types of dialysers is bound to be very limited. The same problem exists in the re-use of the Hollow Fibre Artificial Kidney (HFAK) where cellulose acetate capillaries with a diameter of 25 to 30 μm are used instead of cellulose tubing or sheets. During dialysis treatment the overall performance is reduced proportionally to the loss of membrane area, resulting more likely from thrombus formation in fibre occlusion than from protein layers, according to our studies.

It is interesting to note, as can be seen from figure 6, that there is no significant reduction in the

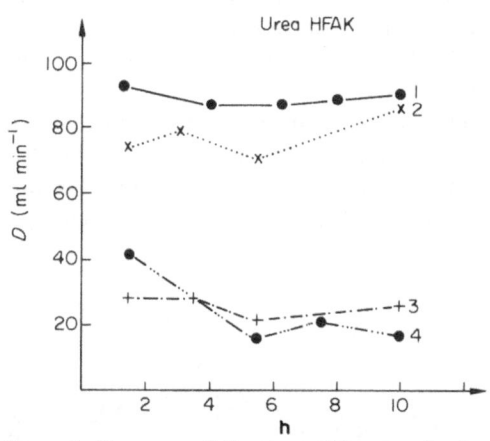

Figure 6. Decrease of the urea dialysance in the HFAK used four consecutive times for dialysis.

dialyser performance from first to second dialysis. However allowing a clotting time in the beginning of the third dialysis for only a short period of about 30 min immediately causes a significant drop in clinical performance.

Another interesting result, that as yet we cannot explain, remains to be reported. By measuring inside the HFAK the electric potential between the blood compartment and the dialysate compartment we found a remarkable change towards the end of the 10 hour dialysis with a reversion of the electric potential from positive to negative (figure 7). Studies to correlate the reduction in dialyser performance to the reported electric phenomenon have been initiated.

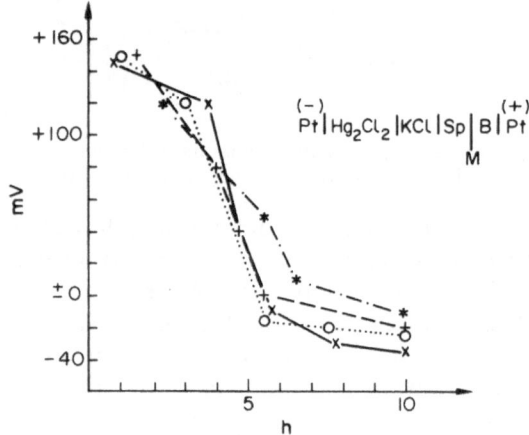

Figure 7. The change of the electric potential in the HFAK during a 10 hours dialysis. (Pt = platinum electrode, M = membrane, B = blood, SP = dialysis fluid, $Hg_2Cl_2 KCL$ = calomel electrode).

NEW ATTEMPTS TO IMPROVE CELLULOSE MEMBRANES

Many groups are working on new optimised haemodialysis membranes with selective or improved permeability. The fabrication of ultrathin cellulose membranes reinforced with cellulose fibres led to a dialysis membrane (Neflex) with satisfactory mass transfer rates and good mechanical strength (Klinkmann *et al.*, 1970) (see figures 8 and 9).

Lately one of our group used ultrasound for altering the pore structure and consequentially the diffusion characteristics in cellulose membranes (Sehner, 1972). Figure 10 shows the changing pattern

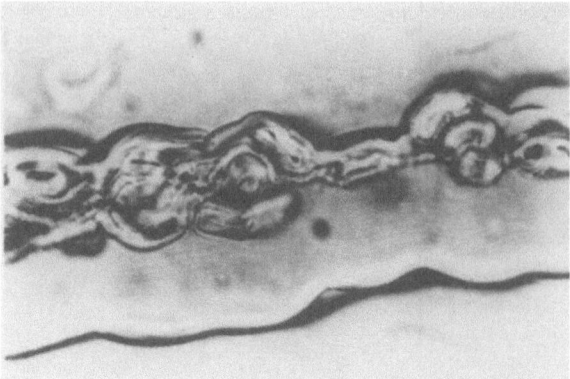

Figure 9. Electronmicrograph of a section through Neflex DVF 30 B with fibre reinforcement (magnification ×600).

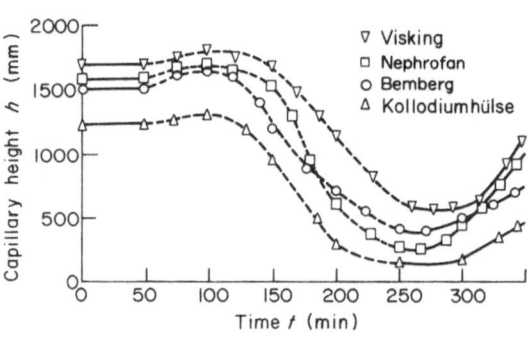

Figure 10. Changes in the osmotic gradient before (————), during (- - - -) and after (————) the application of ultrasound (intensity 10 W/cm² from a distance of 8 cm) to different cellulose membranes (according to Sehner, 1971).

in the osmotic gradient before, during and after the application of ultrasound (10 W/cm²) to different membranes from a distance of 8 cm. Scanning electronmicrographs demonstrate a visible change in the surface structure of the Cuprophane membrane during the treatment with ultrasound (figures 11, 12, 13).

Figure 8. Scanning electromicrograph of Neflex DVF 30 B (magnification ×300).

Figure 11. Surface of Cuprophane-membrane before treatment with ultrasound (magnification ×600).

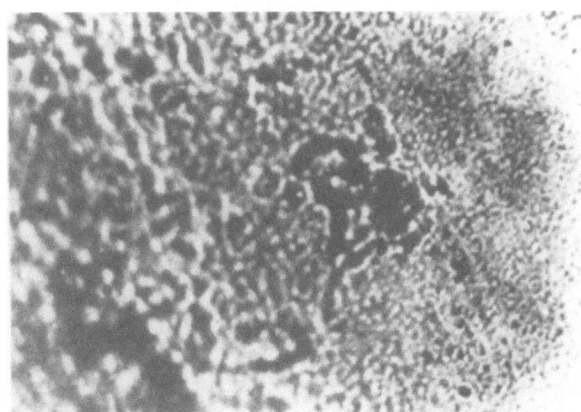

Figure 12. Surface of Cuprophane-membrane during treatment with ultrasound (magnification ×600).

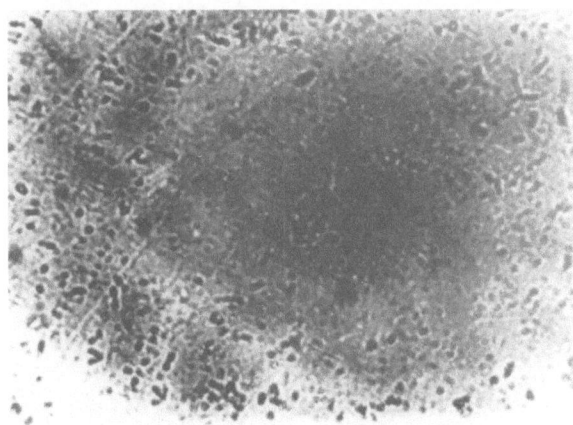

Figure 13. Surface of Cuprophane-membrane after treatment with ultrasound (magnification ×600).

SUMMARY

The past has seen a tremendous increase in the influence and importance of Bioengineering in medical research and patient care. Being medical men with limited knowledge of, but great admiration for, Bioengineering, our impression has always been that to work in this field involves more than merely being a master of only one discipline. It must involve committed enthusiasm for the subject, ability to communicate between people of different interests and with vastly different professional backgrounds. Its practitioners must also show sympathetic interest and be able to listen patiently to the not always specific ideas and individual problems of their medical colleagues.

In our view Strathclyde University has taken in a beautifully conventional way a very unconventional and fundamental step by realising very early a most important aspect: engineers engaged in Biomedical Engineering must be professionals in their own right and not juniors to the medical staff. We are, therefore, convinced that the Strathclyde Bioengineering Unit will continue to be one of the leading centres in the world and make fundamental contributions to improving conventional and unconventional dialysis, and so contribute in relieving the frustrations, difficulties, tensions and inconveniences of thousands of patients suffering from kidney diseases.

REFERENCES

ACCHIARDO, S., and WALKER, B. 1972: *Abstracts of the American Society for Artificial Internal Organs*. Eighteenth Annual Meeting, Seattle, 1.

EASTERLING, R. E. 1972: *Abstracts of the American Society for Artificial Internal Organs*. Eighteenth Annual Meeting, Seattle, 14.

GUTCH, C. E., KOPP, K. F., VAN DURA, D., and KOLFF, W. J. 1972: *Abstracts of the American Society for Artificial Internal Organs*. Eighteenth Annual Meeting, Seattle, 20.

HAMMIL, F. S. 1970: *Thesis*. University of Utah, Division of Artificial Organs.

KLINKMANN, H., ANDRADE, J. D., KIRKHAM, R. L., and LYMAN, D. J. 1970: *Proceedings European Dialysis and Transplant Association* VII: 446.

KLINKMANN, H., HOLTZ, M., WILLGERODT, W., WILKE, G. and SCHOENFELDER, D. 1968: *Proceedings European Dialysis and Transplant Association* V: 78.

KLINKMANN, H. and KOLFF, W. J. 1970: Present Concepts in Internal Medicine III: 6: 437. Letterman General Hospital, San Francisco.

LAVENDER, A. R., and MARKLEY, F. W. 1970: *Proceedings Third Annual Contractors' Conference*. Argonne National Laboratory, 21. U.S. Department of Health, Education and Welfare.

MUIR, W. M. 1971: *Proceedings European Dialysis and Transplant Association* VIII: 359.

MUIR, W. M., and ROSS, D. S. 1967: *In* 'Digest of the VIIth International Conference on Medical and Biological Engineering', Stockholm, 14.

NIAMD Registry 1971: National Institute of Arthritis and Metabolic Diseases, Bethesda, U.S.A.

NOSÉ, Y., MRAVA, G. L., WEBER, D. C., KON, T., NAKAMOTO, S., POPOWNIAK, K. L., and KURUVILA, K. C. 1969: *Transactions of the American Society for Artificial Organs* 15: 118.

PARSONS, F. M., BRUNNER, F. P., GURLAND, H. J., and HARLEN, H. 1971: *Proceedings European Dialysis and Transplant Association* VIII: 3.

SEHNER, J. 1971: *Diplomarbeit*. Martin-Luther-Universität Halle-Wittenberg, Sektion Physik.

SHALDON, S. 1966: 'Scientific Basis of Medicine'. Annual Review 201. University of London, The Athlone Press.

ROWNTREE, L. G., and TURNER, B. B. 1913/1914: *Journal of Pharmacology and Experimental Therapeutics* 5: 275.

SESSION 'A' DISCUSSION[1]

Mr Bain, initiating the discussion, asked how long prosthetic valves will work when they have been encapsulated by the host. He was specifically concerned with alterations in function rather than material, arising, for example, from tissue growing into the corners of a trileaflet valve and thus preventing it from opening. *Dr Black* responding, indicated that they have not yet obtained long term test results *in vivo*. Accelerated *in vitro* tests in lipid and plasma equivalent to ten years' 'wear' have been carried out with good results. The question of tissue ingrowth would depend on the valve geometry which, if correctly designed, would limit this. This may also be assisted by the valve material itself, which could be made to inhibit tissue ingrowth. It had to be accepted that the whole area of prosthetic valves has to date been empirical, and it is sobering to reflect that after some twenty years of effort in the field, the valves used are still very different from their biological counterparts.

Mr Longmore commented that one of the fundamental problems of bioengineering is the inadequacy of the specifications provided. This is particularly so for prosthetic valve design. In this connection it is as well to remember that during the embryonic development of the heart the valves and the blood are differentiated from the same mass of tissue. In fact, to some extent, the blood flowing in the foetus moulds the valve. Consequently, he considered that transplantation of 'biological' valves is likely to be more successful than their mechanical counterparts, not the least because of the self-repairing, self-sustaining characteristics of living tissue. *Dr Wright* raised the question of the long term effectiveness of prosthetic versus biological valves. He pointed out that Starr–Edwards valves implanted in 1960 were still working, while it was accepted at the present time that biological valves fail after 5–6 years. In general he thought that the long-term prospects of patients are much better with prosthetic than with biological valves. *Mr Longmore* considered this a difficult point but felt that he had to disagree. He and his co-workers have compared prosthetic and biological valves and have come down slightly in favour of the latter

although the differences were marginal. He emphasised that in a number of instances the comparison was that of surgical expertise rather than that of the valves themselves.

Professor Mann, in relation to Dr Nosé's contribution, drew attention to the work of Dr William Bernhardt at the Massachusetts Institute of Technology in Cambridge, U.S.A., who, in cardiac assist pumps in calves, spread bovine foetal cells over a silicon Dacron mesh, and using nuclear tagging has shown clearly that the foetal cells persist for months in the *in vivo* situation. *Dr Nosé* agreed that foetal cell seeding seems a most promising approach. He was, however, concerned at the unpredictability of the results. Only with 'biolised' surfaces did he and his co-workers obtain good reproducibility of results for long-term patency.

Professor McGirr commented that Dr Chang's methods appeared to be efficient, giving rapid and adequate clearance. He was concerned, however, about the possible physiological hazards, such as electrolyte imbalance. Additionally, he wondered what effect the use of the microcapsular system has on peripheral resistance to blood flow and high blood pressure. *Dr Cattell* also asked for information from Dr Chang as to how his method normalises electrolytes, copes with water balance and deals with fluid retention. He went on to comment that many of his patients appear to get better as the years go by. This is a source of fascination since they are losing amino acids. He felt that a progressive policy of dietary replacement is the keystone of haemodialysis and the major problem now in dialysis is to make it more acceptable and tolerable. Dr Chang has produced an alternative to the conventional means but he was rather worried by his data. It seemed to him unlikely that Dr Chang would be treating the same kind of patients as his group were dealing with (with a filtration rate of less than 1–2 ml/min for whom water balance was a major problem) or the hypertensive patient or the patient who becomes hyperkalaemic. It seemed to him unlikely that patients of this kind could be managed with one microcapsular session and one conventional dialysis per week. He invited Dr Chang to comment on the possible differences in the patient populations. *Dr Chang*, replying, emphasised that the microencapsulated activated charcoal system is not yet a complete artificial kidney. A system for fluid and electrolyte removal is still to be developed. Thus at present the standard artificial kidney can be completely

---

[1] *Editorial Note.* The discussions at the Symposium were conducted in an informal manner, participant's questions or contributions being followed immediately by the relevant author's answer or comment. The verbatim records have been assembled into 'discussion reports'. In these, contributors and authors are referred to by surname only. Full names, addresses and affiliation are given in the complete alphabetic list of Symposium participants on page vii.

replaced only in those patients who still retain good fluid and electrolyte excretion. Patients with only moderate fluid and electrolyte excretion can be treated by the microcapsule system only if this is supplemented with haemodialysis on the standard artificial kidney for the removal of electrolytes and fluid as required. Finally for those patients with severe problems of electrolyte and fluid retention, the microencapsulated system might be attached in series to the standard haemodialyser, since it is extremely efficient in removing larger molecules and may thus cut down the time required for the standard treatment.

Commenting on the action of the present microencapsulated activated charcoal, Dr Chang continued by saying that it removes only creatinine, uric acid, middle-size and other larger molecules which are not present in high osmolar concentration and do not contribute osmotically very much. In the short term, after a two hours' perfusion which makes his uraemic patients clinically as well as if they had undergone six hours' standard haemodialysis, there appeared to be no significant changes in their electrolyte levels. In the longer term his laboratories' work and that of a number of others have demonstrated the feasibility of a combination of oral ingestion and haemoperfusion. In the former the ingestion of urea or ammonia absorbents and ion exchange resins might remove urea and help also in the removal of electrolytes.

Turning now to the possible hazards of peripheral resistance to flow and consequential high blood pressure, Dr Chang stated that even in patients who tend to have hypertension with standard haemodialysis, no problems have been encountered when using the microcapsular system. With such coated granules, the flow and flow resistance depends on the particle size. The present system has a very low flow resistance of only 10 mm mercury at 2 ml/min flow and this means that no pumps are needed in patients with external AV shunts. As far as the long term effects on peripheral resistance and high blood pressure are concerned, there are no answers available to this question at the present.

Professor Klinkmann referred to an alternative way of using activated charcoal and that is to regenerate dialysate in the conventional system. In this way the required dialysate quantity can potentially be reduced from 300 l to one l per m² of membrane. This permits considerable miniaturisation of the dialysate supply equipment, the problem remaining being the need to reinfuse electrolytes which have been removed. There are probably other essential compounds that are lost through the cellulose membrane. A great deal of work needs to be done in this apparently neglected field. The loss of amino acids has been investigated but little

is known about the loss of other life-sustaining compounds. Until this information is available, we do not know how to sustain life in patients undergoing long-term dialysis.

Responding to a question from Dr Cattell regarding the present state of the artificial kidney programme at Strathclyde, *Dr Gilchrist* briefly outlined the difficulties encountered in attempts to use the standard groove support system for the hydrophilic copolymer membranes developed. A successful support system has now been devised consisting of a plastic mesh similar to that used in the coil kidneys. The use of this kind of support system in the flat bed machine has a number of other advantages such as better distribution of dialysate on the underside. Clinical trials of the membranes are now in progress using this improved support system. He then went on to ask Dr Cattell if he had any comparative costs on preventive medicine as opposed to conventional haemodialysis. Further, how would he envisage the introduction of the necessary mass screening procedures in a country such as Britain were the precepts of preventive medicine to be totally accepted.

In reply *Dr Cattell* indicated that the question of cost of prevention can be broken down into two: the obvious financial cost, and the social and psychological costs. He believed that the cost of a widespread surveillance would probably be about equal to that of total renal replacement. He did not think that in the foreseeable future it will be possible to prevent progressive renal failure, because the methods of doing so effectively—compulsory screening, compulsory treatment, eugenic breeding—represent a 'Big Brother' situation which fortunately would not be accepted by a free society. *Professor McGirr* commented on the possible lag period even if a programme of preventive screening is instituted. Many of the patients who will appear in the next 10–15 years as 'clients' for the various dialysis systems under discussion, will already have kidney damage initiated. *Professor Kenedi and Dr Meeke*, as bioengineers, raised the question of future development and asked for guidance. They invited, in particular, the clinicians to outline what developments they foresee in conventional and unconventional techniques and what shortcomings do they feel require attention.

Dr Nosé, being the first to respond, emphasised that developments should be oriented to simple, economic, easy-to-use, cheap and disposable haemodialytic techniques. In fact, there are in conventional haemodialysers two areas needing special attention: the membrane support systems and the membrane blood port assemblies. He instanced the very substantial programme changes that are taking place in the United States at the present time. The majority of the membrane development programmes have been cut and some three membranes only which are about ready for mass production are being concentrated on. Additionally, dialysate preparation machines are being phased out and development is oriented to dialysate regeneration. *Dr Cattell* referred to the European Dialysis and Transplant Association's recent conference in Florence where five-year survival figures of 80–90 % were quoted for regular dialysis. He felt that these were very good indeed and he saw no reason why such figures should not be extended to the country as a whole with more effective regionalisation of dialysis facilities. He felt that the major problem for engineers is to refine and miniaturise current equipment. He agreed with Dr Nosé that ideally the primary need is the development of a cheap, economic, disposable dialysis unit. *Dr Briggs* felt that while conventional haemodialysis equipment enabled clinicians to treat a large number of patients, there are considerable problems which are partly economic and arise partly from the fact that patients are not being dialysed efficiently enough. He thought that the question of cost might be solved using the microcapsular method of Dr Chang permitting the perfusion of patients not only more efficiently but also more cheaply. *Dr Gibb* commenting on costs pointed out that results from renal transplants are improving to the point that they now compare favourably with dialysis. The cost of a transplant, although large, occurs once and for all and thus now can be considered to compare favourably in economic terms with a long period of dialysis. While it is difficult to decide priorities of money allocation, he felt that there is a case for devoting attention to improving renal transplantation because apart from economic advantages, there are advantages in terms of life quality as well. A successful renal transplant he considered to be infinitely superior to even a very efficient dialysis system. *Professor Klinkmann* also referred to the recent E.D.T.A. Conference in Florence where the latest results showed a clear cut-down in the effect of transplantation and tended to support the superior survival rate of home dialysis. He believed that the success of transplantation depends on the artificial kidney. One of the reasons there wasn't the same success with heart transplants is because there is not a good working artificial heart avail-

able. Putting somebody on to the kidney machine means preparing him for transplantation and giving him also the security that he can come back on the machine if his transplant fails in 3–5 years. Figures from the National Institutes of Health, U.S.A. show about the same cost for the relatively small number of transplants and for the large number of dialysis treatments. He emphasised that if the immunological problems in relation to transplants are not solved, there is really no advantage to be gained at present in using transplants. In consequence, he felt that one cannot divide transplantation and dialysis, they are really just the one treatment with dialysis always remaining the last resort. *Mr Longmore* commented that heart transplants in the few good centres have had results comparable with those of renal transplants. He found it disturbing that in attempting to give people the chance of a few additional years of life, it was necessary to consider costs together with the willingness of the community to provide the relevant finance. He felt that at present dialytic treatment is cost ineffective and instead of looking at refinements and new techniques, the logistics of care provision should be closely examined to improve its efficiency. Bioengineering could take the lead here. For example, if we would put dialysis experts in mobile units, say in double-decker buses centrally controlled and capable of being plugged in to hospitals around the country, then the present resources in staff and money might possibly treat all potential patients. *Professor Klinkmann*, commenting on the bioengineer's role in future development, suggested that their efforts should be in two directions. In the field of conventional analysis there is a need to simplify the equipment, to miniaturise the kidney and to make it cheaper. He thought that conventional dialysis will be with us for at least another 10 years and in consequence bioengineering effort in these directions would be rewarding. In the area of unconventional dialysis, Dr Chang's work is of particular interest and bioengineers will find it rewarding in assisting developments along his lines. *Dr Chang* said that biomedical engineering, as in the field of medicine, should not concentrate on only one specific method. The future, he felt, will develop on the bases of the combination of different approaches such as preventive screening, medical treatment, dialysis and transplantation.

SESSION 'A' SUMMARY

J. M. A. LENIHAN[1]

It would be difficult to summarise a session so full of information and ideas—but the effort is unnecessary, because the contributors themselves have so skilfully demonstrated the broad outlook in their clear statements of problems and principles.

From what they have told us, we may draw two main conclusions and one serious lesson. Firstly, biomedical engineering is a life science. The composition of this morning's list of speakers, with six doctors to one engineer, reflects the common observation that doctors have so far contributed more than engineers; this situation is not surprising, since medicine is much older than science or engineering.

Secondly, we have been reminded that the meaningful problems in biomedical engineering come from the bedside and not from the 'think tank'. A serious obstacle to the union of engineering and medicine is the universal belief that technology follows science. This belief, propagated by scientists and accepted by politicians, leads us to spend great amounts of money on dubious scientific adventures in the hope that useful technology will follow. But history shows that science more often follows technology—thermodynamics followed the steam engine, electrical science grew out of the telegraph industry and most of the major biological problems of today have come from doctors puzzled at the bedside or the operating table.

Mr Longmore told us that, through failure to understand the relationship between science and technology, we have been asking the wrong questions; consequently much costly effort, though based on sound engineering principles, has been wasted.

In other words, biomedical engineering has not been close enough to the clinical scene. The same message was offered more than two centuries ago by Giorgi Baglivi, another pioneer in building bridges between engineering and medicine. Almost as outspoken as Mr Longmore, he advised:

'We must not be surpriz'd to find that the true and genuine Cause of Diseases can never be found by Theoretick Philosophical Principles.'

Another of his precepts might also strike a chord in the present company:

'To heap up great Numbers of Books without using them, to make a Figure at Universities and have a name celebrated in the modern Journals of the Learned; this, I say, will contribute nothing towards the appeasing of the Pain of Diseases. But that End will be effectually compass'd, if you frequently visit the Publick Hospitals and the nasty Beds of the Sick.'

The lesson of this morning's proceedings is that the engineer's repertoire of ideas, instruments and techniques will not make much impact on medicine without his physical presence near the clinical battlefield. Intellect is not enough; the perspective that biomedical engineers should follow is the path that leads them out of the academic realm and into the hospital service. This change of direction may mean abandoning well-founded habits of thought and practice—but it is braver to retreat from a false start than to charge recklessly into a dead end.

---

[1] Western Regional Hospital Board, Glasgow, Scotland.

SESSION B

BIOMECHANICS

Chairman: Professor Carl Hirsch, D.Sc.

Associate: Professor D. C. Simpson, M.B.E., B.Sc., Ph.D., F.R.S.E.

BIOMECHANICS IN THE CLINIC

C. A. McLAURIN[1]

BIOMECHANICS AND THE PATIENT

One of the first questions I always ask when introduced to someone working on biomechanics is, 'What good will it do the patient?' So far I haven't received a satisfactory answer.

Activities in biomechanics are two-fold. One occurs in a biomechanics laboratory equipped with force-plates, electric goniometers, strain gauges, multiple electro-myographic recorders, cinematography and an elaborate battery of computers. In this scientific environment, specialists labour to develop theories to explain in terms of force, displacement and time the performance of existing devices in straight and level walking or some similar routine.

The other activity consists of teaching the information gained by the laboratories to the physicians, orthotists and therapists who are engaged in the clinical care of patients. During the teaching session I suppose most of these clinicians gain some understanding of the principles involved but it is doubtful whether there is really any carry-over into clinical practice.

In short, at the present time, the practice of biomechanics has not yet arrived on the clinical scene and one must first ask if it actually does belong to or is it just so much gobbledegook added to a problem already fraught with complexity and frustration.

My own limited experience outside the clinic has been as an aeronautical engineer where the problems of structural strength, energy and power, simply could not be solved adequately without constant reference to the principles of mechanics as developed from the laws of Isaac Newton over two centuries ago. Mechanics is the common language of all those engaged in structural design and is the basis for estimating performance of all kinds of engines and machines. Why then can we not apply the same powerful tool to solve problems of the motor system in humans?

One reason why biomechanics has not flourished is that the human body does not resemble man-made machines. Indeed, most machines have been designed to suit the mechanical principles on which they are based. Structures have uniform cross-sections, materials are homogeneous and motions are linear, circular or harmonic. We build systems that are readily identifiable with mechanical theory. There are exceptions: a modern aircraft wing and a turbine blade both have organic characteristics, but even here the materials are usually homogeneous and their properties predictable.

It is true that the human body does contain some examples of mechanics—the lever systems illustrated by a muscle acting on a jointed segment are often cited in literature. The fact that the joints are not simple, and the muscle pull varies with displacement, are complications that must be dismissed if we are to stay within the realm of clinical practicality. The human body, however, does have one factor in its favour, although there are individual variations (or tolerances): the basic design has remained unchanged for thousands of years and once we understand the mechanics we will not find our information obsolete in the next few years. Therefore, there is hope that in time we will be able to analyse the human body and its motion patterns in a practical clinical way.

However, there exists still another reason why biomechanics is not practical in the clinic: the form of clinics themselves. First of all, no one in the clinic truly understands biomechanics and, secondly, the patient management process does not lend itself to systematic method. The fact that physicians, therapists and orthotists have little patience with biomechanics is understandable. Very little of what they have been taught is practical—and the clinic is a very practical place. Problems must be faced and solutions sought in a matter of minutes, thus precluding all but the most common methods.

Such measures may be warranted when immediate treatment is essential to prevent contractures or provide support for healing bone. However, an orthotic system is often indicated for chronic problems and it is unfair and unnecessary to ask a patient to live with a device that has not been carefully designed for his individual needs.

Chronic cases present the major area where biomechanics can contribute in a real way to clinical practice, but only if we reinforce our clinics with people who are well trained in biomechanics and restructure our clinic procedures or patient management systems to allow time for adequate assessment,

[1] Ontario Crippled Children's Centre, Toronto, Ontario, Canada.

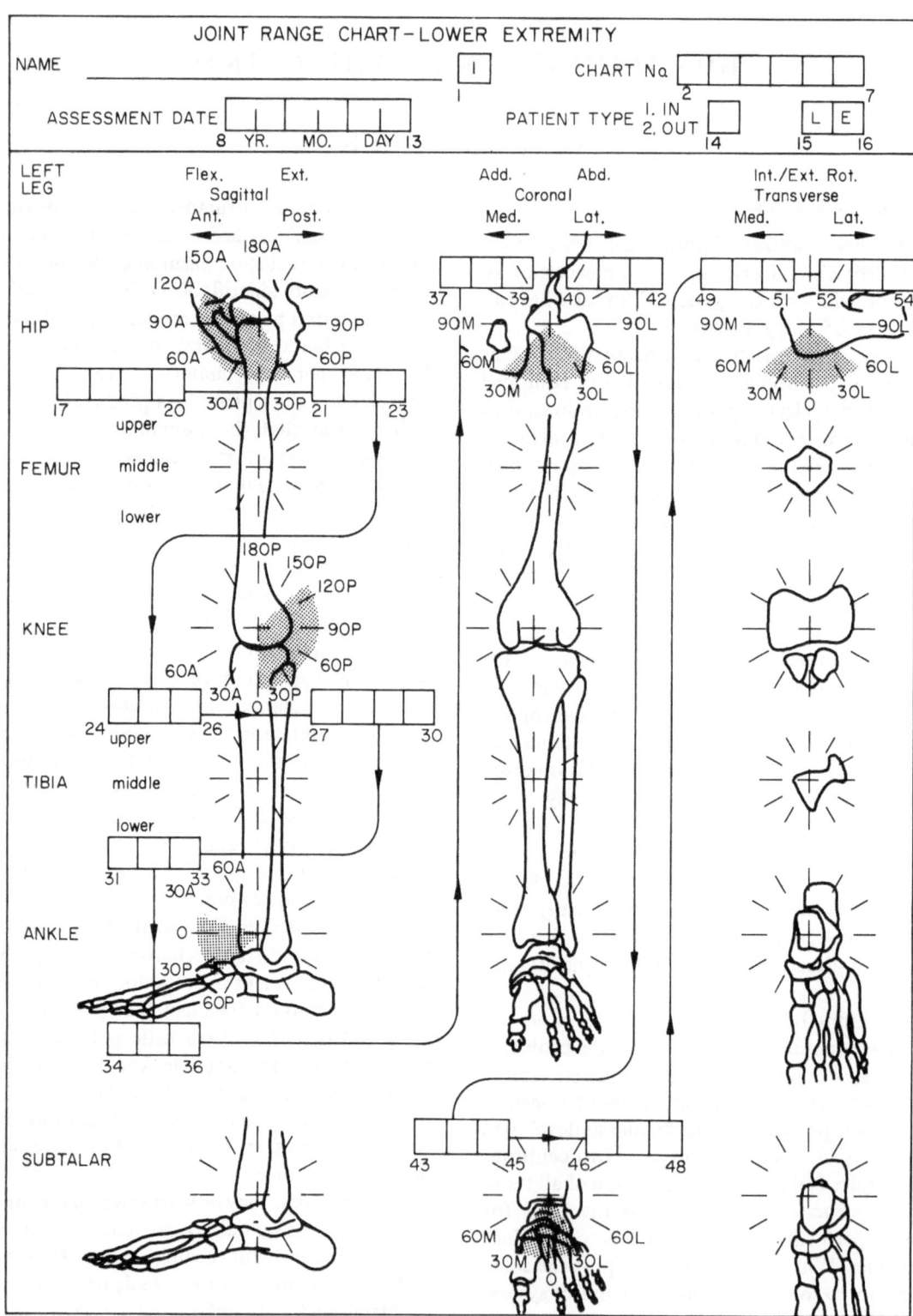

Figure 1. Joint range chart for the left leg—adapted for computer recording.

design, construction and evaluation of the orthotic system.

TRAINING IN CLINICAL BIOMECHANICS

Let us first look at the training of experts in clinical biomechanics. Such an expert might well be an engineer, since engineers normally receive a good grounding in mechanics and problem solving and, at least in Canada and the United States, it is probably cheaper to train an engineer than it is to train an orthotist. But an engineer without further training in anatomy and clinical practice is almost useless—one might even say dangerous, in the same way that any professional is dangerous if called upon to exercise his talents beyond his training and experience. How then does he get this training and experience?

The training might well belong in a school for clinical biomechanics. Does such a school exist? I don't think so. Could such a school exist? Definitely yes! There are schools such as that run at Strathclyde University where biomechanics is well studied. What is needed is the clinical input, a difficult requirement to meet in an engineering school. True, local physicians can refer patients and discuss problems; but to be effective some assessment of the overall problem in biomechanical terms should be made. This can be done by using standard assessment forms in the major clinics and summarising their findings. The functional analysis form for the lower extremities developed by the American Academy of Orthopaedic Surgeons and the National Academy of Sciences in Washington is an excellent example of how this information might be recorded. If these forms are designed for computer use their summary and analysis is greatly simplified. Examples of such forms used at the Ontario Crippled Children's Centre are shown in figure 1. The purpose of the information as summarised is to indicate the areas of major concern and the associated biomechanical problems. Once identified, these problems can be studied in depth at such centres as Strathclyde, and solutions sought using real patients and working in close co-operation with physicians, therapists and orthotists. This would serve not only as a problem-solving centre but would provide a fertile environment for the training of experts in biomechanics.

The essential ingredients are:

(1) access to technical information and instrumentation;
(2) intimate co-operation with physicians, orthotists and therapists;
(3) continuous contact with real patients presenting major problems.

CLINICAL APPLICATION AND SERVICE

Once we have people suitably prepared to practice biomechanics, how can they be effectively employed in the clinics? Let us consider the chronic cases where it is possible to establish goals that can be achieved through long-term programs. First of all, a careful assessment must be made, not just in biomechanics but in all the pathological, physiological, psychological and practical environmental factors effecting the attitudes and life styles of the patients. These factors must then be discussed in depth by the group of experts who comprise (not compromise!) the clinic team. Only then can adequate decisions be made concerning the orthotic and other aspects of the total treatment programme. The 'biomechanics' will require time for solution and would have application not only in the design of orthotic devices but in orthopaedic surgery, physiotherapy and functional training.

To suggest who is going to pay for the biomechanical assessment and recommendations would be a ridiculous presumption, but there is no question that if we are to employ such a service, then it must be paid for. In Canada and the U.S.A. I know of no way this can be done other than through certain government-sponsored research projects. Perhaps in Scotland or other parts of the United Kingdom the monetary problems are more easily solved; but in any event, it is only prudent to ensure that the costs related to biomechanics either result in lower overall costs for the total treatment programme or in increased benefit to the patient. I would challenge the accountants of our bureaucracy to acquire competence in at least the cost side of this cost-benefit equation.

I would caution that progress in clinical biomechanics may be painfully slow, although some occasional dramatic examples can be expected. Hardly any of the problems are simple and a lot of fundamental knowledge is lacking. Let us take for example a child who has lost partial function of his lower extremities due to spina bifida. Figure 2 shows the legs of a five-year-old boy. There are hip flexion deformities—worse on the left than on the right. Both legs are asensory.

The left leg has reasonable muscle power in hip flexion but not in extension. There is a flexion contracture in the left knee with some muscle control.

The right leg is almost totally paralysed. Both feet are paralysed and tend to roll out into valgus, and both Achilles tendons are tight.

The first question in biomechanics is: 'Does the boy have enough motor ability to make walking a worthwhile goal?' If the answer is yes, then the following questions are pertinent:

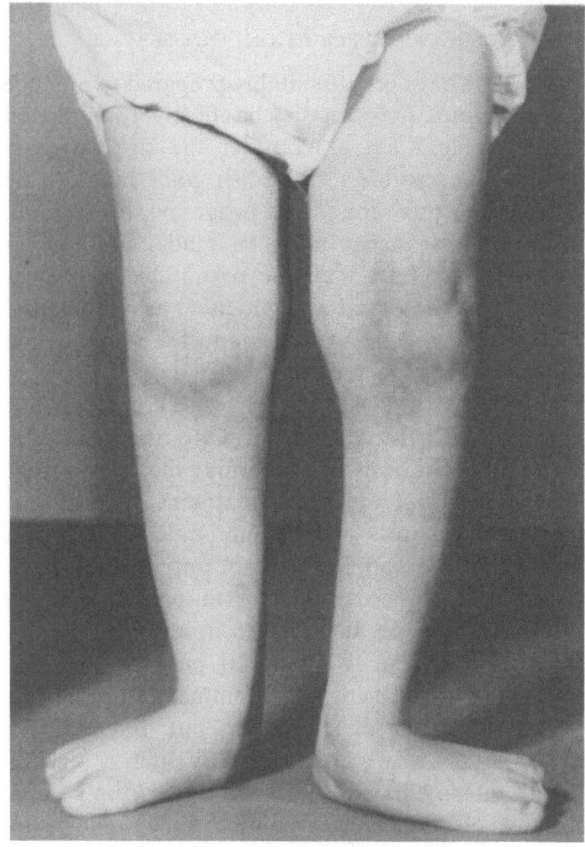

Figure 2. The legs of a five-year-old boy presenting
many clinical problems in biomechanics.

1   What bending moments occur at the hip in
    attempting to stand on either leg?
2   Can some external support or brace be provided
    for this purpose?
3   What effort would be required to extend the hip
    joints of the brace?

4   Is the muscle power adequate to control the
    left knee?
5   If not then what forces will be required to
    control it and the right knee when standing or
    walking.
6   Are locks necessary with the braces at either
    knee joint? If not can stability be gained by
    alignment, if so what alignment?
7   If contractures at hip and knee can be alleviated
    through surgery or therapy, to what extent
    would the bracing requirement be modified?
8   How can the feet be supported to minimise the
    valgus deformity while utilising the intrinsic
    structure of the foot if possible?
9   In what manner and to what extent can the
    sole of the shoe be shaped to minimise the
    valgus deformity and the external rotation?
10 What loads will be imposed on the foot by a
    brace while standing and sitting; and can the
    tissues tolerate these loads?
11 Will prolonged use of the orthotic device tend
    to increase or decrease the deformities?
12 What will be the most acceptable and efficient
    gait pattern with the orthotic device?

When questions such as these can be answered
for each case presented to the clinic, and when a
corresponding orthotic prescription can be detailed,
then we can accept clinical biomechanics as a
reality and claim that some part of our 20th century
technology is being used to benefit the physically
disabled. Otherwise we are but adding credence to
da Vinci's observation: *The greatest misfortune is
when theory outstrips practice* . . .

CLINICAL AND BIOMECHANICAL ASPECTS OF CURRENT PROSTHETIC PRACTICE

GEORGE MURDOCH[1] AND JOHN HUGHES[2]

The earliest recorded instance of prosthetic replacement is described by Herodotus referring to Hegosistratus who, when imprisoned and condemned to death in 484 B.C. escaped by cutting off the foot by which he was secured in the stocks. He subsequently fashioned himself a wooden foot with which functional replacement he served as a soothsayer to the Persian Army until he was recaptured and executed by the Spartans: an event which suggests that Hegosistratus would have been better employed as a limb fitter than a soothsayer and also underlines the inevitable retribution wrought by a frustrated administration.

There is no coherent record of the development of prosthetic practice through the ages, but it is known that it was largely influenced by the surgeons, armourers, and craftsmen involved, by the reasons for amputation and by the social condition of the amputee. Until relatively recent times advance depended upon individual interest and effort. There are recorded instances where this individual brilliance afforded opportunity for advancement which for various reasons could not be achieved. One such individual was Ambroise Paré, the noted French surgeon of the sixteenth century, who for the first time carried out amputation at elected levels using techniques similar to those in use today (Paré, 1540). He provided after-care for his patients, demonstrating an interest in their total rehabilitation, which regrettably is yet often absent. Many of the devices constructed for him by a locksmith with whom he worked closely display a surprising similarity to those in current use (figure 1). His awareness of the need for an overall treatment from amputation to social re-integration and his ability to work constructively with the craftsmen were, in social terms, far in advance of his time and for this reason unacceptable.

Often, however, the barrier was the lack of technology to implement the idea—a lack of material or a suitable manufacturing process. An interesting example is illustrated (figure 2) and described in the U.S. Patent by Parmelee (1863):

> 'the first part of this invention relates to the bucket or socket of the artificial legs or arms intended to receive the stump and it consists in the fastening of such bucket to the stump by means of atmospheric pressure in such a manner that the straps usually employed for this purpose can be dispensed with, and at the same time a perfect fit of the bucket is attained'.

This patent was registered in February of 1863, but it was not until nearly a hundred years later, with

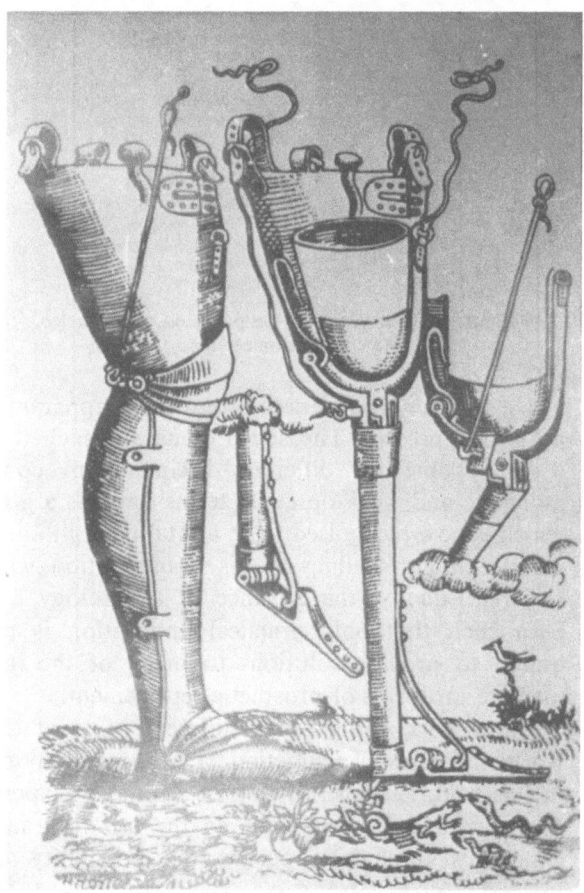

Figure 1. Prostheses by Paré. (From Paré, 1540.)

[1] Dundee Limb-Fitting Centre, Dundee, Scotland.
[2] University of Strathclyde, Glasgow, Scotland.

the application of laminated plastic construction, that it became possible to apply this idea to the treatment of the patient.

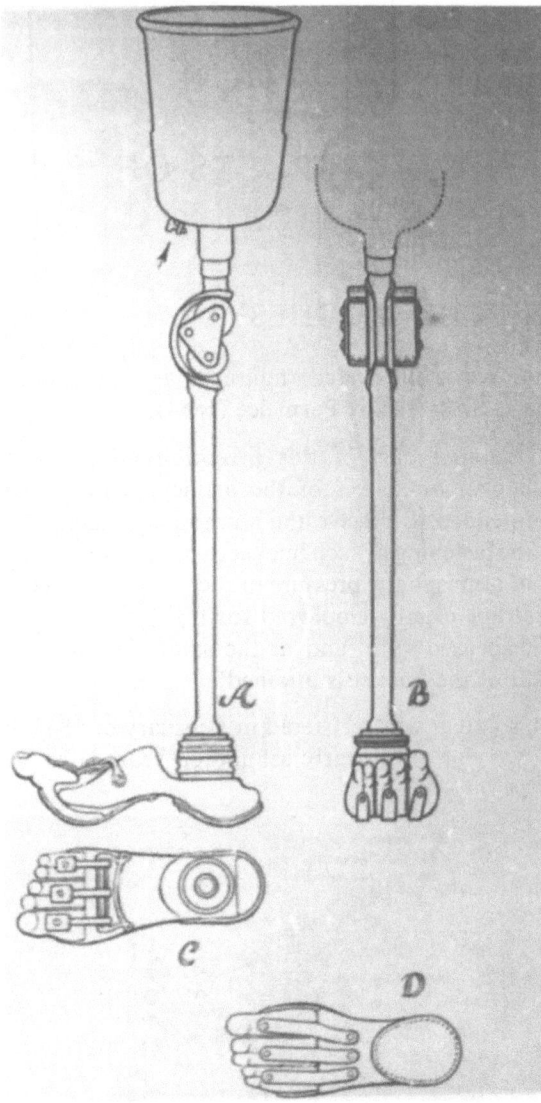

Figure 2. Skeletal prosthesis patented by Parmelee. (From Parmelee, 1863.)

These various barriers to progress are apparently no longer present. The social climate is such that it is possible for different disciplines to communicate and work on equal terms towards a goal which is now recognised to be a total rehabilitation of the patient leading to his re-integration with society. Equally, the advance of technology has been such that only practical application is required to provide solutions to many of the immediate problems of prosthetic replacement.

Some work going on in different parts of the world indicates the direction of forward thinking in this field. Esslinger (1970), in a series of experiments with dogs, has explored the use of transcutaneous implants, opening the possibility of skeletal attachment of the prosthesis. The basic problem is to establish compatibility, in terms of

both physical and biological properties, between the implant material and the body tissues, thus avoiding eventual rejection of the implant. Since the Thalidomide tragedy considerable attention has been paid to external power systems and control theory in an effort to provide useful function to high double-arm deficiencies (Simpson, 1968; Montgomery, 1970; figure 3). Muscle bulge and the electrical potential developed in functioning muscle have both been used to provide volitional control signals (Reswick and Vodovnik, 1967; figures 4 and 5). In contrast

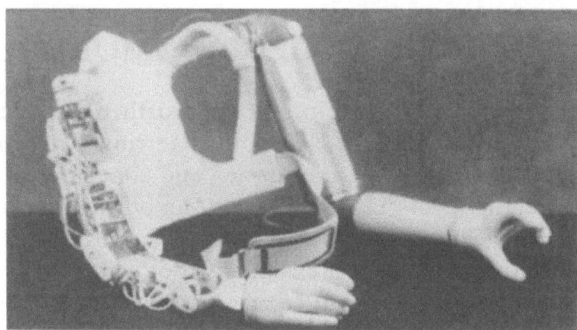

Figure 3. The right arm is an e.p.p. controlled arm with position servo control of r, θ, ϕ and wrist rotation and with stabilisation of hand angle. The left arm carried the CO_2 tank.

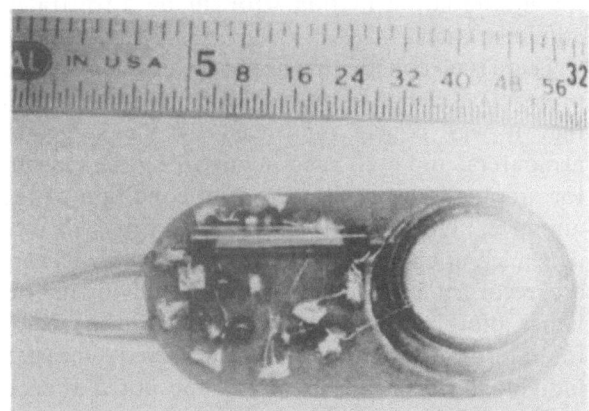

Figure 4. Miniature FM Radio Transmitter used to obtain electromyographic signals by complete implantation. (From Reswick and Vodovnik, 1967.)

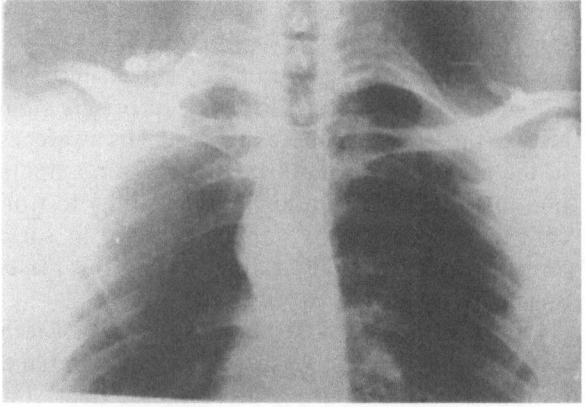

Figure 5. Transmitters implanted in a human and attached to the trapezius muscles. (From Reswick and Vodovnik, 1967.)

Tomovic (1970) has advocated a system of hierarchal control where all subsequent actions are 'automatic' once the movement intention has been initiated. Both approaches may well have application in lower extremity prosthetics with regard to 'knee' and 'ankle' control. It is much less likely that external power can be usefully employed to provide locomotion, although portable power sources are now being developed for the less demanding requirement of the artificial arm (Davies, 1972).

All these advances have highlighted the need for an integrated approach to amputee rehabilitation. The problem is no longer seen as a surgical exercise plus the construction of a limb replacement. A more complete knowledge of human locomotion, a better understanding of the biomechanics in the normal and abnormal and the consequent sophistication of socket and device design, all reveal the need for cross-disciplinary understanding. The surgeon can no longer regard amputation as ablation of a part of the body but must see his operation as a design problem albeit restricted by the available 'material' and the necessary 'structural' characteristics. His solution must meet, so far as is possible, the biomechanical requirements of the interface (the socket) and the related prosthesis. In the same way the prosthetist must express in his work a recognition of the need to match the prosthetic device with the stump; the therapist must design a programme of physical rehabilitation to optimise the efforts of both surgeon and prosthetist; and the social worker must provide the appropriate link between patient and society. Against this background the responsibilities of the engineer are clear. His designs must reflect the clinical and social requirements of the patient and not the abstractions of laboratory theory.

This concept of management is now seen as total integrated progressive patient care bringing with it the recognition of the need for a team approach— the clinic team. In practice the reality of the clinic team is often difficult to achieve. It should however at least function in major decision and prescription making, so that each member may make his professional contribution to the patient's welfare. Essential to the success of the operation of a clinic team is the need for each member to have a sufficient knowledge of the other disciplines to provide a basis for communication.

MAN/MACHINE MATCHING

Interface problems

One of the more obvious responsibilities of the clinic team is to achieve the best attainable match of the patient and his prosthesis. If the prosthesis is truly to be an extension of the skeleton, and if for the moment skeletal attachment is not possible,

then the connection problem lies in achieving an intimate relationship between stump and socket. With a realisation of the requirements of locomotion in a deficit situation, the surgeon of today now performs in his amputation surgery a reconstruction procedure to produce an organ of locomotion. For the present the connection with the device— the socket—is of fixed volume and configuration and consequently the stump construction must reflect these requirements. The result is that level selection and tissue management have received the attention they deserve. The surgery is more precise, less destructive and a specific role is accorded to each tissue. The skin is treated to ensure a thin scar which is mobile. Muscles are attached firmly to the end of the stump recognising physiological requirements for increased power and improved control with the secondary benefit of a stump with a more constant volume and shape. Division of nerve always results in a neuroma and the surgery attempts to ensure that it finds an environment where it will not be the subject of unnecessary or painful stimuli. The medulla of the bone is closed off to improve drainage and the bone end is sculptured to prevent high load areas in contact with the socket (Murdoch, 1967). The net result is an organ in which the bone is firmly located in relation to the other tissues in the stump and an optimal situation with regard to residual muscle power is established. Where these principles have been applied the result has been earlier maturation of the stump and a more stable relationship between stump and socket, permitting the prosthetist and engineer to further sophisticate the design of the socket. The prosthetist and engineer have in turn recognised the necessity of providing a physiologically acceptable environment for the stump.

This total design process is exemplified by the development of the patellar-tendon-bearing prosthesis for the below-knee amputee. The requirement of maximum functional restoration implies the removal of restriction on the knee joint. This in turn places rather exacting requirements on the members of the clinic team. The osteomyoplastic procedure described by Ertl (1949) enables the surgeon to produce a stump of relatively stable volume with firm underlying tissue and good muscle function, around an intact knee joint (Murdoch, 1970; figures 6 and 7). A specification by the engineer of the force patterns imposed on the stump during normal activity coupled with an indication of the likely behaviour of the loaded tissue allows the prosthetist to produce a functional socket. This he does by studying the individual stump in terms of its capacity to sustain load in specific areas, and designing the socket to apply the forces involved with acceptable and desirable tissue deformation (figure 8). The well-recognised success of this

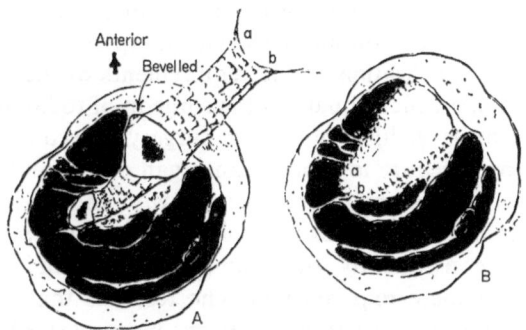

Figure 6. Osteomyoplasty showing formation of bridge between tibia and fibula constructed by osteo-periosteal flaps. (From *Artificial Limbs* (1962) **6** (2).)

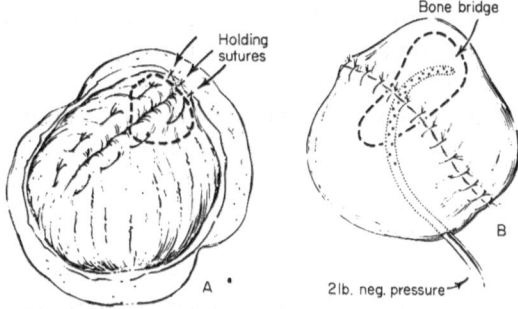

Figure 7. Osteomyoplasty demonstrating antero-lateral muscle group sutured under tension to the posterior muscles and secured by anchor sutures to the osteoperiosteal bridge. (From *Artificial Limbs* (1962) **6** (2).)

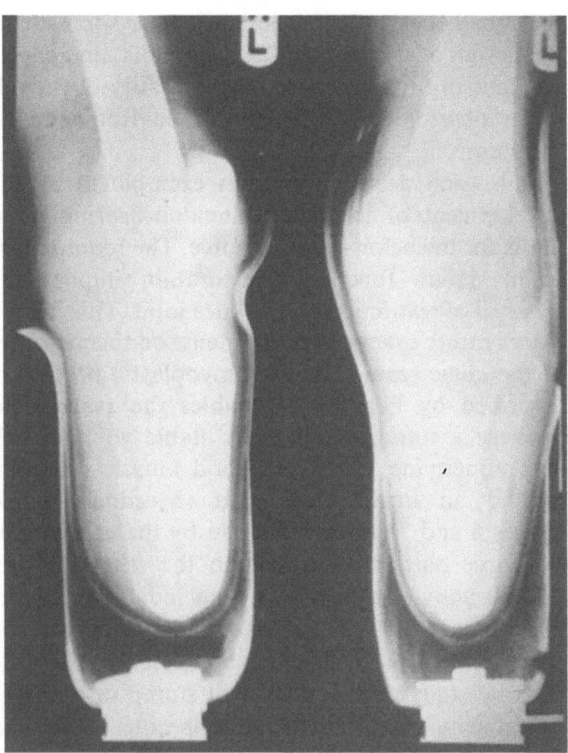

Figure 8. Radiograph of below-knee stump demonstrating bone bridge and showing total contact fitting in patellar-tendon-bearing socket.

procedure has clearly demonstrated the validity of the approach (Radcliffe and Foort, 1961).

Devices

At this point in time, at least in the practice of lower-limb prosthetics, no serious attempt has been made to apply a rational approach to the matching of man and machine. Prescription of the device and the design of the device itself has been a purely subjective procedure related to experience, individual expertise, and intuition, and the results have never been evaluated in objective terms. This statement may also be generally applied to upper-limb prosthetics although the pneumatic-powered prosthesis described by Simpson (1968) attempts to provide

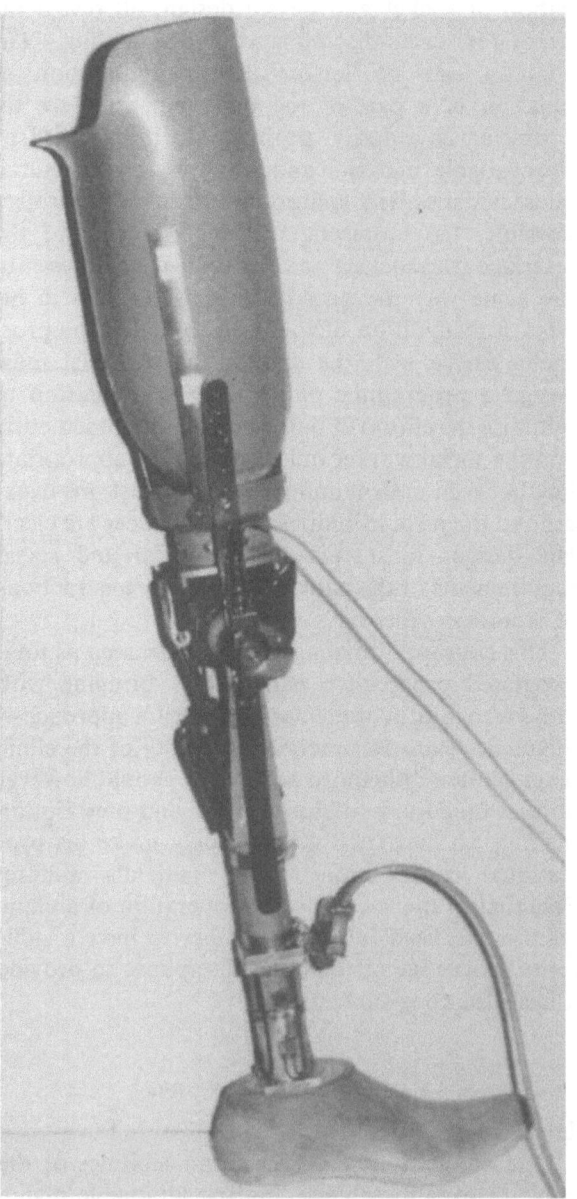

Figure 9. Modular prosthesis for above-knee amputee. Goniometer at knee and dynamometer in shin allow calculation of forces exerted by amputee.

'proprioceptive matching' in terms of shoulder/prosthesis end-point position. Even in this instance however an objective assessment is not at present available.

There is no basis for the prescription of prosthetic devices in terms of restoration of functional loss. However, measurement of the amputee's remaining functional capability, compared objectively with that in the normal subject, should provide the necessary information related to both prescription of the device and the design of the device required. Functional performance in the normal subject has now been widely studied (Eberhart, 1947; Hughes, Paul and Kenedi, 1970). Dynamic studies of amputee performance have also been carried out using conventional walkpath techniques and pylon dynamometer techniques (Eberhart, 1947; Lowe, 1969; Hughes, Lowe and Paul, 1970; figure 9). If it is possible to show that the dynamic performance of the amputee relates to his static capability, assessed for example by isometric measurements of force development of remaining musculature, then the possibility for a clinical prescription system is open (Hughes and McGechan, 1971). Similarly a comparison of static measurement and dynamic performance in an enlarged study would provide the necessary design information to ensure that the devices produced were objectively related to the requirements of the wide variety of amputee types encountered. This study requires relatively simple technology in terms of measurement but sophisticated means of data handling for the study of significant samples of the amputee group.

The primary aim is to provide the clinic team with an objective method of prescription and the complementary system which will satisfy the clinical needs. It should also be possible to measure the extent to which matching has been achieved, i.e. the extent to which the patient's functional deficit has been supplemented to provide optimal restoration of function. These objectives place rather special requirements on the system of prosthetic supply. The system must be flexible in terms of incorporation of the total range of devices required and available. The possibilities must be available for interchange and adjustment to meet, firstly, prescription and, secondly, modification possibly shown to be necessary by the evaluation procedure. It is fortunate that at this time when the problems of man/machine matching have been recognised and are being tackled, other considerations have given rise to the evolution of an entirely new system of construction of prosthetic devices. The necessity to move away from craft construction and to utilise the latest developments in modern production technology have given rise to the modular system of prosthetic assembly—a system of assembly from premanufactured standardised components. If the

modules are designed in accordance with the approach outlined, the modular system provides the sort of facility visualised—the patient's capability is measured, the appropriate combination of devices chosen, the prosthesis constructed and the outcome evaluated (Hughes, 1970).

It is perhaps important to realise that to take the best advantage of modularisation a system of international standardisation is required. Only in this way will it be possible to use and apply the results of world-wide research effort. It should be remembered that conventional systems of construction, e.g. the crustacean metal limb, with the inherent difficulties of incorporating different types of devices and materials have been one of the biggest inhibiting factors to progress in this country in the last twenty years. It is also fascinating to compare the design of Parmalee with a modular limb of the present generation and to speculate on the possible progress which might have been achieved if this philosophy had been adopted in 1863, capable as it is of simple modification and redesign.

Restoration of appearance

It is clearly desirable to restore to the patient an appearance of normality in both the dynamic and static sense although there may often be a compromise dictated by functional considerations. This is particularly true in the dynamic situation where the movement of limb segments will be primarily dictated by requirements of support and control, but it is also true in the static condition where there may be requirements such as matching with standard components, e.g. feet. The move away from craft construction towards modular assembly has highlighted this very large problem. The modular prosthesis is skeletal in appearance and must be covered to simulate the normal covering, not only in terms of basic shape but also feel, colour, texture, thermal properties, and the changes which take place with movement and the different postures adopted by the patient.

At present the patient has the choice of an individually sculptured cover, which negates the philosophy behind modular assembly, or one selected from a range of ready-made sizes. One manufacturer, for example, provides over three hundred such covers in an attempt to satisfy the individual patient's requirements. Neither alternative does more than attempt to meet the needs with respect to basic shape, feel, and colour, and both carry different penalties such as delay and expense. One approach which may improve this situation, by providing individual shape without recourse to manual skill, is an adaptation of the method described by Cruikshank (1969), which employs a photographic technique to produce a replica. A specification of material properties and a search of available materials is a pressing need before the other criteria such as changes in shape, texture, etc., can be satisfied. The priority which should be accorded to this problem has now been internationally recognised (National Academy of Sciences, 1971).

EDUCATION

Having accepted that those concerned with the treatment of the patient, from before amputation till return to employment, should work together as a clinic team, it becomes obvious that each member should have professional status. This includes the prosthetist, who must be given the necessary training and education to enable him to attain this status. This is recognised in a recent report, 'The Future of the Artificial Limb Service in Scotland' (Scottish Home and Health Department, 1970) which in a unique and exhaustive investigation examines the treatment of the amputee and makes a number of important recommendations. The same report outlines in detail a specimen course of training and education including scientific, technical, medical and social elements. It also envisages training for other members of the team, e.g. the nurse, the therapist, and the doctor. It recommends that all prosthetic specialists should have formal courses of instruction in amputation surgery, biomechanics, and prosthetics; that all orthopaedic surgeons in training should have experience in prosthetics; and implies that the medical student should have a knowledge of biomechanics. The report also suggests how the clinic team concept can be implemented. It has received the approval of the Secretary of State for Scotland who has accepted it and instructed his officers to implement it in its entirety.

THE PATIENT

It is proper that we should finish this paper by referring to the first and most important member of the clinic team, the patient. His requirements are the proper basis for all design, surgical philosophy, surgical and prosthetic technique, and prescription criteria. His problems as they are presented to the clinic should also form the basis for research effort. These problems may present themselves at an individual level; they may expose general areas of ignorance requiring long-term basic research; or they may indicate the need for a new device or system of management. In this context and working within this philosophy it becomes possible to identify areas of priority of effort: the development of a light-weight functional and cosmetic prosthesis for the geriatric; the development of cosmetic finishing methods acceptable to the patient in both static and dynamic situations; intensified studies on appropriate amputation designs to produce desirable stump characteristics such as end bearing and

perhaps fenestration to allow closer integration of stump and socket; and the development of means of objective assessment of functional capability and functional restoration after amputation.

One conclusion is absolute. The clinic-team concept points the way to future development and every effort must be made to ensure that the whole team is placed in an environment where their different disciplines may collectively operate to the benefit of the patient.

REFERENCES

CRUIKSHANK, J. 1969: Light beam profiling and reproduction. U.K. Patents 31386, 7 and 8/69.
DAVIES, B. L. 1972: A portable hydraulic power unit for prostheses. *Eng. Med.* **1** (2).
EBERHART, H. D. (Ed.) 1947: 'Fundamental Studies of Human Locomotion and other Information relating to Design of Artificial Limbs', Subcontractors' Report to the Committee on Artificial Limbs National Research Council, Prosthetic Devices Research Project. College of Engineering, University of California, Berkeley, Serial No. CAL 5, Vol. 1.
ERTL, J. 1949: Uber Amputationsstumpf, *Chirurg.* **20**: 218.
ESSLINGER, J. O. 1970: A basic study in semi-buried implants and osseus attachments for application to amputation prosthetic fitting. *Bull. Prosthetic Res.* **10-13**: 219-225.
HUGHES, J., PAUL, J. P., and KENEDI, R. M. 1970: Control and movement of the lower limbs. Pp. 147-179 in 'Modern Trends in Biomechanics' (Ed. D. Simpson). Butterworths, London.
HUGHES, J., LOWE, P. J., and PAUL, J. P. 1970: Dynamic assessment of above-knee prostheses, Yugoslav Com. for Electronics and Automation. *Proc. Symposium on Advances in External Control of Human Extremities.*
HUGHES, J. 1970: Interchangeability, standardization and the modular concept. Pp. 417-420 in 'Prosthetic and Orthotic Practice' (Ed. G. Murdoch). Edward Arnold, London.
HUGHES, J., and McGECHAN, M. B. 1971: The objective assessment of amputee gait. *In* 'Symposium on Human Locomotor Engineering'. *Inst. Mech. Eng.*
LOWE, P. J. 1969: 'Knee mechanism performance in amputee activity'. Ph.D. Thesis, University of Strathclyde, Glasgow.
MONTGOMERY, S. R. 1970: External power systems for the upper extremity. Pp. 387-398 in 'Prosthetic and Orthotic Practice' (Ed. G. Murdoch). Edward Arnold, London.
MURDOCH, G. 1967: Levels of amputation and limiting factors. *Ann. R. Coll. Surg. Engl.* **40**: 204-216.
MURDOCH, G. 1970: The surgery of the below-knee amputation. Pp. 45-60 in 'Prosthetic and Orthotic Practice' (Ed. G. Murdoch). Edward Arnold, London.
NATIONAL ACADEMY OF SCIENCES 1971: 'Cosmesis and Modular Limb Prostheses'. Report on Conference, C.P.R.D. San Francisco.
PARÉ, A. 1840: 'Œuvres Completes', *Edition Malgaigne, Paris* **1**: 616-621.
PARMELEE, D. D. 1863: U.S. Patent 37 637.
RADCLIFFE, C. W., and FOORT, J. 1961: The Patellar-Tendon-Bearing Below-Knee Prosthesis. Biomechanics Laboratory, Univ. of California.
RESWICK, J. B., and VODOVNIK, L. 1967: External power in prosthetics and orthotics, an overview. *Artificial Limbs,* **2** (2): 5-21.
SCOTTISH HOME AND HEALTH DEPARTMENT 1970: 'The Future of the Artificial Limb Service in Scotland'. Report of a Working Party set up by the Secretary of State for Scotland. H.M.S.O.
SIMPSON, D. C. 1968: 'An Externally Powered Prosthesis for the Complete Arm'. *Institution of Mech. Eng. Symposium on Basic Problems of Prehension, Movement and Control of Artificial Limbs (1968-1969)* **183** (3J): 11-17.
TOMOVIC, R. 1970: Multi-level control of human extremities. Pp. 190-200 in 'Modern Trends in Biomechanics', Vol. 1 (Ed. D. Simpson). Butterworths, London.

TRADEOFFS AT THE MAN-MACHINE INTERFACE IN CYBERNETIC PROSTHESES/ORTHOSES

ROBERT W. MANN[1]

Although contemporary research efforts in prostheses and ortheses have yet to have much general impact on the rehabilitation of humans with neuro-muscular-skeletal defects, some research results and, to a much lesser extent, clinical applications of advanced concepts and techniques have progressed to the point where an overall codification of the field is possible. An attempt to assess the advantages and liabilities of different approaches may prove useful.

In such a comprehensive assessment it is, of course, impossible to deal with detail, but in fact this limitation may prove beneficial. In prosthetic/orthotic research, as in medicine, there is a tendency to parochialise disabilities into very specific categories. This approach is of course understandable in medicine given the incomprehensible complexity of human biology and the unlimited varieties of disease and trauma. This diversity produces a centrifugal tendency resulting in the many specialties and subdivisions within medicine.

In approaching the classification and assessment task of this discourse I prefer the integrating approach to nature familiar to the physical scientist or engineer. We interpret nature, and synthesise, by defining and applying the fundamental laws which control the interaction of matter; each separate research study or design project becomes but an individualised expression of the whole.

Employing this philosophical stance, I will now attempt to coalesce into a simple schemata the many and particular aspects of augmenting an amputee or providing a paralytic with a supportive or substitute appendage.

THE SYSTEM

Let me first observe that conceptually all such systems can be categorised into three entities: the man, the machine and the interface (figure 1).

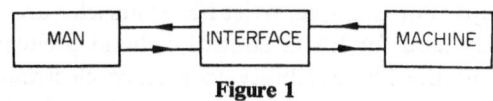

Figure 1

The man

The man, of course, is the object of our attention just as he is the patient of the medical doctor.

Although he suffers a neuromuscular skeletal defect which can range over a broad and complex spectrum from simple distal amputation to high cervical spinal injury, we always deal with a human with the major part of his central nervous system and brain intact, and with significant (though different from case to case) portions of his spinal cord, peripheral nervous system, and musculature viable. The same brain and central nervous system, whose activity must certainly be the operational definition of life, is ultimately the means by which control will be achieved over any prosthetic/orthotic machine.

The machine

It is not my intent herein to dwell on the machine. As an artificial limb or orthotic brace it can assume manifold forms dictated by the complication of its function and the sophistication of the technology applied. It will of course reflect the nature of the disability and attempt to mimic the natural situation. But whether it is, for example, an artificial limb substitute for the upper extremity or an orthosis to support a flail arm, considered in generalised terms the machine's functions are similar though certainly its appearance differs. Thus in subsequent discussion herein 'machine' subsumes prostheses and/or orthoses of whatever nature or complexity.

As a design engineer with now thirty years of experience (and still designing) I am confident that all of the really significant technological problems of adequate prostheses/orthoses, including such considerations as minimum weight, adequate power and energy, responsiveness of control, cosmesis, etc., are achievable within the present state-of-the-art or with moderate specialised developmental efforts, limited only by the financial and manpower resources brought to bear. Therefore, in this paper, I dismiss further consideration of the machine with the caveat that satisfactory machines can be achieved when we can specify their functional requirements and when we are prepared to commit resources to their realisation.

The interface

Between the substantially intact human and the machine lies the interface. This I submit is the heart of our problem, to which I devote the remainder of

[1] Department of Mechanical Engineering, Massachusetts Institute of Technology, Cambridge, Massachusetts, U.S.A.

this discussion. For the prostheses/orthoses problem in my view reduces to explicating the optimal means for bilaterally interfacing the human and the machine so as to regain for the human as normal an appendage or neuromuscular control as he previously enjoyed, or would have enjoyed if normal.

In addressing ourselves to the interface we recognise it as the cybernetic element, the communication and control link between the man and his machine (Weiner, 1948). The ultimate control source, whose signals the interface will process, is the brain and central nervous system of the human; but the physical interface between the man and the machine, at least conceptually, can occur at different levels of the neuromuscular skeletal system. For the purposes of this discussion I select as levels the brain, the peripheral nervous system, and the musculature. To assess the advantages and liabilities of intervention with the human system and interconnection with the machine at these different locations we must establish a basis for evaluation.

ASSESSMENT CRITERIA

I propose two basic measures—access and innateness. Under 'access' I include the initial and/or long-term hazard associated with the level of intervention, and the relative convenience of the interface for regular, daily use. My 'innateness' measure connotes a scale which puts a particular interface between, at one extreme, communication and control of a natural extremity, and at the other extreme, an interface which demands extensive training and considerable attention from the human for control of his appendage. Accessibility and innateness will frequently trade off one against the other; and subdivisions of the access measure, hazard and convenience may also be polarised.

Specific examples may serve to rationalise my choice of criteria, e.g. skin electrodes for electromyographic signal detection pose a very low level of hazard but do represent some inconvenience in daily use. By dramatic comparison, the detection of electroneurological activity on the surface of the cortex involves surgical procedures which present risk and long-term electrode contact raises serious questions of biomaterial compatibility in the presence of electrical activity. But were such implants ultimately proved useful and safe, their permanent nature could eliminate daily inconvenience.

Let us now proceed to inspect the range of interface possibilities, considering for each the measures of access and innateness.

THE TRADITIONAL INTERFACE

Contemporary clinical practice in prostheses and orthoses is our base-line for comparison. The traditional control concept, whether we consider elbow or knee prostheses for the upper and lower extremity amputee or an aid for the paralysed such as a feeder, stresses access with low or minimal hazard at the expense of inconvenience and results in an innateness level which demands much training and a high level of attention. Gross, unrelated, muscular skeletal motion is harnessed through mechanical connections to flex the prosthetic joint or support and move the flail extremity (Wilson, 1970).

EXTERNAL POWER

The introduction of external power as a substitute for human muscular effort increases the interface options since now the new independent energy source can be modulated by signals of much lower power level. However, in many cases this new opportunity has been only modestly exploited, using the same gross, unrelated, muscular skeletal contortions to operate say an electrical switch rather than pull the more traditional cable. Essentially no improvement on the innateness scale is achieved (Sell and Merten, 1969).

When the bulging of an anatomically relevant muscle is used to control a switch, some innateness gain can be achieved, but the external artifact of bulge is not directly related to the level of muscle contraction, and the range of control transferable to the external power source is very limited, usually to off-on control.

ELECTROMYOGRAPHY

The electromyographic signal, the electrical activity associated with the contraction of skeletal muscle, has more recently been used as a means of modulating an external power source. Electrodes in contact with the skin detect the EMG signal. The hazard is nil, especially if the electronics are designed to permit the use of dry electrodes, thereby eliminating the dermal irritation which can accompany the repeated use of electrode paste. The electrodes can be integrated into the prosthesis socket reducing inconvenience to a minimum.

In most cases, little improvement on the innateness scale has resulted since the EMG signals have been derived from anatomically unrelated muscle groups. For example, wrist flexor muscles are commonly used for EMG control of hand prehension, due to the inaccessibility to surface electrodes of the deep, natural, prehension muscles (Skachkov, 1966). Or the patient is required to train his muscles to generate non-anthropomorphic responses which satisfy the control characteristics of the prosthesis/orthosis system (Childress, 1969).

In a very few cases, notably the 'Boston Arm' (Rothchild and Mann, 1966; Mann, 1968), innateness possibilities with EMG have been exploited. Here the residual bicep and tricep muscles, the natural flexors and relaxers of the anatomical elbow joint provide a natural source of EMG for control of an artificial elbow joint. With this approach innate control of the prosthetic elbow is virtually automatic.

Surface electrodes on the skin, of course, respond to any and all sources of electrical activity in the musculature and may therefore include the EMG generated by irrelevant muscle groups. By employing relatively minor surgery, either transcutaneous electrodes or implanted electrodes with telemetry (Reilly, 1968) can provide more localised access to specific muscle bodies, including deep muscles such as, for example, the finger prehensors in the forearm. With fine enough electrodes the activity of individual motor units in the muscle can be detected (Basmajian *et al.*, 1965).

The tradeoff between surface and deep electrodes on the innateness scale can be complex. Surface electrodes inevitably collect or 'integrate' the muscular activity from a volume of tissue. In a relatively simple anatomical situation like the upper arm for elbow control only, this 'integration' can be used to advantage for natural control. In other cases, like the complex forearm, volume conduction confuses the sources of signals reaching the electrodes, thereby diminishing the opportunity for natural control. On the other hand, transcutaneous or implanted electrodes can detect activity in very specific regions. Whether innateness improves depends on an adroit balance of isolation achieved through localisation while avoiding inadequate statistical information on the overall muscle performance due to restricted sampling of EMG activity.

Simultaneous, volitional control of multiple motor unit sites has been demonstrated; but very high levels of conscious attention are demanded from the human, and the muscle in which the particular units are located must be maintained close to the relaxed state so as not to generate too much background EMG. It would appear that motor unit control, while expanding our understanding of volitional control in the nervous system, is not directly relevant to the control of practical prosthese or orthoses.

POSTURAL EMG

The statics and dynamics of normal neuromuscular physiology are of course not only modulated by the efferent or motor information emanating from the brain and spinal cord but also reflect the sensory or afferent feedback which proprioceptive sensory organs in the muscles, tendons, and joints return to the spinal cord and brain. Note in figure 1

that bilateral communication and control paths interconnect the man, the interface, and the machine. The normal system, and therefore any prosthetic/ orthotic system, must recognise and replicate to the same extent possible sensory feedback paths. The effectiveness of the 'Boston Arm' derives not only from the natural source of efferent signals but also from the deliberate design of a force feedback system in the machine's servo mechanism, which exacts from the residual but dysfunctional muscles increased muscle tone to handle greater loads. The load on the artificial device communicates with the central nervous system through the intact muscle spindle and tendon sensory organs and the afferent sensory system.

But where muscles anatomically related to the substitute prosthetic or orthotic joints are either ablated or flail, innate control through direct access to those muscles is no longer possible. Even here electromyography can yield useful control information by exploiting the roles of proximal muscles in the natural postural control system.

The intact neuromuscular-sensory-skeletal system in normal posture and in coordinated motion is an exquisitely orchestrated ensemble (Roberts, 1967; Mountcastle, 1968). Simple motions, such as elbow flexion with the upper arm in a horizontal plane, automatically invoke subtle dynamic responses from the complex musculature of the shoulder, back, and chest to maintain the upper arm horizontal as the moment of inertia of the upper extremity changes with elbow flexion. The detailed muscular coordination demanded in the normal range of mundane, everyday activities is itself extraordinary, not to mention the absolutely remarkable manifestations of muscular interaction exhibited in sports, gymnastics and other examples of maximum effort and highly coordinated human skills.

The Temple arm (Wirta and Taylor, 1969) is an example of utilised EMG signals from ten muscle groups in the torso. This activity, recorded during the isometric flexion of the upper extremity joints of a normal human, provides the basis for the design of a pattern recognition and weighting circuit. Then the EMGs from the amputee, processed through the circuit, control three joints of an upper extremity prosthesis.

Enhanced physiological replication resulting from maximising the proprioceptive sensory feedback from the prosthesis to the human should achieve more natural coordinated control over distal ablated or paralysed joints (Jacobsen, 1971).

Research and development in the field of remote manipulation and teleoperators (Johnsen and Corliss, 1967; Whitney, 1969) addresses the problem of simultaneous coordinated control of devices with multiple, sometimes redundant, degrees of freedom. Although the human controller here is intact, trained, and committed, improvements in control algorithms which reduce or eliminate human attention on the control of individual joints so as to permit concentration on the task are germane to multiple-degrees-of-freedom prostheses and orthoses.

PERIPHERAL NERVES

More direct control of distal joints is conceptually possible in direct access to activity in peripheral nerve bundles which, prior to amputation or paralysis, serviced the muscles to be replaced or motivated. A typical peripheral nerve, say the ulnar in the upper arm, has in the order of 10^5 separate fibres, afferent and efferent, with different propagation velocities. The correlation of nerve signals in such bundles with motor and/or sensory activity is being explored (Rothchild, 1968). This approach must be considered speculative, for not only must the information be interpretable, but also long-term, indwelling, biocompatible electrodes, appropriately shielded from the surrounding EMG activity must be developed. Were these formidable tasks to be realised, the level of innateness would be very satisfactory. Access would require surgery, and long-term biocompatibility must be assured, but with appropriate telemetry convenience the idea should be very acceptable.

THE BRAIN

Our review from gross, unrelated control motions through the refinements of electromyographic and electroneurological interfaces have progressed towards the ultimate source of human sensorimotor activity, the brain.

Psychophysiological electroencephalographic research has demonstrated the topological relationship between cortex site and sensorimotor function. One might argue that electrical detection of activity at this level would represent the most innate source. While orthogenetic arguments posit the cortex as the source of volitional commands in the muscular sensory system, the integrating roles of the basal ganglia, cerebellum, and brain stem are extremely complex and quite incompletely understood. The activity of interest is somehow distributed through the volume of the cortex, and not even all of the area of the cortex surface is accessible due to surface convolutions. Even with direct access under the *dura mater*, resolution is still subject to electrode size limitations.

Electroencephalographic signal detection by electrodes on the skin over the brain poses no hazard. The several rhythms of brain activity, alpha, beta, theta, and mu or arcuate, and the contingent negative variations of Gray Walter have been explored. While interesting, these are most in-adequate on our innateness scale since very substantial training and/or attention on the part of the subject is required. Phase-locked loop techniques have been employed to extract from the 'noise' of electrodes on the skin over the motor cortex, the signals associated with gross muscular activity in the extremities (Nirenberg *et al.*, 1972). These results so far do not demonstrate anything like the reliability mandatory for a prosthesis/orthosis.

In animal experiments the electrical activity in an individual pyramidal nerve cell has been correlated with a monkey's movement of its upper extremity (Evarts, 1967). More recently Humphrey *et al.* (1970) have simultaneously monitored the activity in several cells in the motor cortex. An analysis of the information in the parallel pathways led them to believe that 'the information about a given movement is carried not simply in the discharge patterns or spike trains, but to a significant extent by the temporal relations between them'. Although the surgical and long-term biocompatibility aspects of such procedures for humans raises many questions, the approach suggests an extraordinarily high level of innateness for the particular joint flexion considered.

Whether at the brain, peripheral nerve, or muscle-motor unit level, the choices lying between more gross detection of the bioelectric signal, so as to exploit physiological integration *versus* fine-scale detection, e.g. of the individual neuron which may preclude assembling information reflecting the extraordinarily graduated and flexible response capabilities of the total neuromuscular system, are formidable.

SUMMARY

In this necessarily short attack on so comprehensive a question it has not of course been possible to deal in anything like adequate detail with very many subtleties of each level of interface intervention. However, I believe that even this rather superficial treatment can serve to support my conviction that the central issue of prostheses/orthoses is the choice of interface. I have tried in these brief remarks to stress and elaborate the tradeoffs between access and innateness and thereby encourage the consideration of these evaluative concepts.

Quite understandably any one investigator with a particular therapeutic or rehabilitative goal must make a great many decisions to narrow his investigation to a manageable scope. If in doing so we at least identify whether our level of interface is compatible with our aspirations for access and innateness, thereby making our choice consistent with our goals, perhaps this discussion will have served its purpose.

REFERENCES

BASMAJIAN, J. V., BAEZA, M., and FABRIGAR, C. 1965: Conscious control and training of individual spinal motor neurons in normal human subjects. *J. New Drugs* **5** (2): 78-85.

CHILDRESS, D. S. 1969: A myo-electric three-state controller using rate sensitivity. P. 5 *in Proc. 8th Int. Conf. Med. Bio. Eng. and 22nd Ann. Conf. Eng. Med. Bio.* (IEEE publication no. 69 C 20-JCMBE, New York).

EVARTS, E. V. 1967: P. 215 *in* 'Neurophysiological Bases of Normal and Abnormal Motor Activities'. Raven Press, New York.

HUMPHREY, P. R., SCHMIDT, E. M., and THOMPSON, W. D. 1970: Predicting measures of motor performance from multiple cortical spike trains. *Science* **170**: 758-762.

JACOBSEN, S. C. 1971: 'Prosthetic Control'. Unpublished Doctoral Thesis Proposal, Department of Mechanical Engineering, M.I.T., U.S.A.

JOHNSEN, E. G., and CORLISS, W. R. 1967: Teleoperators and human augmentation. *NASA* **SP-5047**.

MANN, R. W. 1968: Efferent and afferent control of an electromyographic, proportional-rate, force sensing artificial elbow with cutaneous display of joint angle. *Proc. Inst. of Mech. Eng.* **183** (3J): 86-91.

MOUNTCASTLE, V. B., Ed. 1968: *in* Neural control of movement and posture, Mosby, St. Louis. 'Medical Physiology', Vol. II, Part XI.

NIRENBERG, L. M., HANLEY, J., and STEAR, E. B. (In press): EEG motor signal tracking with an adaptively designed phase-locked loop. *IEEE Trans. Bio. Eng.*

REILLY, R. E. 1968: Emgor: an implantable sensor for myo-electric signals. *Proc. Inst. of Mech. Eng.* **183** (3J): 109-116.

ROBERTS, T. D. M. 1967: 'Neurophysiology of Postural Mechanisms'. Plenum Press, New York.

ROTHCHILD, R. D. 1968: 'Decoding of Electrical Noise from Nerve Bundles'. Unpublished Doctoral Thesis Proposal, Department of Mechanical Engineering, M.I.T., U.S.A.

ROTHCHILD, R. D., and MANN, R. W. 1966: An EMG controlled, force sensing, proportional rate elbow prosthesis. *Biomedical Engineering Symposium, June 1966*, Marquette University.

SELL, G., and MERTEN, H. 1969: An electrically operated hand. *Bio-Med. Eng. (GB)*, 4: 563-565.

SKACHKOV, A. N.: Bioelectrically controlled artificial hands. Pp. 97-101 *in* 'External Control of Human Extremities'. *Proc. Int. Symp., Dubrovnik, 1966* (Belgrade: Yugoslav Committee for Electronics and Automation, 1967).

WEINER, N. 1948: 'Cybernetics'. The Technology Press of M.I.T. and Wiley, New York.

WHITNEY, D. E., 1969: Resolved rate control of manipulators and human prostheses. *IEEE Transactions, Man-Machine Systems*, **MMS-10** #2.

WILSON, A. B. 1970: Limb prosthetics—1970. *Artificial Limbs* **14**: 1-52.

WIRTA, R., and TAYLOR, D. 1969: Development of a multiaxis myoelectrically controlled prosthetic arm. *3rd Int. Symp. Ext. Control Human Extremities, Dubrovnik*.

FURTHER READING

ALLES, D. S. 1970: Information transmission by phantom sensations. *IEEE Transactions, Man-Machine Systems*, **MMS-10** #1.

DORCAS, D. S., and SCOTT, R. N.: A three state myo-electric control. *Med. Biol. Engng.* **4** (1966): 367-370.

FULFORD, G. E., and HALL, M. J. 1968: 'Amputation and Prostheses'. Wright, Bristol.

MANN, R. W., and REIMERS, S. D. 1970: Kinesthetic sensing for the EMG controlled 'Boston Arm'. *IEEE Transactions, Man-machine Systems*, **MMS-10** #1.

RESWICK, J. B., and VODOVNIK, L., External power in prosthetics and orthotics, an overview. *Artificial Limbs* **11** (2).

VODOVNIK, L. *et al.* 1969: Development of orthotic systems using functional electrical stimulation and myoelectric control. *SRS-7060-23-68, Progress Report #1, Ljubljana, Yugoslavia*.

MODULAR CONCEPTS IN PROSTHETICS/ORTHOTICS[1]

D. S. McKENZIE[2]

GENERAL

The Concise Oxford Dictionary defines a module as 'standard, unit, for measuring; (archit.): unit of length for expressing proportions, usu. semidiameter of column at base'. In prosthetics/orthotics, as indeed in many other engineering fields, we use the word to denote a component or subassembly of a whole product, each module being complete in itself and attached to adjacent modules at a standard interface by releasable fastenings. It is therefore a misnomer and perhaps 'unitary construction' would be a better term. By whatever the name, it was evident that a prosthetic system of this sort could provide a solution for the most pressing problem of the British Limb Service, namely the unacceptably long time taken to make and to repair limbs.

Since the 1920s the artificial leg supplied in Britain was a light metal stressed skin structure riveted to various forgings, castings or pressings comprising the joints. The processes required to assemble and prepare the limb for fitting and for finishing it thereafter were, of necessity, central factory procedures; while fittings and other work on the patient had to be done at peripheral centres. It was found impossible, despite continual effort, to achieve a significant reduction in the time taken. It therefore became necessary to adopt a new philosophy as a basis for entirely new design criteria. We therefore visualised a comprehensive modular system through which the temporary and definitive needs of virtually all amputees, whatever the site of amputation, whatever the age, whatever the level of activity could be met in a matter of a few days rather than months. Stocks of centrally manufactured components and subassemblies would be held at peripheral centres and provision made for the fabrication of sockets there. The fastenings between modules would be such as to make assembly and disassembly easy and quick. Provision would be made for static and dynamic alignment, the latter not readily achieved in the conventional prosthesis. The system would maintain all the prescribing options currently available to clinicians, and have no relatively adverse features. It would, it was to be hoped, be compatible with mechanisms and devices of merit developed in other countries, this implying a degree of acceptance of internationally agreed standards of geometry and interfaces.

PATIENT BENEFITS

I have emphasised that the primary objective of embarking on a modular system was speed but it was proper also to consider what further benefits might accrue to the patient from the modular concept and the opportunity to redesign, so that the lower limb prosthesis might be improved.

For example, the modular nature of the assembly would make it possible for the patient to try various prescribable alternatives such as knee mechanisms, without incurring delays, or the novice could start with a quite simple knee mechanism and graduate to a more sophisticated device without interrupting his rehabilitation as he progressed. If the alignment device could be designed so that it could stay in the prosthesis, adjustments could be made at a later time as, for example, the new patient gaining confidence may get improvement from a narrower walking base. If the design was such that the finishing process could be done in an hour or so, the patient could take delivery on the same visit at which he has his fitting. Repairs, based on unit replacement, could be done on a 'while-you-wait' basis. The design could be such that the vertical space within the limb occupied by joints and mechanisms was minimised so as to leave the operating surgeon the greatest freedom to choose the level of bone section. If, as seemed inevitable, a central strut structure was adopted, it should be possible to devise a soft cosmetic cover of acceptable shape, appearance and texture, odour, thermal effect, noise etc. Factors of this sort could be and were included in the design criteria.

The old stressed skin structure had one particular virtue, namely that failure, if it occurred, was a gradual collapse with little hazard to the patient. It seemed likely that the sort of structures which would emerge in the new designs could be at risk of catastrophic failure. It was therefore necessary to determine values for the various modes of loading which would provide a basis for type approval testing and for quality assurance in production both in terms of static and fatigue testing.

---

[1] Crown copyright.
[2] Biomechanics Research and Development Unit, Roehampton, London, England.

79

LOGISTICS

Finally the logistic implications in a service like ours must be considered at a very early stage in planning a major programme of conversion. In the central factories, there will have to be a drastic change in manufacturing methods. The possible reactions of the labour force will need careful negotiation. It is recognised that there will be a period of a number of years during which the existing limbs will be phased out and replaced by modular systems and this may well cushion any possible disquiet among the labour force. Capital expenditure on plant and tooling will be considerable as will that on the building up of the initial stocks of components. The peripheral centres will each require to be reviewed in relation to bench space for assembly, socket making facilities etc. and for storage of components. New building may be required in certain instances and some redeployment of labour to carry out the new processes to be undertaken at peripheral centres will be inevitable. Such considerations tend to require a considerable lead time and most of them have some implication on the design criteria of the system itself.

Such, then, in the very superficial way that my time permits, were the considerations which formed the basis of our concept of the modular prosthetics system for lower limbs when we started the programme five years ago. It is of interest that the only significant addition to the design criteria is the emergence of the following concept: that it is desirable for the mechanical structures to be sufficiently slender to allow the soft cosmetic cover to envelop them completely as a continuum from the foot to the upper part of the socket.

ORTHOTICS

Orthotic, as opposed to prosthetic, devices in general lend themselves less readily to modularity. There is so often the need for a custom-built service. Nevertheless, there are already systems of hand splints which may be regarded as being modular, and a number of drop foot appliances are more or less available 'off the shelf'.

Orthotic devices are used to assist joints which are unstable or which lack power. I am attracted by the design philosophy which states that the basic orthosis should provide freedom to move in all the axes of the natural joint concerned. Constraints or power assistance are then only provided in relation to those degrees of freedom needing assistance in the individual case. It seems to me that this approach lends itself to a modular system although it is probable that considerable anthropometric data will be required to determine the range of sizes and other variables which will need to be met.

There also seems to be some possibility of developing modular systems for a number of spinal braces.

Acknowledgment

I am grateful to Dr R. H. L. Cohen, C.B., Department of Health & Social Security, for permission to present this paper, which expresses the views of its author and does not necessarily represent those of the Department.

FOOTPRINT PATTERNS DURING WALKING

J. D. CHODERA AND R. W. LEVELL[1]

INTRODUCTION

The development of Man's striding gait occurred well over a million years ago and is a characteristic of the human being. No other mammals, even primates, have an habitual bipedal gait as a natural method of propulsion. The structure of the human body is adapted in form and function to this very special kind of locomotion, which is performed by genetically imprinted movement stereotypes (Napier, 1970). The adaptability of the neuro-muscular control system to changes in walking conditions produces large variations in the walking pattern from strolling or walking on a slippery surface, through the rolling gait of a sailor, to the highly elaborate motion of the athlete. But all these variations contain some basic common features of the 'normal human walk'.

The only contact with the support surface during walking is through the soles of the feet; consequently the most significant interactions with the physical environment take place at or through this surface. These interactions can be separated into two groups of phenomena: force vectors, and the shape and localisation of the contact areas. The present work belongs to the second group, being a study of the geometry of the footprint patterns.

During walking the centre of mass of the body is moved along a spatial trajectory. Fischer, in his classic study (Fischer, 1899), has shown that the vertical projection of this trajectory onto the walking surface results in a sinusoidal pattern. This arises from the alternate transfer of weight from one leg to the other. The direction of propagation of the body mass following such a sinusoidal path may be defined by the mean line between the excursions. From common experience it is known that a man is able to walk from point A to point B. The simplest line joining two such points on a plane is a straight line. When examining a straight walk most authors use just such a straight line as a co-ordinate basis

[1] Biomechanical Research and Development Unit, Roehampton, London, England.

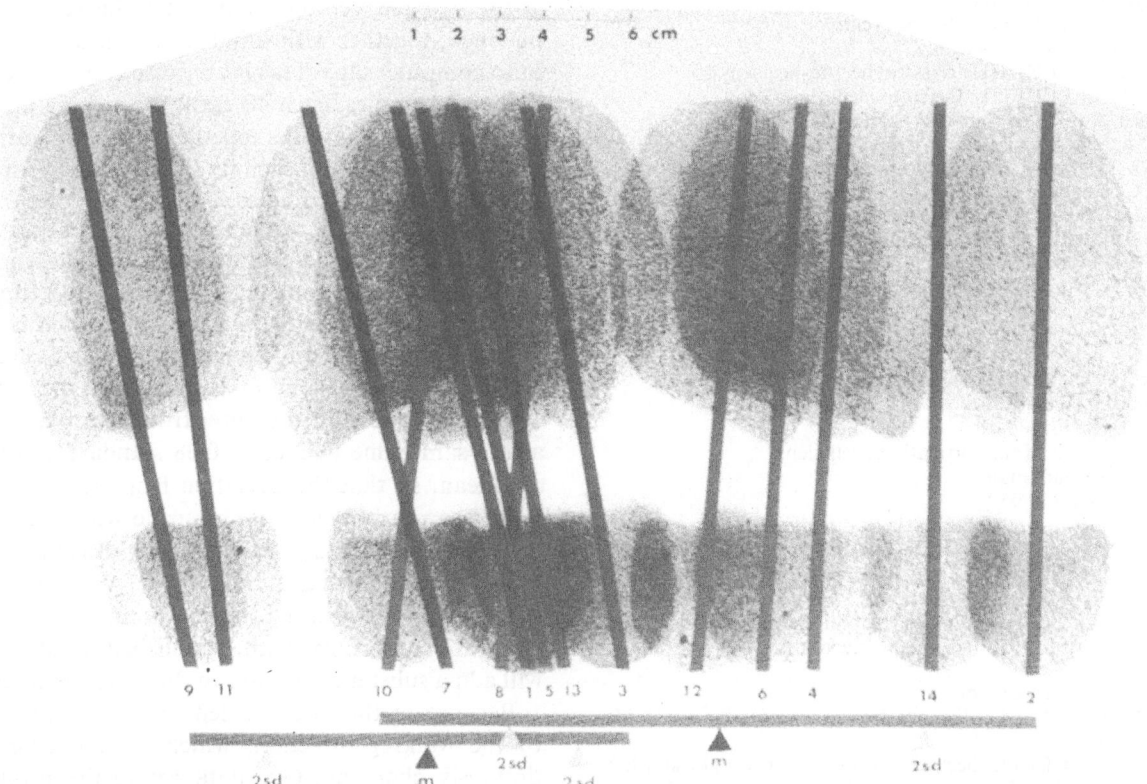

Figure 1. Footprints overlapped by compression of longitudinal axis. (Right-handed, female subject of 43 yr; height 165 cm; weight 69 kg; normal walk).

for their measurements. Initially we too used this straight line basis and we compressed the longitudinal axis so that the positions of the footprints overlapped and we found a scatter between these footprints to the sides, as shown in figure 1. Using the accepted methods of averages between footprint pairs we found that the walking direction changes smoothly from step to step. Indeed, not one of the 320 experimental records we have collected from our 40 subjects (both normal and amputees or

disabled in the age range 4 to 60 years) showed a straight line walk. Straight line walk, then, does not appear to exist.

It is known that if we walk in an environment where visual feedback of directional information is absent (e.g. walking with closed eyes, or in a desert or wood) then we tend to walk in curves. At the other extreme, for example when walking along a single contrasting line for short distances, we produce less scatter to the sides.

ANGLE A: angle between two centroid-to-centroid directions
ANGLE B: angle between the mean footprint and the particular walking direction at the centroid
ANGLE RT: angle between actual walking direction and the co-ordinates of the walkway at the centroid
B mean: mean of footprint angles
C: centre or centroid
CC: correlation coefficient
CV: coefficient of variation
DISTR. DISTL: half stride length right and left
EXP: experiment
F, FST: fast walk
FPANGR, FPANGL: footprint angles right, left
FPL: footprint length
FPT: footprint
H: heel
HEELX, HEELY: heel co-ordinates
HH: heel-to-heel distance
L1, R1: walk with load of 350 g on left (right) ankle, subject's own speed
L2, R2: as above but 700 g
M: mean
MIDX, MIDY: centroid co-ordinates
N: normal walk, subject's own speed
N1: normal walk at the beginning of the experiment
N2: repeated normal walk
NOEQ: walk without equipment
NOSL: bare feet walk
NROF, EVT, NE: number of events
RC: regression coefficient
RIGHT or LEFT HH: classical heel-to-heel length
RIGHT or LEFT TT: toe-to-toe length
R, L: right, left or right (left) handed
SD: standard deviation
SEBM: standard error between means
SECC: standard error of the coefficient of correlation
SEM: standard error of the mean
SIGN: P-value of the significance from Student's Tables
SLW: slow walk
SQSUM, MEANSQ: summation terms
SUM: total of variables
SUMSQ· sum of squares
T: toe
TOEX, TOEY: toe co-ordinates
TT: toe-to-toe distance
T.T: T-test (Student)
TTCC: T-test of the correlation coefficient
W, M: woman, man
X: significant at 95%
XHH: heel-to-heel width in the X direction
XHR, XHL: heel co-ordinates left and right in the X direction
XHTR, XHTL: heel-to-toe width right and left in the X direction
XTR, XTL: toe co-ordinates right and left in the X direction
XTT: toe-to-toe width in the X direction

Some Combined Groups:
CCR (or CCL)—L: distance between successive centroids right and left
HCR (or HCL)—L: heel to centroid distance right and left
TCR (or TCL)—L: toe to centroid distance right and left

List of Abbreviations used in Graphs and Tables

EXPERIMENTAL TECHNIQUE

In our experimental technique, which has already been described (Chodera, in press) we are using 48 cm wide strips of aluminium foil for recording the footprint patterns. This foil is attached to a walkway which is 120 cm wide and has any kind of rough surface such as a wire mesh, or a rubber mat with a surface profile. Metal foil has been used so that we may also record heel and toe contact times. The foil preserves fairly well the position and shape of the footprints of several (in our experiments up to 10) experimental walks and permits their evaluation at any later time. A straight line representing the longitudinal axis of the walkway is recorded on the foil, using steel tape, as a series of distinct marks 40 cm apart. This permits any deformation of the foil (e.g. stretching or wrinkling) to be accounted for in the processing. Heel and toe positions are recorded by metal markers fixed to the shoes or bare feet; these points may be recovered with an accuracy of ± 1 mm. A semi-automatic X–Y co-ordinate reader is used to recover the points and the results are punched, together with suitable identification flags, onto computer tape. This is carried out in a series of sequential frames, each 40 cm long, picking up the values from all 10 walks together in one operation. The software sorts out each single walk and produces a table of X–Y co-ordinates as shown in figure 2.

Because the width of the foil was 48 cm, the subject was constrained to walk within that width. Under these conditions for example in experiment No. 16.1 (see figure 3), we found the width co-ordinate of walking direction to change 11·5 cm in a recorded length of 8 m, taken from a 12 m normal walk. The stride width measured from heel to heel was at the same time 6·02 cm $\pm$ 0·38 standard error of the mean, so that the deviation from straight line walk was nearly twice that of stride width. This is not a negligible value. If the stride widths and foot angles are calculated in relation to the supposed straight line between the starting point and the end point in conformity with established practice, it will add a substantial scatter to the resulting figures.

Because of this, we decided to relate all figures to the walking direction, which is of course continuously changing. Our data define the position of each footprint by two fixed points. During the bipedal stance phase the body weight is transferred

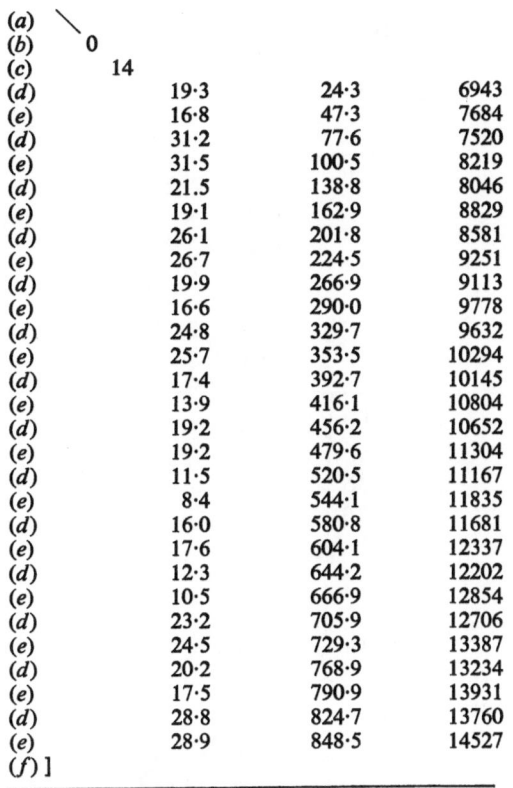

(a)	\		
(b)	0		
(c)	14		
(d)	19·3	24·3	6943
(e)	16·8	47·3	7684
(d)	31·2	77·6	7520
(e)	31·5	100·5	8219
(d)	21.5	138·8	8046
(e)	19·1	162·9	8829
(d)	26·1	201·8	8581
(e)	26·7	224·5	9251
(d)	19·9	266·9	9113
(e)	16·6	290·0	9778
(d)	24·8	329·7	9632
(e)	25·7	353·5	10294
(d)	17·4	392·7	10145
(e)	13·9	416·1	10804
(d)	19·2	456·2	10652
(e)	19·2	479·6	11304
(d)	11·5	520·5	11167
(e)	8·4	544·1	11835
(d)	16·0	580·8	11681
(e)	17·6	604·1	12337
(d)	12·3	644·2	12202
(e)	10·5	666·9	12854
(d)	23·2	705·9	12706
(e)	24·5	729·3	13387
(d)	20·2	768·9	13234
(e)	17·5	790·9	13931
(d)	28·8	824·7	13760
(e)	28·9	848·5	14527
(f)	1		

Figure 2. Computer tape showing input data for footprint studies (a) Data separator; (b) First foot (0 = left, 1 = right); (c) Number of footprints; (d) Heel *X*, heel *Y* co-ordinates (cm), heel-on time (ms); (e) Toe *X*, toe *Y* co-ordinates (cm), toe-off time (ms); (f) End of input data.

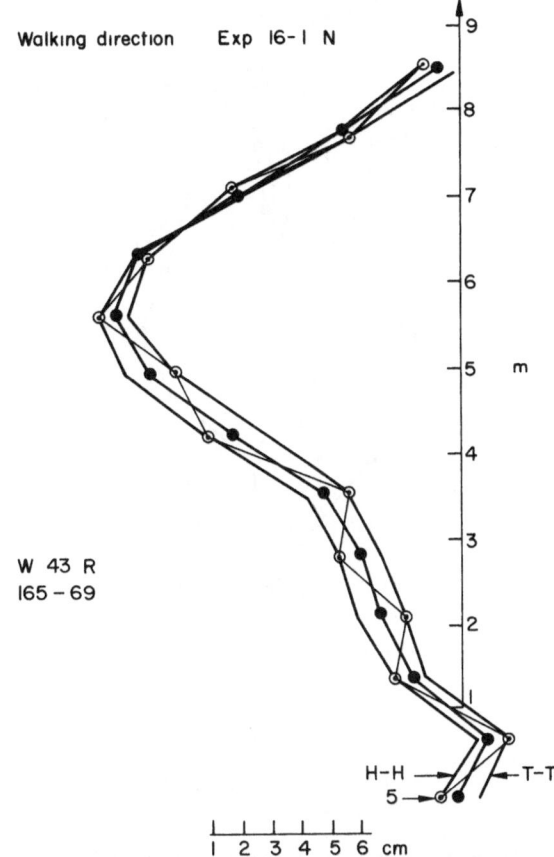

Figure 3. Walking direction of one walk for different weightings (same subject and walk as in figure 1).

from one leg to the other. At the instant of equal loading the projection of the centre of gravity of the body onto the walking surface was assumed to lie at the centroid of the four points comprising the two footprints. Figure 3 shows the walking direction line derived from the same experiment as above, using selected different point weighting factors, i.e. between heels only, between toes only, between heels and toes equally, and between ½ of the heel-to-toe distance, this being the approximate projected position of the ankle. The choice of weighting is arbitrary and each choice has almost the same probability of finding a point which lies approximately on the line produced by projecting the trajectory of the centre of mass of the body onto the plane of walking. For example, line No. 5 in the picture is that due to toe-to-heel weighting.

We are using, at present, equal weighting between heels and toes because that is the most similar to the methods used by other authors of footprint studies; and we need to compare results. But the mean line between the extreme weightings corresponds more nearly to the weighting between ankles. This can be appreciated from the anatomy of the foot structure as well as from the function of the foot during walking.

The line joining the weighted centroids of each footprint pair changes in direction from pace to pace. The angle, alpha, between successive segments of this line varies with directional changes about 180°, being less than this when the direction of the segment changes to the right. We found that the direction of the segments changed smoothly over the whole of the recorded walk (figure 4) and conformed with one of three basic shapes; a curve to the left, a curve to the right, or an S-shaped curve. The S-shaped curve was observed more frequently at slow walking speeds. We believe this smooth curve to be due to deliberate directional corrections carried out in response to feed-back from a higher level, there being more time available at slow speeds to effect such corrections over the recorded distance.

We recorded the footprint pattern of our subjects under varying conditions. We asked them to walk comfortably at their own natural speed, to walk slowly, and fast; we also put weights of 350 and 700 g on the ankles successively on the left and then the right side. Immediately after these loading experiments we recorded another normal walk, then one with bare feet and finally one without the foot contact cables which are servo-driven and follow the subject automatically. This latter experiment was performed to check any influence of the recording equipment on the footprint pattern.

We chose the mean direction of the two line segments meeting at the weighted centroid of a pair

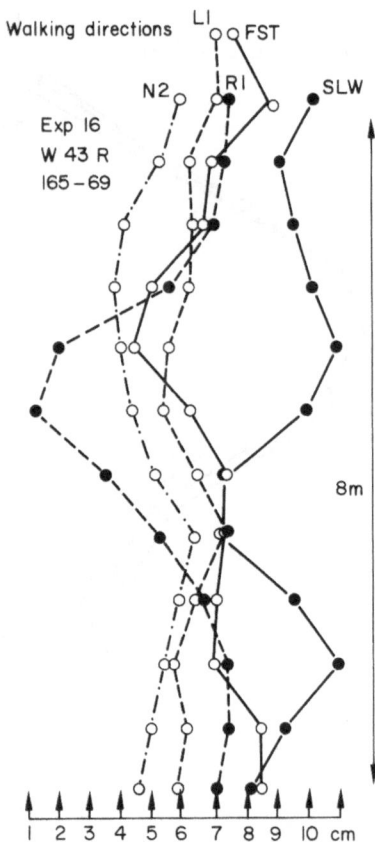

Figure 4. Walking direction curves for different walking conditions (all equal weighting; experiment No. 16; subject as in figure 1).

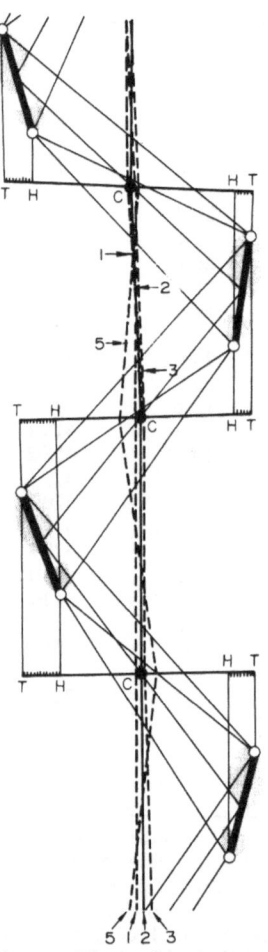

Figure 5. Diagram illustrating the processing technique and the effects of different weighting factors on the derived walking direction.

of footprints to represent the instantaneous walking direction appropriate to that footprint pair. We derived this by producing an orthogonal set of coordinates located at the weighted centroid such that the transverse axis was the bisector of the angle between the successive centroid-to-centroid directions alpha (figure 5). These new instantaneous direction co-ordinates are related to the walking or measurement co-ordinates by simple translation and rotation operations, the rotation angle RT being that between the longitudinal measurement axis and the mean, instantaneous walking direction. By measuring the heel-and-toe contact points of the two footprints with reference to this new set of axes the variations due to changes in walking direction are removed. The heel-to-heel distance, toe-to-toe distance and foot angle, each related to the particular walking direction at this weighted centroid and all other parameters of the stride, are expressed in terms of these new co-ordinates and are thus independent of the circumstances of the walk.

The time relationships are also well defined because the timing and duration of the bipedal stance phase is recorded separately. The observation and evaluation of all the data is thus carried out in a defined manner and under precisely defined conditions.

PROCESSING

To perform the necessary calculations on 320 test walks, each containing 10 to 20 single steps, we used two stages of computer processing. The first processes the input data from the foil, compensating for the possible distortions of the recorded pattern and separating them out into sets of X and Y co-ordinates referred to the walkway and expressed in centimetres and accurate to ± 1 mm (figure 2). The second programme processes the input data in phases. First it calculates the length of the footprints and simple scatter statistics of the input data. The coefficient of variation is used as a measure of the accuracy of the experiment.

At the very beginning of the project we used dust prints for recording purposes and the spread of footprint lengths varied around 5%. First attempts using the aluminium foil recording technique led to an improved accuracy in the range of 3%. The present technique supplements this with special markers and input data checks and the precision of the method is now in the range of 0·5–1·5%. The accuracy of the technique is one order higher than the spread of the biological parameters, a necessary condition for statistical significance of the results when there are a relatively small number of events in each single walk. In the second phase, the X and Y co-ordinates of the weighted centroids, which are also the origins of the new co-ordinate systems, are determined. The distances between these centroids are calculated and printed in table 4 with the classical stride lengths (figure 6b).

Next the angle alpha and the instantaneous walking direction are found. This instantaneous walking direction is the basis of the translation and rotation of the co-ordinate system in which the relationships between successive footprints are expressed. The new transverse co-ordinates of the heels and toes of both left and right feet are determined separately. The transverse difference between heel and toe on each side are divided by the particular footprint lengths to form the cosine of the footprint angle. All stride parameters are grouped into tables. Table 1 (figure 6a) gives input co-ordinate data, footprint lengths and the co-ordinates of the weighted centroids from which the walking direction is found. The mean values, standard deviations, standard errors of the mean, confidence belt about the mean in the range of two standard errors of the mean, and the coefficient of variation are printed together with the number of events. These simple statistics are printed for all arrays of variables appearing in each table. In table 1 we also display the differences between successive positions of the same point on the foot and their scatter statistics.

Table 2 (in figure 6b) shows the angles. The angle alpha gives the changes of walking directions and angle RT the difference between the original and transformed co-ordinate systems step by step. The next two columns give the footprint angles right and left and their mean is given in column 4. This mean value differs slightly from the angle of the mean footprint because of slight scatter in the footprint lengths. The mean footprint angle beta has been calculated as the angle between the line connecting centre points of the lines joining heel to heel and toe to toe of the original input data and the instantaneous walking direction at the particular centroid. It is printed to permit a visual check of the footprint angle. These arrays are one expression of the asymmetry of the footprint angles against the instantaneous walking direction.

Table 3 displays stride widths. In addition to the usual heel-to-heel width, the transverse values of right and left heel co-ordinates are given, showing the relation between the heels of each side and the instantaneous walking direction line. A similar display is given for both toes. The difference between these has some relation to the side stability of the walking body.

Table 4 shows the heel-to-heel distances, and toe-to-toe distances, processed in the conventional way. Additionally a further centroid-to-centroid distance is calculated. This length is the distance between two subsequent points of the defined walking direction and corresponds to the half-stride length. It may be related to the foot-in-stance phase, which is responsible for the transfer of the body mass between those two points. We distinguish, therefore, the right and the left half-stride lengths.

The sum of both our half-stride lengths and the classic heel-to-heel distances are approximately equal. The correlation coefficient for a theoretically straight walk equals 1·0. In practice the correlation in our experiments is around 0·9 and will decrease with any deviation of the walking direction from a straight line. In such a case the foot inside of the curve makes shorter strides and there is a difference between left and right stride lengths expressed by the conventional measurement method. Evaluating the line of progression by our method results in only one value, which includes any circumstance during walking.

The differences printed in Table 4 represent the differences between two subsequent steps of the same side and between steps of the opposite sides. The scatter of values is not significant in totals between left and right side, but the variance of differences is larger on one side in all records.

The end of Table 6 evaluates the significance of differences between left and right side values. Tests are calculated on standard error of the difference between means (SEBM) tested for significance at 95 % (in column X, 0 = not significant, 1 = significant), Students T-test (T.T.) with its probability value P (Sign), correlation coefficient, tested by T-test and standard error of correlation, regression coefficient and summation terms.

In the first part of Table 6 are printed means, variances, standard deviations, co-efficients of variation, standard errors of the mean and simple summation terms together with the number of events for all variables of the single experimental run.

RESULTS

Equal weighting of toes and heels has been used for processing two experiments, the first on a normal, healthy woman, age 46, right handed, the other on a below knee left-side amputee, right-handed and well fitted with an experimental PTB prosthesis. This sample from our experimental material involved: speed changes, loading the ankles on each side separately with approximately 1 % and 2 % of the body weight, a walk in bare feet and a walk without any electrical recording equipment attached. Standard conditions with adaptation walking, 10 minute rest periods and other precautions for ensuring homogeneous results have been fulfilled (figures 6A, 6B and 7).

TABLE 1: EXPERIMENT 16; RUN 1; FPT No. 14; FIRST FOOT LEFT

WEIGHTING : 0.25 0.25 0.25 0.25, W 43 R 165 69 17.11.1970

FPT	HEELX	HEELY	TOEX	TOEY	FPL	CENTRE	MIDX	MIDY
1	19.3	24.3	16.8	47.3	23.1		24.7	o
2	31.2	77.6	31.5	100.5	22.9	1	24.7	62.4
3	21.5	140.8	19.1	162.9	22.2	2	25.8	120.4
4	26.1	201.8	26.7	224.5	22.7	3	23.3	182.5
5	19.9	266.9	16.6	290.0	23.3	4	22.3	245.8
6	24.8	329.7	25.7	353.5	23.8	5	21.7	310.0
7	17.4	392.7	13.9	416.1	23.7	6	20.4	373.0
8	19.2	456.2	19.2	479.6	23.4	7	17.4	436.1
9	11.5	520.5	8.4	544.1	23.8	8	14.6	500.1
10	16.0	580.8	17.6	604.1	23.4	9	13.4	562.4
11	12.3	644.2	10.5	666.9	22.8	10	14.1	624.0
12	23.2	705.9	24.5	729.3	23.4	11	17.6	686.6
13	20.2	768.9	17.5	791.9	23.2	12	21.3	749.0
14	28.8	824.7	28.9	848.5	23.8	13	23.8	808.5

	HEELX	HEELY	TOEX	TOEY	FPL		MIDX	
MEANS	0.73	61.57	0.93	61.63	23.25		20.05	
STAND.DEV.	7.45	3.40	10.78	3.34	p.47		4.21	
S.E.M.	2.07	0.94	2.99	0.93	0.13		1.17	
MEAN+2*SEM	4.86	63.46	6.91	63.48	23.50		22.39	
MEAN−2*SEM	−3.40	59.68	−5.05	59.78	23.00		17.72	
COEFF.VAR.	1018.97	5.53	1158.45	5.42	2.02		21.01	
NR OF EVT.	13	13	13	13	14		13	

DIFFERENCES OF THE COORDINATES

11.90	53.30	14.70	53.20
−9.70	63.20	−12.40	62.40
4.60	61.00	7.60	61.60
−6.20	65.10	−10.10	65.50
4.90	62.80	9.10	63.50
−7.40	63.00	−11.80	62.60
1.80	63.50	5.30	63.50
−7.70	64.30	−10.80	64.50
4.50	60.30	9.20	60.00
−3.70	63.40	−7.10	62.80
10.90	61.70	14.00	62.40
−3.00	63.00	−7.00	62.60
8.60	55.80	11.40	56.60

TABLE 6 STRIDE STATISTICS

VARIABLE	NUMBER	MEAN	VARIANCE	S.D.	C.V.	SEM	SUM	SUMSQ
FOOTPRINT LENGTH AND DIRECTION								
FPL	14	23.25	0.22	0.47	2.0	0.13	325.51	7571.23
MIDX	13	20.05	17.74	4.21	21.0	1.17	260.70	5440.96
FOOTPRINT ANGLES								
ANGLE A	11	179.88	2.91	1.71	0.9	0.51	1978.70	355963.03
ANGLE RT	11	−0.20	4.12	2.03	−999.9	0.61	−2.23	41.69
FPANGR	11	2.24	1.94	1.39	62.2	0.42	24.65	74.67
FPANGL	11	−6.75	1.84	1.36	−20.1	0.41	−74.27	519.92
BMEAN	11	−2.25	1.34	1.16	−51.5	0.35	−24.75	69.12
ANGLE B	11	−2.24	1.32	1.15	−51.2	0.35	−24.67	68.54
STRIDE WIDTHS								
XHR	11	3.47	0.52	0.72	20.7	0.22	38.12	137.27
XHL	11	−2.56	0.37	0.61	−23.7	0.18	−28.15	75.70
XTR	11	4.37	0.20	0.45	10.2	0.13	48.12	212.46
XTL	11	−5.28	0.34	0.59	−11.1	0.18	−58.09	310.15
XHTR	11	0.91	0.33	0.57	62.8	0.17	10.00	12.34
XHTL	11	−2.72	0.31	0.56	−20.5	0.17	−29.94	84.60
XHH	11	6.02	1.55	1.25	20.7	0.38	66.27	414.75
XTT	11	9.65	0.87	0.93	9.7	0.28	106.20	1034.05
STRIDE LENGTHS								
DIST R	6	62.34	2.46	1.57	2.5	0.64	374.06	23332.75
DIST L	6	62.09	4.57	2.14	3.4	0.87	372.51	23150.29
RIGHT HH	6	124.62	9.55	3.09	2.5	1.26	747.72	93227.41
LEFT HH	6	124.17	16.00	4.00	3.2	1.63	745.04	92595.23
LEFT TT	6	124.76	10.24	3.20	2.6	1.31	748.55	93437.96
LEFT TT	6	124.17	20.82	4.56	3.7	1.86	745.01	92610.31

GROUPS	SEBM	X	T.T. SIGN.	C.C.	TTCC/NE SIGN	SECC	SQSUM	MEANSQ	RC *
CCR-L	1.08	0	0.24 > .50	0.15	0.31/4 > .50	.45	23259.1	1057.23	.11
YHHR-L	2.06	0	0.22 > .50	0.12	0.24/4 > .50	.45	92975.2	4226.15	.09
YTTR-L	2.28	0	0.26 > .50	0.19	0.38/4 > .50	.45	93101.8	4231.90	.13
HCR-L	0.28	1	21.26 < .01	−0.77	3.61/9 < .01	.32	110.9	2.64	−.91
TCR-L	0.22	1	43.47 < .01	−0.63	2.42/9 < .05	.32	264.0	6.29	−.48
HTR-L	0.24	1	15.07 < .01	0.34	1.09/9 < .50	.32	51.7	1.23	.35
HH-TT	0.47	1	7.73 < .01	0.86	5.08/9 < .01	.32	736.5	17.54	1.15
ANGR-L	0.59	1	15.33 < .01	0.42	1.40/9 < .20	.32	316.2	7.53	.43

Figure 6A. Computer output tables 1 and 6 from a single walk (subject, experiment and walk as in figures 1, 2 and 3; second series of experiments; accuracy 3%)

TABLE 2 STRIDE ANGLE

CENTRE	ANGLEA	ANGLERT	FPTANGR	FPTANGL	FPTMEAN	ANGLEB
1	183.39	−0.59	1.35	−5.62	−2.13	−2.08
2	178.64	−1.61	3.13	−4.60	−0.73	−0.69
3	179.59	−0.72	2.24	−7.42	−2.58	−2.65
4	180.67	−0.85	3.02	−7.29	−2.13	−2.08
5	181.56	−1.96	4.14	−6.55	−1.20	−1.19
6	179.81	−2.65	2.66	−5.87	−1.60	−1.63
7	178.55	−1.83	1.84	−5.67	−1.91	−1.95
8	178.22	−0.21	4.15	−7.28	−1.56	−1.62
9	177.45	1.95	1.99	−6.49	−2.25	−2.20
10	179.81	3.32	−0.15	−7.86	−4.00	−3.94
11	181.01	2.91	0.28	−9.62	−4.66	−4.64
MEANS	179.88	−0.20	2.24	−6.75	−2.25	−2.24
STAND.DEV.	1.71	2.03	1.39	1.36	1.16	1.15
S.E.M.	0.51	0.61	0.42	0.41	0.35	0.35
MEAN+2*SEM	180.91	1.02	3 08	−5.93	−1.55	−1.55
MEAN−2*SEM	178.85	−1.43	1.40	−7.57	−2.95	−2.94
COEFF. VAR.	0.95	−999.92	62.18	−20.10	−51.52	−51.22
NO. OF EVT.	11	11	11	11	11	11

TABLE 3 STRIDE WIDTHS

FTP	CENTRE	XHR	XTR	XHL	XTL	XHTR	XHTL	XHH	XTT
2	2	4.94	5.47			0.53			
3	2			−4.12	−6.29		−2.17	9.05	11.76
3	3			−3.02	−4.80		−1.78		
4	3	3.29	4.53			1.24		6.31	9.32
4	4	3.22	4.11			0.89			
5	4			−2.16	−5.17		−3.01	5.38	9.28
5	5			−2.49	−5.45		−2.96		
6	5	3.34	4.59			1.25		5.83	10.04
6	6	2.86	4.58			1.71			
7	6			−2.37	−5.07		−2.70	5.24	9.65
7	7			−2.03	−4.45		−2.42		
8	7	2.70	3.78			1.08		4.73	8.23
8	8	3.22	3.97			0.75			
9	8			−2.42	−4.77		−2.35	5.65	8.74
9	9			−2.03	−5.04		−3.01		
10	9	2.69	4.38			1.69		4.73	9.43
10	10	3.37	4.17			0.81			
11	10			−2.49	−5.06		−2.57	5.85	9.23
11	11			−2.86	−5.97		−3.11		
12	11	4.45	4.39			−0.06		7.31	10.36
12	12	4.04	4.15			0.11			
13	12			−2.16	−6.02		−3.86	6.19	10.17
MEANS		3.47	4.37	−2.56	−5.28	0.91	−2.72	6.02	9.65
STAND.DEV.		0.72	0.45	0.61	0.59	0.57	0.56	1.25	0.93
S.E.M.		0.22	0.13	0.18	0.18	0.17	0.17	0.38	0.28
MEAN+2*SEM		3.90	4.64	−2.19	−4.93	1.25	−2.38	6.78	10.22
MEAN−2*SEM		3.03	4.10	−2.92	−5.63	0.56	−3.06	5.27	9.09
COEFF.VAR.		20.75	10.21	−23.66	−11.09	62.81	−20.54	20.69	9.67
NO. OF EVT.		11	11	11	11	11	11	11	11

TABLE 4 STRIDE LENGTHS

	Y-HEELS DISTANCE Y-TOES				HEELS < DIFFERENCES > TOES			
	LEFT	RIGHT	LEFT	RIGHT	1–3	2–4	1–3	2–4
	116.52	124.30	115.62	124.09	−7.78	0.00	−8.47	0.00
	126.11	127.91	127.12	129.00	−1.80	−1.81	−1.88	−3.03
	125.82	126.62	126.13	126.27	−0.80	2.08	−0.14	2.87
	127.94	124.64	128.12	124.51	3.30	−1.31	3.61	−1.85
	123.70	125.31	122.82	125.39	−1.60	0.94	−2.57	1.69
	124.95	118.93	125.20	119.28	6.02	0.36	5.91	0.19
MEANS	124.17	124.62	124.17	124.76	−0.45	−1.89	0.04	−0.06
STAND.DEV.	4.00	3.09	4.56	3.20	4.75	4.39	1.44	2.44
S.E.M.	1.63	1.26	1.86	1.31	1.94	1.96	0.59	1.09
MEAN+2*SEM	127.44	127.14	127.89	127.37	3.43	2.04	1.22	2.12
MEAN−2*SEM	120.91	122.10	120.44	122.14	−4.33	−5.82	−1.13	−2.24
COEFF.VAR.	3.22	2.48	3.67	2.57	****	−232.18	3319.41	****
NO. OF EVT.	6	6	6	6	6	6	6	5

	LEFT CC	RIGHT CC	LCC-LCC	RCC-RCC	LCC-RCC
	58.04	62.10			−4.06
	63.31	64.23	5.27	2.13	−0.92
	62.99	63.22	−0.32	−1.01	−0.23
	64.01	62.29	1.03	−0.94	1.73
	61.63	62.67	−2.38	0.39	−1.04
	62.54	59.55	0.91	−3.12	2.98
MEANS	62.09	62.34	0.90	−0.51	−0.26
STAND.DEV.	2.14	1.57	2.80	1.94	2.45
S.E.M.	0.87	0.64	1.25	0.87	1.00
MEAN+2*SEM	63.83	63.62	3.41	1.22	1.74
MEAN−2*SEM	60.34	61.06	−1.61	−2.24	−2.26
COEFF.VAR.	3.44	2.51	311.34	−380.35	−947.24
NO. OF EVT.	6	6	5	5	6

Figure 6B. Computer output tables 2, 3 and 4 from a single walk (subject, experiment and walk as in figures 1, 2 and 3; second series of experiments; accuracy 3%)

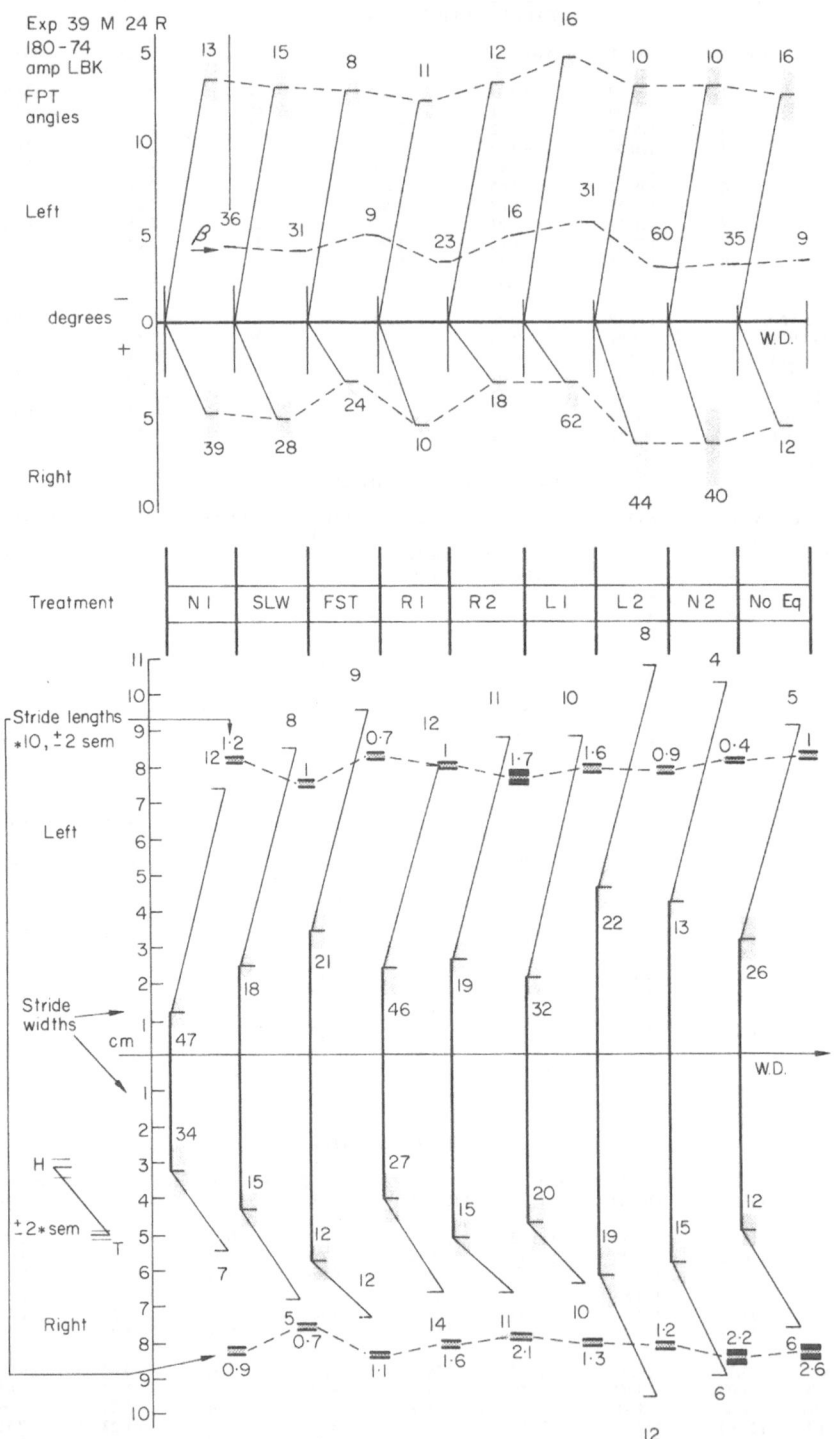

Figure 7. Graphic display of the statistical results of one complete experiment comprising ten distinct walks under different conditions. Angles at the top, and stride widths and lengths at the bottom. (Male subject aged 24 yr; right-handed; left leg amputated below knee.)

The direction of the walk of the first of these samples changed in the range of s.d. ±4·2 cm when the average stride width was found to be 6·02 cm ± 1·25 cm standard deviation. Similarly in all other walks the excursions from the straight line are not negligible in comparison with the stride width. Each single walk is represented on the graph by mean values of the stride widths between heels and toes (figure 7). The scatter of each set of values is displayed as a zone representing twice the standard error of the mean on each side of each mean. The numbers printed alongside are the co-efficients of variation. The line in the centre represents the walking direction. The stride widths are significantly asymmetrical about this line. We have found this asymmetry of the stride parameters in all of our records. This asymmetry is typical of the human walking pattern. For right-handed subjects the asymmetry is such that the mean value of the left heel is placed nearer to the line of propulsion and the left foot is rotated more laterally and shows less scatter of footprint angles than the right.

To express this asymmetry we used the angle of the mean footprint at each weighted centroid. The sign of this angle beta is generally preserved during changing walking conditions. However, there are individual differences between subjects in the values of mean angles and their scatter. In our material we have not found any single walk which does not have significant asymmetry.

At different walking speeds the footprint angles change. We believe that this effect is related to a rotation of the pelvis, but we have not at this stage proved this. Increasing the walking speed also caused the stride width to increase. At slow walking speeds we found differences between individuals; for instance amputees walked with a larger stride width than the others.

Loading the ankles produced generally more effect with the load on one side than on the other. The side affected most is the one furthest from the line of propulsion and having a smaller angle. Subjectively, the patients felt the load on this side more distinctly. The changes produced include widths, angles and scatter. In nearly all normal (i.e. without changes in walking conditions) walks made immediately after a loading experiment some effects due to loading were observed. This memory phenomena disappeared after some rest and free walking.

Bare feet walking was found to be different from walking in shoes. The asymmetry was less noticeable, stride widths being narrower with less scatter on all the data. Walking without the electrical recording equipment being attached was found in some cases to be surprisingly different. This seemed to be due to individual sensitivity to the walking conditions rather than to any physical influence.

Long term amputees showed some rigidity and inflexibility in their patterns. Also they were more prone to tiredness.

The stride lengths of all normal speed walks were found to be nearly the same, having a range of variation of less than 5%. The bare feet walks were usually slightly faster.

DISCUSSION

The asymmetry of all stride parameters, the non-symmetric response to experimental conditions and of amputees on different sides have been demonstrated. This suggests that each human leg has a slightly different function to perform. We may suggest for instance that the one which is nearer to the line of propulsion plays a larger part in

load bearing action and body balance and maintains the larger foot angle with a smaller scatter than the other foot to create more regular conditions for the transfer of body weight. In the meantime the other foot changes the narrower foot angle more and is more involved in steering the direction of the walk. Whatever explanation of this asymmetry of function is eventually established, it must result in a different prosthetic treatment for each leg.

We now have values of stride-widths and other asymmetries measured under defined conditions. We think that this data may be a useful basis for calculation and correction of the vertical alignment of prostheses. The footprint angles show a definite range of scatter which is not as large as reported in previous literature (Murray, 1964). Our data may also be used for setting the prosthetic foot in the necessary position within the patient's own functional tolerance band. This may improve the present practice of setting the foot-angle on the prosthesis by guesswork, either symmetrically or by reliance on the patient's subjective assessment which we have shown to have an unexpected dependence on previous changes.

Perhaps this may also help to reduce the influence on the motion of the patient's pelvis and limit the later appearance of rotoscoliosis of the spine, observed in a high percentage of patients and causing later problems with slipped discs for example.

It is known that every patient/prosthesis combination has a most suitable walking speed. The stride length results, which are related to the walking speed (Grieve, 1968) seem to confirm the existence of a patient's own natural speed of walking in a range which is narrower than the estimates obtained from metabolic rate changes or other observed parameters. Given specific data for a particular patient the stride length and swing phase time may be used for considering the weight and mass distribution of the prosthetic pendulum and to fit it better to the remaining stump capabilities.

There are more facts and relationships latent in the data already collected; for example, a considerably larger scatter in heel position than in toe position can be seen and this suggests that the foot angle fulfils some definite function. The anthropometric relations and time records have not yet been studied fully. Some of the basic data we have collected are now being assembled for publication.

The method of recording and analysis produces data with a high accuracy and is a useful tool for comparative and diagnostic purposes. In the future we hope to be able to use it for modelling the gait of individual subjects and to forecast the results which will be produced by changes in the walking conditions and prosthetic parameters.

REFERENCES

CHODERA, J. D.: The geometry of human foot print pattern during walking. *Bio-Medical Engineering* (in press).

FISCHER, O. 1899: Der Gang des Menschen, II Theil. Die Bewegung des Gesamtschwerpunktes und die ausseren Krafte. *Abhandlungen d. Math. Phys. Cl. d. Koenigl. Sachsich. Gesellsh. der Wissenschaft*, Bd. **25**: 1–130.

GRIEVE, D. W. 1968: Gait patterns and the speed of walking. *Bio-Medical Engineering*. (March): 119–122.

MURRAY, M. P., DROUGHT, A. B., and KORY, R. C. 1964: Walking patterns of normal men. *J. Bone Joint Surg.* **46**-A: No. 2.

NAPIER, J. 1970: 'The Roots of Mankind'. Allen and Unwin, London.

DESIGN ASPECTS OF ENDO-PROSTHESES FOR THE LOWER LIMB

JOHN P. PAUL[1]

BACKGROUND

In current clinical practice hip arthroplasty devices are being fitted to patients of progressively younger age groups (Charnley, 1971). Considerable research has been undertaken into the friction and wear characteristics required of the bearing surfaces of joint replacements (Duff Barclay, 1967; Scales, 1969). Attention has been directed principally to the design of the acetabular component of the hip joint and problems of attachment of this to the pelvic structure. There has been little research or analysis of the interface between the prosthesis and the femur, largely because clinically this region rarely gives rise to problems. The form of the femoral component of the prosthesis is generally a curved intramedullary stem of approximately rectangular cross-section secured either with bone chips or methyl methacrylate cement, a typical average design of stem being shown in figure 1.

By comparison the knee arthroplasty is an operation more frequently undertaken than the hip and the clinical end results are less predictably satisfactory. It is of interest, therefore, to look at the mechanical aspects of knee joint design, loading and performance. Basically, the available prostheses fall into two distinct categories: those providing mechanical restraint against dislocating forces, exemplified by figure 2; and those providing replacement joint surfaces with possibly restraint against medio-lateral linear dislocation. It will be seen in figure 2 that the assembly comprises a single axis hinged connection with an extension stop. The end pieces fit into formed recesses in the femur and tibia and the intramedullary stems are straight and of cross-section which is either of one form or tricuspid. The devices may be fitted using methyl methacrylate cement. If cement is not used the stem may have fenestrations formed in it.

Studies of the loads transmitted at the hip joint have been reported by Inman (1947) and Kummer (1972) and many others whose analysis is based on a plane force system and who have taken no account of the characteristic pattern of variation of ground to foot force in locomotion. Intravital measure-

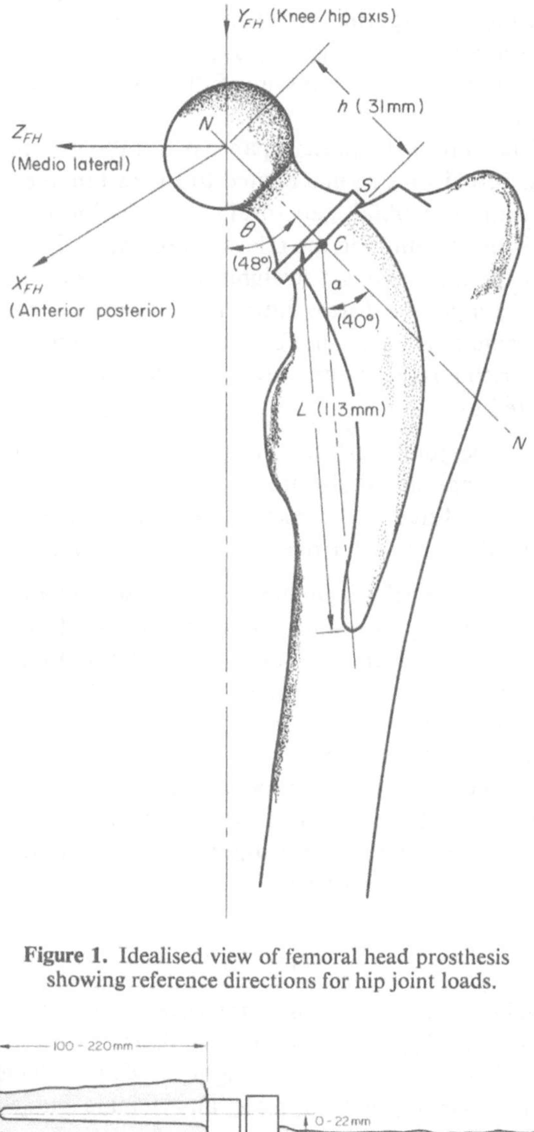

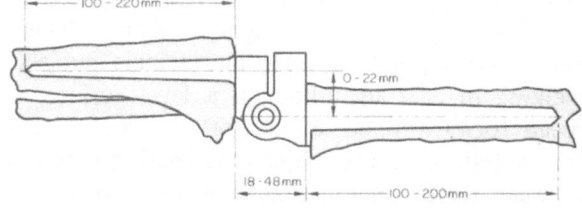

Figure 1. Idealised view of femoral head prosthesis showing reference directions for hip joint loads.

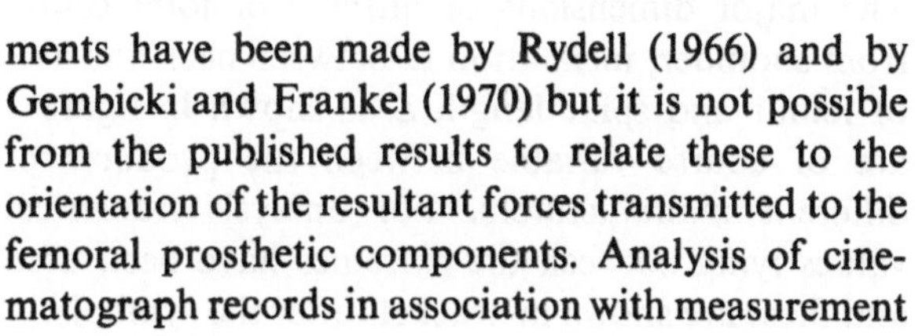

Figure 2. Idealised view of knee prosthesis.

ments have been made by Rydell (1966) and by Gembicki and Frankel (1970) but it is not possible from the published results to relate these to the orientation of the resultant forces transmitted to the femoral prosthetic components. Analysis of cinematograph records in association with measurement

[1] University of Strathclyde, Glasgow, Scotland.

John P. Paul

of ground-to-foot forces is the basis of the results reported for hip joint forces by Paul (1967a, 1967b) and for knee joint forces by Morrison (1967). Paul's results describe the variation of hip joint force components relative to vertical and horizontal axes and are not relevant to the loading of femoral components. Morrison's results are referred to orthogonal axes related either to the vertical or to the instantaneous tibial ones and cannot easily be interpreted to give the loading on the distal part of the femur. Experimental work by Carlsson (1970) should allow further knowledge of the mechanics of hip function and analysis of the ankle is being conducted by Condie (1973).

The aim of the present paper is to present Paul's and Morrison's results related to the instantaneous femoral and tibial axes to allow discussion of the interface loading on the relevant prosthetic devices. For the hip there is a straightforward resolution of the component forces into the appropriate components relative to the femoral axes. For the knee Morrison reports the values of the forces transmitted by

1   the joint surfaces in the medio lateral and longitudinal directions;

2   the force in the relevant cruciate ligament;

3   the force in the relevant collateral ligament.

A knee prosthesis similar to that shown in figure 2 will certainly transmit the forces referred to in items 1 and 2 above. The loading transmitted by the prosthetic replacement due to valgus/varus moment action will depend on the amount of the joint capsule resected and the pessimistic design assumption is that the device transmits all this moment. Morrison's results have therefore been used to calculate the resultant force and moment components on this basis.

RESULTS

The components of hip joint force in the reference directions X_{FH} Y_{FH} Z_{FH} shown in figure 1 vary with time in the manner shown in figures 3, 4 and 5. These curves correspond to male and female subjects walking in a straight line on a level surface at mean forward velocities between 2·1 and 1 m/s (mean 1·46 m/s). The major load-bearing junction between femur and prosthesis is the shoulder S at the neck (figure 1). Patently the hip forces exert more than simple compression at this interface. The major dimensions of offset h of joint centre from shoulder, inclination Θ between neck and axis of femur and spike length L as shown in figure 1 are of course variable between the products of different manufacturers. For analysis numerical values lying between the extremes have been used and these are shown in brackets after the symbols.

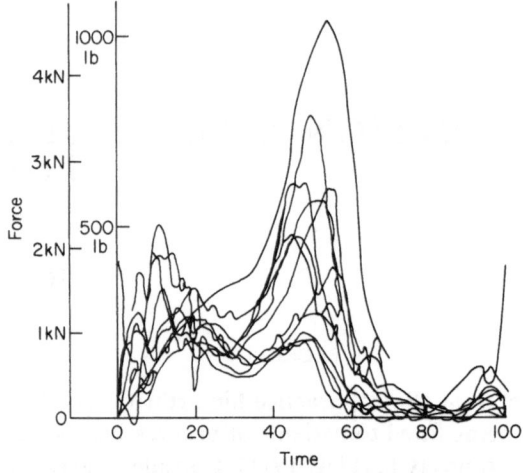

Figure 3. Variation with time of the component force Y_{FH} described in figure 1.

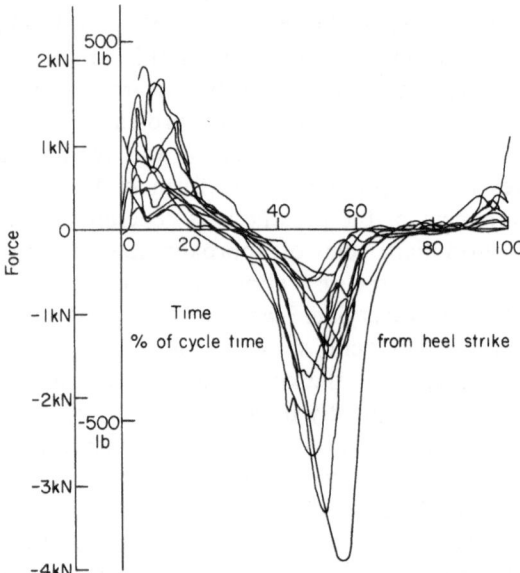

Figure 4. Variation with time of the component force X_{FH} described in figure 1.

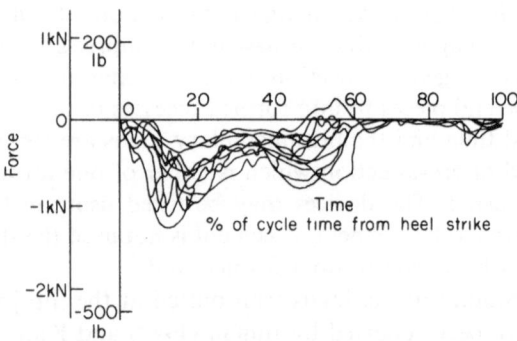

Figure 5. Variation with time of the component force Z_{FH} described in figure 1.

Taking account of the orthogonal forces at the head and the dimensions h and Θ the force system can be expressed as the equivalent actions at S as shown in figure 6 where

$$P = Y_{FH} \cos \theta - Z_{FH} \sin \theta$$
$$F_1 = Y_{FH} \sin \theta + Z_{FH} \cos \theta$$
$$F_2 = X_{FH}$$
$$M_1 = (Y_{FH} \sin \theta - Z_{FH} \cos \theta)h$$
$$M_2 = X_{FH} \times h$$
$$T = 0$$

(neglecting angle of femoral ante-torsion)

Although this indicates no tendency to rotate the prosthesis about the neck axis through C there will be a component of M_2 namely T acting about

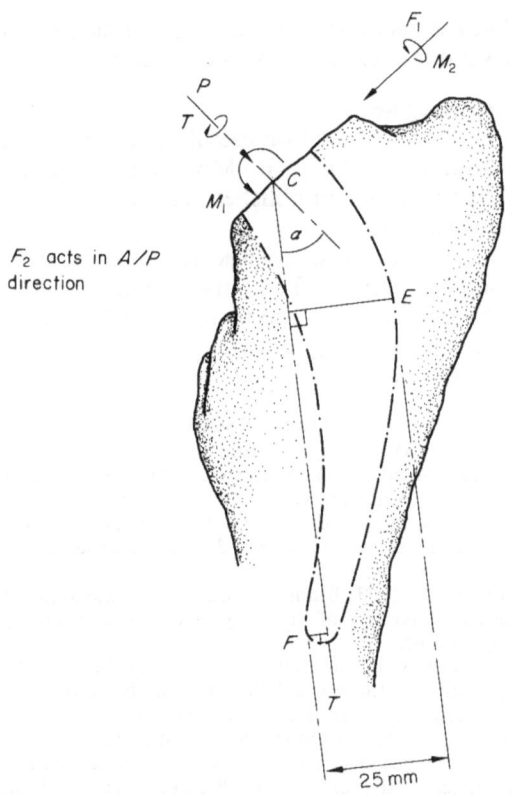

F_2 acts in A/P direction

Figure 6. Typical load actions acting on femur referred to prosthesis shoulder surface.

axis CT on a parallel axis which will tend to loosen the bond between prosthesis and the shaft by a rotational action where:

$$T^1 = M_2 \sin \alpha$$

Taking typical results from the graphs it is apparent that worst conditions occur at approximately 10% and 50% of cycle after heel strike (toe-off and heel strike of the contra-lateral foot respectively).

Taking middle range values from the curves on figures 3, 4 and 5, the magnitudes of load actions at the worst conditions are shown in table 1. The

moments M_1 and M_2 can be presumed to be transmitted by variation in the lateral pressure between prosthesis and the bone. As an extreme design assumption this may be considered to be two equal and opposite forces at the shoulder and at the tip, 113 mm apart. For example, for M_2 at 50% of cycle the maximum value of this force is given by 47 Nm/0·113 m = 0·41 kN. If torsion is assumed to be prevented by similar forces acting tangentially at the points E and F at the extreme distances from axis CT, for the 25 mm dimension shown the torque T^1 would imply forces of 1·2 kN. It is apparent that loading conditions are more severe in torsion than in the other rotational systems.

For the knee joint the loading conditions differ considerably from those at the hip. Figure 7 shows the variation of valgus/varus knee moment relative

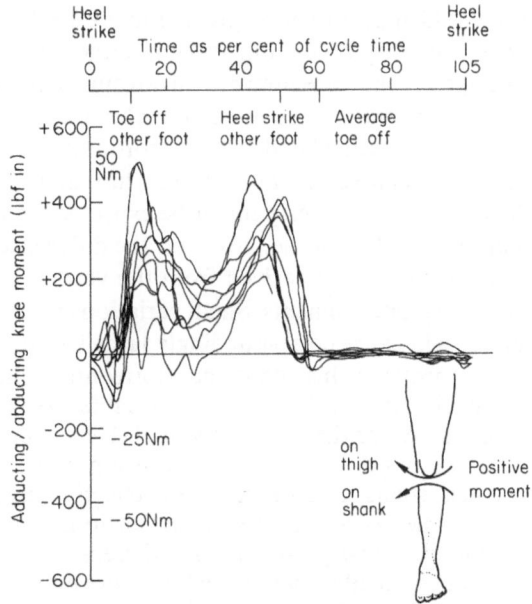

Figure 7. Valgus/varus moments about a horizontal axis through the knee joint. Derived from the results of Morrison (1967).

to fixed vertical and horizontal axes for the test subjects of figures 3–5 (drawn from the results of Morrison (1967)). These do not correspond to the loading on the prosthetic components due to the varying inclination of the limb segments in space. These curves have maxima of 30 Nm at the 10 and 50% cycle points and there is simultaneously a moment about the vertical axis of typical magnitude ±7 Nm. At these instants the thigh may be inclined

Cycle point	P	M_1	M_2	F_1	F_2	T^1
10%	1·25(280)	15(130)	28(250)	0·49(110)	0·90(200)	18(160)
50%	1·64(370)	34(300)	−47(−420)	1·08(240)	−1·5(−340)	−30(−270)

TABLE 1: Magnitudes of load actions at femur/femoral hip prosthesis interface. Units N or Nm (lb and lb in. values in parentheses)

John P. Paul

Cycle point	Femur			Tibia		
	Axial force	Adducting moment	Torque	Axial force	Adducting moment	Torque
10%	1·46(330)	30(270)	6(50)	1·53(340)	31(280)	1·7(15)
50%	1·61(360)	31(280)	0·5(5)	1·70(380)	30(270)	6·8(60)

TABLE 2: Values of load actions at the interfaces between tibia, femur and knee prostheses. Units N and Nm (lb and lb in. values in parentheses)

at 24° in front or 14° behind the vertical respectively while the corresponding shank angles are 10° and 26°. When the load actions are resolved into the appropriate axes relative to the long bone the major values are shown in table 2.

Since the lengths of stems of the knee prostheses are comparable with those for the hip and the moment values are of the same order of magnitude there would appear to be no great difference between them in intensity of loading at the interface. Indeed the design for the knee stem has an advantage in that its flexural stiffness more nearly matches that of the bone shaft. The calculated values of torque appear to be so low relative to those at the hip that little difficulty in torsional grip might be expected. In the calculation of this torque, however, the difference is taken between terms derived from the vertical and horizontal plane moments. Small variations in either quantity will cause correspondingly large variations in the torque value and the horizontal plane moment is known to be very variable between different steps for the same individual (Harper *et al.*, 1961). The major difference between hip and knee prosthesis stems is, however, their effective cross-section. The knee devices have sections which approximate closely to circles, whereas the hip devices are basically rectangles whose sides may be in the ratio 2·5/1·0. The tangential force to resist the maximum torque on a knee prosthesis calculated on the same basis as that for the hip has a value of 0·7 kN compared with 1·2 kN for the hip prosthesis. In view of the fact that the area of material in shear resisting this load may be much less than half that at the hip this appears to be a critical design feature. Two of the available devices have additional pins or flanges formed on the seating shoulder. The pins would appear to assist the grip, but, since the flanges are tangentially situated relative to the axis of the stem, they would appear to be much less efficacious than similar material distributed as radial fins.

SUMMARY

The loading on some hip and knee joint prostheses is analysed.

The load actions at the interfaces indicate that the likely mode of failure is by axial rotation.

Acknowledgments

This work is based on calculations performed by the author and Dr J. B. Morrison and access to Dr Morrison's unpublished results is gratefully acknowledged.

The author also wishes to acknowledge the assistance received from many manufacturers of prosthetic devices in the form of the loan of prostheses and working drawings and the supply of relevant literature.

REFERENCES

CARLSEN, C. 1970: Experimental measurement of pressure distribution on the human hip joint. SESA Spring Meeting, Huntsville, Alabama, U.S.A.

CHARNLEY, J. 1971: The present status of total hip replacement. *Internal Publication No. 30*. Centre for Hip Surgery, Wrightington Hospital.

CONDIE, D. N. 1973: Biomechanics of the ankle joint. Ph.D. Thesis, University of Strathclyde, Glasgow, Scotland (in preparation).

DUFF BARCLAY, I., and SPILLMAN, D. T. 1967: Total human hip joint prostheses—a laboratory study of friction and wear. *Proc. Inst. Mech. Eng.* **181**: 3J: 90.

GEMBICKI, F. W., FRANKEL, V. H., and BURSTEIN, A. H. 1970: AM/FM telemetry of load on a femoral lead fixation appliance. *Internal report*. Biomechanics Laboratory, Case Western Reserve University.

HARPER, F. C., WARLOW, W. J., and CLARKE, B. L. 1961: The forces applied to the floor by the foot in walking *National Building Studies Research Paper 32*. H.M.S.O., London.

INMAN, V. T. 1947: Functional aspects of the abductor muscles of the hip. *J. Bone Jt. Surg.* **39** (3): 607.

KUMMER, B. 1972: Biomechanics of bone. *In* 'Biomechanics: its Foundations and Objectives' (Ed. Y. C. Fung). Prentice Hall, New Jersey.

MORRISON, J. B. 1967: The forces transmitted by the human knee joint during activity. Ph.D. Thesis, Univ. Strathclyde, Glasgow, Scotland.

MORRISON, J. B. 1968: Bioengineering analysis of force actions transmitted by the knee joint. *Bio-Med. Eng.* **3**: 4: 164.

PAUL, J. P. 1967a: Forces transmitted by joints in the human body. *Proc. Inst. Mech. Eng.* **181** (3J): 8.

PAUL, J. P. 1967b: Forces at the human hip joint. Ph.D. Thesis, Univ. Glasgow, Scotland.

RYDELL, N. W. 1966: Forces acting on the femoral head prosthesis. *Acta. Orthop. Scand. Suppl. 88* **37**: 1.

SCALES, J. T., KELLY, P., and GODDARD, D. 1969: Friction torque studies of total joint replacements—the use of a simulator. *In* 'Lubrication and Wear in Joints' (Ed. V. Wright). Sector, London.

BIOMECHANICAL ASPECTS OF ORTHOPAEDIC IMPLANTS

I. L. PAUL, E. L. RADIN, AND R. M. ROSE[1]

INTRODUCTION

Total joint replacement prostheses for implantation into human beings have been developed for the hip, knee, shoulder, elbow, metacarpophalangeal and metatarsophalangeal joints. New designs and modifications of old designs are presently appearing at an increasing rate. There are well over thirty different hip prostheses (and almost as many knees) implanted in patients.

The Swanson metacarpophalangeal joint replacement is one-pieced, silicone-rubber, double stemmed device, the stems of which are loosely inserted into the medullary canals of the bones after the joint has been excised. Because of the loose insertion, sliding takes up some of the bend at the 'joint'. When inserted into patients with rheumatoid arthritis whose grip and function are limited by the multiple-joint involvement of the disease, the results are excellent (Swanson, 1972; Rhodes et al., 1972).

Of total knee joint replacement prostheses, only the metal hinged design has been in use long enough to have an adequate follow-up. Young of the Mayo Clinic reported on a ten-year experience. In the 38 patients, who had this operation, there was a 100% failure rate (Young, 1971).

Since total hip joint replacement has taken place much more extensively than the others, we shall concentrate on this class of prostheses. Insufficient clinical experience is available at this time to permit meaningful comment on metatarsophalangeal, shoulder and elbow replacements. However, the general principles developed should be applicable to all firmly anchored articulated total joint replacements.

INFECTION AND LOOSENING OF TOTAL HIP PROSTHESES

Great Britain has been paramount in the development of total hip replacement techniques. The presently popular all-vitallium McKee–Farrar prosthesis had its origins in the 1940s, and was last revised by its manufacturer in 1968. The first successful metal on plastic total hip replacement was designed by Charnley. His initial experiences, in the late 1950s, with stainless steel on Teflon proved disastrous and all these devices had to be removed. His later experiences with stainless steel on high density polyethylene in 379 hips, which were done as primary procedures over seven years ago, incurred a 4% infection rate, no significant wear, the average being 0·13 mm per year and a 1·3% incidence of mechanical failure. Loosening in one patient was ascribed to excessive use of the reamer in softened bone; in the other instance the patient engaged in athletic activities. It should be stressed that in this early series of Mr Charnley's cases the patients were generally physiologically old and debilitated, and that this series excluded operations done as secondary (salvage) procedures. The results in his patients who have had multiple operations on their hips are less successful (Welch and Charnley, 1970).

In another study of 138 hips replaced by Charnley with his prosthesis, some of which did include revisions of other types of hip surgery, five had infections and three showed wear greater than 3 mm after seven years (Eftekhar and Charnley, 1971).

The experience of others with total hip replacement has not been as promising. Wilson et al. (1971) reported a 21% incidence of complications in 100 Charnley hip joint replacements. Loosening and infection were the two major complications. The incidence of complications in that series increased with time after surgery. Roles (1971) reported a 25% complication rate in his series of Charnley hip joint replacements.

Patterson and Brown (1972) reporting on 386 vitallium on vitallium hip joint replacement prostheses (McKee–Farrar) with an average follow-up time of 1·4 years reported a 15% complication rate, mainly infection and loosening. Again, late infection seemed a persistent problem. Amstutz (1970), Galante (1971), Lazansky (1970) and Morris and Nicholson (1970) have reported similar findings in shorter-term series.

Wilson and Scales (1970), using a hip joint simulator, studied the methacrylate cement fixation of McKee–Farrar prostheses. Their results suggest that high loadings through the hip joint may, on some metal on metal prostheses, result in the development of frictional torque values of such magnitude as to disrupt the bone at the bone-acrylic interface. When this occurs clinical 'seizing' of the articulation is apparent. They referred to the 10% incidence of loosening at Stanmore. On the basis of this work, the surface configuration of the

[1] Engineering in Living Systems Laboratories, Massachusetts Institute of Technology, Cambridge, Massachusetts, U.S.A.

McKee–Farrar design has been altered to prevent 'seizing'.

Late infections are troublesome in that many of them are asymptomatic for a long period of time. The suspicion must be entertained that the cause of these late infections, which are usually aseptic, is the methylmethacrylate cement degrading with time. Certainly, liberated monomer from the acrylic could be sufficiently toxic to the tissues to cause inflammation and this would also explain the relationship between the late infection and loosening which has been reported with fair consistency.

Another problem which appears with time is the release of constituents of the prosthesis throughout the body. It has been shown that one can find quite significant concentrations of cobalt in the hair of patients who have a vitallium on vitallium hip joint inserted (Scott, 1971). Particles made *in vitro* from such prostheses have been shown to be carcinogenic in rats (Heath *et al.*, 1971). The variables of particle size and species sensitivity make this finding not necessarily applicable to man, however.

OPERATING FORCES AND THE EFFECTS OF FRICTION

Assuming biological compatibility (which cannot, at this time, be taken for granted in the long-term situation), success or failure of hip joint replacements will be determined by their long-term response to the forces they operate under. *In vitro* testing in our laboratory of the most popular currently available models (including the revised McKee–Farrar) suggests that lubrication is adequate and that surface effects are manageable (Weightman *et al.*, 1972). Representative friction curves determined in our laboratory (Radin *et al.*, 1970; Weightman *et al.*, 1972) using Paul's (1965) loading profile and by others (Duff-Barclay and Spillman, 1967; Walker, 1971) are represented in figure 1. They show coefficients of friction of the three most commonly used types of hip joint replacements prostheses to be below 0·20 when lubricated with serum or synovial fluid.

Mr Charnley's low friction hip joint replacement device has significantly lower frictional characteristics than the McKee–Farrar and slightly lower friction than the Charnley–Muller. Charnley has designed his prosthesis to accomplish this purpose by limiting the femoral head diameter to 22 mm (Charnley, 1972). It should be obvious that although the frictional forces generated by the prosthesis will be almost independent of this contact area, the frictional torque generated by the prosthesis will be higher with increasing head diameter. However, with increasing head size the contact stresses will be reduced correspondingly.

Frictional torque is of itself of little consequence in normal gait. With the highest measured frictional

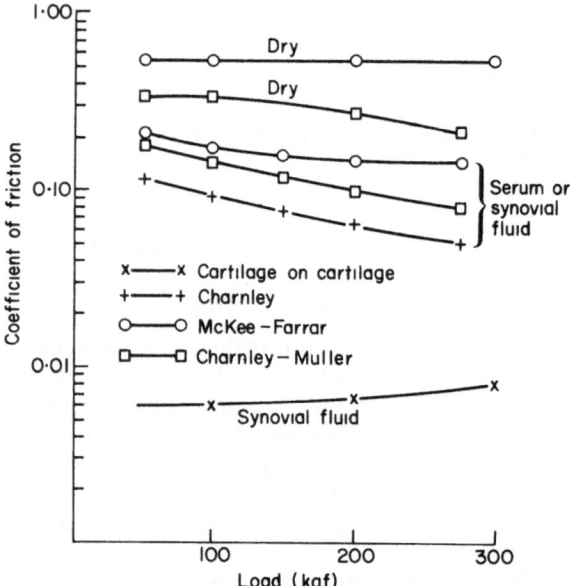

Figure 1. Frictional characteristics of the three most common hip prostheses compared to normal joint friction.

coefficients already quoted above, the additional energy required in gait will be of the order of 2–5%. The frictional torque also has relatively small effects on the stresses on the femoral prosthesis shaft and on the interface between the 'cement' and the bone. If one considers the geometry of the proximal femur operating under the conditions quoted in figure 2, the resulting bending moments, shear

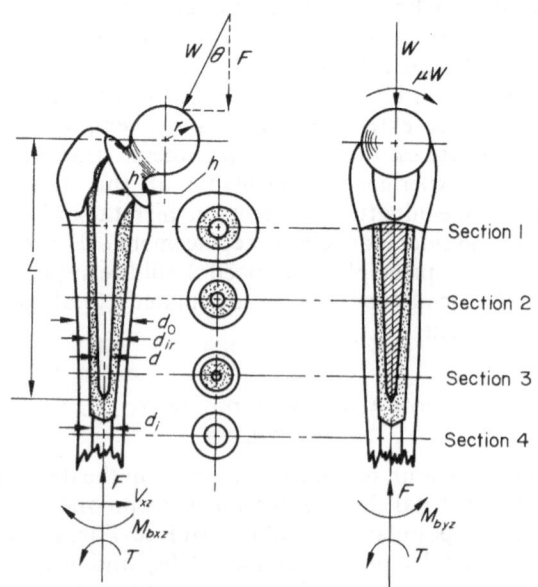

Figure 2. Idealised configuration of a 'cemented' hip prosthesis shown with forces, moments and torques arising in the femoral shaft due to joint reaction forces (normal and tangential to the surface).

forces, and torsion which must be transmitted by the upper femor or prosthetic appliance to the femoral shaft are shown in figure 3. An equivalent force $W = 100$ kg is used to obtain relative values.

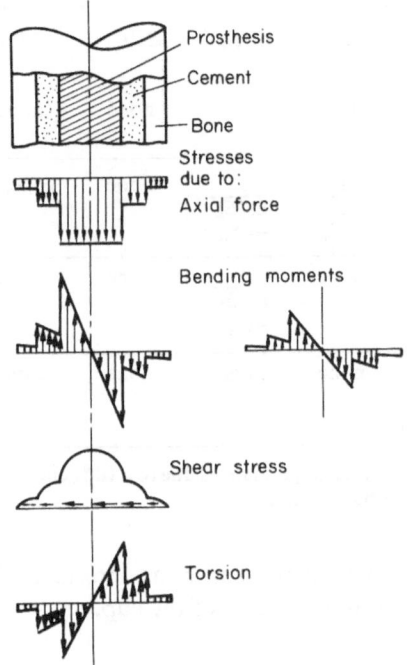

Figure 3. Forces, moments and torques transmitted by a normal femur or by a femur with a total hip prosthesis when loaded as in figure 2.

The contribution to these loads by the frictional forces at the sliding surfaces are shown separately (figures 4 and 5) in order to demonstrate the effects of the higher friction coefficients normally encountered in prostheses as compared to joints.

The resulting stress distributions at four sections of a normal femoral shaft are shown in figure 4. The tensile and compressive stresses at the outside

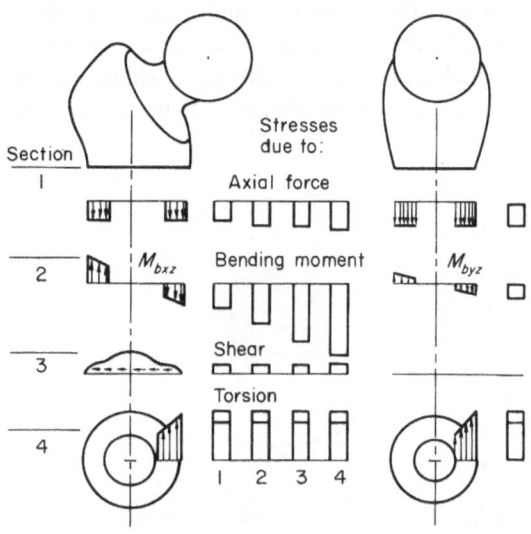

Figure 4. Resulting stress distributions at four sections along the length of a normal femur.

and inside of the femoral shaft represent the sum of the stresses due to axial forces and the bending moments created by the offset of the femoral head from the shaft axis. The tensile and compressive stresses at the anterior and posterior aspects of the

femoral shaft are due to the axial force and to the bending moment created by the frictional force of the femoral head. It should be noted that shear stresses in the femoral shaft arise not only from torsion transmitted by the shaft but also from the bending moments in the two planes (due to the offset femoral head and to frictional forces). These shear stress contributions are maximal at the anterior and posterior aspects and lateral and medial aspects of the shaft respectively. Since these represent shear between adjacent fibres parallel to the short axis, they are critical in determining the bonding strength required at the prosthesis-'cement'-bone interfaces.

Figure 5 shows the corresponding stress distributions for a shaft with a total hip prosthesis bonded

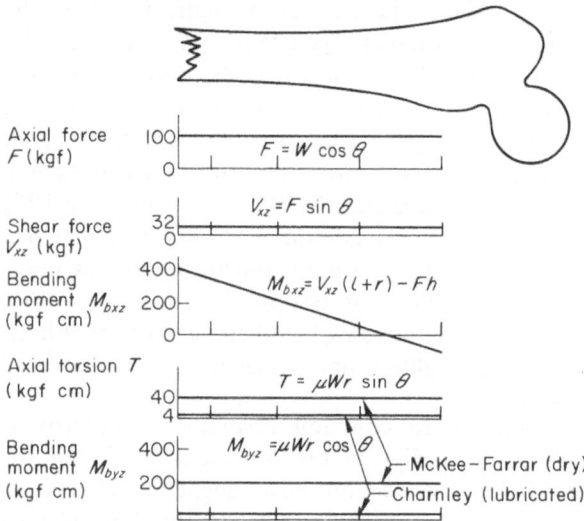

Figure 5. Resulting stress distributions at four sections along the length of a femur with a total hip prosthesis.

in place. The numerical values are for a load of $W = 100$ kg, using the standard mechanical properties for vitallium, polymethylmethacrylate, and bone. Friction coefficients representative of metal-on-plastic and metal-on-metal prostheses are both shown.

The very pronounced effects on the stresses in the remaining femoral shaft due to the introduction of the prosthesis can be seen by comparison with figure 4.

In terms of clinical application the stresses of primary interest are the maximum stress in the remaining bone, the maximum shear stresses at the prosthesis-'cement' and the 'cement'-bone interfaces, and the maximum stress in the prosthesis stem. The effects of the various parameters on these quantities were evaluated in terms of their functional effects as well as percentage changes of these stresses due to a 10% increase in a given parameter. These results are tabulated in figure 6. The functional expressions give an indication of the relationship

	μr	r	d_1/d_{1r}	d	L	E_c/E_B	E_p/E_B	E_c/E_p	h
Max. tensile stress in bone shaft	none	+1%	$(d_1/d_{1r})^3$ -4%	none	$1/L$ -10%	none	none	none	h $+10\%$
Max. shear stress at prosthesis–cement interface	none	none	none	$\dfrac{1}{d^3}$ -16%	$\dfrac{1}{L}$ -6%	$\dfrac{1}{E_c/E_B}$ -4%	none	E_c/E_p $+6\%$	h $+10\%$
Max. shear stress at cement–bone interface	none	none	$\dfrac{1}{(d_1/d_{1r})^3}$ -6%	none	none	$\dfrac{1}{E_c/E_B}$ -6%	none	none	h $+10\%$
Maximum stress in prosthesis stem	none	none	none	$\dfrac{1}{d^3}$ -16%	none	none	none	none	h $+10\%$

Figure 6. Effects of parameter changes on maximum stresses and percentage changes due to a 10% change in parameter (using McKee-Farrar for initial values).

of the effect. The percentage figure indicates the magnitude of the change due to a 10% increase in a parameter. These values can be used to quickly evaluate the effects on the various maximum stresses when a change in design parameter is contemplated.

The values in figure 6 indicate that lengthening of the prosthesis stem by 10% will result in a 10% decrease in the maximum bone stress and a 6% decrease in the maximum prosthesis-cement interface stress. Since it has no effect on the other two calculated stresses, lengthening of the prosthesis stem is desirable if one wishes to lower the overall stress.

Contrary to common assumptions (Charnley, 1972), the radius of the prosthesis head and the coefficient of friction have negligible effects on all of the meaningful stresses in the femoral part of the prosthesis. It should be noted, however, that the friction coefficient times the radius of the head has a direct influence on the maximum shear

stress felt at the cup-bone interface. This may be related to loosening of the cup.

LONG-TERM WEAR OF THE ARTICULATING SURFACES

It would seem, therefore, that friction *per se* is not a crucial consideration, but that other prosthetic characteristics have greater effects on the relevant maximal stresses. The radius of the femoral head component does have a significant effect on the contact stresses which are crucial in determining the wear-tear characteristics of the prosthesis. Wear data from our laboratory (Weightman *et al.*, in press) for the three most commonly used types of hip joint replacements prostheses are represented in figure 7. Assuming a linear rate of wear for the life of the prosthesis, wear would not seem to be a limiting factor. None of these *in vitro* experiments have included impact loading, which has been shown in our laboratories to be an extremely

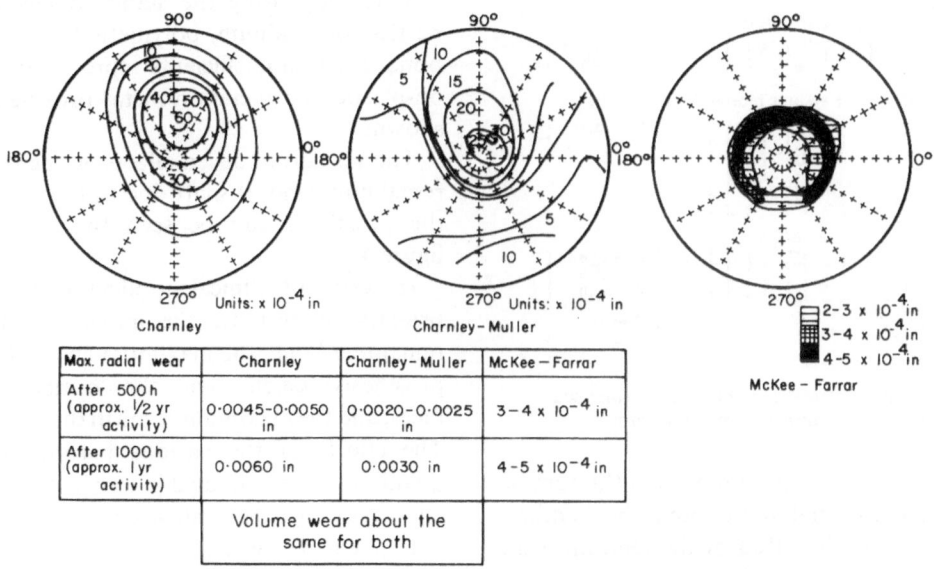

Max. radial wear	Charnley	Charnley–Muller	McKee – Farrar
After 500 h (approx. ½ yr activity)	0·0045–0·0050 in	0·0020–0·0025 in	3–4 x 10^{-4} in
After 1000 h (approx. 1 yr activity)	0·0060 in	0·0030 in	4–5 x 10^{-4} in
	Volume wear about the same for both		

Figure 7. Wear data for the three most common kinds of prostheses obtained on a walking simulator.

$$\tau_{max} \leq \left(\frac{2 \times 10^3}{N}\right)^{1/9} \gamma_R \tau_y$$

Charnley–Muller	McKee–Farrar
$\tau_{max} = \frac{12P}{\pi D^2}\sqrt{\left(\frac{1}{2}\right)^2 - \mu^2}$	$\tau_{max} = 0{\cdot}2 \sqrt[3]{PE^2\left(\frac{D_1 - D_2}{D_1 D_2}\right)^2}$
$\mu = 0{\cdot}05$. $P = 500$ lb $\tau_y = 0{\cdot}054 \times 10^3$ lb in^{-2} $\gamma_R = 0{\cdot}2$	$E = 36 \times 10^6$ lb in^{-2} $\tau_y = 82\,000$ lb in^{-2} $\gamma_R = 0{\cdot}2$
$\tau_{max} = \frac{960}{D^2}$	$\tau_{max} = 6{\cdot}7 \times 10^3 \times \sqrt[3]{\frac{1}{D^4}}$
$\therefore \frac{1}{D^2} \leq 0{\cdot}15$	$\frac{1}{D^{3/4}} \leq 2{\cdot}3$

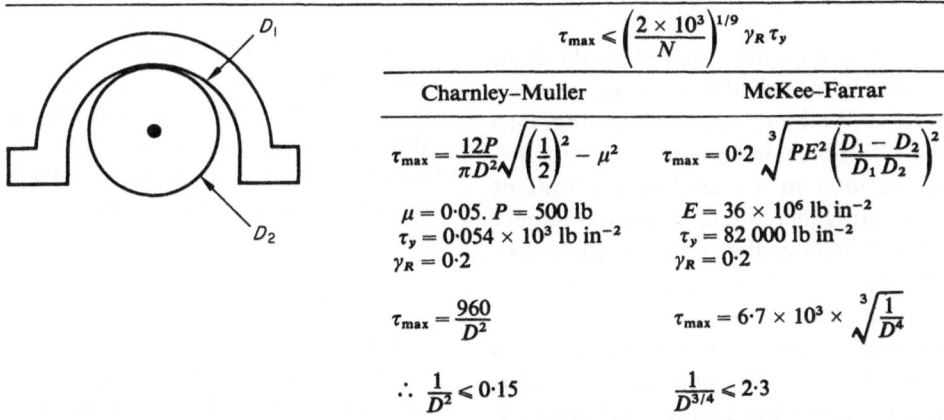

Figure 8. Definition of parameters and calculation of zero wear performance for McKee-Farrar and Charnley-Muller prostheses.

significant determinant in the wear of normal joints, both *in vitro* (Radin and Paul, 1971) and *in vivo* (Simon *et al.*, in press). This may explain the clinical reports citing marked wear in certain individual cases. Long-term effects of the wear debris on the surrounding tissue are still questionable and cannot be easily related to basic principles.

An estimate of the effects of prosthesis parameters on the threshold of wear can, however, be formulated from basic wear principles. One approach which appears to have good experimental and analytical basis is to look at conditions which will result in zero wear. This approach was introduced by the Endicott Development Laboratories of IBM (MacGregor, 1964) and adopted by us (Weightman and Paul, in press) to the particular case of a spherical surface prosthesis. Figure 8 defines these parameters and gives the analytic expressions for this model. The two examples in figure 8 for a metal-on-metal prosthesis can be easily chosen within the practical range to theoretically meet the zero wear conditions while the metal-on-plastic prosthesis is unable to meet the criteria. Effects of parameter changes important in wear considerations are tabulated in figure 9. Although consideration of wear using the 'zero wear' criteria yields some insight into the effects on wear of various parameter changes, experience both in the laboratory (figure 7) and on prostheses which have been examined after use in patients demonstrate that the metal-on-metal prosthesis does show some slight wear. It is too early on the basis of clinical experience to dismiss wear as an insignificant parameter particularly since long-term body reaction to wear debris has not been adequately documented.

In addition to the material wear properties the most important parameters for wear are the radii of the prosthesis head and acetabulum, since these determine the critical shear stresses. Thus, given two prostheses designs of the same materials, the larger the radius of curvature the less the wear.

Parameter	Symbol	Functional effect	Percentage change
Friction coefficient	μ	$\frac{1}{(1/4 + \mu)^2}$	−14%
Head radius	r	$1/r^{3/4} - 1/r^2$	−22%
Yield stress in shear of weaker material	τ_y	τ_y	+10%
Young's modulus	E	None for metal on plastics for conforming geometries $1/E$ for metal on metal (Hertzian contact)	+7%
Difference between unloaded seat diameter and head diameter	D_1/D_2	None for metal on plastic for conforming geometries $1/(D_1/D_2)$ for metals	−6%

When one parameter is changed the others are assumed constant

Figure 9. Effects of parameter changes on zero wear threshold and percentage changes due to a 10% increase in parameter.

THE EFFECT OF NORMAL IMPACT LOADING

Under normal conditions, the relatively high stresses applied to joints will almost always be of an impulsive nature. Walking, running, getting into or out of a chair, climbing, hammering, shovelling—all will load the joint in a repetitive intermittent fashion. The resulting increase in maximum impact force due to the stiffening effect of a prosthesis can be estimated assuming linear force-deflection behaviour. This has been done, and the effects of changing parameters on the maximum impact force transmitted by a femur with a prosthesis are tabulated in figure 10. From these functional effects it can be appreciated that the introduction of a

Parameter	Symbol	Functional effect	Percentage change
Friction coefficient	μ	none	none
Radius of prosthesis	τ	none	none
Distance from head center-line to femur axis	h	$1/h^2$	-14%
Ratio of inside diameter of femur to reamed inside diameter	d_i/d_{ir}	$\dfrac{1}{(d_i/d_{ir})}$	-1%
Prosthesis stem diameter	d	d	$+1\%$
Length of prosthesis stem	$1/L$	$1/L$	-10%
Ratio of cement modulus to bone modulus	E_c/E_B	E_c/E_B	$+10\%$
Ratio of prosthesis modulus to bone modulus	E_p/E_B	E_p/E_B	$+12\%$
Ratio of cement modulus to prosthesis modulus	E_c/E_B	E_c/E_B	$+10\%$

Figure 10. Effects of parameter changes on peak joint forces and stresses during axial impact including percentage changes due to a 10% increase in parameter.

relatively compliant 'cement' between the prosthesis and bone would significantly influence the attenuation of peak dynamic forces. Attempts to design fixation mechanisms to replace the potentially labile polymethylmethacrylate 'cement' now in common use must consider the effects of relative stiffness on the magnitude of the peak dynamic stresses. In addition to acting to attenuate impact, this 'cement' also acts as a stress distributing medium between the relatively stiff prosthesis and the relatively elastic bone (Wagner and De Marnefee, 1968). Metal-to-bone interfaces have potentially high local stress concentrations which can cause resorption of bone and subsequent loosening.

SUMMARY

Of the biomechanical aspects of total hip replacement, loosening of the prosthesis would seem to be a more serious problem than are friction and wear, at least in the context of present technology and practice. However, as younger and more active patients have joint replacements performed, and static and peak dynamic forces on the implants are increased, the significance of friction, wear, and resistance to repetitive impact loading must be kept in mind.

Acknowledgments

The work described in this report was supported by the United States Department of Health, Education and Welfare (Social and Rehabilitation Service) under award RD-3341 MPO, and the Division of Sponsored Research at the Massachusetts Institute of Technology.

REFERENCES

AMSTUTZ, H. C. 1970: Complications of total hip replacement. *Clin. Orthop.* **72**: 123.

CHARNLEY, J. 1972: The long-term results of low-friction arthroplasty of the hip performed as a primary intervention. *J. Bone Joint Surg.* **54B**: 61.

DUFF-BARCLAY, I., and SPILLMAN, D. T. 1967: Total human hip joint prostheses—A laboratory study of friction and wear. *Proc. Instit. of Mech. Eng. London* **181**: 3J: Paper 10.

EFTEKHAR, N., and CHARNLEY, J. 1971: Low friction torque arthroplasty—a study of long-term results. *Clin. Orthop.* **81**: 93.

GALANTE, J. 1971: Total hip replacement. *Orthop. Clin. N. Amer.* **2**: 139.

HEATH, J. C., FREEMAN, N. A. R., and SWANSON, S. A. U. 1971: Carcinogenic properties of wear particles from prostheses made in cobalt-chromium alloy. *Lancet:* 564.

LAZANSKY, M. G. 1970: Complications in total hip replacement with the Charnley technique. *Clin. Orthop.* **72**: 40.

MACGREGOR, C. W. 1964: 'Handbook of Analytical Design for Wear'. Plenum Press, New York.

MORRIS, J. B., and NICHOLSON, D. R. 1970: Total prosthetic replacement of the hip joint in Auckland. *Clin. Orthop.* **72**: 33.

PATTERSON, F. B., and BROWN, C. S. 1972: The McKee-Farrar total hip replacement: preliminary results and complications of 308 operations performed in five general hospitals. *J. Bone Joint Surg.* **54A**: 257.

PAUL, J. P. 1965: Bioengineering studies of forces transmitted by joints. In 'Biomechanics and Related Bioengineering Topics' (Ed. R. M. Kenedi). Pergamon Press, Oxford.

RADIN, E. L., and PAUL, I. L. 1971: The response of joints to impact loading I: *In vitro* wear. *Arthritis Rheum.* **14**: 356.

RADIN, E. L., PAUL, I. L., and POLLOCK, D. 1970: Animal joint behavior under excessive loading. *Nature* **226**: 554.

RHODES, K., JEFFS, J. V., and SCOTT, J. T. 1972: Experiments with silastic prostheses in rheumatoid hands. *Ann. Rheum. Dis.* **31**: 103.

ROLES, N. C. 1971: Complications of total prosthetic replacement of the hip and knee joints. *J. Bone Surg.* **53B**: 760.

SCALES, J. T., KELLY, P., and GODDARD, D. 1969: Friction torque studies of total joint replacements: the use of a simulator. *Ann. Rheum. Dis.* **28**: Supplement, p. 30.

SCOTT, K. T. B. 1971: Forum on metallic surgical implants. *J. Bone Joint Surg.* **53B**: 346.

SIMON, S. A., RADIN, E. L., and PAUL, I. L. in press: The response of joints to impact loading: II. *In vivo* behavior of subchondral bone. *J. Biomech.*

SWANSON, A. B. 1972: Flexible implant arthroplasty for arthritic finger joints: rationale, technique and results of treatment. *J. Bone Joint Surg.* **54A**: 435.

WAGNER, J., and DE MARNEFEE, R. 1968: Mechanical study of the cemented Austin Moore hip prosthesis. *Acta Orthop. Belgica.* **34**: 253.

WALKER, P. S. 1971: The friction of internal artificial joints. *In Proc. Conf. on Human Locomotor Engng.* Instit. of Mech. Eng. London. p. 123.

WALKER, P. S., WILSON, P. D., Jr., and GOLD, S. L. 1971: The tribology (friction, lubrication and wear) of artificial hip joints. *J. Bone Joint Surg.* **53A**: 1660.

WEIGHTMAN, B. O. and PAUL, I. L. (in press): An analytical model for wear applied to the design of total hip prostheses.

WEIGHTMAN, B. O., PAUL, I. L., ROSE, R. M., SIMON, S. R., and RADIN, E. L. (in press): A comparative study of total hip replacement prostheses.

WEIGHTMAN, B. O., SIMON, S. R., PAUL, I. L., ROSE, R. M., and RADIN, E. L. 1972: Lubrication mechanisms of hip joint replacement prostheses. *J. Lubric. Tech.* **94**: 131.

WELCH, R. B., and CHARNLEY, J. 1970: Low friction arthroplasty of the hip in rheumatoid arthritis and ankylosing spondylitis. *Clin. Orthop.* **72**: 22.

WILSON, J. N., and SCALES, J. T. 1970: Loosening of total hip replacements with cement fixation: clinical findings and laboratory studies. *Clin. Orthop.* **72**: 145.

WILSON, P. D., AMSTUTZ, H. C., CZERNIECKI, A., SALVATI, E. A., and MENDES, D. G. 1971: Total hip replacement with fixation by acrylic cement: a preliminary study of 100 consecutive McKee–Farrar prosthetic replacements. *J. Bone Joint Surg.* **53B**: 760.

YOUNG, H. H. 1971: Use of a hinged vitallium prosthesis (Young type) for arthroplasty of the knee. *J. Bone Joint Surg.* **53A**: 1658.

THE LUBRICATION OF HUMAN JOINTS

D. DOWSON[1] AND V. WRIGHT[2]

The remarkable performance of load-bearing human joints is well known, but it has to be admitted at the outset that our knowledge of the mode of operation is far from complete. It might be thought that an answer to the question 'Is the human joint protected by boundary or fluid-film lubrication?' would be readily available, but there is no clear response. The reason for this might be that we are asking the wrong question.

Tribologists find it convenient to classify the mode of lubrication of machine elements as 'boundary' or 'fluid-film', since such representations aid design, lubricant selection, and diagnosis of failures. Yet many lubricated components encounter both forms of lubrication at different times and under different conditions of operation. A journal or thrust bearing might operate under full fluid-film conditions for most of the time, but on starting and stopping, and perhaps during periods of excessive loading, the surfaces come so close together that boundary lubrication dictates the friction and wear characteristics.

The human joint is a dynamically loaded bearing which operates intermittently and it would indeed be surprising if a single mode of lubrication accounted for its tribological characteristics at all times. In this paper the basic properties of the bearing components, the range of operating conditions, and the evidence of particular modes of lubrication under specific conditions will be reviewed.

If the bearing material, articular cartilage, shows signs of distress it becomes necessary to consider the possibility of relieving the condition by improving or changing the lubricant. This is perhaps the most obvious remedial procedure in tribological terms and it is an approach which is frequently adopted in engineering situations. Some progress has been made in recent years in investigating the efficacy of synthetic lubricants in the biological situation, and the present position will be reviewed.

When joint disorder becomes so severe that total replacement is justified, the tribologists' main concern is to find materials with adequate wear resistance and satisfactory strength. It is understandable that the general engineering properties of the materials and overall joint performance in a biological environment should have occupied the major studies of artificial joints, but it is surprising that understanding of the lubrication of such joints within the body is so sparse. Brief mention will be made of the lubrication of artificial joints.

Studies of natural load-bearing joints and the design and development of artificial joints forms an exacting yet exciting field of endeavour for the tribologist working in bioengineering. The pace has quickened considerably in the last decade, but in spite of a considerable increase in knowledge, there are many unresolved problems.

THE LUBRICATION OF NATURAL HEALTHY JOINTS

Figure 1 is a representation of the synovial bearing. It seems probable that both 'fluid-film' and 'boundary' lubrication can be encountered under different conditions and it has to be noted that the range of loads and speeds in the dynamically loaded human joint is considerable. It further appears that once a fluid film has been formed between the cartilage surfaces it provides exceptional resistance to the close approach of the solids during squeeze-film conditions by either the 'boosted' (Dowson et al., 1970) or 'weeping' (McCutchen, 1959) mechanisms, or even both.

Direct measurement of the film thickness is difficult and the conclusion that coherent fluid-films can form under some conditions is based upon calculations (Dowson, 1967) and interpretations of friction measurements (Dowson et al., 1968). It seems to be widely accepted that elastic distortion of the surface will aid the formation of fluid films in many activities, and the form of fluid-film lubrication must therefore be elastohydrodynamic. This mode of lubrication will certainly provide the low friction characteristics typical of many human joints. A range of typical dimensions in joints based upon measurements and calculations is shown in figure 2.

[1] Department of Mechanical Engineering, The University, Leeds, England.

[2] Rheumatism Research Unit, The University, Leeds, England.

Under prolonged loading it seems that the surfaces will come so close together that boundary lubrication must occur. Indeed, it appears from the lack of sudden change in friction characteristics as the severity and time of loading is increased,

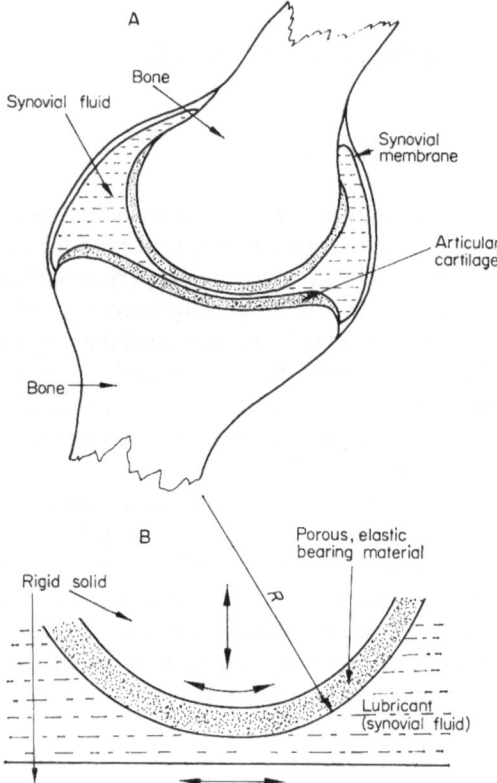

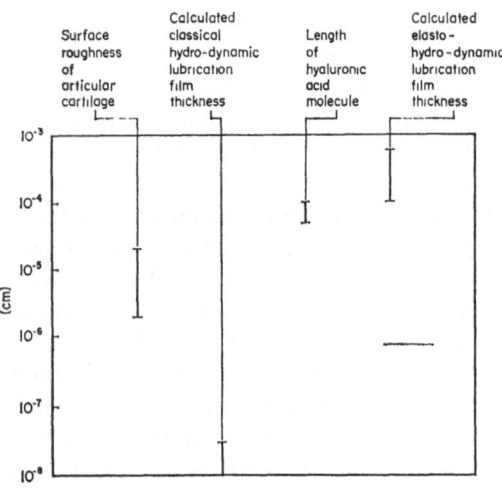

Figure 1. (A) Diagram of a human joint. (B) Model of equivalent bearing.

Figure 2. Typical dimensions of human joints.

that the efficacy of boundary lubrication in human joints is remarkable. It is known (Maroudas, 1967) that gels tend to form on the surfaces of cartilage, and although our knowledge of the surface interactions between synovial fluid and cartilage is increasing, a definitive understanding of the boundary mode of lubrication in synovial joints is still

awaited. This is likely to be a most active area of future research on joint lubrication.

Some of the basic features of load-bearing synovial joints and the conditions under which they operate which are of relevance to studies of both fluid-film and boundary lubrication are now briefly reviewed.

Synovial fluid

The joint lubricant is a dialysate of blood plasma with the addition of a non-sulphated mucopolysaccharide, hyaluronic acid. The chemical and physical properties of the fluid have been discussed by Davies (1967). It appears that the viscous properties of the fluid are determined by its hyaluronic acid content; the coefficient of viscosity being related to hyaluronic acid concentration in an approximately linear manner. The lubricant is highly non-Newtonian, as figure 3 shows. It has been suggested (Charnley, 1959; McCutchen, 1962) that the chemical affinity of the hyaluronic acid-protein complexes for cartilage is a dominant

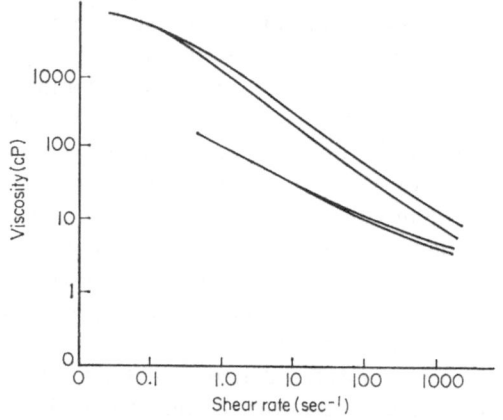

Figure 3. Viscosity-shear rate.

feature of the lubrication of human joints, and this must undoubtedly be true in boundary lubrication conditions. It has also been suggested (Walker et al., 1968) that the migration behaviour of these complexes plays an important role during squeeze-film action and that the surface roughnesses and permeability of the cartilage are also important in 'boosted' lubrication. The gels formed on cartilage surfaces have been studied by Maroudas (1967), and the behaviour of synovial fluid on surfaces of articular cartilage and the aggregation of hyaluronic acid – protein complex on such surfaces has been studied by Walker et al. (1969, 1970). It has been suggested that the structural form and alignment of these aggregates plays an important role in squeeze-film and possibly boundary lubrication.

Articular cartilage

The bearing material in synovial joints has been studied extensively in recent years. It is a permeable, non-homogeneous elastic material. At first (Davies *et al.*, 1967), it was thought that the surface of cartilage was exceedingly smooth, but more recent studies have raised doubts about this view (Walker *et al.*, 1968). It is now thought that the surface quality of normal healthy cartilage lies in the range 20 to 250 millionths of an inch on the c.l.a. scale. Edwards (1967) has drawn attention to the fact that fully saturated cartilage contains 70 to 80 per cent liquid and McCutchen (1962) has deduced that the matrix form of cartilage can be represented by holes of diameter about 60 Å (6×10^{-7} cm). The effective elastic modulus varies with the time of application of the load but McCutchen (1962) has quoted values in the range 10^6 to 10^8 dyn/cm^2.

Joint geometry, loads and speeds

A most important feature of joints in relation to lubrication studies is the degree of geometrical conformity. Geometrical conformity is expressed as the radius of the equivalent cylinder or sphere near a plane. It has been noted (Dowson, 1967) that in natural knee and hip joints this radius lies in the range 2–100 cm.

Paul (1967) has carried out a comprehensive study of loads on the articular surfaces of joints in a walking cycle. The results show that loads up to three or four times body weight can easily be encountered. More recent work at Leeds has shown that forces encountered by the knee joint in simple athletic activities can be much in excess of these figures.

The speeds of movement in human joints appear to lie in the range 0 to 10 cm/sec for normal activities; the motion often being a combination of rolling and sliding.

These values of radii, loads, and speeds show the wide range of conditions encountered in joints. Their importance must be carefully considered in studies of joint lubrication.

It appears that, in common with many intermittently operated bearings, the human joint experiences both fluid-film and boundary lubrication. The form of fluid-film lubrication has been analysed extensively and it appears to be a combination of elastohydrodynamic action with a remarkable ability to prolong squeeze-film times. It is important to note that hydrodynamic, elastohydrodynamic, squeeze-film, boosted, and weeping are all terms associated with features of a single mode of lubrication: fluid-film. The friction of joints under fluid-film lubrication is determined by the viscous characteristics of the synovial fluid. Surface contact and wear is thus minimal under these conditions.

An exact description of the boundary-lubrication mechanism in joints is still awaited. It appears to be related to the hyaluronic acid – protein complex and its attachment and orientation on the cartilage

surface. Radin *et al.* (1970) have suggested that the lubricating fraction of bovine synovial fluid is a protein or glycoprotein.

SYNTHETIC LUBRICANTS

If a synthetic lubricant is to be used to relieve the symptoms of early osteoarthrosis it should possess the necessary rheological characteristics, be resistant to degradation, tolerable within the joint space, and retained there for periods acceptable to the patient. A number of workers have considered the potential of this approach to the lubrication of osteoarthrosic joints, and Helal and Karadi (1968) have reported favourably on the role of silicone fluid in this connection. Nuki *et al.* (1969) have drawn attention to the promising features of polyvinylpyrrolidone whilst Radin *et al.* (1971) have compared silicone, methyl cellulose, and polyvinylpyrrolidone with buffer, serum, and synovial fluid in a friction-measuring machine known as an arthrotripsometer. None of the synthetic lubricants performed as well as synovial fluid, but polyvinylpyrrolidone was slightly superior to both buffer and serum. Silicone fluid gave the highest friction in all cases, a result which is difficult to reconcile with the clinical success with this fluid reported by Helal and Karadi (1968).

Wright *et al.* (1971) have evaluated silicone as an artificial lubricant in osteoarthrosic joints. A pilot study carried out on five inpatients suggested that some relief of pain was produced by silicone 300 cs, and the authors proceeded to a controlled trial on twenty-five outpatients with a total of forty osteoarthrosic knees. In this trial 10 ml of silicone 300 cs was injected into one knee and the results compared with those from a similar injection of 10 ml of normal saline. The results were assessed with no knowledge by the patient or the examining physician of the substance injected. After one week one pair showed a preference for silicone, ten for saline, and nine indicated no preference. The results are shown in figure 4A. After one month the results failed to reach an agreed level of statistical significance. Five patients showed a preference for saline, one for silicone, and fourteen showed no preference, giving the representation in figure 4B.

Knee stiffness was recorded throughout these studies by means of a knee arthrograph, and the authors reported that there was no significant difference between saline and silicone as lubricants. Figure 5 shows an interesting record of the variation with time of the stiffness of the knees of a patient admitted to hospital as part of this study on the effect of injections of the two substances. It shows that admission to hospital itself had a marked effect in reducing joint stiffness and that there was no difference in the effect of the injections.

The reasons for the superior performance of saline over silicone are not clear. It is difficult to see how any direct effective lubricating action could be produced by the saline, although Roberts and Tabor (1971) have reported an interesting electrical

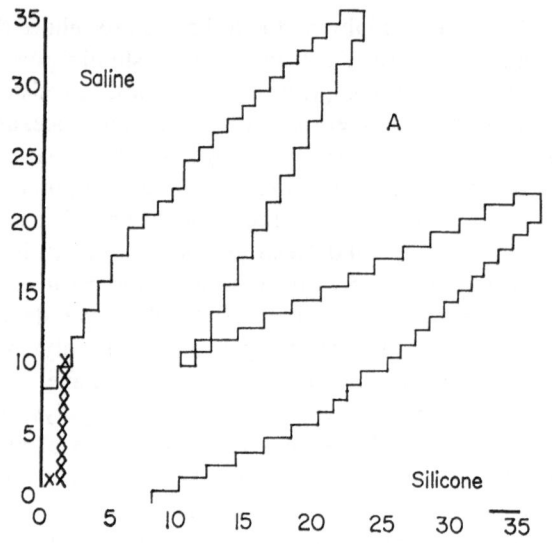

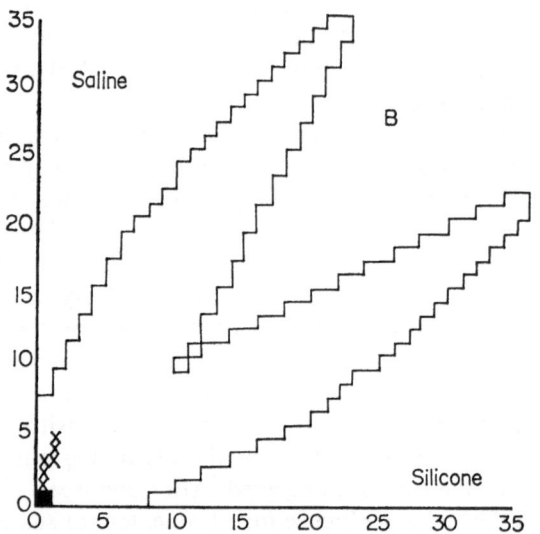

Figure 4. Sequential analysis after treatment with saline and silicone. (A) After one week. (B) After one month.

double-layer effect in the elastohydrodynamic lubrication of soft solids. It may be that the saline, if it is effective at all, stimulates the production of more effective synovial fluid in the joint capsule. At present, prospects for the development of satisfactory synthetic lubricants for osteoarthrosic joints do not appear to be good, although several patients think that they benefit from injections of some fluids. There is a need for further careful study in this field.

THE LUBRICATION OF ARTIFICIAL JOINTS

The combinations of materials employed in load-bearing joints such as the hip and the knee

fall into two categories: metal-on-metal and metal-on-plastic. If both components are made of metal it is generally thought that similar materials should be used to minimise corrosion resulting from electro-chemical action in the body environment. Such an arrangement is normally highly undesirable from the tribological point of view, owing to the ease with which like metals adhere and weld at asperity junctions. The materials most commonly used are cobalt-chrome-molybdenum and stainless steel (En 58J), and ultra-high molecular weight high-density polyethylene, in metal-on-metal and metal-on-plastic prostheses respectively.

Very little work has been reported on the lubrication of these artificial joints *in vivo*. Dowson and Wright (1971) have considered the lubrication of artificial hip joints and have concluded that fluid-film lubrication is unlikely to occur. The surface

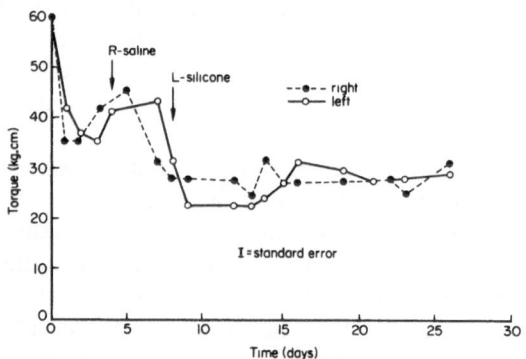

Figure 5. The stiffness of knees of a patient in hospital.

quality of the components suggests that films of thickness greater than 0·3 μm ($1\cdot2 \times 10^{-5}$ in) are required. The relatively rigid materials are unlikely to assist the development of such films under normal operating conditions and the great potential of normal joints for elastohydrodynamic operation is lacking.

The problem of lubrication of artificial joints thus resolves itself into a case of boundary or, perhaps, mixed-film action, but what is the nature and quantity of lubricant available in such joints? There is very little information available on these questions. *Post mortem* examinations have shown that there is usually some fluid present and that it sometimes has the appearance of healthy synovial fluid. However, in some cases known to the authors the joints appear to be operating under 'dry' conditions.

A good boundary lubricant might well reduce the friction and wear in metal-on-metal prostheses, but the situation is more complicated in metal-on-plastic arrangements. It is considered that a transfer film of plastic is usually formed on the metallic component in the latter: a view supported by the observation that the coefficient of friction for a metal-on-plastic is little different from the coefficient for the plastic's rubbing on itself. The metal-on-plastic prosthesis is thus capable of satisfactory operation in the dry condition and it might even prefer dry situations to wet since there is some evidence that fluids can inhibit the formation of transfer films.

SUMMARY

In healthy synovial joints, both fluid-film and boundary-lubrication conditions contribute to the remarkable performance of the bearing. In the fluid-film situation elastohydrodynamic and squeeze-film actions are important. The 'boosted' and 'weeping' lubrication actions which have been proposed for the squeeze-film phase appear to be unique to the biological bearing. The aggregation of gel-like structures on cartilage appears to play an important role in the boundary lubrication of joints and it has been suggested that the efficacy of such lubrication is related to a protein constituent of the synovial fluid.

Attempts to employ synthetic lubricants to relieve the early signs of osteoarthrosis have yet to achieve results which would lead to the wide acceptance of such procedures, but further studies are required before a definitive answer to this issue can be presented.

Little is known about the rheological nature and quantity of lubricant present in artificial joints. Such knowledge is urgently required for the development of materials and designs and the better understanding of the tribological characteristics of replacement bearings for the human body.

REFERENCES

CHARNLEY, J. 1959: The lubrication of animal joints. *Proc. Symp. Biomech.* **12.**

DAVIES, D. V. 1967: Properties of synovial fluid. *Proc. Inst. Mech. Eng.* **181** (3J): 25-29.

DOWSON, D. 1967: Modes of lubrication in human joints. *Proc. Inst. Mech. Eng.* **181** (3J): 45-54.

DOWSON, D., LONGFIELD, M. D., WALKER, P. S., and WRIGHT, V. 1968: An investigation of the friction and lubrication in human joints. *Proc. Mech. Eng.* **182** (3N): 68-76.

DOWSON, D., UNSWORTH, U., and WRIGHT, V. 1970: Analysis of 'boosted lubrication' in human joints. *J. Mech. Eng. Sci.* **12** (5): 364-369.

EDWARDS, J. 1967: Physical characteristics of articular cartilage. *Proc. Inst. Mech. Eng.* **181** (3J): 16-24.

HELEL, B., and KARADI, B. S. 1968: Artificial lubrication of joints: use of silicone oil. *Ann. Phys. Med.* **9**: 334-340.

MAROUDAS, A. 1967: Hyaluronic acid films. *Proc. Inst. Mech. Eng.* **181** (3J): 122-124.

McCUTCHEN, C. W. 1959: Sponge-hydrostatic and weeping bearings. *Nature, Lond.* **184**: 1284.

McCUTCHEN, C. W. 1962: The frictional properties of animal joints. *Wear* (5): 1.

NUKI, G., FERGUSON, J., BOYLE, J. A., and BODDY, K. 1969: Rheological simulation of synovial fluid by a synthetic polymer solution. *Nature, Lond.* **224**: 1118-1119.

PAUL, J. P. 1967: Forces transmitted by joints in the human body. *Proc. Inst. Mech. Eng.* **181** (3J): 8-15.

RADIN, E. L., PAUL, I. L., and WEISSER, P. A. 1971: Joint lubrication with artificial lubricants. *Arthritis Rheumatism,* **14** (1): 126-128.

RADIN, E. L., SWANN, D. A., and WEISSER, P. A. 1970: Separation of a hyaluronate-free lubricating fraction from synovial fluid. *Nature, Lond.* **228**: 377-378.

ROBERTS, A. D., and TABOR, D. 1971: The extrusion of liquids between highly elastic solids. *Proc. R. Soc. Lond. A,* **325**: 323-345.

WALKER, P. S., SIKORSKI, J., DOWSON, D., LONGFIELD, M. D., and WRIGHT, V. 1969: Behaviour of synovial fluid on surfaces of articular cartilage; a scanning electron microscope study. *Annals of the Rheumatic Diseases* **28** (1): 1-14.

WALKER, P. S., UNSWORTH, A., DOWSON, D., SIKORSKI, J., and WRIGHT, V. 1970: Mode of aggregation of hyaluronic acid protein complex on the surface of articular cartilage. *Annals of the Rheumatic Diseases* **29** (6): 591-602.

WRIGHT, V., HASLOCK, D. I., DOWSON, D., SELLER, P. C., and REEVES, B. 1971: Evaluation of silicone as an artificial lubricant in osteoarthrosic joints. *Br. Med. J.* **2**: 370-337.

A NOTE ON WEEPING LUBRICATION

C. W. McCUTCHEN[1]

Professors Dowson and Wright say that joints enjoy 'either the boosted or weeping mechanisms, or even both'. It would be diplomatic for me to join the professors in their pan-theorism. But boosted and weeping lubrication are based on diametrically opposed deductions about the flow of liquid through the rubbing surfaces of the cartilages. They cannot coexist. If joint lubrication is worth studying at all, if learning about it is to benefit future sufferers from arthritis, the Leeds group and I should clearly delineate our differences rather than paper them over.

When boosted lubrication was first described (Walker *et al.*, 1968), the Leeds group said that it occurred 'by fluid entrapment between irregularities of the cartilage surface, the squeezing of water through the small pores of the cartilage, and the production of concentrated synovial fluid on the cartilage surface.'

The theory is compounded of three elements: surface irregularities, escape of fluid through the cartilage pores, and the presence of molecules too big to go through the pores. But we know that when cartilage is lubricated by physiological saline rather than synovial fluid friction is still very low, though not so low as before. The big molecules that were one of the theory's three legs are now missing. Surely the theory falls.

But the Leeds group explained (Longfield *et al.*, 1969), that 'It is only in the final stages of "squeeze-out" that the presence of hyaluronic acid molecules becomes absolutely essential...' so the big molecules were not very important after all, and the irregularities of the cartilage surface which used to entrap fluid now let it escape '... through the fine network of passages formed between the contours of the surface of the cartilage and the flat slideway.' The only part of the theory to remain steadfast is the escape of liquid through the cartilage.

But speeding up this presumed escape by using cartilage in a thin layer supported on a porous backstop hurts rather than helps lubrication. The Leeds group says (Longfield *et al.*, 1969) that this results from '... a more rapid passage of fluid through the cartilage and a more rapid rise in the viscosity of the remaining fluid in the trapped pools...' which cause '... a quicker rise in the frictional force.' The flow through the cartilage is now bad rather than good, and so are the big molecules, because without them the viscosity would not rise.

Each of the three elements of the theory has somewhere or other been said to produce contrary effects. To borrow a simile from nuclear physics the theory has met its anti-theory and the two should have annihilated each other. They did not, no doubt because the contradictory interpretations were never present at once, but followed each other as the occasion demanded.

Has boosted lubrication any property that does not change? It has. Note that fluid flow through the surface of loaded cartilage is assumed to be from the squeeze film into the cartilage.

Flow in this direction would be disastrous for lubrication because it would rapidly deflate the squeeze film. But the direction is wrong. Consider what happens when two cartilages are squeezed together. Very soon the high spots on one rubbing surface meet the high spots on the other. Liquid flows laterally along the partial squeeze film surrounding the high spots and also through the bulk of the cartilage. If the former flow is faster than the latter the pressure in the squeeze film will drop below that in the bulk cartilage and liquid will flow from bulk to film.

This automatic replenishing of the partial squeeze film is called weeping lubrication (McCutchen, 1959). It makes the partial squeeze film last far longer than it otherwise would. For it to happen, the resistance of the crack between cartilages to flow parallel to the cartilage surfaces must be less than that of a slice of bulk cartilage material of the same thickness. By the Leeds group's own data (Longfield *et al.*, 1969) the undulations of the cartilage surface are 200 or more times bigger than the pores in cartilage are wide (McCutchen, 1962). The flow resistance of the crack is at least $(200)^2 = 40\,000$ times that of an equally thick slice of bulk cartilage so the condition is satisfied. The liquid that flows through the crack is more viscous than that which flows through the bulk, but not much, especially at the high shear rates obtaining in the crack.

For the flow to go from crack to bulk as boosted lubrication demands, the surface would have to be at least 200 times smoother than it is. Ergo the fluid goes the way that weeping lubrication says it does.

[1] National Institutes of Health, Bethesda, Maryland, U.S.A.

By ignoring this free transfusion to the squeeze film the Leeds group misses the real cause of the long squeeze film times.

To be sure, their idea of which way the liquid flows has an intuitive appeal. In common experience liquids move and solids stand still. If one cartilage pushes on another will it not push the other's liquid back into the cartilage sponge instead of the other way round? No, because the sponge is deformable and is pushed back into the liquid as we have just worked out; as a demonstration illustrates.

A cellulose sponge is tightly encased in a strong polyethylene bag. The sponge is soaked full of water. One face of the bag is perforated so it can weep. The other is not. Both faces are, in effect, roughened by encasing the bag in a nylon stocking. This effectively provides a weeping and an ordinary bearing, back to back so they can be compared one against the other. The double bearing is placed between glass plates and one of these plates is twisted relative to the other. As both bearings are soaking wet, the sliding takes place at the bearing that has the lower friction. This is seen to be the weeping bearing.

It is also seen (by the increasing puddle) that water is flowing *from* within the imitation cartilage (the cellulose sponge) *toward* the mating glass plate as the weeping theory suggests it should. Boosted lubrication would have it flow away from the glass into the cartilage.

Eventually the supply of liquid becomes exhausted and the weeping bearing loses its advantage. If now the load is lowered the weeping bearing has for a while the greater friction as its sponge sucks water out of the squeeze film, making the hydrostatic pressure there negative and clamping the rubbing surfaces together. This behaviour also obtains with cartilage.

(See Professors Dowson's and Wright's comments in the Discussion, p. 124.)

REFERENCES

LONGFIELD, M. D., DOWSON, D., WALKER, P. S., and WRIGHT, V. 1969: *Bio. Med. Eng.* 4: 512.
McCUTCHEN, C. W. 1959: *Nature* **184**: 1284.
McCUTCHEN, C. W. 1962: *Wear* 5: 1.
WALKER, P. S., DOWSON, D., LONGFIELD, M. D.. and WRIGHT, V. 1968: *Ann. Rheum. Dis.* 27: 512.

INTRAVITAL MEASUREMENT OF FORCES IN THE HUMAN SPINE: THEIR CLINICAL IMPLICATIONS FOR LOW BACK PAIN AND SCOLIOSIS

ALF NACHEMSON AND GÖSTA ELFSTRÖM[1]

The orthopaedic surgeon works with the supportive system of the human body and from this it follows that he must think along the same lines as many engineers—in terms of angles, of stability and motion, of static and dynamic loads. He is basically best trained in biology, but many advances in this field of medicine in the past, and with certainty also in the future, have and will be achieved by close cooperation with engineers who have become interested in orthopaedic problems.

LOW BACK PAIN

Among the most common disabling diseases in the modern societies today is low back pain (Hult, 1954; Horal, 1969). Society's economic loss from this relatively benign ailment is enormous. Much has been written about this disease, yet the cause of it and henceforth the proper treatment is largely unknown (Armstrong, 1952; Hanraets, 1959; Naylor, 1962; Peyron, 1967).

Of all the different structures low in the human back which have at various times and by various authors been blamed for pain, the intervertebral disc alone (figure 1) seems to be the offending structure. The evidence is only indirect, however, and is as follows.

1. Sciatica, which is known to be caused by disc materials herniating out from the disc and pressing on the nerve root, is always preceded by one of several low back pain attacks (Hult, 1954; Hanraets, 1959; DePalma and Rothman, 1970).

2. Patho-anatomical studies have revealed that of all the structures in the back, the intervertebral disc alone shows signs of degenerative changes, and these occur at the same periods of life as low back pain, i.e. between the ages of 25 and 50 years (Friberg and Hirsch, 1949; Hirsch and Schajowicz, 1952; Naylor, 1962; Peyron, 1967).

3. It is often possible in patients who have had low back pain to cause pain, of the same type as experienced earlier, by injections into the disc (Hirsch, 1948; Lindblom, 1950; DePalma and Rothman, 1970).

4. Experiments performed during surgery under local anaesthesia (Hirsch *et al.*, 1963; DePalma and Rothman, 1970) and after surgery (nylon threads tied around different structures in the back (Smyth, 1958) were pulled after two to three weeks) have demonstrated that when the posterior part of the disc is touched or pulled, pain of the same type as experienced earlier is elicited.

Since the majority of our patients claim that the onset of their symptoms occurs in connection with lifting heavy weights or similar activity (Hult, 1954;

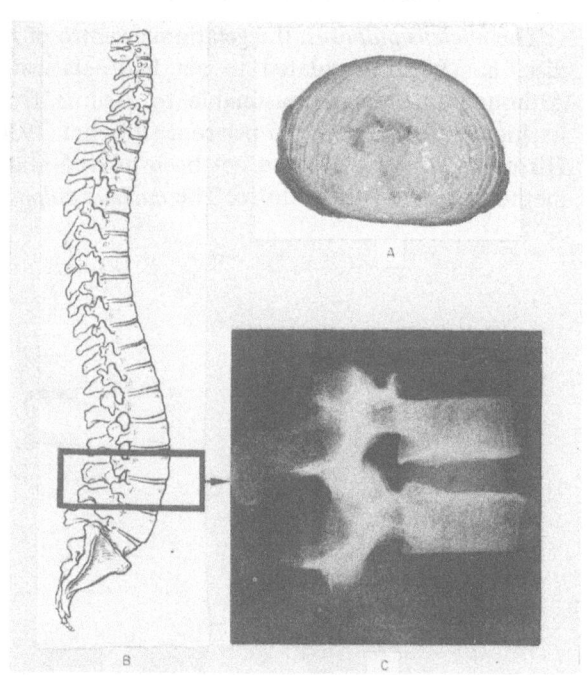

Figure 1. (A) Lumbar disc, transversally cut, with the gelatinous centre, the *nucleus pulposus* and the surrounding fibrous structure, the *annulus fibrosus*. (B) Schematic drawing of the lumbar spine. (C) Roentgenogram of a motion segment (L3).

Hanraets, 1959; Horal, 1969), and since all the sufferers claim an increase in severity of their symptoms following such tasks, it is easily understood that doctors have been interested in the loads to which the lumbar spine is subjected.

It has been said that low back pain is the price we have paid for becoming erect. The veterinaries

---

[1] Department of Orthopaedic Surgery I, University of Göteborg, Sweden.

have, however, demonstrated the occurrence of macroscopic and microscopic disc degeneration and syndromes similar to sciatica in many animal species (Olsson, 1951; Hansen, 1959). Obviously the quadruped animals also suffer. Mechanical factors are however of importance since in man as well as in animals those regions of the spine that are subjected to the heaviest mechanical stresses are those that most often show symptoms.

Theoretical calculations of the load on the lumbar spine have been much too high; forces above the known fracture load of cadaveric and living lumbar spine have been calculated (Armstrong, 1952; Bayer, 1954; Perey, 1957; Hirsch and Nachemson, 1961). The high values have been observed by a number of investigators (Bartelink, 1957; Morris *et al.*, 1961; Eie, 1966), during demonstration of the load-relieving effect of increased intra-abdominal and intra-thoracic pressure. Nevertheless, their calculated loads were still excessive. It is therefore clear that direct measurements should be of definite value to solve this problem.

MEASUREMENTS OF DISC PRESSURE

The *nucleus pulposus*, the gelatinous centre of the disc, has been postulated to act hydrostatically. Although this seems reasonable to assume from its high water content, 85 per cent (Püschel, 1930; Hirsch *et al.*, 1952), it had not been proved and a method was evolved to do so. The *nucleus pulposus*

was punctured with a specially constructed hollow needle (figure 2) connected to an electro-manometer by elastic polyethylene tubing. By turning the opening of the needle into the three directions of principal stress in the loaded disc it was possible to prove that the nuclei of normal and nearly normal discs behave hydrostatically (Nachemson, 1960, 1962, 1963). Pressures in the normal discs were 1·5 times the vertical load applied per unit area and there was a linear increase in pressure for external loads up to 300 kgf. It was also possible to relate the forces acting in the nucleus to the forces acting in the *annulus fibrosus*, the fibrous structure surrounding it. The vertical stress in this structure is approximately 50 per cent of the applied external load per unit area while the tangential, tensile strain is four to five times the applied external load (figure 3). In the cadaver disc there exists an intrinsic pressure of about 1 kgf/cm^2 and it was also demonstrated that the *ligamentum flavum*, situated between the posterior arches and facets, prestresses the disc by a force of around 1·5 kgf (Nachemson and Evans, 1968). Because this ligament is situated at a distance from the motion centre of the disc (Rolander, 1966) it creates an intradiscal pressure of about 1 kgf/cm^2. In this manner the disc is prestressed, providing some intrinsic stability of the spine.

The above-mentioned *in vitro* experiments provided a basis for intravital disc pressure studies (figure 2). In the first series of experiments in 50 subjects a needle built on the above-described

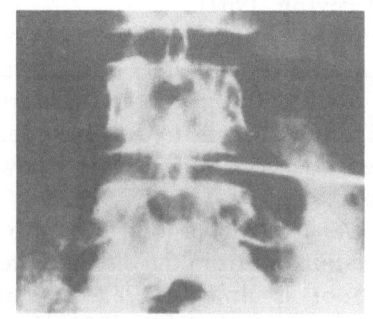

Figure 2. (A) Schematic drawing of method for measuring intradiscal pressure *in vitro*. (B) X-ray showing needle in position for *in vivo* tests. (C) Schematic drawing of method for measuring intradiscal pressure *in vivo*.

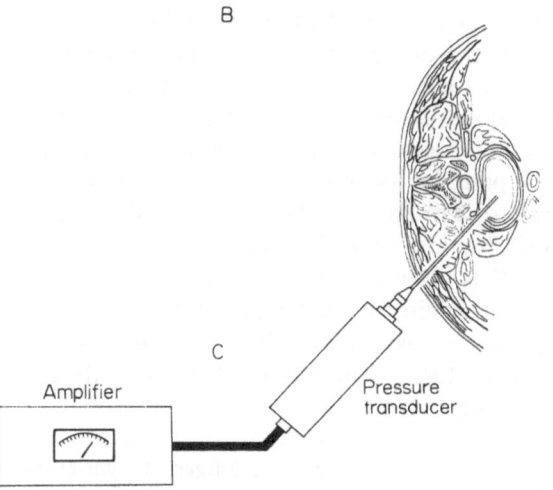

principles was used, and different static positions of the body were studied (Nachemson and Morris, 1964). A summary of the results is seen in figure 4, where the total load of the third lumbar disc in a

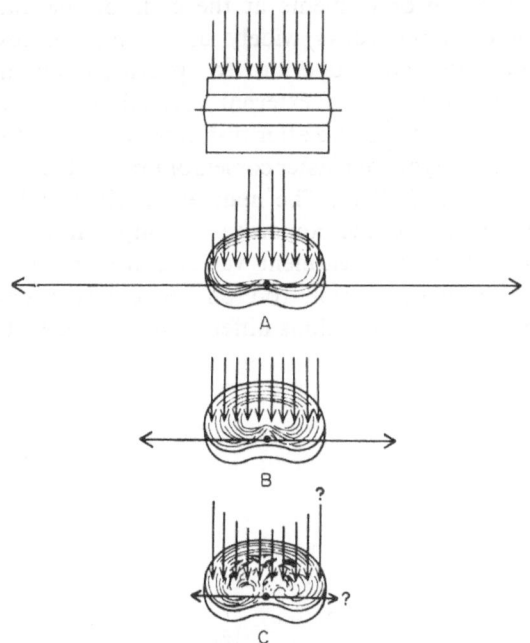

Figure 3. Approximate relationship between vertical stress and tangential strain in different parts of (A) normal, (B) degenerated, and (C) very degenerated lumbar discs.

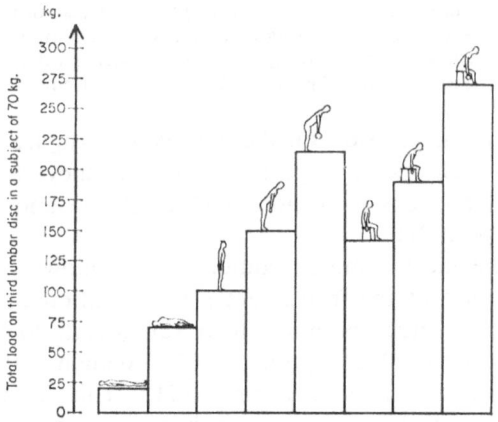

Figure 4. Total load on the L3 disc in different positions in a subject weighing 70 kg. Positions shown are (1) reclining (relaxed, supine), (2) reclining (lateral decubitus), (3) standing upright, (4) standing and leaning 20° forward without and (5) with a 20-kg load in arms, (6) sitting upright, arms and back unsupported, (7) sitting and leaning 20° forward without and (8) with a 20-kg load in arms.

subject of 70 kg has been calculated from the data obtained.

The system used did not allow for any dynamic measurements because of the poor dynamic characteristics of the polyethylene membrane. With the help of Toyota Research and Development Com-

pany a new pressure gauge has been developed. With this the intradiscal pressures are measured by means of a subminiature pressure transducer, the operation of which is based on the piezoresistive effect of semiconductor strain gauges (figure 5). The transducer is connected as a Wheatstone bridge. The change in resistance due to pressure changes on the diaphragm causes an out-of-balance current from the bridge. The output from the bridge is amplified in an integrated circuit amplifier which is in turn connected to a recorder.

With this subminiature pressure transducer it has been possible to measure the pressures in the disc and calculate the loads on the lower lumbar spine in some common movements, manoeuvres, and therapeutic exercises (Nachemson and Elfström, 1970). Some of the results are seen in tables 1 and 2.

Such information is of direct value when treating

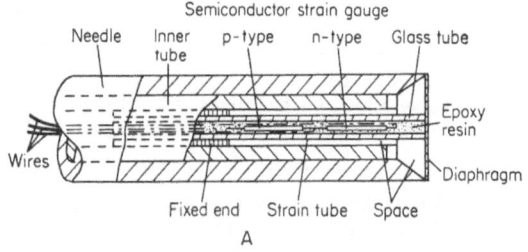

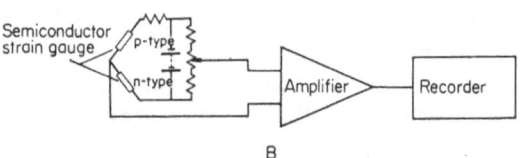

Figure 5. (A) Schematic drawing of the subminiature pressure transducer. (B) Principal diagram of the pressure-measuring system.

the patient with low back pain, since we now know more about the forces to which the patients' discs are subjected when they perform various tasks and exercises (Lidström and Zachrisson, 1970). Also the data could help to clarify the aetiology of low back pain, since with the knowledge gained from the autopsy experiments the tensile stresses in the posterior part of the *annulus fibrosus* can be calculated (figure 3). The strength of these posterior fibres is in some instances as low as 100 kgf/cm² (Galante, 1967). As shown in figure 6, lifting a weight the 'wrong way' will increase the pressure to 35 kgf/cm², which gives a tensile strain in the posterior part of the *annulus* of around 105 kgf/cm². At present it is not known whether such ruptures, which can obviously occur in everyday life, in themselves cause pain or if they allow irritating materials to leak out from inside the disc (Naylor, 1962;

Feffer, 1963; Hirsch *et al.*, 1963; Peyron, 1967; Nachemson *et al.*, 1968; Nachemson, 1969).

SCOLIOSIS

Another classical orthopaedic problem is scoliosis, i.e. a lateral curvature of the spine (figure 7), which occurs most frequently in adolescent girls (Bouillet and Vincent, 1967; James, 1967). Until we know

Position/movement/manoeuvre	Load (kgf)
Supine	50
Standing	100
Upright sitting, no support	140
Walking	115
Twisting	120
Bending sideways	125
Coughing	140
Jumping	140
Straining	150
Laughing	150
Bending forward 20°	150
Lifting of 20 kg, back straight, knees bent	185
Lifting of 20 kg, back bent, knees straight	390

TABLE 1: Approximate load on L3 disc in 70-kg individual in different positions, movements, and manoeuvres

more about the etiology of this disease, we are limited, with more severe curvatures, to relatively crude methods of treatment. These consist of different mechanical ways of correcting the spine and holding it so that a posterior or anterior long-fusion mass heals over the corrected curvature (Moe, 1958; Cotrel, 1966; Hall, 1970; Stagnara *et al.*, 1970).

Position/movement/manoeuvre	Load (kgf)
Standing	100
Bending forward 20° with 10 kg in each hand	215
Supine	50
Supine in traction (30 kg)	35
Bilateral straight leg raising, supine	150
Sit-up exercise with knees bent	210
Sit-up exercise with knees extended	205
Isometric abdominal muscle exercise	140
Active back hyperextension, prone	180

TABLE 2: Approximate load on L3 disc in 70-kg individual in different positions and exercises

If this bone heals it holds the spine in a straighter position for the future. Smaller curvatures for which no operation is needed can be somewhat corrected by the use of external braces or plaster casts, of which the Milwaukee brace is regarded the most satisfactory (Blount, 1964; Moe and Kettleson, 1970; Stagnara *et al.*, 1970). These external braces,

however, have to be worn all day and all night until the patient is skeletally mature around the age of 17 years.

The Harrington method (Harrington, 1962, 1967) utilising an internal distraction rod applied to the posterior elements at the ends of the curved spine, is nowadays widely used, and the results obtained are superior to previous operative methods, in which external correction and maintenance of the curve after fusion were achieved with different types of plaster corsets or braces (Goldstein, 1969; Hall, 1970). The manner in which different surgeons use the apparatus during surgery and post-operative treatment varies considerably. The use of a distraction rod on the concave side is universal, but opinions differ as to the necessity of

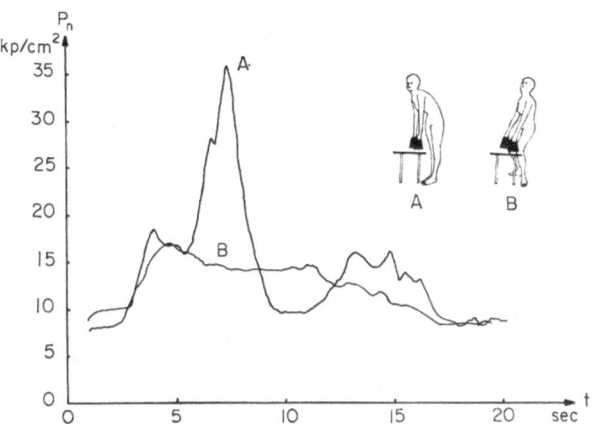

Figure 6. Pressure recorded from the L3 disc in a 25-year-old male lifting 20 kg (A) with bending of back and knees straight, (B) with back straight and bending of knees. (1 kp/cm² = 1 kgf/cm²).

using compression on the convex side (figure 10). The post-operative period of recumbency ranges from 6 weeks to 6 months (Nachemson and Elfström, 1971).

The most common complication, following the insertion of the instruments, is dislocation of the upper distraction hook with subsequent loss of correction (Harrington, 1962; Waugh, 1966; Nachemson and Elfström, 1971). This is most frequently due to the relatively poor resistance to axial load of the upper purchase site.

AXIAL FORCE MEASUREMENT IN HARRINGTON RODS

Autopsy studies have provided approximate figures for the maximum load the different purchase sites can withstand without fracturing (Harrington, 1962; Waugh, 1966). Measurements of the forces occurring in the Harrington distraction rod during operation have been made using rods with wire-connected, built-in strain gauges, but because technical difficulties, post-operative measurements have been unsuccessful (Harrington, 1962; Waugh, 1966; Hirsch and Waugh, 1968). From these

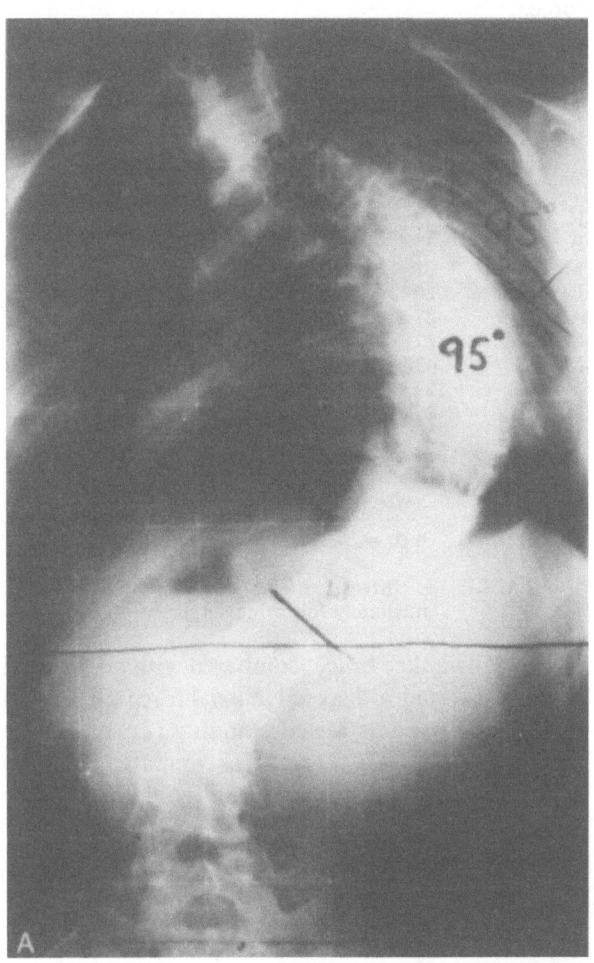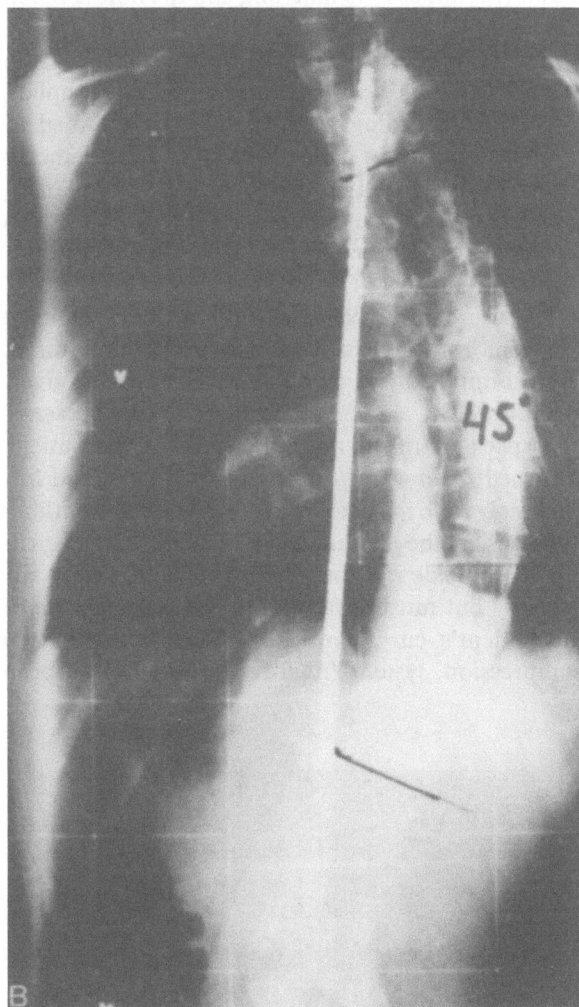

Figure 7. (A) 19-year-old female with a thoracic idiopathic scoliosis of 95°. (B) Result three years following correction with the Harrington distraction rod and fusion.

measurements an upper tolerance limit for the distracting force applied has been evaluated, and information has been obtained on some important steps of the correction. Many other questions, especially about what happens during the post-operative period, were not answered.

There has been a definite need to follow the axial load decline in the rods as well as to determine what loads are induced by different types of movements and exercises which are regularly performed post-operatively by these patients.

As part of our equipment for intravital measurements we used a 'pressductor', invented by O. Dahle of the Swedish Company ASEA, the functioning of which is based on changes in mutual inductance between two coils (Nachemson and Elfström, 1971). The 'pressductor' is a transformer in which the primary and secondary windings are so placed that the coupling between them is zero in the mechanically unloaded state. If the 'pressductor' is subjected to compressive stress, the permeability in the direction of the stress is reduced, and the magnetic flux is deviated to the transverse

direction by an amount proportional to the impressed force. If an alternating current is applied to the primary winding, a voltage approximately proportional to the impressed force is obtained from the secondary winding (figure 8).

The transducer developed for our experiments is a stainless-steel cylinder in which one part of the rod slides to apply axial forces to the 'pressductor'

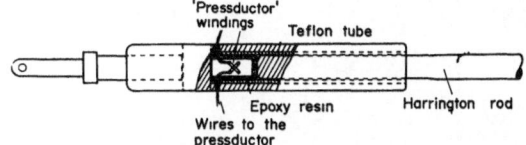

Figure 8. Schematic drawing of the force gauging 'pressductor' in the modified Harrington distraction rod. The upper part (right) with the notches (not shown) slides down towards the 'pressductor' and is available in different lengths.

(figure 8) encapsulated in the plastic, Araldite D. In order to minimise losses by friction between the rod and the steel cylinder, the inside of the cylinder is lined with a layer of Teflon. Since the sleeve of the steel cylinder is 45 mm long and the hole is made with a small tolerance, bending of the rod in the cylinder can be neglected. The transducer itself (the 'pressductor') is sensitive only to axially applied force. The recordings can therefore be assumed to be measurements of axial load alone. The telemetry system consists of two main parts: the implanted transmitter and the external receiver, which also includes a generator for power transfer to the implanted transmitter (figure 9).

The use of this telemetry system on 11 patients has demonstrated many points which have direct clinical implications, and a few examples are given in tables 3, 4, and 5. From these it can be seen: that special care should be taken when the patient is turned from the operating frame to the bed immediately following surgery (table 3); that breathing exercises can be instituted early (table 4); that turning to the side can be performed with safety the first day but must always be to the concave side of the patient's curvature (table 5). The use of the compression system (figure 10), which has been

discussed, is probably not indicated (table 6). Figure 11 shows the gradual decrease with time of the axial force.

Patient	Maximum load during distraction (kgf)	After 60 minutes (kgf)	Lift and turn (kgf)	Coughing (kgf)
K. E.	42	31	43	n.p.
B. A.	25	18	32	27
M. S.	21	12	16	19
L. K.	20	14	23	48
A. J.	27	14	n.p.	19
E. J.	22	8	12	n.p.
M. E.	43	28	31	n.p.
Y. N.	29	16	21	n.p.
R. B.	25	11	n.p.	n.p.
A. Å.	39	15	17	n.p.
E. M. J.	37	24	20	n.p.

n.p. = Test not performed.

TABLE 3: Measurements during and immediately following surgery

The Milwaukee brace, compared with no external support, exerted a distractive axial force of 4–6 kgf in the four patients tested standing (figure 12). In the supine position there was, compared with no external support, a reduction in axial force of 2–4 kgf wearing the brace.

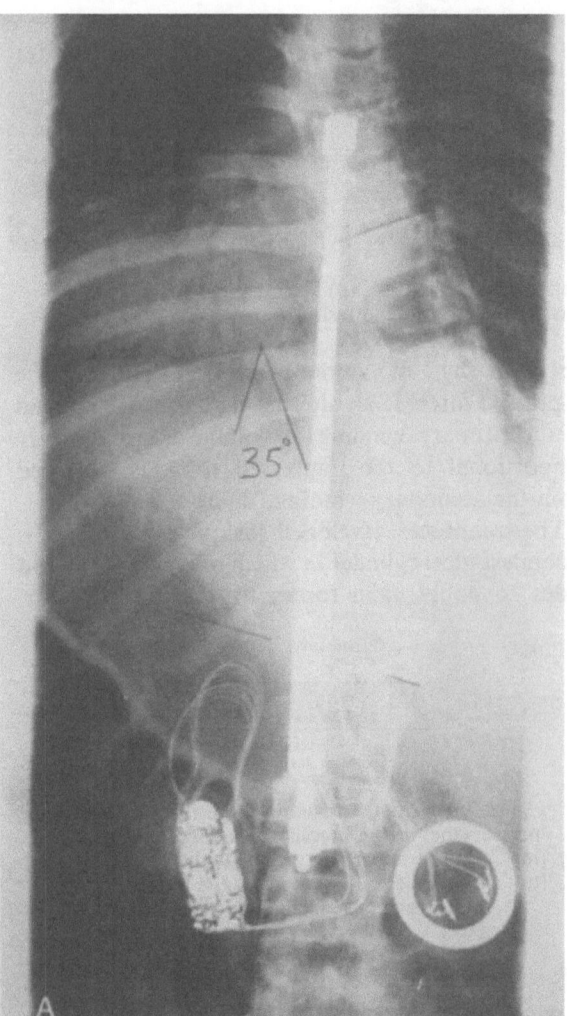

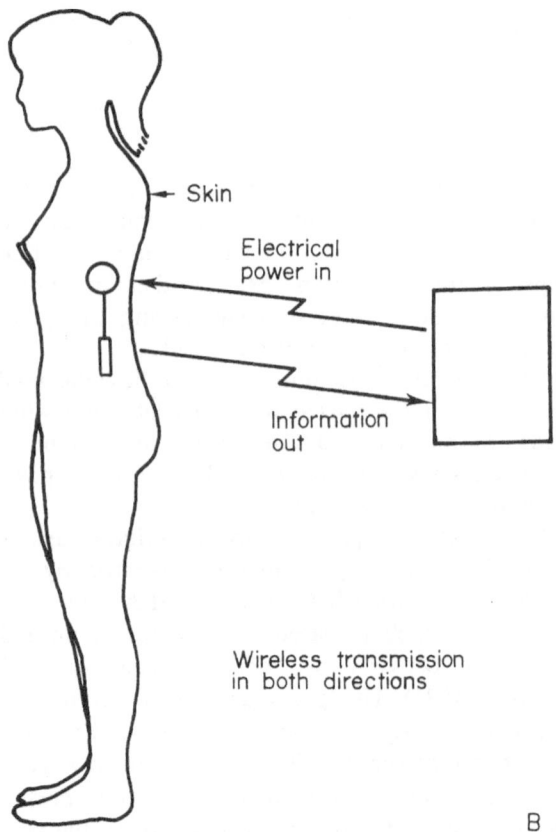

Figure 9. The principle of the telemetry system used: (A) Roentgenogram; (B) a 14-year-old patient showing the modified Harrington distraction rod with wires and telemetry units.

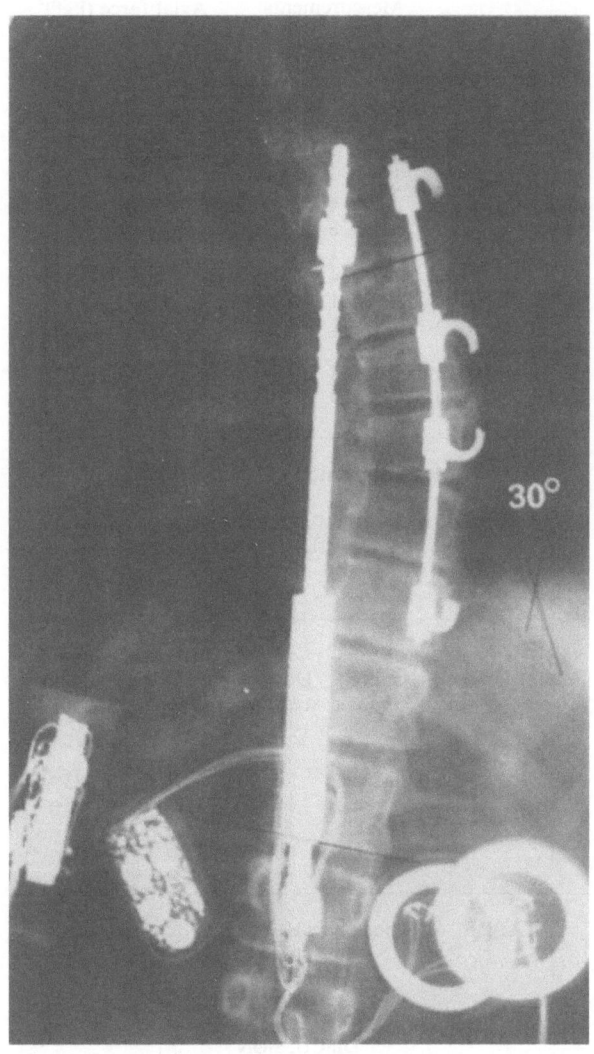

FIG. 10

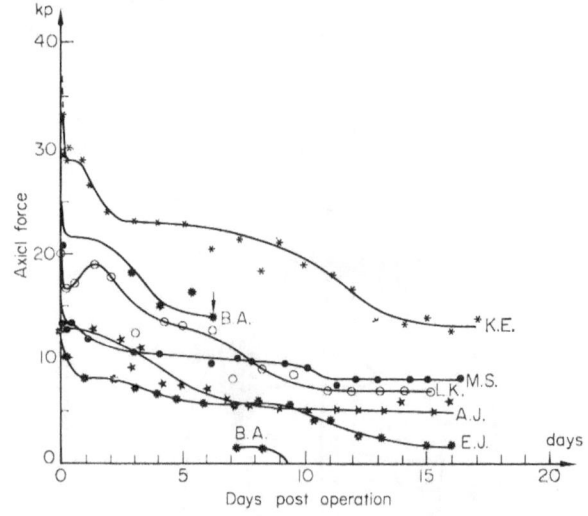

FIG. 11

Figure 10. Roentgenogram from a 15-year-old patient, showing the modified Harrington distraction rod with telemetry units, as well as the Harrington compression system on the convex side of the curve. (1 kp = 1 kgf).

Figure 11. The post-operative decline with time of the axial force on the Harrington rod in the supine position in six of the patients studied. Patient B. A. had an upper hook displacement after the sixth post-operative day (arrow). The displacement was actually detected from force recordings and was verified by a roentgenogram. The stable force values observed in the other patients after 12 days are about one-third of the maximum distraction force obtained during operation.

Patient convexity	Measurements performed days post-operative	Load supine, relaxed (kgf)	Bicycling in bed (kgf)	Vital capacity determination (kgf)	Coughing (kgf)
K. E. right	10	21	right 18 left 23	24	23
M. S. right	8	10	right 9 left 12	13	16
L. K. left	12	8	right 13 left 7	11	9
A. J. right	7	6	right 5 left 6	8	11
E. J. right	12	4	right 3 left 6	5	6
M. E. double	2	26	n.p.	n.p.	27
Y. N. right	11	2	right 3 left 3	3	n.p.
A. Å. right	14	1	right 3 left 3	n.p.	n.p.

n.p. = test not performed.

TABLE 4: Load effect of various exercises

Evaluation, from 'standing' results, of the different parts of a Milwaukee brace demonstrated the importance of a well-fitted pelvic cage, of either chin

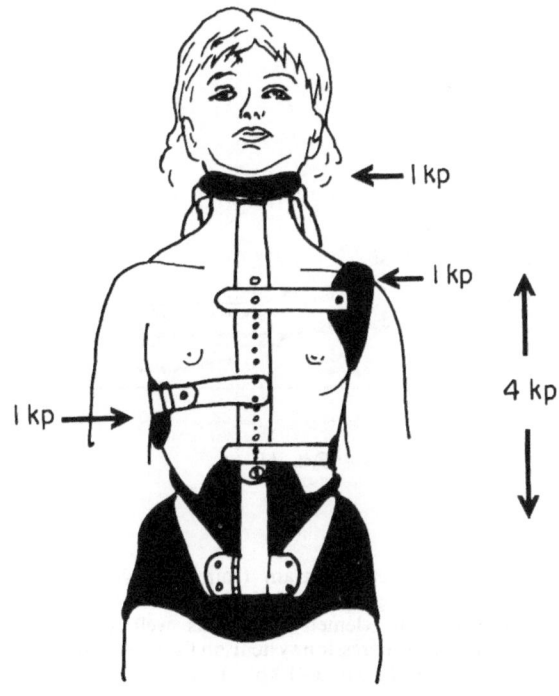

Patient	Convexity	Measurements performed days post-operative	Axial force (kgf)		
			Supine	Right side	Left side
K. E.	right	7	22	25	22
B. A.	left	5	16	16	18
M. S.	right	4	10	12	10
L. K.	left	8	9	10	13
A. J.	right	1	18	23	19
E. J.	right	4	7	8	9*
M. E.	double	1	24	29	29
Y. N.	right	1	14	16	14
R. B.	right	6	1·5	4	2
A. Å.	right	4	3	6	5
E. M. J.	right	5	12	14	13

* Compression system added.

TABLE 5: Axial forces in the supine, the right and the left positions

Patient	Age (years)	Curve type	Degree standing	Distraction rod (kgf)	Compression system (kgf)
A. J.	18	Thoracic	70°	27	33
E. J.	12	Thoracic	55°	22	18
Y. N.	20	Thoracic	55°	30	34
R. B.	13	Thoracic	80°	18	23

TABLE 6. Effect of Harrington compression system

DISCUSSION

We have described two areas in which intravital measurements of forces based on methods provided by engineers have been applied to common orthopaedic problems. Only by the cooperation of orthopaedic surgeons and biologically oriented engineers can such research be performed. It is true that they solve the etiological problems of neither low back pain nor scoliosis, but they are certainly helpful in the everyday treatment of these patients. The data presented also provide the orthopaedic surgeon with knowledge that can be used in future research.

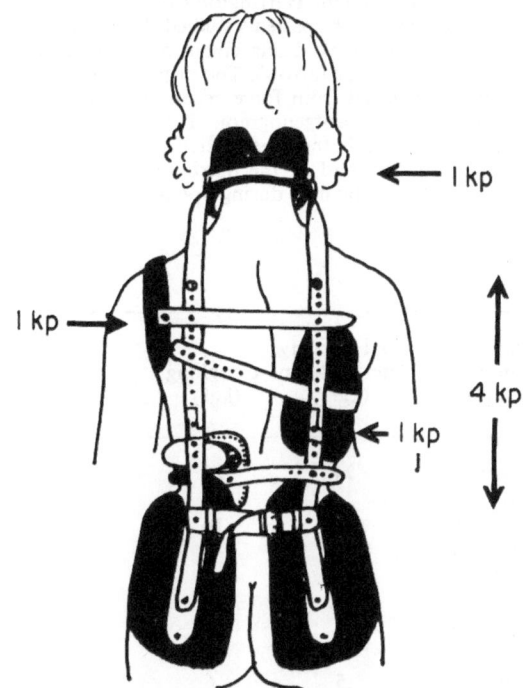

Figure 12. Approximate distracting forces on the scoliotic spine resulting from different parts of the Milwaukee brace. (1 kp = 1 kgf).

or occiput pads and especially of two side supporting pads (figure 12).

Mechanically these last-mentioned results should be valid also for non-operated patients with moderate curves of idiopathic origin for whom this brace is commonly used.

REFERENCES

ARMSTRONG, J. R. 1952: 'Lumbar Disc Lesions. Pathogenesis and Treatment of Low Back Pain and Sciatica'. Livingstone, Edinburgh and London.

BARTELINK, D. L. 1957: The role of abdominal pressure in relieving the pressure on the lumbar intervertebral discs. *J. Bone and Joint Surg.* **39(B)**: 718-725.

BAYER, H. 1954: Mit welchen Kräften wirken die Rückenstrecker auf die Lendenwirbelsäule ein. *Z. Orthop.* **84**: 607-615.

BLOUNT, W. P. 1964: The Milwaukee brace in the treatment of the young child with scoliosis. *Arch. Orthop. Unfallchir.* **56**: 363-369.

BOUILLET, R., and VINCENT, A. 1967: La scoliose idiopathique. *Acta Orthop. Belgica* 33: 96-388.

COTREL, Y. 1966: The end results in the treatment of idiopathic scoliosis. *Congr. SICOT, Paris. In Excerpta Medica, Int. Congress Series* 116: E77-8.

DEPALMA, A. F., and ROTHMAN, R. H. 1970: 'The Intervertebral Disc'. Saunders, Philadelphia.

EIE, N. 1966: Load capacity of the low back. *J. Oslo City Hosp.* 16: 73-98.

FEFFER, H. L. 1963: A physiological approach to lumbar intervertebral disc derangement. Pp. 111-126 in 'Current Practice in Orthopedic Surgery', Vol. 1 (Ed. J. P. Adams). Mosby, St Louis.

FRIBERG, S., and HIRSCH, C. 1949: Anatomical and clinical studies on lumbar disc degeneration. *Acta Orthop. Scand.* 19: 222-242.

GALANTE, J. O. 1967: Tensile properties of the human lumbar annulus fibrosus. *Acta Orthop. Scand., Suppl. 100.*

GOLDSTEIN, L. A. 1969: Treatment of idiopathic scoliosis by Harrington instrumentation and fusion with fresh autogenous iliac bone grafts. *J. Bone and Joint Surg.* 51(A): 209-222.

HALL, J. E. 1970: The management of scoliosis. *J. Bone and Joint Surg.* 52(A): 408.

HANRAETS, P. R. M. 1959: 'The Degenerative Back and Its Differential Diagnosis'. Elsevier, Amsterdam.

HANSEN, H.-J. 1959: Comparative views on the pathology of disc degeneration in animals. *Lab. Invest.* 8: 1242-1259.

HARRINGTON, P. R. 1962: Treatment of scoliosis. Correction and internal fixation by spine instrumentation. *J. Bone and Joint Surg.* 44(A): 591-610.

HARRINGTON, P. R. 1967: Instrumentation in structural scoliosis. Pp. 95-123 in 'Modern Trends in Orthopaedics', Vol. 5. Butterworths, London.

HIRSCH, C. 1948: An attempt to diagnose the level of a disc lesion clinically by disc puncture. *Acta Orthop. Scand.* 18: 132-140.

HIRSCH, C., and SCHAJOWICZ, F. 1952: Studies on structural changes in the lumbar annulus fibrosus. *Acta Orthop. Scand.* 22: 184-231.

HIRSCH, C., PAULSON, S., SYLVÉN, B., and SNELLMAN, O. 1952: Biophysical and physiological investigations on cartilage and other mesenchymal tissues. VI. Characteristics of human nuclei pulposi during aging. *Acta Orthop. Scand.* 22: 175-183.

HIRSCH, C., and NACHEMSON, A. 1961: Clinical observations on the spine in ejected pilots. *Acta Orthop. Scand.* 31: 135-145.

HIRSCH, C., INGELMARK, B.-E., and MILLER, M. 1963: The anatomical basis for low back pain. *Acta Orthop. Scand.* 33: 1-17.

HIRSCH, C., and WAUGH, T. 1968: The introduction of force measurements guiding instrumental correction of scoliosis. *Acta Orthop. Scand.* 39: 136-144.

HORAL, J. 1969: The clinical appearance of low back disorders in the city of Gothenburg, Sweden. *Acta Orthop. Scand., Suppl. 118.*

HULT, L. 1954: Cervical, dorsal, and lumbar spinal syndromes. A field investigation of a non-selected material of 1200 workers in different occupations with special reference to disc degeneration and so-called muscular rheumatism. *Acta Orthop. Scand., Suppl. 17.*

JAMES, J. I. P. 1967: 'Scoliosis'. Livingstone, Edinburgh and London.

LIDSTRÖM, A., and ZACHRISSON, M. 1970: Physical therapy on low back pain and sciatica. *Scand. J. Rehab. Med.* 2: 37-42.

LINDBLOM, K. 1950: Technique and results of diagnostic disc puncture and injection (discography) in the lumbar region. *Acta Orthop. Scand.* 20: 315-326.

MOE, J. H. 1958: A critical analysis of methods of fusion for scoliosis. An evaluation in two hundred and sixty-six patients. *J. Bone and Joint Surg.* 40(A): 529-554.

MOE, J. H., and KETTLESON, D. N. 1970: The end results of Milwaukee brace treatment of 169 patients with idiopathic scoliosis. *In* Onzième Congrès de Chirurgie Orthopédique et de Traumatologie, Mexico, 6-10 Octobre 1969. Bruxelles, Imprimerie des Sciences, S.A.

MORRIS, J. M., LUCAS, D. B., and BRESSLER, B. 1961: Role of the trunk in stability of the spine. *J. Bone and Joint Surg.* 43(A): 327-351.

NACHEMSON, A. 1960: Lumbar intradiscal pressure. Experimental studies on post mortem material. *Acta Orthop. Scand., Suppl. 43.*

NACHEMSON, A. 1962: Some mechanical properties of the lumbar intervertebral discs. *Bulletin of the Hospital for Joint Diseases,* 23(2).

NACHEMSON, A. 1963: The influence of spinal movements on the lumbar intradiscal pressure and on the tensile stresses in the annulus fibrosus. *Acta Orthop. Scand.* 33(3).

NACHEMSON, A., and MORRIS, .J. 1964: *In vivo* measurements of intradiscal pressure. *J. Bone and Joint Surg.* 46(A): 1077-1092.

NACHEMSON, A., and EVANS, J. H. 1968: Some mechanical properties of the third human lumbar interlaminar ligament (Ligamentum flavum). *J. Biomechanics* 1: 211-220.

NACHEMSON, A., DIAMANT, B., and KARLSSON, J. 1968: Correlation between lactate levels and pH in discs of patients with lumbar rhizopathies. *Experientia* 24: 1195-1196.

NACHEMSON, A. 1969: Intradiscal measurements of pH in patients with lumbar rhizopathies. *Acta Orthop. Scand.* 40: 23-42.

NACHEMSON, A., and ELFSTRÖM, G. 1970: Intravital dynamic pressure measurements in lumbar discs. A study of common movements, manoeuvres and exercises. *Scand. J. Rehab. Med., Suppl. 1.*

NACHEMSON, A., and ELFSTRÖM, G. 1971: Intravital wireless telemetry of axial forces in Harrington distraction rods in patients with idiopathic scoliosis. *J. Bone and Joint Surg.* 53(A): 445-465.

NAYLOR, A. 1962: The biophysical and biochemical aspects of intervertebral disc herniation and degeneration. *Ann. R. Coll. Surg. Eng.* 31: 91-114.

OLSSON, S.-E. 1951: On disc protrusion in dog. *Acta Orthop. Scand., Suppl. 8.*

PEREY, O. 1957: Fracture of the vertebral end-plate in the lumbar spine: an experimental biomechanical investigation. *Acta Orthop. Scand., Suppl. 25.*

PEYRON, J.-G. 1967: Biologie du disque intervertébral. *Sem. Hop. Paris* 43: 3318-3335.

PÜSCHEL, J. 1930: Der Wassergehalt Normaler und Degenerierter Zwischenwirbelscheiben. *Beitr. z. path. Anat. u. allg. Path.* 84: 123.

ROLANDER, S. D. 1966: Motion of the lumbar spine with special reference to the stabilizing effect of posterior fusion. *Acta Orthop. Scand., Suppl. 90.*

SMYTH, M. J., and WRIGHT, V. 1958: Sciatica and the intervertebral disc. An experimental study. *J. Bone and Joint. Surg.* 40(A): 1401-1418.

STAGNARA, P., DU PELOUX, J., FAUCHET, R., MAZOYER, D., and CALLAY, C. 1970: Distraction cast for the treatment of scoliosis. *J. Bone and Joint Surg.* 52(A): 406.

WAUGH, T. R. 1966: Intravital measurements during instrumental correction of idiopathic scoliosis. *Acta Orthop. Scand., Suppl. 93.*

RECENT MEASUREMENTS OF THE INTRA-ABDOMINAL PRESSURE

N. EIE[1]

INTRODUCTION

The relieving effect of the intra-abdominal pressure is an important factor in the mechanics of the lumbo-sacral spine. By counteracting the compression produced by the contraction of the erector spinae muscles and by tending to elongate and straighten the lumbar spine anteriorly, the intra-abdominal pressure assists significantly in the prevention of both compression fractures and bending lesions of the lumbar spine. The high intra-abdominal pressure which occurs during heavy physical exercises explains why muscular individuals may expose their back to extremely heavy loading without causing damage to the spine. The maximum spontaneous-intra-abdominal pressure of an individual may be considered a measure of the load which the individual's back may tolerate.

EQUIPMENT FOR MEASUREMENT OF THE INTRA-ABDOMINAL PRESSURE

Measurement of the intra-abdominal pressure was first carried out in 1957 by Davis and Bartelink using the balloon-catheter-manometer method, which however did not register the high peaks of the pressure. In 1962 Eie and Wehn used electrical recording of the intra-abdominal pressure variations and were able to register pressures up to 300 mm of mercury. By means of a small pressure transducer, a so-called 'electric pill', a modulator and a pocket tape recorder, it has also been possible to register the intra-abdominal pressure during motion in a variety of activities such as skijumping, jetflying etc. The results of these investigations were published (Eie, 1966), and showed that the intra-abdominal pressure varies according to demand.

Recently the Central Institute for Industrial Research in Oslo has built a very small radio transmitter, weighing only 43 g, shown in figure 1. This can be comfortably carried by any person and the impulses from the electric pill swallowed by the test subject are recorded through a radio receiver. A typical record is shown in figure 2. This is a sample of measurements of the intra-abdominal pressure in three female members of the Norwegian

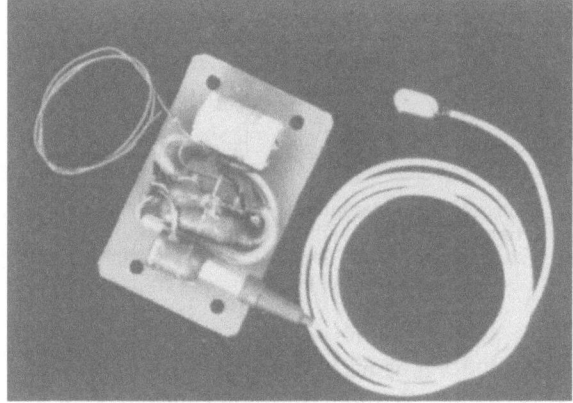

Figure 1. Radio-transmitter and 'electric pill' for intra-abdominal pressure measurements.

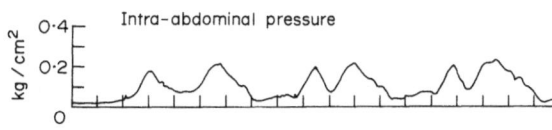

Figure 2. Intra-abdominal pressure during gymnastic exercises.

Figure 3. Heavy weight lifting, 305 kgf.

[1]Neurological Department, Ullevaal Hospital, Oslo, Norway.

International Gymnastic Team taken during the performance of exercises such as headvaults, free cartwheels, handsprings etc. Another significant example is that of weight lifting. At present the world record in heavy-weight lifting is approximately 375 kgf. During such heavy lifts the intra-abdominal pressure must be between 300 and 400 mmHg. Figure 3 shows the Norwegian champion in discus throwing lifting 305 kgf. During this lift his intra-abdominal pressure rose to 340 mmHg or 0·462 kgf/cm² (see record on right of figure 4).

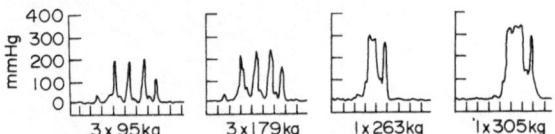

Figure 4. Intra-abdominal pressure during weight lifting records left to right: 3 successive lifts of 95 kg, 3 successive lifts of 179 kg, 1 lift of 283 kg and 1 lift of 305 kg.

This pressure corresponds to a load of approximately 400 kgf as his abdominal cross-section area was 857 cm². Without the relieving effect of the intra-abdominal pressure, the vertebral compression would have been 1114 kgf, which is well above the fracture limit of a lumbar vertebra. When the same man was asked to produce his maximum intra-abdominal pressure by muscle contraction alone and without lifting, the pressure recorded rose to 420 mmHg = 0·571 kgf/cm². This would allow him—without damage to his lumbar spine—to beat the weight lift of the world champion!

SUMMARY

These tests indicate that the lumbar spine has efficient protection in well-developed abdominal muscles, which in turn may be an important factor in the prevention of the low back pain syndrome.

REFERENCES

Bartelink, D. L. 1957: The role of abdominal pressure in relieving the pressure on the lumbar intervertebral disks. *J. Bone Joint Surg.* **39 B**: 718–725.
Davis, P. R. 1959: The causation of herriae by weight-lifting. *Lancer* **11**: 155–157.
Eie, N. 1966: Load capacity of the low back. *J. Oslo City Hosp.* **16**: 73–98.
Eie, N., and Wehn, P. 1962: Measurements of the intra-abdominal pressure in relation to weight bearing of the lumbosacral spine. *J. Oslo City Hosp.* **12**: 205–217.

SESSION 'B' DISCUSSION

In response to a question from Professor Hirsch, *Professor Mann* defined 'trade-off' as the operation we all go through many times a day when we consider the balance between those desirable ends which to quite an extent are incompatible. He went on to explain that the particular 'trade-off' he was drawing attention to in his paper was that, on the one hand it would be desirable to access the central nervous system in a way which had minimal long and short term effects on the patient, and on the other hand, so that the information derived was explicitly characteristic of the particular control that the central nervous system was exerting in relation to a natural appendage. As an example of the minimal patient effect he instanced the use of surface electrodes for electromyographic (EMG) sensing—these obviously do not interfere with the patient if, for example, in the case of an amputee they are incorporated in the socket. The patient puts these electrodes on and takes them off with the socket. However, the signal that they pick up is an abbreviated and filtered representation of the basic control information which originates in the central nervous system. An example of direct accessing of the central nervous system is demonstrated by the work of Humphries *et al.* (quoted in Professor Mann's paper) who in fact inserted electrodes into the motor cortex of animals and thereby derived electrical information which they showed to be well correlated with the animals' motor activity. This kind of very direct access is obviously not achievable in the human patient, certainly at the present time. *Professor Hirsch* asked if it was hoped to extract from muscle EMG signals, say in amputee stumps, information that would reflect some sort of gait pattern similar to what would be expected in normals, or possibly utilise the information to direct an appropriate gait. *Professor Mann* had relatively little information to offer as his group's lower extremity studies were at a very early stage. They know that the EMG pattern of activity in amputees is different from that in the normal. However, he hoped it would be possible to take the biological information which the amputee's deficient system can generate (and which he thought, and he emphasised this as being the important point) the central nervous system generates in response to the conscious desires of the patient. If this is in fact the case, then it will be possible to convert this information to control and operate a prosthesis in a way which would

either be akin to normal, or would at least correspond to what the patient would like to do. *Mr Condie* referred to studies carried out on Mr Murdoch's patients, where in some instances it was found that the stump's muscles may continue to contract and provide an EMG system which corresponded to normal. In other cases, however, they have found a quite clear conversion of the function of muscles as a result of amputation surgery. This was paralleled in the experience of surgical treatment of poliomyelitis where, for example, the hamstring tendons were transferred to perform the function of the quadruceps muscles. Mr Condie's and his co-workers' interest in studying the process of how muscles 'learn' their new function, is oriented to develop treatment patterns leading to more selective and more constructive lines of surgery. This in turn should evolve to an understanding of the most desirable pattern of muscle activity from the point of view of rehabilitation. *Professor Mann* concurred, with the qualification that, as implied in his paper, the more training can be minimised, the closer the desirable optimum of rehabilitation is approached. *Mr Murdoch* expressed his agreement with Professor Mann's thesis. He suggested that at times intuitive solutions at below-knee amputations involving a specific kind of surgery, a certain kind of socket and a corresponding suspension which evolved, backed by theoretical biomechanical analysis, produced a situation where there has been little effective intervention and where, as demonstrated by EMG studies, the action approaches normal. *Mr Condie*, enlarging on this at Mr Murdoch's invitation, instanced below-knee amputations where the calf muscles have been re-attached to the stump using myoplastic procedures, and in particular patients it was found that these muscles continued to contract in precisely the phasic manner in which they would have were the ankle joint and foot still present. There were several instances recorded in medical experience of this kind of result.

Dr Maroudas opening the discussion on the papers by Dowson, Wright and McCutchen referred to her own introduction of the idea of ultrafiltration some years ago in relation to the concepts of squeeze film versus booster lubrication. Her idea of ultra-filtration suggested that when cartilage surfaces approach each other very closely, the resistance in the gap between these surfaces to the outflow of fluid increases and some of the fluid seeps through the cartilage thus leading to the formation of a hyaluronic acid/water film between the two surfaces. This concept, in her view, holds only for very small gaps and she arrived at the figure of 200 Angstroms for the thickness of the equilibrium configuration of this hyaluronic acid/water film. Professor Dowson, who incorporated this ultrafiltration idea in his development of a booster lubrication concept, extrapolates to a very much thicker film. Dr Maroudas felt that the mechanism, by which hyaluronic acid molecules are retained within large gaps of something like 1 mm, is unclear since it is well known that hyaluronic acid will escape sideways through a gap of this thickness, fluid flow being very rapid in such films. She expressed her agreement with Dr McCutchen that when larger gaps are generated through asperities coming into contact, there are likely to be pools between them.

Professor Dowson, responding, commented that in the analysis of boosted lubrication[1] it was necessary to draw upon published data about the increase in viscosity with the concentration of hyaluronic acid (H/A) and the relationship employed does, of course, extend to large film thicknesses.

They did not wish to imply that this mathematical relationship is an accurate representation of the physical events at large gaps, and they agreed that the process of concentration normally assumed to be operative in the boosted lubrication concept was likely to be restricted to small separations. This should not, however, be seen as a serious limitation upon the theory since, in common with all squeeze-film actions, the spectacular increase in the time of approach of the surfaces becomes evident only when the gaps are small. It is thus only in the latter stages of the process that a significant difference develops between squeeze-film times for 'normal' and 'boosted lubrication' actions.

Professor Wright and he did not believe it necessary to think that the 'weeping' and 'boosted' lubrication concepts were diametrically opposed and unable to co-exist, as stated by Dr. McCutchen. The boosted lubrication mechanism would be valid only if, by some mechanism, the viscosity of the lubricant increased as the surfaces came closer and closer together. They envisaged from the outset the restrictions which could be placed upon the migration of large H/A molecules both into the porous cartilage and by side-leakage between the relatively rough bearing surfaces. It was later found, and clearly stated in the theoretical analysis of the boosted lubrication concept,[1] that side-leakage of the low-viscosity constituent of the lubricant would far exceed the flow into the porous cartilage.

They have never stated, that 'weeping' lubrication does not operate in a human joint. The aspect of the concept which appeared to worry many people when it was first proposed was the idea that fluid would be trying to flow back into the region of high pressure in the film, rather than sideways to regions of lower pressure. They believe that the problem of flow through a porous, elastic bearing material like cartilage during squeeze-film action is exceedingly complex and difficult to analyse in the present state of knowledge. In some circumstances, fluid flows into the cartilage, while in others, fluid may be expressed into the clearance space to provide the basis for weeping lubrication. A start has been made on the analysis of this situation by Norman and Higginson,[2] but until more is known about the effective elasticity of the cartilage matrix, the compressibility of the fluid within the cartilage, and the permeability of the cartilage, it is extremely difficult to reach firm conclusions.

They were entertained by Dr. McCutchen's delightful little demonstration of weeping lubrication, and he has to be congratulated upon his enterprise. They thought, however, that it was dangerous to base conclusions upon illustrations with models which do not simulate exactly the properties and geometry of the materials involved in synovial joints. They are not yet convinced that the demonstration is relevant to the problem of the mode of lubrication of animal joints.

Dr Edwards questioned the need in weeping lubrication for a fluid film mechanism to separate the articular surfaces completely. He thought that the main point in weeping lubrication was that fluid pressures at the interface reduced the pressures at the boundaries. In consequence, postulating this kind of mechanism for weeping lubrication, the phenomena would not depend on the presence of a fluid film separating the two surfaces, and in fact it wouldn't even matter in what direction the fluid might flow. *Dr Swanson* highlighted a gap in the discussion on the mechanism of synovial joint lubrication. This arose when one attempted to relate the various theories to the observed magnitudes of friction in synovial joints. Basing his calculations on the kind of frictional moment which he and other investigators have measured in cadaveric human hips for example, and supposing that this was to be provided entirely by the shearing of the fluid film, he estimated the relationship between the rate of shear and the viscosity of the film to correspond to the measured frictional moment. Supposing that his calculations were correct, he found that the viscosity required at any given shear rate would be at least 100 times greater than

[1] DOWSON, D., UNSWORTH, U., and WRIGHT, V. 1970: Analysis of 'boosted lubrication' in human joints. *Journal of Mechanical Engineering Science*, **12**: (5): 364–369.

[2] NORMAN, R. 1971: The lubrication of porous elastic solids with reference to the functioning of animal joints, *Ph.D. Thesis*, University of Durham.

that ever detected in synovial fluid. This, on the face of it, seemed to be arguing more in favour of water flowing out of the film than of water flowing in. He liked Dr McCutchen's demonstration and he certainly could not refute it *as applied to the friction between two glass plates lubricated by a cellulose sponge;* but he had to state that he was not very happy about the idea of water coming into the synovial fluid film. It seemed to him that this can only reduce, rather than increase, the viscosity. *Dr McCutchen*, responding to Professor Dowson's remarks, accepted that by reasons of practicality he discussed only one feature of boosted lubrication, and also that there existed a whole spectrum of uncertainties.

The discussion from this point continued with exchanges on a variety of details. This the Chairman, *Professor Hirsch* soon closed, commenting that obviously the discussion of this particular topic could go on indefinitely, indicative of a major need for more basic and reliable information. He invited all participants concerned to apply themselves to meeting this need.

SESSION 'B' SUMMARY

D. C. SIMPSON[1]

The interest which the papers have aroused has been shown in the discussion and the points of difference have emerged, so I think it only falls to me to make a few general remarks in conclusion rather than to attempt in any way to summarise the session.

Firstly, however, I would like to ask the organisers to send our good wishes to Colin McLaurin who was unfortunately prevented from coming to give his paper by illness in his family. In his paper, which John Hughes read as the opening communication of the session, he reminded us that the problems which are met with in the clinic are very rarely straightforward, that seldom is there a straightforward solution and that the step from bioengineering theory in the laboratory to practical application in a clinic is a very much bigger one than is often realised. I think we need this reminder.

It is difficult and no doubt unfair in the very short time at my disposal to single out particular papers for comment, but nevertheless I feel I must say how much I was interested in the simple and elegant techniques of Chodera and Levell for collecting data about gait without apparently prejudicing that data by imposing restrictions or restraints on the subject. Only too often in making clinical measurements we record what is conscious arm or leg action, when we are really interested in the unconscious, and this is forgotten in assessing the results.

At meetings like these and particularly at inter-national gatherings, I think we sometimes make things much more difficult for ourselves than we have to, because we have not defined sufficiently precisely what we mean by the terms we use; I was therefore very pleased to hear McKenzie start his paper with a definition of his subject, a feature which was lacking in the paper which preceded his, to the loss of the listeners. A further lesson perhaps for all of us, but I think for myself especially, came from the extra contribution from McCutchen. He is obviously a brave man but it was not only an excellent demonstration of how his ideas of joint lubrication contrasted with those of Dowson, but also a demonstration that the working of a model also illustrates an idea and will convince sceptics about what the model itself will do and will not convince them about its application to the original system under discussion.

As one who had to sit through a session only too aware that he has a commitment hanging over him to comment at the end, I would say that, although one of the contributors had some of the best presented and most informative slides I have seen for years, the session included some quite appalling ones indeed. This is unfortunate because bioengineers often seem already to have communication problems which arise from the interdisciplinary nature of the work; could I therefore make a strong plea as an associate chairman, not only for clarity and conciseness in language, but also for considerably more care in the preparation of visual aids so that we can all appreciate the work which is being described.

[1] Princess Margaret Orthopaedic Hospital, Edinburgh, Scotland.

SESSION C

TISSUE MECHANICS

Chairman: Professor T. Gibson, M.B., Ch.B., D.Sc., F.R.C.S.(Edin.), F.R.C.S.(Glasg.).

Associate: Dr L. Sokoloff, M.D.

FAILURE CHARACTERISTICS OF BONE AND BONE TISSUE

A. H. BURSTEIN, D. T. REILLY AND V. H. FRANKEL[1]

The failure characteristics of the structural elements of the human body have interested men for centuries. During the last several decades the question 'how strong?' has been asked and answered repeatedly. As in other areas of mechanics the answers to the questions have varied, depending both upon the technique that the observer has used and the astuteness of the observer.

The study of the mechanical properties of bone tissue has progressed through several plateaux. The first plateau occurred with the work of Messerer (1880), who investigated some of the simpler mechanical properties of bone tissue. These early investigators treated the bone tissue as a linear elastic material. Later work by other researchers (Evans, 1958) recognised bone as an anisotropic elastic material. A third plateau was reached with the work of McElhaney and Byars (1965) when it was recognised that bone was a viscoelastic material. Still more recently the composite structure of bone has been considered by such researchers as Currey (1964) and Katz (1971), and the two-phase nature of the material has been incorporated into its conceptual modelling. Recently Piekarski (1970) examined crack propagation and energy of fracture of bone tissue.

Our own recent work (Burstein et al., 1972) has led to the discovery of the plastic nature of bone behaviour. Using bovine bone we have found that, under tensile loads, specimens whose load axis coincides with the long axis of the bone exhibit marked plastic deformation (figure 1). If the mater-

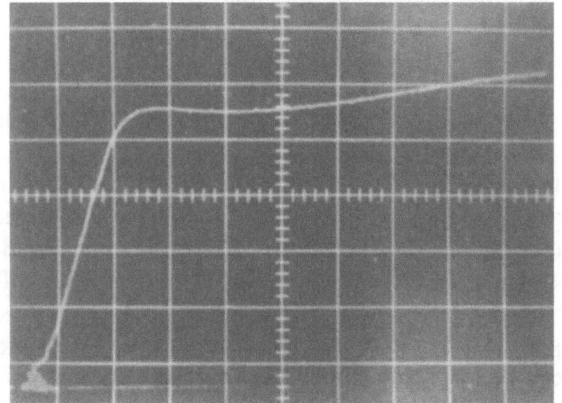

Figure 1. Load v. deformation oscillograph record for bovine femur specimen loaded in uniaxial tension.

ial tested were homogeneous, the plastic portion of the load deformation curve would suggest a material that was work hardening. However, since the material is a multi-phase substance rather than a homogeneous substance, the portion of the yield curve which is horizontal probably represents a pull-out or separation type of phenomenon, while the final portion of the yield curve with its gradually increasing load probably represents final fracture or failure of some of the remaining portions of the two-phase material.

That this behaviour is truly plastic is shown in figure 2 which represents several loading and unloading cycles of a bovine tension test specimen.

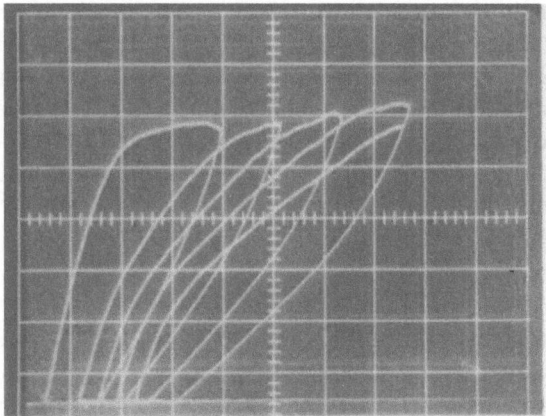

Figure 2. Cyclic tensile load in bovine specimen. Note how envelope of curves approximates curve of figure 1.

Several minutes were allowed to elapse between each of the successive loading cycles which were of $\frac{1}{2}$-second duration. Some of the deformation produced during each portion of the loading cycle beyond the yield point is permanent and irreversible, and successive deformations are additive. The viscous nature of the material is also shown in this figure which shows some recovery of the deformation. In addition, a change in the stiffness with succeeding load cycles can be noted. In virtually all of the tension tests with the load coincident with the long axis of the bone, failure occurred by a transverse fracture. Although the failure was obviously ductile in nature, there was no permanent reduction in cross-sectional area as would be found in a material such as steel. Standard histological examination techniques did not show micro-fracture, shear separation, or any other type of

[1] Case Western Reserve, Cleveland, Ohio, U.S.A.

microdiscontinuity either in the region of the fracture or in the region of yielding. For bovine bone the mean ultimate stress was

$$172 \text{ MN/m}^2 \pm 22 \text{ MN/m}^2$$

with a total strain of $2.9\% \pm 0.8\%$ and an elastic modulus of $24\,500 \text{ MN/m}^2 \pm 5100 \text{ MN/m}^2$.

In order to determine the effect of anisotropy on the yielding characteristics of bovine bone, specimens were cut so that their load axis would lie along a line perpendicular to the long axis of the bone and tangent to the circumference. As expected, the mechanical properties of these specimens differed grossly from the elastic properties of those specimens whose load axis coincided with the long axis of the bone.

Our findings were similar to those of Sweeney *et al.* (1965) in that these specimens had a lower elastic modulus and a lower ultimate tensile strength. Twenty-five specimens showed an ultimate stress to failure of $52 \text{ MN/m}^2 \pm 8 \text{ MN/m}^2$ and an elastic modulus of $11\,100 \text{ MN/m}^2 \pm 1770 \text{ MN/m}^2$. In addition, these specimens exhibited little or no plastic deformation and had a very much lower total strain to failure (mean of $0.7\% \pm 0.1\%$). When additional specimens were prepared with load axis at an angle of 30 degrees to the long axis of the bone and again tangential to the circumference, the expected mid range of properties was encountered. The mean ultimate stress was $111 \text{ MN/m}^2 \pm 9 \text{ MN/m}^2$ with an elastic modulus of

$$18\,000 \text{ MN/m}^2 \pm 1900 \text{ MN/m}^2.$$

While the mechanical response of human bone is very similar to that of bovine bone, it is certainly not identical. This may be due, in part, to histological

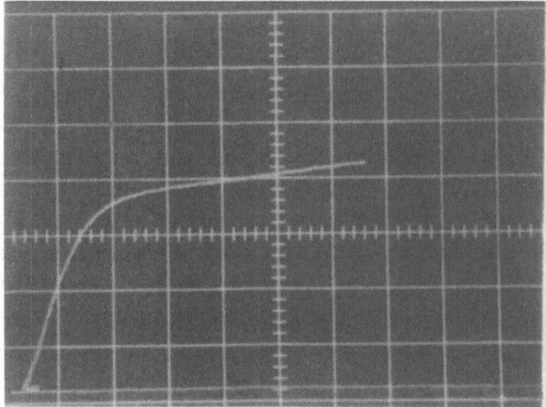

Figure 3. Load *v.* deformation oscillograph record for a human femur specimen loaded in uniaxial tension.

structure (viz., plexiform structure *v.* haversian). To clarify the differences in mechanical response of these tissues, human bone specimens were prepared in identical manner to those of bovine. At present

we have completed our tests only on those specimens whose load axis is parallel to the long axis of the human bones. The specimens were from both tibiae and femora and were tested to failure using loading times of well under 1 second. Extreme care was taken to have the exposed surfaces of the specimens completely saturated with saline at all times before and during the test.

The load deformation curve for bone tissue taken from a 57-year-old male is shown in figure 3. The primary difference between human and bovine bone tissues appears to be that the human bone tissue exhibited a so-called work-hardening region of plastic deformation immediately upon yielding, whereas the bovine tissue displayed a perfectly plastic region before entering into this so-called

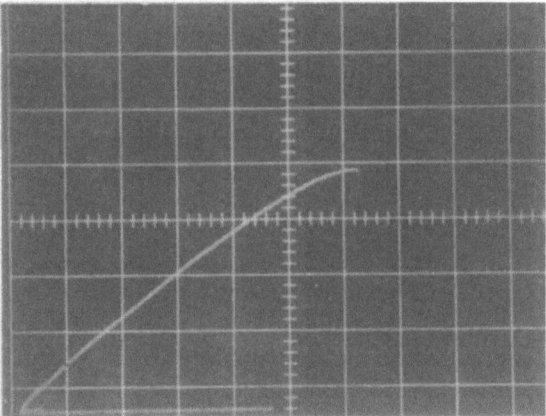

Figure 4. Load *v.* deformation oscillograph record for bovine femur specimen loaded in uniaxial compression.

work-hardening region. For 30 specimens from four subjects the mean yield strength was

$$107 \text{ MN/m}^2 \pm 10 \text{ MN/m}^2$$

while the mean ultimate strength was

$$151 \text{ MN/m}^2 \pm 18 \text{ MN/m}^2.$$

The initial elastic modulus was

$$14\,100 \text{ MN/m}^2 \pm 2260 \text{ MN/m}^2$$

with a total strain to failure of $4.6\% \pm 1.2\%$.

The elastic and plastic behaviour of bovine tissue was also investigated under compressive loadings. Specimens identical to those used in the tension tests were prepared, placed between parallel jaws, and loaded until failure occurred. The load-deformation behaviour of bovine specimens differed markedly from those tested in tension (see figure 4). These specimens exhibited little or no detectable plastic behaviour, and the failure surfaces were consistently at oblique angles. This is an indication of shear initiation of failure, and again, unlike materials such as steel, the shear mode of failure was not accompanied by significant plastic deformation.

It is interesting to note that under compressive loading, histological examination revealed the presence of what appeared to be microfractures at an angle of approximately 45 degrees to the loading axis. Such histologically apparent features have been noted by Chamay (1970). These cracks, which appear in directions of principal shear planes, appear to be more numerous if the stress field is non-uniform, e.g. column buckling, beam bending, than if the stress field is uniform, e.g. a direct, compressive loading. The findings of both tension and compression testing are consistent with a multi-phasic material which exhibits yield by means of pull-out which creates voids and crazes. This hypothesis is strengthened by a phenomenon noticed during our tension tests. When the specimen began its plastic deformation the test section became lighter if viewed by reflected light and darker if viewed by transmitted light. This behaviour is reminiscent of the crazing phenomenon in many polymers.

The ability of bone tissue to yield under tensile stress can, of course, greatly enhance its value as a structural material. The strength of a simple beam in bending may be increased by as much as 85 to

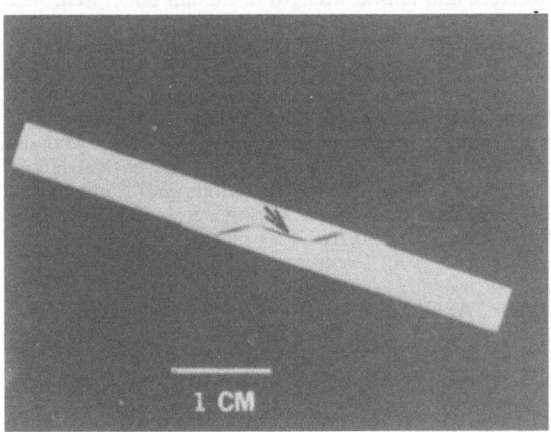

Figure 5. Torsional fracture of bovine specimen. Arrow shows shear crack at point of highest shear stress and in principal shear plane.

100 per cent if the material of which it is constructed is ductile. The same holds true for bones (see Burstein *et al.*, 1972). While knowledge of the tensile and compressive properties of bone tissue is important, it does not in and of itself answer all questions regarding the elastic, plastic, and failure characteristics of bone tissue. Multi-dimensional stress fields must be used to establish a more complete understanding of the ultimate behavioural characteristics of bone tissue.

One of the more important and more complex stress fields is pure shear. This stress configuration exists when long bones are loaded in torsion. We have machined square cross-section specimens of bovine bone and subjected them to rapid torsional loadings. While square cross-section specimens do not have a uniform shear stress they were used

instead of round cross-section specimens since they have the distinct advantage of allowing the determination of exact point of initiation of the fracture. The stress distribution in a square section is such that the maximum shear stresses occur at the midpoint on each of the four faces. Bovine bone was tested in this manner and produced the typical

Figure 6. Vascular network in bovine plexiform bone fractured by initiating shear crack. ×100 magnification.

fracture pattern shown in figure 5. Fracture initiation occurred on a plane of maximum shear stress parallel to the axis of the specimen. Figure 6 shows the surface of fracture initiation and that it is in fact the vascular network between lamellae. This vascular network is the triangular plane in the centre of the scanning electron micrograph (figure 6), while the remainder of the field shows the spiral portion of the fracture sloping towards the viewer (in the lower part) and away from the viewer (in the upper part).

In order to verify the nature of fracture initiation of a whole bone under a pure shear stress field in

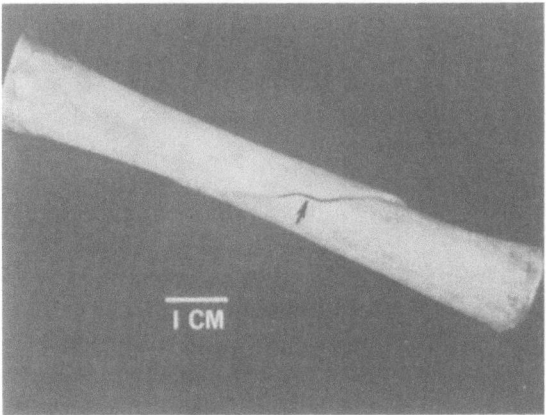

Figure 7. Dog femur fractured by torsional load. Arrow points to initiating shear crack.

which principal shear planes coincide with and are perpendicular to the long axis of the bone, additional torsion tests were conducted on dog femora and tibiae. Careful observation of the fracture surface invariably located a surface with the appearance of that in figure 7. It is our opinion that this surface, which is parallel to the long axis of the bone, represents the point of initiation of the torsional fractures. The spiral fracture, which accounts for the gross appearance of the classic torsional failure, appears to propagate from the initiating shear crack.

These preliminary data, and recognition of the plastic properties of the tissue allow us to speculate on the type of investigations and conceptual modelling required in the future. We have already begun to measure modulus, extent, and strain rate sensitivity of one plastic or work-hardening region. The problem of stress concentrations is being looked at again in the light of the material's ability to plastically deform and alleviate the stress concentration effect. The histological structure of bone tissue (especially the plexiform nature of bovine bone) seems to lend itself most easily to modelling as an orthotropic material.

This type of model can be formulated by five independent material constants. All of these constants must of necessity be determined from specimens of bone material which have meticulously been kept wet through all processes of fabrication and testing if the elastic-plastic nature of the material is to be measured. Also a strain rate representative of physiological loading should be used since the viscoelastic nature of the material will alter the material constants with different strain rates. Lang (1969) determined these constants for bovine phalanx using an ultrasonic technique.

The multi-phasic nature of bone material and its influence on fracture mechanics, plastic deformation, and mechanical properties needs further elucidation. The phenomenon seen in our tensile tests which resembles crazing, suggests that the mechanism for the plastic deformation of bone tissue is similar to that in polymers and depends on pullout or void formation. This might be further substituted by performing tension tests in an environment which suppresses pull-out and the formation of voids. Such an environment, for example, would be superimposed hydrostatic pressure. Under superimposed hydrostatic pressure some polymers which are extremely ductile in a uniaxial tension test become brittle (Christiansen et al., 1971).

REFERENCES

BURSTEIN, A. H. CURREY, J. D., FRANKEL, V. H., and REILLY, D. T. 1972: The ultimate properties of bone tissue: the effects of yielding. *J. Biomechanics* **5**: 35.

CHAMAY, A. 1970: Mechanical and morphological aspects of experimental overload and fatigue in bone. *J. Biomechanics* **3**: 236.

CHRISTIANSEN, A. W., BAER, E., and RADCLIFFE, S. 1971: Mechanical behaviour of polymers under high-pressure. *Philos. Mag.* **24**: 451.

CURREY, J. D. 1964: Three analogies to explain the mechanical properties of bone. *Biorheology* **2**: 1.

EVANS, F. G. 1958: Relations between the microscopic structure and tensile strength of human bone. *Acta Anat.* **35**: 285.

KATZ, J. L. 1971: Hard tissue as a composite material: 1, bounds on the elastic behaviour. *J. Biomechanics* **4**: 455.

LANG, S. B. 1969: Elastic coefficients of animal bone. *Science* **165**: 287.

McELHANEY, J. H., and BYARS, E. F. 1965: Dynamic response of biological materials. *ASME 65-WA/HUF-9.*

MESSERER, O. 1880: 'Uber Elasticitat and Festigkeit der Menschlichen Knocchen'. Verlag der J. G. Cotta' schen Buchlandlung, Stuttgart.

PIEKARSKI, K. 1970: Fracture of bone. *J. Appl. Physics* **41**: 215.

SWEENEY, A. W., KROON, R. P., and BYERS, R. K. 1965: Mechanical characteristics of bone and its constituents. *ASME 65-WA/HUF-7.*

A NOTE ON THE HISTOLOGY OF CEMENT LINES

L. SOKOLOFF[1]

The so-called 'cement line' that demarcates the edge of osteones is a well-documented *locus minoris resistentiae* in cortical bone (Dempster and Coleman, 1961; Piekarski, 1970). Unlike certain other histological features of bones and joints, it has not received rigorous anatomical scrutiny. It seems worthwhile to call attention to the structure of the cement line in view of the mechanical importance attributed to it in promoting or arresting crack propagation.

Cement lines are not native features of bone. They do not exist in all mammalian species; and when they do, they arise as secondary osteones develop during periods of physiological remodelling (Amprino, 1967). They thus have characteristics of so-called 'reversal lines' where new bone is laid down on a site of previous physiological resorption.[2] Like osteones themselves, then, they do not appear to be primary parts of the general mechanical design of bone. The extent to which the remodelling process—and hence cement line formation—is governed mechanically is a contentious matter (Johnson, 1966; Amprino, 1967). In the formation of new bone during remodelling, a preliminary ground substance is deposited. One type of thinking about the cement is that it represents a residuum of this ground substance.

From what is known of the chemical composition of bone, the principal candidate constituents of cement lines are: (1) collagen, the fibrous tension-resisting material; (2) mineral, principally in the form of hydroxyapatite crystallites, but also some amorphous calcium phosphate; and (3) protein-polysaccharide and other proteinaceous materials corresponding to what is called 'ground substance' in other connective tissues.

In physical terms two rather separate functions have been ascribed to the protein-polysaccharides. They are polyanionic macromolecules that exist in viscous solutions. Approximately 8 per cent by volume of mineralised bone matrix is water (Robinson, 1960). When the space the ground substance sols occupy is relatively large, they contribute to the flow characteristics of the tissue

as in cartilage (Ogston, 1970). By contrast, where they abut on other charged groups, their function may be quite different. Because of their polarity, they likely interact with collagen or mineral (Herring, 1972) and may thereby cause adhesion and stiffness.

Bone, anatomically, is an obviously composite material. The terminology applied to its components by histologists does not always coincide with that of material scientists. Anatomically the word 'matrix' is applied to the entire extracellular material, soft and hard, of the osseous tissue proper. When bone is fractured in shear, osteones are 'pulled out' of the cortex, apparently along the cement line. By analogy to certain two-phase materials, the osteone has been considered accordingly as 'fibre' and the cement line as the weak 'matrix' (Piekarski, 1970; Welch, 1970). In other physical analyses of bone, the mineral crystallites have been treated as the 'fibres' and the collagen fibres, because of their low elastic modulus, as the 'matrix' (Currey, 1969). In cartilage, where the system is simpler, the collagen behaves appropriately as the fibre, and the ground substance as the 'matrix' (Ogston, 1970).

Cement lines appear as refractile bands perhaps 1 μm or so wide in unstained sections. The question has been raised whether they are real structures rather than optical artefacts arising where the edges of osteones overlap or their constituents change direction (Schmidt, 1959). They ordinarily are recognised in histological preparations by appearing blue when stained with haematoxylin, particularly Delafield's haematoxylin. Haematoxylin staining is non-specific and its precise mechanism is not known (Pizzolato and Lillie, 1968). It depends on mordanting by metallic components with anionic groups in the tissues. The same stain reacts with sulphated ground substance components and calcific deposits. Inasmuch as the cement lines are readily visible in decalcified specimens, the presumption is that the haematoxylin staining of the cement line is not related to a mineral salt proper. (Haematoxylin also stains the prefailure craze-lines that develop when bone is compressed excessively (Chamay, 1970). These do not correspond to cement lines and the basis for their staining is obscure. Perhaps as with the anomalous staining properties of stressed soft connective tissues, interstitial components are displaced or different

[1] Laboratory of Experimental Pathology, National Institutes of Health, Bethesda, Maryland, U.S.A.
[2] Interosteonal cement lines resemble but are not identical with the 'resting' lines that develop where bone formation halts for a time and then is renewed.

reactive groups on the surface of the fibres are exposed.)

Two other thin haematoxyphil lines of some special mechanical interest are seen in relation to the articular cartilage (Green et al., 1970). One of these is located at the junction of the calcified layer of articular cartilage with bone. It has similarities to the interosteonal cement lines of the bone proper but is a little thicker. The other line is seen at the interface of the calcified with the non-calcified layers of the cartilage. Its affinity for haematoxylin is much greater than that of the cement lines and it is more refractile. If often appears reduplicated as advancing waves of calcification occur in the cartilage; hence the name, the 'tide-mark'. Although the latter and cement lines are both thin, haematoxyphilic lines that demarcate the limit of a process of mineral deposition in tissue, their constitution and physical behaviour show several distinct differences.

That cement lines are authentic structures is demonstrated by their affinity for haematoxylin and certain other stains. Bodian's copper-protargol stain was originally applied to demonstrating neuraxons in paraffin-embedded tissues but it does not have a known specific chemical substrate. Grimley noted that it also stains the canaliculi of decalcified bone. It also brings out the cement lines very clearly (figure 1). The tidemark is not stained by this reagent although the osteochondral

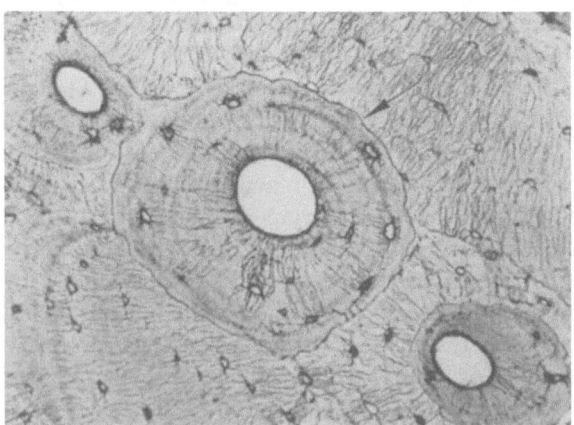

Figure 1. Cement line (arrow) in paraffin embedded section of decalcified bone. Bodian stain (× 320).

junction takes it up strongly. There is thus a difference in the tissue components reacting with haematoxylin and Bodian's stain respectively.

In microradiographs of the osteochondral cement line, considerable quantities of mineral are described (Green et al., 1970), so there is a mineral as well as an organic component. An apparent clear zone immediately external to the protargol-stained line suggests the possibility that these two components may not occupy the same space. It has been reported that the interosteonal cement line also

appears more radio-opaque than the surrounding bone (Smith, 1963). Furthermore, less carbon residue was seen in the cement lines than with osteones after controlled microincineration. From this the conclusion was drawn that there is a higher ratio of mineral to organic material in the cement line than in osteones. Nevertheless in the usual microradiographs, the interosteonal cement lines are not consistently heavily radio-opaque, indeed they are sometimes relatively lucent. The amount of mineral found by X-ray absorption in one study of sections 40 μm thick (Philipson, 1965) was comparable to that of heavily mineralised osteones. The electron probe analysis reported by Mellors (1964) also showed no increased quantity of Ca or P in this position. Technical factors may contribute to the contradictory data here. Osteones are not simple geometric cylinders but branch and pursue variably tortuous courses (Cohen and Harris, 1958). In the thick sections used for microradiography, the surface of the osteone is infrequently oriented over any length in the direction of the X-ray beam with the result that it is unlikely to present sufficient overlay for radio-opacity. Apparent radiolucence may also at times be an artefact arising from separation of osteones during preparation of the undecalcified sections. It must also be noted that Fawns and Landells (1953) found no calcific material in cement lines using alizarin or von Kossa stains.

Whether collagen is present in cement lines is not clear from light microscopy. It has been known for a long time (Weidenreich, 1930; Weinmann and Sicher, 1947) that silver impregnation methods employed for demonstrating fine collagen or reticulin fibres are rejected quite selectively by the cement lines unlike bone or cartilage matrix. The usual conclusion that collagen is not present in the cement line goes beyond this evidence. Birefringence comparable to that of the light bands of lamellar bone is not ordinarily found. The usual stains for collagen are not helpful. Cement lines cannot be distinguished from the rest of the bone matrix in the Masson trichrome or Van Gieson stains. In an X-ray diffraction study of two species (whale and orang-utang) in which cement lines appeared particularly thick (5 μm) in microradiographs, collagen patterns were obtained (Philipson, 1965). These cement lines were much thinner when examined with phase contrast microscopy and one wonders whether they could have been located with sufficient precision to allow the interpretation offered.

The diffuse metachromasy and alcian blue staining of osteoid and other ground substances are not found in mature bone or cement lines. The acid mucopolysaccharide content of mature bone is relatively low (Herring, 1972). Histochemical staining indicates that much of that mucopolysaccharide is located within lacunae and canaliculi rather than the bone matrix or cement lines. While these histochemical methods by no means exclude the presence of protein-polysaccharides from bone matrix or cement lines, the amount present must be very small, in fine dispersion and closely associated with other matrix components. The situation is thus very different from that of cartilage. Bone matrix is diffusely stained by the periodic acid-Schiff reagent which depends for a positive reaction on vicinyl hydroxyl groups in the substrate. It may well be staining the glyco- and sialoproteins, at least some of which are intimately associated chemically with the collagen (Herring, 1972). There is a slight accentuation of the periodic acid-Schiff stain at some of the cement lines but not a uniform or pronounced one to indicate a special compartmentalisation here. There are no stainable neutral lipids or phospholipids (Baker's stain).

The tidemark of articular cartilage is a weak point. In fractures (Fawns and Landells, 1953) and degenerative joint disease of man (Pommer, 1915) and laboratory animals (Sokoloff, 1956) minute separations appear at the junction of the calcified and non-calcified cartilage (figure 2.) That they develop during life is demonstrated at times

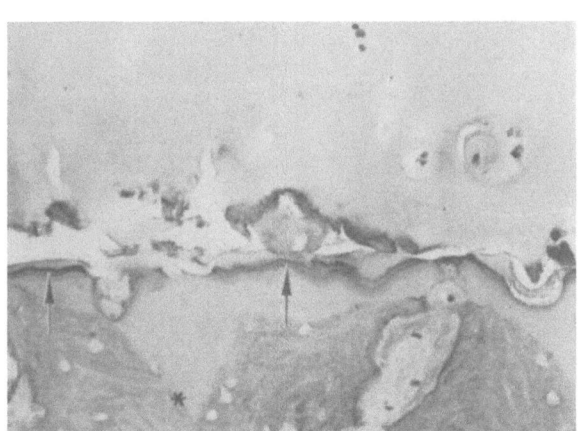

Figure 2. Separation of the non-calcified from the calcified layer of articular cartilage at the tidemark (arrows), degenerative joint disease, human patella. The tidemark is reduplicated in places. The cement line is not visible without a special stain at the osteochondral junction (*) (Hematoxylin and eosin, × 250).

by proteinaceous exudates in the clefts and by cellular and matrix changes in the immediate vicinity. Collagen fibrils course uninterrupted through the tidemark into the calcified layer. We must conclude that the abrupt differences in the elastic moduli of the two layers of the cartilage at the tidemark are responsible for this failure.

The cement line at the base of the calcified layer, by contrast, appears tightly bonded to the bone. Conventional histological methods have not revealed a collagenous bridge between the two tissues. In scanning electron micrographs, Ohnsorge, Schütt

and Holm (1970) have described collagen fibrils in the calcified layer as being closely packed, arranged perpendicular to but terminating at the osteochondral junction. Mital (1970) noted that the fibrils of the cartilage gather into bundles which blend with calcific material near the osteochondral junction. The bundles penetrate the subchondral bone in places but intimate details of collagen in the cement line could not be discerned in these undecalcified preparations. The tightness of the junction has traditionally been attributed to the coarse serration and interdigitation of the bone and cartilage. There is, in addition, as in the interosteonal cement lines, a marked finer irregularity. The asperities in the cement lines are measured in μm while the first order irregularities of the osteochondral junction are in mm.

From this brief review we see that there is no concensus on the structure or composition of cement lines. Their very existence as specialised structures has been questioned. The presence of each of the constituents has variously been affirmed and denied. Electron microscopy might offer useful information about the collagen and mineral in cement lines and so let us have a better idea of whether they cement osteones together or allow them to slip by each other. Recent excellent reviews of transmission (Cameron, 1972) and scanning electron microscopic (Boyde, 1972) characteristics of bone do not, however, address themselves to this subject.

Acknowledgments

I am happy to acknowledge helpful discussions with Drs L. C. Johnson and C. W. McCutchen.

REFERENCES

AMPRINO, A. 1967: Bone histophysiology. *Guy's Hosp. Rep.* **116**: 51–69.

BOYDE, A. 1972: Scanning electron microscope studies of bone. *In* 'The Biochemistry and Physiology of Bone' (Ed. C. H. Bourne), 2nd ed. Academic Press, New York and London, pp. 259–310.

CAMERON, D. A. 1972: The ultrastructure of bone. *In* 'The Biochemistry and Physiology of Bone' (Ed. C. H. Bourne), 2nd ed. Academic Press, New York and London, pp. 191–236.

CHAMAY, A. 1970: Mechanical and morphological aspects of experimental overload and fatigue in bone. *J. Biomechanics* **3**: 263–270.

COHEN, J., and HARRIS, W. H. 1958: The three-dimensional anatomy of haversian systems. *J. Bone Joint Surg.* **40A**: 419–434.

CURREY, J. D. 1969: The relationship between the stiffness and the mineral content of bone. *J. Biomechanics* **2**: 477–480.

DEMPSTER, W. T., and COLEMAN, R. F. 1961: Tensile strength of bone along and across the grain. *J. Appl. Physiol.* **16**: 355–366.

FAWNS, H. T., and LANDELLS, J. W. 1953: Histochemical studies of rheumatic conditions. I. Observations on the fine structures of the matrix of normal bone and cartilage. *Ann. Rheumat. Dis.* **12**: 105–113.

GREEN, W. T., Jr., MARTIN, G. N., EANES, E. D., and SOKOLOFF, L. 1970: Microradiographic study of the calcified layer of articular cartilage. *Arch. Path.* **90**: 151–158.

HERRING, G. M. 1972: The organic matrix of bone. *In* 'The Biochemistry and Physiology of Bone' (Ed. C. H. Bourne), 2nd ed. Academic Press, New York and London, pp. 127–189.

JOHNSON, L. C. 1966: The kinetics of skeletal remodeling. *Birth Defects, Original Article Series* **2**: 66–142.

MELLORS, R. C. 1964: Electron probe microanalysis. 1. Calcium and phosphorus in normal human cortical bone. *Lab. Invest.* **13**: 183–195.

MITAL, M. A. 1970: Biomechanical characteristics of the human hip joint and a technique of homotransplantation of its articular cartilage. *M.Sc. Thesis, University of Strathclyde, Glasgow.*

OGSTON, A. G. 1970: The biological functions of the glycosaminoglycans. *In* 'Chemistry and Molecular Biology of the Intercellular Matrix' (Ed. E. A. Balazs). Academic Press, New York and London, Vol 3, 1231–1253.

OHNSORGE, J., SCHÜTT, G., and HOLM, R. 1970: Rasterelektronenmikroskopische Untersuchungen des gesunden und des arthrotischen Gelenkknorpels. *Z. Orthop.* **108**: 268–277.

PHILIPSON, B. 1965: Composition of cement lines in bone. *J. Histochem.* **13**: 270–281.

PIEKARSKI, K. 1970: Fracture of bone. *J. Appl. Physics* **41**: 215–223.

PIZZOLATO, P., and LILLIE, R. D. 1968: The impregnation of bone and pathologic calcification by metal salts and their recognition by unoxidized hematoxylin. *Histochemie* **16**: 333–338.

POMMER, G. 1915: Zur Kenntnis der Ausheilungsbefunde bei Arthritis deformans, besonders im Bereiche ihrer Knorpelsuren, nebst einem Beitrag zur Kenntnis der lakunären Knorpelresorption. *Virchows Arch.* **219**: 261–278.

ROBINSON, R. A. 1960: Chemical analysis and electron microscopy of bone. *In* 'Bone as a Tissue'. (Eds. K. Rodahl, J. T. Nicholson, and E. M. Brown Jr.). McGraw-Hill, New York, pp. 186–250.

SCHMIDT, W. J. 1959: Grenzscheiden der Lakunen und Kittlinien des Knochengewebes. Polarisationsoptische Analyse kollagenfreier Kongorotgefärbter Schliffe. *Z. Zellforsch.* **50**: 275–296.

SMITH, J. W. 1963: Age changes in the organic fraction of bone. *J. Bone Joint Surg.* **45B**: 761–769.

SOKOLOFF, L. 1956: Natural history of degenerative joint disease in small laboratory animals. 1. Pathologic anatomy of degenerative joint disease in mice. *A.M.A. Arch. Path.* **62**: 118–128.

TONNA, E. A. 1959: The histochemical nature and possible significance of the subperiosteal reversal lines of aging rat femora. *J. Gerontol.* **14**: 425–429.

WEIDENREICH, F. 1930: Das Knochengewebe. *In* 'Handbuch der mikroskopischen Anatomie des Menschen'. (Ed. W. von Möllendorff). Springer, Berlin. Vol. **2**, Part 2, pp. 391–520.

WEINMANN, J. P., and SICHER, H. 1947: 'Bone and Bones. Fundamentals of Bone Biology'. C. V. Mosby, St. Louis, pp. 464.

WELCH, D. O. 1970: The composite structure of bone and its response to mechanical stress. *Recent Adv. Eng. Sci.* **5**: 245–262.

THE STRUCTURE OF THE BONE CARTILAGE JUNCTION

A. J. PALFREY[1]

INTRODUCTION

There have been a number of studies of the ultra-structure of articular cartilage, one of the most recent being that by Palfrey and Davies (1966) and the subject has recently been reviewed by Ghadially and Roy (1969). On the other hand there have been relatively few studies on the fine structure of bone, the most recent review being that of Hancox (1972).

Many of these studies have implied that the line of junction between articular cartilage and the underlying bone is a relatively straight line. Furthermore, no clear distinction has been drawn between the calcified part of the cartilage and the bone. It is by no means clear whether the dense lamina, which can so readily be identified either in a radiograph or in a prepared bone, is formed of cartilaginous or of bony matrix.

This gap in our knowledge can be attributed to the difficulties encountered in the preparation of specimens for microscopy. In undecalcified sections the structure of the matrix is obscured by a mass of apatite crystals. Older methods of decalcification involved the use of relatively long immersion often in strongly acid solutions, with resulting loss of detail in both cells and matrix.

This investigation uses a new rapid decalcifying method (Wu and Michaels, 1969; Hoole, 1971) in conjunction with glutaraldehyde fixation (Palfrey, 1972). The resulting preparations show little deterioration in the structure of the matrix and much cytological detail is preserved.

MATERIALS AND METHODS

Specimens were taken from six 150 g white Wistar rats (4 male, 2 female). The distal ends of both femora were excised after either the administration of ether anaesthesia or the intraperitoneal injection of 0·1 ml/100 g of a 6% solution of nembutal.

In the first two animals specimens were fixed by immersion for 4 h in 3% glutaraldehyde in phosphate buffer, pH 7·3 (Sabatini et al., 1963). The distal half of the femur was excised and split longitudinally with bone forceps. Only the peripheral parts of

these preparations showed adequate fixation. Subsequent animals were anaesthetised with nembutal, the abdominal aorta exposed and an intra-aortic infusion of the fixative made distally through a No. 20 hypodermic needle under a constant head of pressure (160 cm of water); the correct position of the needle was confirmed by prior injection of not more than 2 ml of normal saline. After the death of the animal the distal half of the femur was excised, split longitudinally and fixation continued by immersion in a further volume of the same solution for 4 h. Specimens were repeatedly washed in buffer prior to decalcification for 2–5 h in RDC (Bethlehem Instrument Ltd., Hemel Hempstead, Herts.). Thereafter blocks were cut so that no measurement was greater than 1 mm, post-fixed in 1% osmic acid in veronal acetate buffer, pH 7·4, and embedded in araldite (Palfrey and Davies, 1966).

After reorientation, sections were cut from every block at a thickness of 1 μm and stained with Azur II–methylene blue for optical microscopy (Richardson et al., 1960). Thin sections for electron microscopy were cut from selected areas and stained with uranyl acetate (Barnett and Palfrey, 1965). Thereafter further sections were cut at 1 μm, thus giving precise localisation of the areas from which the thin sections were obtained.

RESULTS

The optical microscope preparations show a well marked difference between the staining properties of the cartilage and the bone matrix (figure 1). The cartilage matrix stains a blue or mauve which varies in intensity with the depth from the surface but which is more intense in that part of the matrix which was calcified during life. In contrast the bone matrix is stained a very light blue and can be distinguished readily. The bone can be seen as a number of finger-like processes extending into the cartilaginous matrix, which is present as a series of irregular septa. A similar picture is found in both coronal and sagittal sections, indicating the finger-like form of the bony processes. Blood vessels are conspicuous within the bone although in perfused preparations blood cells are generally not seen within these vessels.

---

[1] St. Thomas's Hospital Medical School, London, England.

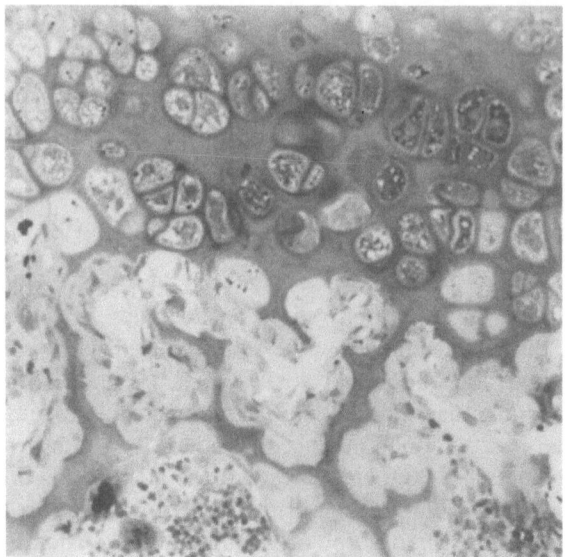

Figure 1. Optical micrograph of the junctional zone between calcified cartilage (above and right) and the finger like processes of bone, seen in oblique and longitudinal section. Araldite embedding, stained with Azur II–methylene blue (magnification × 256).

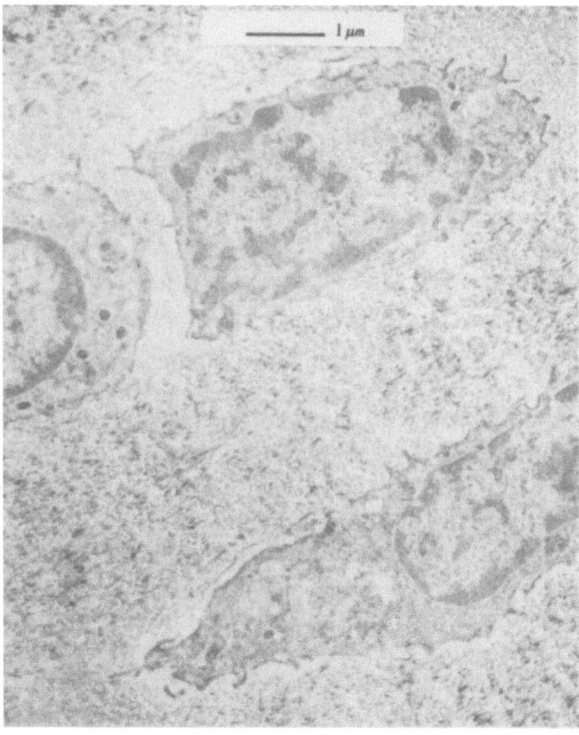

2a

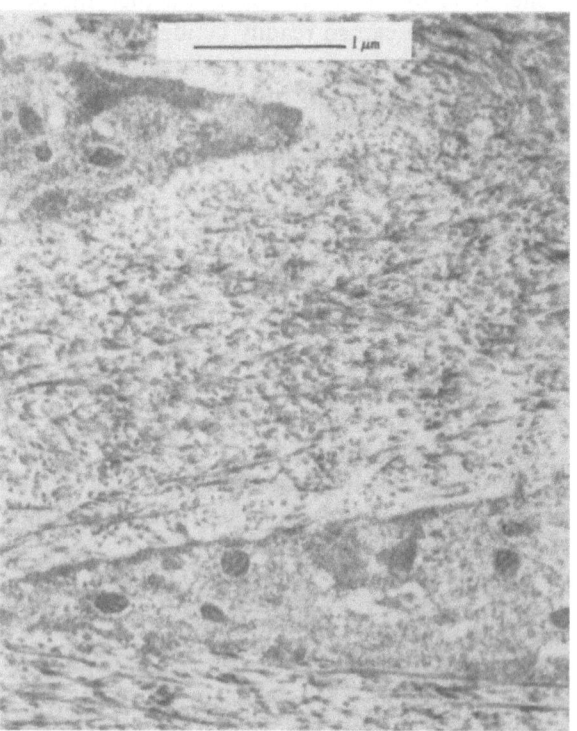

2b

Figure 2. Electron micrographs to show the matrix of the superficial zone of the articular cartilage. Notice that most of the collagen fibres are sectioned obliquely or transversely. Some electron-dense granular material is present in the interfibrillar matrix. Stained with uranyl acetate (magnification (a) × 15 000; (b) × 30 000).

The cartilage matrix

The matrix of the cartilage near the articular surface contains many collagen fibres (figure 2a) most of which are cut transversely or obliquely; at high magnification (figure 2b) these fibres can be seen to vary from 10 to 50 nm in diameter. The proportion of collagen fibres arranged in a parallel array varies widely in different parts of the cartilage. Near the margin of the cartilage (figure 3) a parallel arrangement is almost universal. In such areas the period of the cross banding on the collagen can be measured as 60–70 nm; the diameters of the fibres are more difficult to determine since two fibres often overlie one another, but it probably does not differ from that in more typical parts of the matrix.

In the deeper part of the non-calcified cartilage (the deep zone of Davies *et al.*, 1962) there is some evidence of cell degeneration (figure 4), but the structure of the matrix does not differ from that of the more superficial layers. The collagen fibres are of similar diameter (15–50 nm) and exhibit the same periodicity.

The bone matrix

The structure of the bone matrix after decalcification may be studied in sections taken from the shaft of the femur (figure 5). The collagen fibres are found in more ordered arrangement than those in the cartilage matrix. In sections cut transversely to the long axis of the bone (figure 5a) a high proportion of the fibres are cut transversely; similarly most fibres in longitudinal sections (figure 5b) are cut along their length. The diameters of the fibres are most readily measured in transverse section and vary between 40 and 150 nm; most of the fibres are seen in profile as irregular polyhedra. The distances between adjacent fibres vary from 25–30 nm and are thus generally less than the diameters of the fibres. The periodicity of the

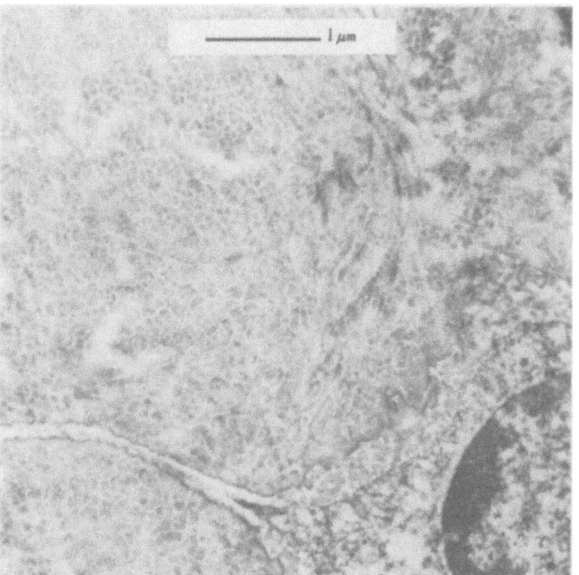

Figure 3. Part of the articular cartilage as seen in an electron micrograph. This specimen was taken from the periphery of the articular surface. Many fibres are seen running in the plane of the section and the characteristic cross banding of collagen is present. Part of a chondrocyte is shown in the lower part of the field (magnification × 20 000).

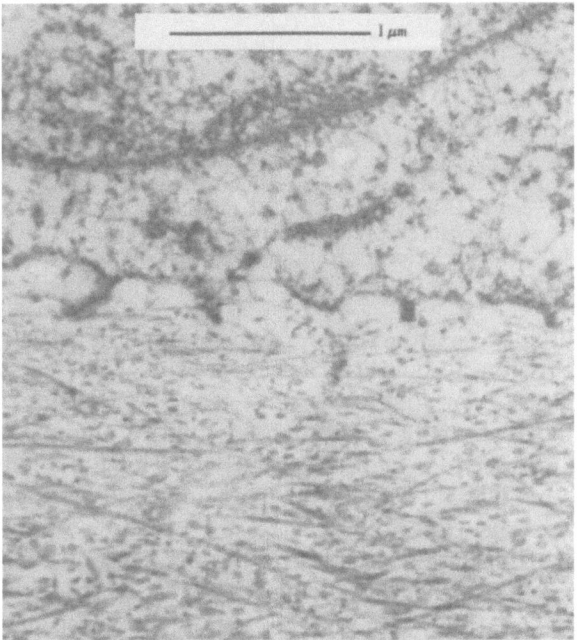

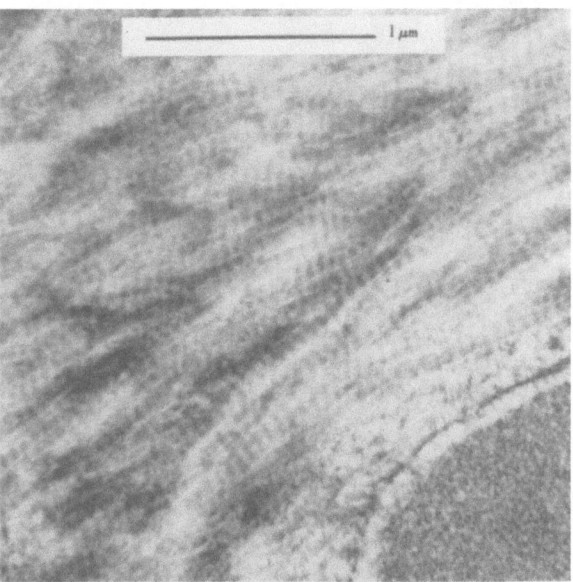

5b

Figure 5. Electron micrographs to show the structure of the matrix of bone in specimens taken from the shaft of the femur. In (a) the bone and most of the collagen fibres are sectioned transversely; the large diameters of the fibres and the small intervals between them can be seen; part of an osteocyte and its processes are present. In (b) the bone and most collagen fibres are sectioned longitudinally and the cross banding can be seen; the fibres run for only about 1 μm in the plane of the section, suggesting that they are inclined to the long axis of the bone (magnification (a) × 22 500; (b) × 45 000).

Figure 4. Electron micrograph to show the structure of the cartilage matrix near one of the chondrocytes in the deep zone of the cartilage. Collagen fibres are present in both transverse and longitudinal section (magnification × 40 000).

fibres may be discerned in longitudinal section, varying between 60 and 65 nm.

Calcified cartilage

The part of the cartilage which was calcified before preparation may be identified by its relation to the degenerate cells immediately superficial to the finger-like processes of bone (figure 6). When this part of the matrix is examined with the electron microscope (figure 7) it is very similar in appearance to the general cartilaginous matrix. The collagen fibres run in a variety of directions and measure from 15–50 nm in diameter; finer fibres predominate and these show no periodic banding. The distance between the individual fibres is generally greater

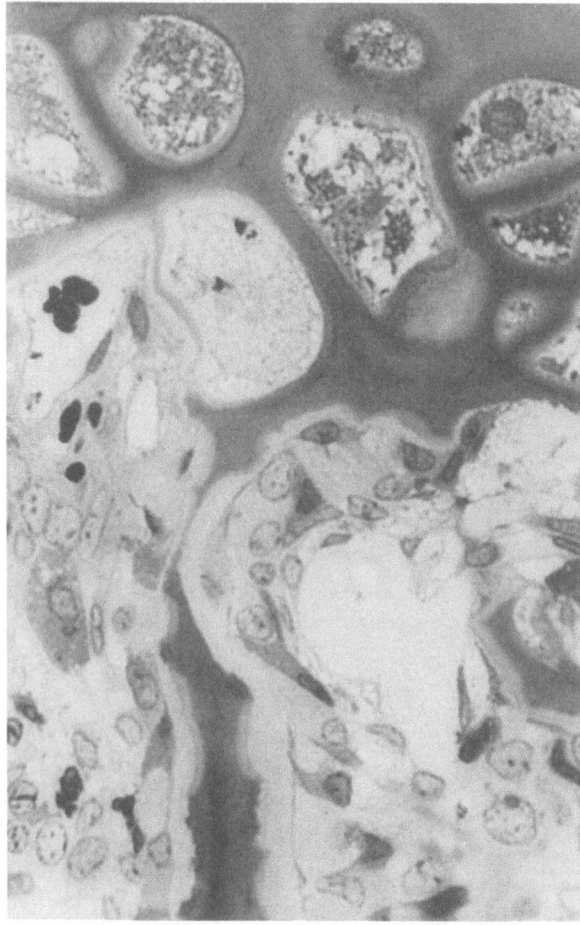

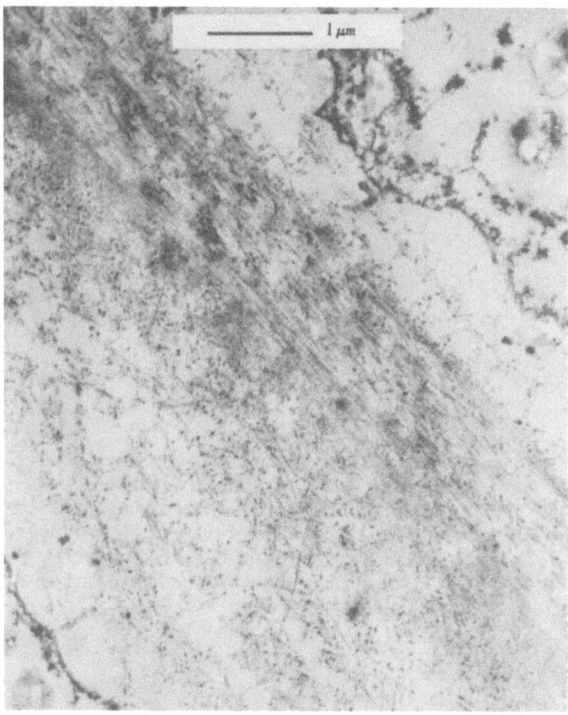

its neighbours by 50 to 300 nm, the interval some-times including granular material.

On the bone side of the demarcation the structure of the matrix is clearly that of bone: the collagen fibres are generally large in diameter (50–150 nm) and many demonstrate the usual periodicity. The distance between the individual fibres is

Figure 6. Optical micrograph taken at the bone cartilage junction. In the upper part of the field the degenerate cells of the calcified cartilage are separated by the deeply stained cartilaginous matrix. Below, two bony processes containing blood vessels are surrounded by the lightly stained bony matrix, with an intervening septum of cartilaginous matrix (magnification × 800).

Figure 7. Electron micrograph of the calcified cartilaginous matrix. Notice the small collagen fibres and the relatively large amounts of dense granular material in the inter-fibrillar matrix, within which translucent areas are present. Parts of two degenerate cells are present at the top right and bottom left (magnification × 20 000).

than the diameter of the fibres and in some areas very much larger, varying up to as much as 300 nm. Some interstices show no apparent structure but in others electron dense granular material is present.

The junctional region

The interface between bone and cartilage matrix varies in structure in different specimens. The most usual appearance is that of a dense line, varying in width between 20 and 40 nm (figure 8a). In other areas (figure 8b) this line is replaced by a broader band varying in width between 200 and 500 nm but which is marginated on both sides by a dense line, some 40 nm wide, blending with the matrix of both the bone and the cartilage; the central band is formed by a reticulum of dense material in a translucent background.

On the cartilaginous side of this junction the fine collagen fibres (10–30 nm in diameter) can be identified, and generally do not cross the line of demarcation. Each collagen fibre is separated from

usually less than their diameter and is usually between 10 and 20 nm. However, this matrix differs from that of the shaft of the bone (figure 5) in that the collagen fibres do not have a preponderant direction, both transversely and longitudinally sectioned fibres occurring in an apparently random mixture. The other interesting feature of the matrix in this situation is the presence of dense material between the collagen fibres; some of this material may represent collagen fibres which have been sectioned obliquely so that their borders are not defined, but in a few areas very fine unbanded fibres (about 5 nm diameter) can be identified. It is impossible to exclude the presence of dense granular material; a few translucent areas up to 100 nm diameter are present.

No cells are seen at the junction between the two types of matrix, and chondrocytes have not been found in the cartilaginous matrix near the line of junction. Occasional processes of osteocytes approach the junction (figure 9). Each process is

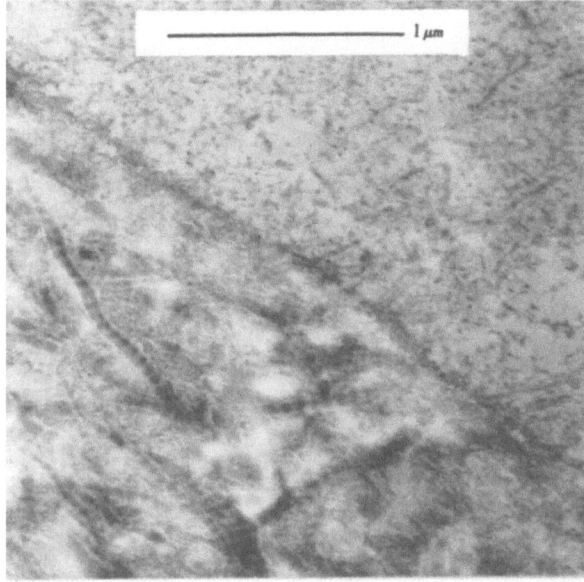

8a

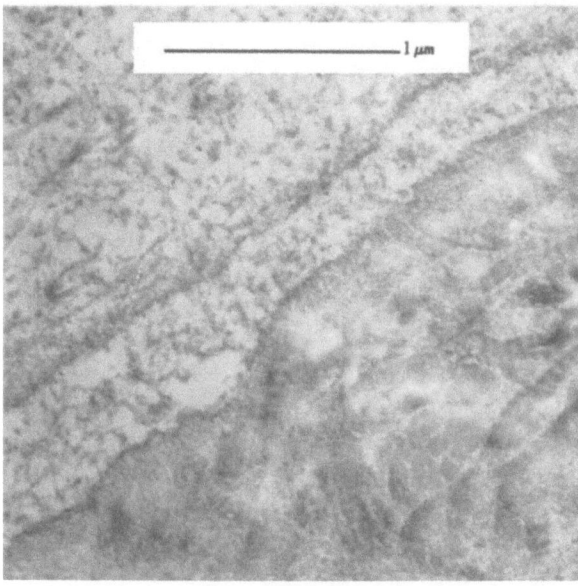

8b

Figure 8. Electron micrographs to show the line of junction between bone matrix (below) and cartilage matrix (above). In (a) the junction is marked by a narrow dense line, in (b) there is a wide zone of dense reticular material, marginated on both sides by a narrow dense line. Notice the irregular arrangement of the collagen fibres in the bone (magnification (a) and (b) × 45 000).

about 100 nm in diameter and is surrounded by a plasma membrane, but shows no internal structure. The process is separated by a translucent and structureless space from the wall of the canaliculus, which is itself demarcated from the general bone matrix by a dense irregular membrane some 10 nm in width. In most parts of this junctional region no collagen fibres cross from bony to cartilaginous matrix, although such an appearance is seen occasionally (figure 10a). It is important to remember that such an appearance will be seen if the collagen fibres on either side end at the junction,

and if the two ends happen to coincide; this coincidence will occur particularly where the plane of the section passes obliquely through the line of junction. This explanation is supported by the presence of relatively small fibres (10–20 nm in diameter) in close association with the junction, particularly on the cartilaginous side. When seen at high magnification (figure 10b) the junctional zone is formed by moderately dense granular material, within which occasional clear areas but no formed elements can be identified.

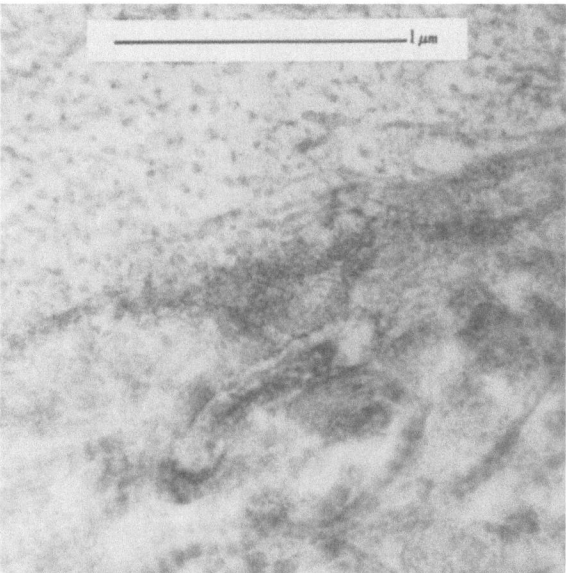

Figure 9. In this electron micrograph an osteocyte process approaches the line of junction between bone and cartilage matrix. A clear space separates the process from the surrounding canaliculus, but the significance of the space is uncertain (magnification × 60 000).

Mechanical stability

It is not, of course, possible to examine the mechanical properties of this junction in this type of specimen. However, it is of interest that in some sections partial separation occurs near the junctional zone between the two types of matrix. An early stage of such a separation is seen as a series of clear areas (figure 11a) which in more extreme instances become confluent (figure 11b); in both examples the clear areas occur within the substance of the cartilage, and some of the cartilaginous matrix adheres to the true line of the junction. It would seem that both the bony matrix and the line of junction are more resistant to this type of trauma than the matrix of the cartilage.

DISCUSSION

The intracellular matrix of both bone and cartilage is formed from collagen fibres, mucopolysaccharides and other interfibrillar material; each of these elements can be identified with the

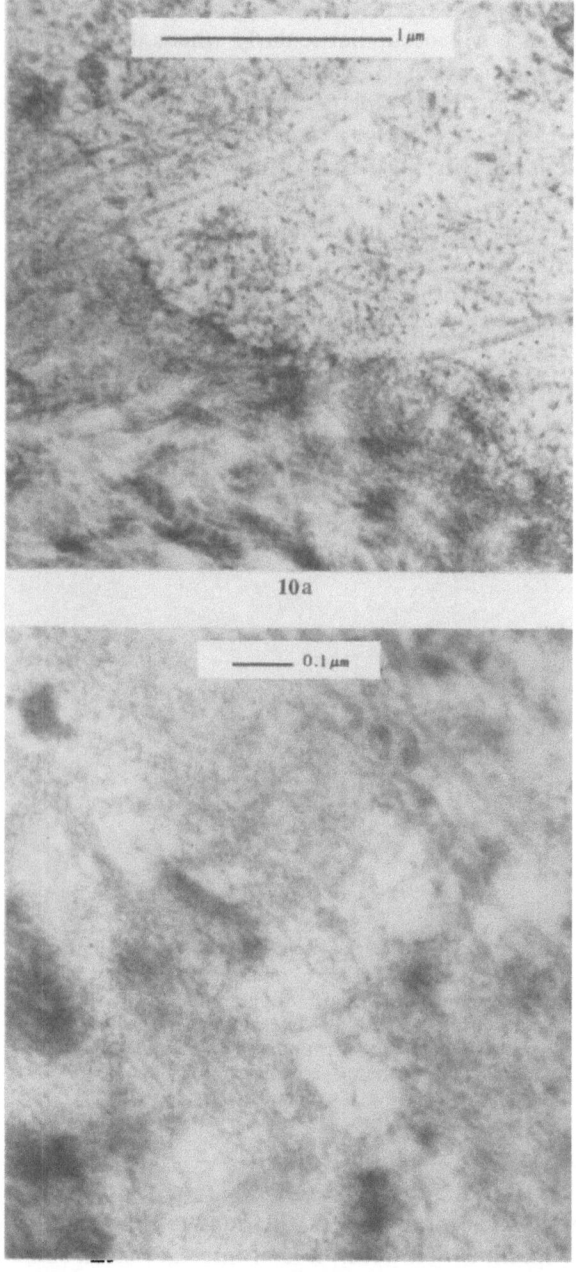

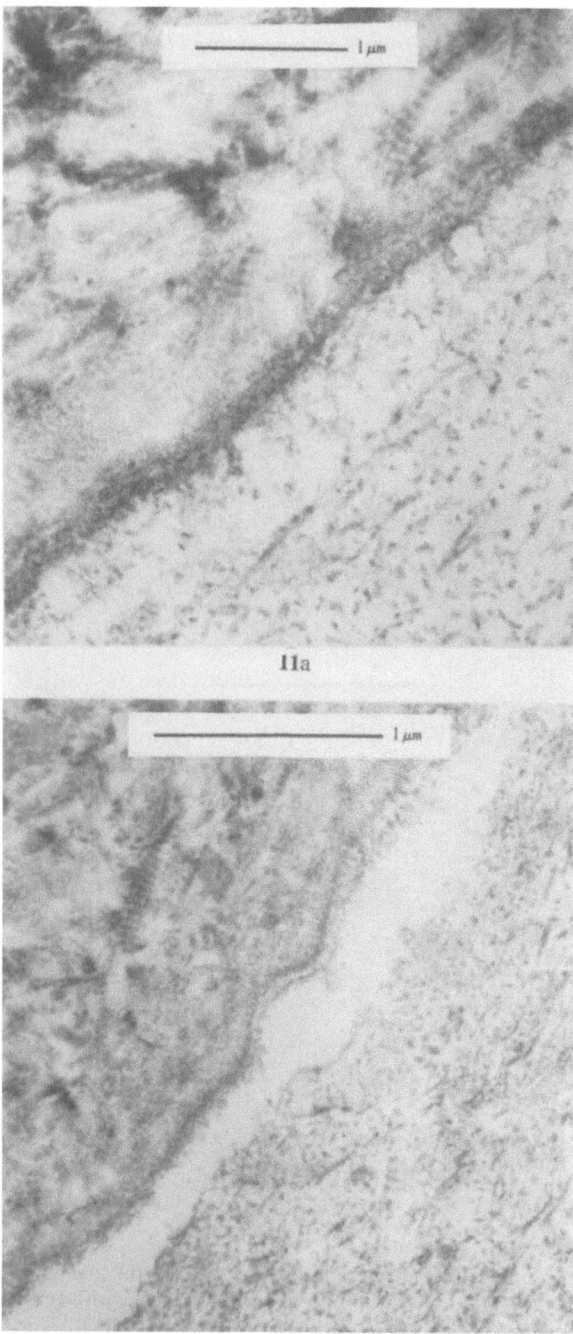

Figure 10. Electron micrographs to illustrate the detailed structure of the junction between the matrices. In (a) four collagen fibres approach and perhaps cross from cartilage to bone. In (b) the more usual structure is seen, with the junction represented by a band of dense granular material; cartilage matrix is present in the upper right corner and bone in the lower left (magnification (a) × 45 000; (b) × 120 000).

Figure 11. Two electron micrographs chosen to illustrate the site of disruption when this occurs near the line of junction. In (a) a series of discrete holes are present, while in (b) there is one continuous line of separation, but in both the rupture is in the matrix of the cartilage and not at the junction (magnification (a) × 45 000; (b) × 30 000).

electron microscope. However, the matrix of bone may be distinguished from that of cartilage by a number of factors.

First, the diameters of the collagen fibres in cartilage vary between 10 and 50 nm, whereas those in bone are usually between 50 and 150 nm. Second, in most parts of the cartilage the distances between the fibrils are generally less than their diameters, whereas in bone the fibres are substan-

tially larger than the interval between them. This difference may also be expressed by saying that the collagen fibres in bone occupy a larger proportion by volume of the matrix than in cartilage.

The third difference between cartilage and bone matrix is that in the former the collagen fibres as seen in transverse section are round, whereas in bone they are in the form of irregular polyhedra. This may be only a reflection of the difference in size of the two groups of fibres, since large irregular fibres have also been seen in older articular cartilage and in synovial membrane (unpublished observations).

The fourth point of difference is that in articular cartilage from a typical section most of the fibres are cut transversely, whatever the plane of the section. In bone taken from the shaft most of the fibres run approximately parallel to the long axis of the bone. In those parts of the bone which are immediately related to articular cartilage an almost random arrangement is seen, perhaps reflecting the complex pattern of forces exerted on the end of the bone.

The junction

The junction between cartilage and bone would seem not to be crossed by collagen fibres, except perhaps in rare instances. The line of junction has been seen with the electron microscope as a mass of electron-dense material, sometimes concentrated into a dense line, sometimes spread into a reticular mass with intervening translucent areas. The most obvious identification of this material is with mucopolysaccharides since these compounds have been thought to show these electron microscope appearances in a number of connective tissues (Ghadially and Roy, 1969; Hancox, 1972). Nevertheless this cannot be a positive identification from the nature of the technique, and it may well be that some other chemical description would be appropriate.

It is a matter of clinical observation that detachment of the cartilage from the underlying bone does not occur as a result of trauma. The complex pattern of finger-like processes of bone extending into the cartilaginous matrix goes some way to explaining this observation, as does the irregular contour of the bony processes. The observations reported here show that the line of junction appears to lack formed elements such as connective tissue fibres. This work also suggests that the line of junction is stronger than the adjacent cartilage matrix, even if only in response to the unnatural trauma produced in these materials during preparation for microscopy.

The tissues which have formed the subject of this study have been notoriously difficult to fix, even for optical microscopy. The present method of fixation seems to compare with other methods that have been tried, but there is still room for improvement. In these circumstances spaces seen in relation to plasma membranes and specifically spaces between osteocyte processes and the walls of their canaliculi should be interpreted with caution.

SUMMARY

1 The structure of the junctional region between articular cartilage and the subchondral bone has been studied after rapid decalcification by optical and electron microscopy.

2 Differences of structure between the cartilaginous and bony matrix have been described.

3 The differences in structure between subchondral and cortical bone are less pronounced than those between the general matrix of these two tissues.

4 The line of junction is formed by a band of electron dense material, presumed to be mucopolysaccharide in nature.

5 Collagen fibres only rarely cross the junction.

6 There is evidence to suggest that the junction is stronger than the adjacent cartilaginous matrix.

REFERENCES

BARNETT, C. H., and PALFREY, A. J. 1965: Absorption into rabbit articular cartilage. *Journal of Anatomy* **99**: 365–375.

DAVIES, D. V., BARNETT, C. H., COCHRANE, W., and PALFREY, A. J. 1962: Electron microscopy of articular cartilage in the young adult rabbit. *Annals of rheumatic Diseases* **21**: 11–22.

GHADIALLY, F. N., and ROY, S. 1969: 'Ultrastructure of Synovial Joints in Health and Disease'. Butterworths, London.

HANCOX, N. M. 1972: 'Biology of Bone'. Cambridge, London.

HOOLE, P. F. 1971: Rapid decalcification of bone for diagnostic histology. *Medical Laboratory Technology* **28**: 201–204.

PALFREY, A. J., and DAVIES, D. V. 1966: The fine structure of chondrocytes. *Journal of Anatomy* **100**: 213–226.

PALFREY, A. J. 1972: Some preliminary observations on the fine structure of decalcified bone. *Journal of Anatomy* **111**: 482–483.

RICHARDSON, K. C., JARRETT, L., and FINKE, E. H. 1960: Embedding in epoxy resins for ultrathin sectioning in electron microscopy. *Stain Technology* **35**: 313–323.

SABATINI, D. D., BENSCH, K. G., and BARRNETT, R. J. 1963: Cytochemistry and electron microscopy. The preservation of cellular ultrastructure and enzymatic activity in aldehyde fixation. *Journal of Cell Biology* **17**: 19–58.

WU, A., and MICHAELS, L. 1969: A new proprietary decalcifying agent; comparison with established methods and use in routine histopathology. *Canadian Journal of Medical Technology* **31**: 224–227.

BIOENGINEERING ORIENTATED STUDIES ON THE STRUCTURE OF HUMAN HIP JOINT CARTILAGE

P. F. MILLINGTON AND I. C. CLARKE[1]

Successful and continued functioning of articular joints is essential to normal life. In many instances, however, joint pain does occur giving rise to considerable discomfort if not complete incapacity. A common cause is the breakdown of cartilage which can result from various diseases, e.g. osteo-arthrosis, a disease of articular cartilage (Wright, 1969). Before we can understand the manner in which cartilage becomes disrupted or begin to interpret mechanical data derived from the study of whole or part joints, it is essential to know the normal structural arrangement of the tissue. It may then be possible to determine the role of cartilage under normal joint function and the mechanism by which the joints are lubricated.

Investigations of the composition and organisation of cartilage have been undertaken for at least 200 years. During the period 1723 to 1950 the light microscope was the only instrument really suitable for this work. Since 1950 the introduction of electron microscopes and the development of microprobe analysers has greatly changed the effectiveness of morphological studies.

Much of the earlier work has been re-presented in a number of reviews during 1971 (Gardner and McGillivray, 1971a; Clarke 1971a, b; Millington et al., 1971a). Since 1968, members of the Bioengineering Unit, University of Strathclyde, have been concerned with both the structural and mechanical aspects of cartilage. This paper is concerned with an assessment of the associated morphological work.

SCANNING ELECTRON-MICROSCOPE STUDIES

In 1968, McCall demonstrated the value of the scanning microscope (SEM) in the study of articular cartilage. He described cell-like lumps on the surfaces of the tissue from young subjects, but ridges on the surface of human adult cartilage. His results did not confirm those previously reported by light and transmission electron-microscope studies, except for one notable paper by Hammar (1894) who described shallow depressions on the articular surfaces (Gardner and McGillivray, 1971b). McCall (1969) also described the internal fibrillar organisation of articular cartilage. He identified three zones: (1) the superficial zone containing tightly packed bundles of fibrils orientated parallel with the surface; (2) the mid zone, composed of loosely packed randomly arranged fibrils; and (3) a deep zone, in which the fibrils were packed together in radially orientated bundles.

Following this preliminary study, a comprehensive research programme was developed to provide further basic information on possible load-bearing structures in the human hip joint, a prerequisite for bioengineering studies of joint lubrication and wear.

Surface studies

Although McCall (1968) and Walker et al. (1969) described large parallel arrays of ridges on the surface of articular cartilage, there remained the possibility that this appearance could have been due to preparation artifact. Clarke (1971c) soon discovered that dehydration of the tissue resulted in shrinkage large enough to produce such distortion. He minimised the shrinkage by preparing comparatively large blocks of cartilage with underlying bone attached. The surfaces of such specimens generally contained arrays of depressions 20 to 40 μm in diameter in an otherwise smooth surface (figure 1 A). The frequency of the depressions varied from 190 to 900 per mm^2, with an average value of 430 per mm^2. It was noted that the shapes of the depressions corresponded to the underlying cells which suggested that the cells had in fact collapsed.

To avoid the dehydration steps used in scanning electron microscopy, Clarke (1971d) studied replicas (figure 1 B) taken from moist articular surfaces in vitro (figure 1 C). Measurement of the replicated contours with a profile recorder indicated that the depths of the depressions on articular surfaces varied from 0·2 to 5 μm (figure 2 A). However, while dehydration had been avoided, the presence of the replicating medium itself may have affected the tissue and altered the spatial characteristics. In addition, it was difficult to estimate how accurately the wet articular contours had been replicated. To overcome these limitations Clarke (1972) applied

---

[1] University of Strathclyde, Glasgow, Scotland.

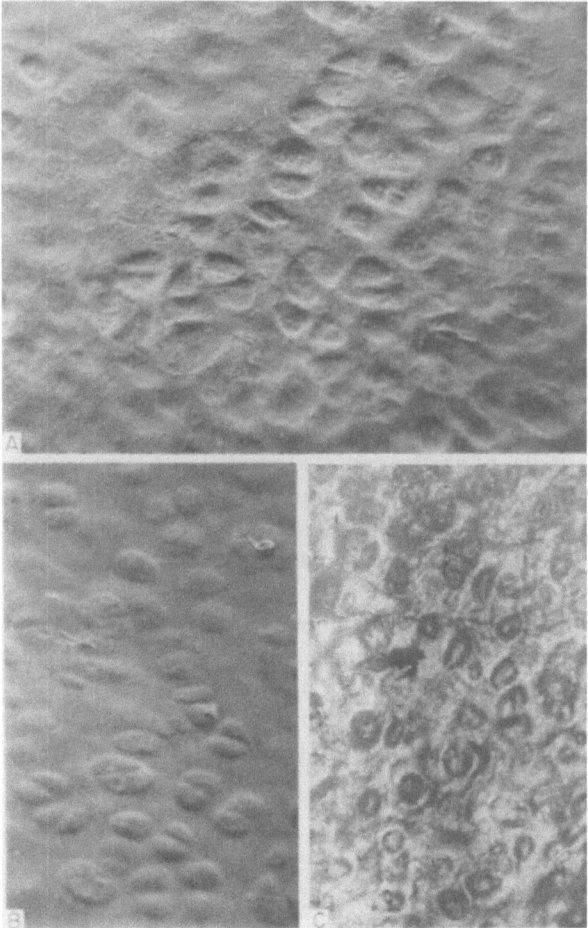

Figure 1. (A) Scanning electron micrograph of dehydrated articular surface, showing typical oval depressions 20 to 30 μm in diameter, magnification × 270. (B) Scanning electron micrograph of acrylic replica taken from a moist articular surface. The 'figure-of-eight' contours are now represented by humps, magnification × 270. (C) Reflected-light micrograph illustrating similar contours as in figure 1, A on a moist articular surface. (Magnification × 270).

stereomicroscopy techniques to both the articular surfaces and their replicas. In this study the replicated contours varied from 0·3 to 9 μm in height and the articular surface contours from a similarly low value to 15 μm at the upper limit (figure 2, B, C). The good correlation in the 0–9 μm range (figure 2) suggested that the replicas had reproduced the wet articular surfaces fairly accurately and furthermore that the SEM preparation techniques had not significantly altered the degree of irregularity of the wet *in vitro* articular surface contours. Of the two methods of contour measurement, the SEM stereomicroscopy was the more satisfactory because there was no stylus to cut into the replica surface. This may have accounted for the generally lower values measured by the profile recorder. Additionally, with SEM stereomicroscopy the operator has a true 3-D image in which he can identify and measure each object. Table 1 sum-

marises the range of values obtained by these methods.

Within the small variations described above it has been possible to measure with considerable accuracy the absolute peak-to-valley heights of the articular surface contours in prepared tissue specimens. However, although it has been shown that similar undulations exist *in vivo* (Gardner and McGillivray, 1971*b*) the possibility exists that such contours could well be formed by humps as opposed to depressions. It must be emphasised that no reliable information is yet available by which we can define the exact nature and magnitude of the contours of joint surfaces *in vivo*.

The Superficial zone

The superficial zone of cartilage is of considerable interest in relation to the normal lubrication and function of the articulating surfaces. Most specimens prepared for transmission electron microscopy have a thin amorphous layer on the outer

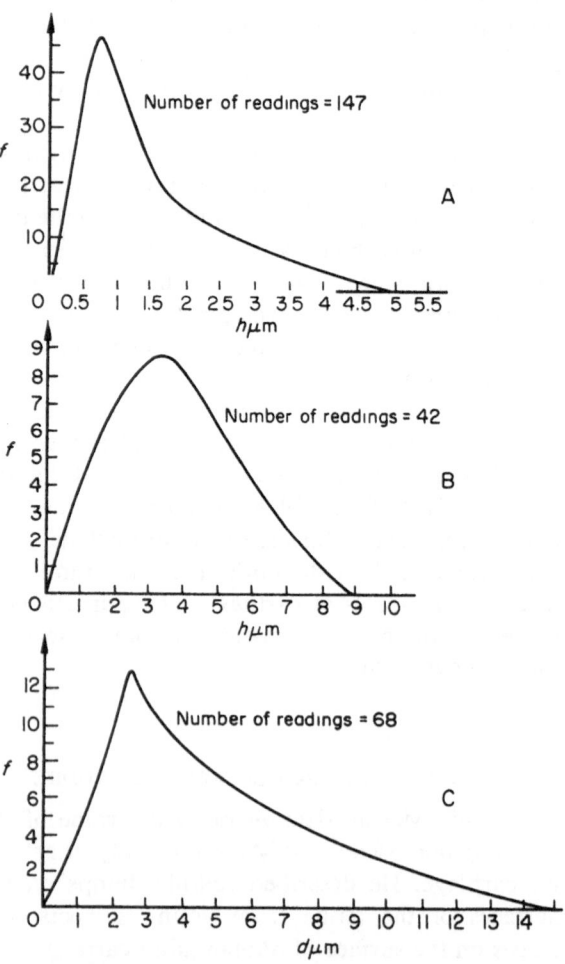

Figure 2. Distribution (f) of depression depths (d) in articular surfaces and corresponding heights (h) on replicated contours, similar to those described in figures 1 A and B. (A) Measured from replicas by means of a profile-recording device. (B) Measured from articular replicas by stereomicroscopy. (C) Measured from articular surfaces by stereomicroscopy.

surface of the cartilage, (Weiss *et al.*, 1968). The cartilage, when viewed by scanning electron microscopy, shows little of the underlying fibrous structure, thus indicating the presence of some amorphous material on the surface. Even uncoated specimens have the same sort of appearance. However, Clarke (1971*a*) has now confirmed that at magnifications above ×9000 in the SEM, the

from the surface of cartilage to expose the underlying fibrous networks. Mital and Millington (1971) studied these peelings in great detail and discovered that the peeled layer often took the form of a wedge-shaped piece of tissue tapering away to a very thin layer close to the articular surface. Careful study of these peelings and the exposed surface led to the suggestion that beneath the thin surface amorphous

A

Study	Technique used	Subject	Number of measurements	Peak-to-valley heights	
				range (μm)	typical values (μm)
Measurement of peak-to-valley heights of articular surface contours	Profile recorder	Articular surface replicas	147	0·2–5·0	0·2–2·0
	Stereomicroscopy (scanning electron microscope)	Articular surfaces	68	0·4–14·9	1·0–5·0
		Articular surface replicas	42	0·3–8·9	1·0–6·0

B

Study	Technique used	Area examined	Number of measurements	Depression or lacunar (number/mm²)		Depression or lacunar diameter ranges (μm)
				range	average	
Assessment and comparison of frequencies of articular surface depressions and underlying lacunae	Scanning electron microscope	Lacunae visible on tangentially-sectioned articular surfaces	24	186–1007	611	10–53
		Depressions visible on articular surfaces	31	189–1070	588	10–68
	Reflected-light microscope	Lacunae visible on tangentially-sectioned articular surfaces	11	191–366	287	13–60
		Depressions visible on articular surfaces	10	200–323	249	12–54
	Transmitted-light microscope	Lacunae visible in tangential sections	22	295–725 (equivalent to 29 500–72 500/mm³)	427 (42 700/mm³)	10–71

TABLE 1: Summary of data obtained *in vitro* from micro-studies of adult human femoral articular surfaces. The degree of irregularity created by depressions on the articular bearing surfaces is given in A, while the frequencies of such contours are listed in B.

surface layer has the appearance of a densely woven, random, fibrillar network which is generally thought to be collagen (Weiss *et al.*, 1968).

From thick specimens the scanning electron microscope gives information relating only to the surface presented to the electron beam. To obtain information relating to the underlying layers, specimens must be fractured or cut. Clarke (1971*a*) found that a layer of tissue could be peeled away

layer there existed a zone in which could be detected a transition from the three-dimensional random arrangement of fibrils in the deep region to parallel fibres very close to the surface. This suggestion was based on an assessment of predominant orientations of the fibrils close to the surface. The suggested arrangements are illustrated in figure 3. In both these studies and those by Clarke (1971*a*) there were no closely packed parallel bundles of fibrils observed which could have represented the orderly array of surface fibres or ridges described by McCall (1968) and Walker and colleagues (1969).

The superficial zone of fractured surfaces, when

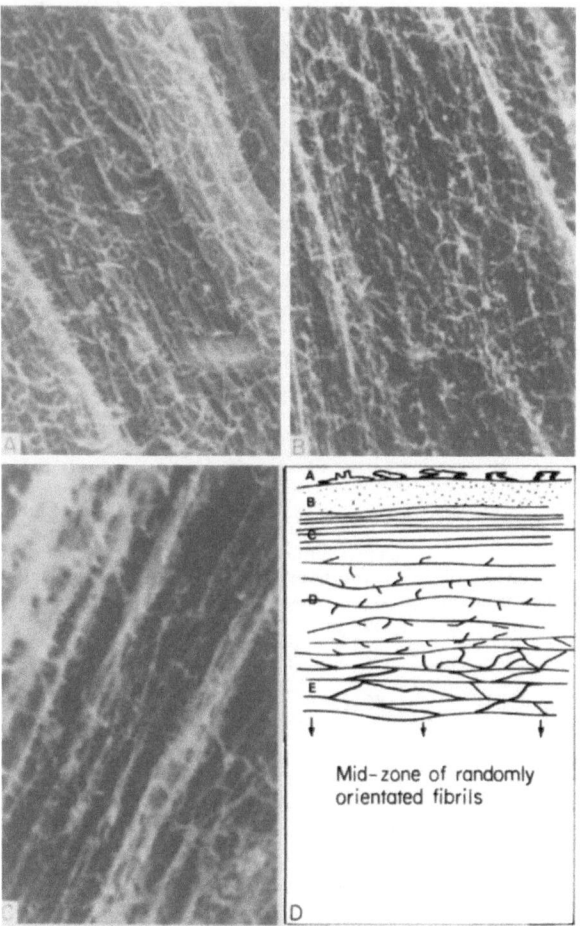

Figure 3. The structure of the superficial region of femoral articular cartilage. (A) Scanning electron micrograph of layer C, showing close packed parallel fibrils (Magnification × 6000). (B) Scanning electron micrograph of layer D showing many fibrils both with cross-linking and many short torn fibrils (fronded). The fronds may have been caused by the tearing action used in the preparation of the tissue wedge (Magnification × 5500). (C) Scanning electron micrograph of layer E in which the cross-linking of the composite fibrils is a prominent feature. This zone merged imperceptibly with the more random arrangement of the mid-zone of articular cartilage (Magnification × 12000). (D) Diagram of suggested transition layers of the superficial region of articular cartilage, where layer A represents the material deposited on the surface during preparation; layer B is the amorphous surface 20 to 30 nm thick; layer C is the unbridged parallel fibril layer less than 50 nm thick; layer D represents the fronded parallel fibrils and may be up to 3 microns thick; and layer E (4 μm) represents the net-like structures overlying the more random arrangement of the mid-zone fibrils.

inspected in the scanning electron microscope, seldom showed cellular structures. Nevertheless Clarke (1972) has been able to show that chondrocytes and their lacunae do occur very close to the articular surface itself. Indeed, the data available indicates a very close correlation between surface depressions and lacunae.

The mid and deep zones

Beneath the thin superficial layer composed principally of parallel layers of fibrils lies a zone in which the fibrillar network has no regular orientation. This random network extends throughout the cartilage to the calcified region with little variation in density or orientation. The thickness of fibrils varied considerably from 40 to as much as 460 nm and strands of over 2100 nm thickness (2·1 μm) have been measured in the matted networks of some fractured specimens.

While the fibrillar organisation at high magnifications appeared generally random below the superficial zone, at low magnifications the network of fibrils took on the appearance of a series of radially orientated overlapping layers. Clarke (1971a) found that such layers turned obliquely at or near the superficial zone or appeared to run parallel with it. The resulting pattern was strikingly similar to the simplified 'arcade' structural model proposed by Benninghoff (1925).

Possibly the appearance of these radially orientated layers in the tissue led to the various reports of 'radially orientated fibrils', particularly in the deeper zones (Benninghoff, 1925; Little et al., 1958; McCall, 1969).

In the basal regions Mital (1970) found that the fibrils became orientated radially and in some instances came together to form bundles that penetrated into the subchondral boneplate. Also in the basal regions of articular cartilage, certain channels have been identified running from the spongy bone through the subchondral boneplate into the cartilage zone (Mital and Millington, 1970). These varied between 50 and 70 μm in width in the subchondral and osteochondral region, and decreased to between 5 and 10 μm wide at the junction of the calcified and deep zones. They contained in their walls what appeared to be pores, 3–5 μm in diameter. An example of such a channel is shown in figure 4.

Articular cartilage is avascular and much experimental work has been done to determine its source of nutrition. Some confusion arose from the early work which did not distinguish between immature and mature cartilage. However, Maroudas et al. (1968) and others found that in pigs the cartilage-bone interface was permeable to water and solutes in the immature, but not the mature tissue. This indicated that the adult cartilage could not receive its nourishment through the bony tissue layers, as was possible in the younger tissue. However, Greenwald and Hayes (1969) suggested that there were distinct species differences. They showed that vascular channels passed through the subchondral plate and penetrated the articular cartilage in the femoral heads of adult human subjects but were absent in adult rabbit tissue. Mital and Millington (1970) suggested that the channels they described by scanning electron microscope study of human articular cartilage were those identified by Greenwald and Hayes.

Role of articular cartilage fibrils in load bearing

Following the identification of fibrillar structures in articular cartilage, there has been considerable speculation about their role in load carrying.

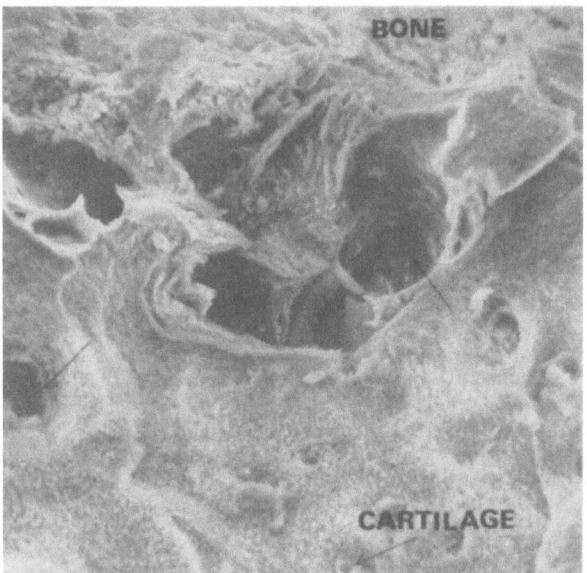

Figure 4. Scanning electron micrograph of channels between bone and cartilage in the human femoral head. The channels (arrows) appear to traverse the region from the subchondral bone plate into the cartilage itself. The walls of the channels appear to be perforated at intervals with pore-like structures. It is suggested that these channels could be concerned with nutrition of cartilage (Magnification × 1200).

Benninghoff (1925) suggested that the load carried by the surface layer was transmitted to the calcified zone by means of radial fibrils while MacConaill (1951) argued that a system of fibrils oblique rather than radial would be more suited to resist the applied forces.

However, it is now known that while both radial and oblique fibrils exist to some extent, the overall structure is that of a random network except for the distinct regions immediately subjacent to the articular surface and near the calcification zone.

Possibly the superficial zone of densely layered fibrils is the main structural entity on which depends the load-carrying ability and ultimately the integrity of the tissue. This surface layer is much less permeable than the underlying zones (Muir *et al.*, 1970) and could therefore function as the envelope or 'membrane' of the fluid-filled system which comprises the cartilage matrix. Clarke (1971*a*) suggested that such a system would permit joint loads to be transmitted through the cartilage to the underlying bone in the same manner that a balloon when stepped upon will transmit the load to the floor. However, as the load is applied, the balloon membrane can freely expand, without detriment to itself, in those regions away from the loaded area; expansion includes a displacement of the membrane outwards and a lateral expansion of the membrane itself. The fibrillar structures of cartilage may well be adapted to withstand just such derangements. The interwoven fibril bundles parallel to and just below the articular surface may resist any lateral expansion of the cartilage, while simultaneously acting as a bearing surface and a semipermeable 'membrane'. The general randomly orientated fibrillar network, enclosed in the calcified zone, may act as a bracing system under tension for the restraint of the surface 'membrane', that is, the superficial zone.

IN-VITRO LOAD STUDIES OF SYNOVIAL JOINTS

Articular cartilage is subjected to cyclic loading throughout the life-span of each individual. Recent work has suggested that synovial fluid and articular cartilage contribute little if at all to the force-attenuating properties of synovial joints (Radin *et al.*, 1969, 1970). It is currently thought that the bony structures themselves absorb most of the energy transmitted across the joints during cyclic loading.

Studies have been made in the Strathclyde Bioengineering Unit to determine the time-dependent deformation response of articular cartilage to static load, and to compare the load-cycle responses of constrained and unconstrained tissue (Graham, 1969). However, the form in which the experiments were conducted did not allow for detailed analysis. One limitation was that small sections of tissue were employed and hence the results could not be meaningfully extrapolated to the situation where complete joints were under load. In the latter example the degree of congruity with which the two components of the joint fit together may be of importance. Hammond and Charnley (1967) suggested that the femoral head could be completely accommodated within the acetabulum under the load of body weight through compliance of the two layers of cartilage. Studies by Bullough *et al.* (1968) indicated that neither the femoral head nor acetabulum was completely spherical, especially towards the tips of the 'horseshoe' acetabular area. They found that there was however a significant increase in acetabular sphericity with age. An investigation by Greenwald (1970) showed that generally, at loads greater than 25 per cent of body weight, the entire cartilage surface of the acetabulum was in contact with the femoral cartilage. With ageing the load required to produce complete contact became much less, confirming that the joints become more congruent with age.

Accepting the need for load studies of complete joints rather than isolated plugs of tissue, Mital (1970) attempted to extend the work of Graham (1969) by loading hip joints. Mital demonstrated that there was a difference between the response of the dominant side joint components ('right-handed') when compared with that of the opposite-side hip. With the data plotted with respect to age there appeared to be a decrease in failure load and the two curves approached one another on ageing, indicating a connection with the reduced activity of the subject (Millington et al., 1971).

CONCLUSIONS

From these and other experiments it was seen that a detailed analysis of the fibrous structure of articular cartilage combined with a study of the load and lubrication properties of the joint was necessary before meaningful attempts at a description of the mechanical properties of the tissue could be made. It is now hoped that through a process of continual refinement it will be possible to integrate both the structural and mechanical aspects of the cartilage studies. To this end studies of the response of articular cartilage to static and eventually dynamic loading in an oscillating rig (a simulated wear programme) have been undertaken (Clarke, 1972; Parker, 1972). Through these and other studies it is hoped that a positive contribution may be made to our understanding of the function of this important tissue in both health and disease.

REFERENCES

BENNINGHOFF, A. 1925: Die Modelleirenden und former haltenden Faktoren des Knorpe Reliefs. Z. Anat. Ent wegesche 76: 43.

BULLOUGH, P., GOODFELLOW, J., GREENWALD, A. S., and O'CONNOR, J. 1968: Incongruent surfaces in the human hip joint. Nature 217: 1290.

CLARKE, I. C. 1971a: Articular cartilage: A review and scanning electron microscope study. J. Bone and Joint Surg. 53(B): 732.

CLARKE, I. C. 1971b: Human articular surface contours and related surface depression frequency studies. Ann. Rheum. Dis. 30: 15.

CLARKE, I. C. 1971c: Surface characteristics of human cartilage; a scanning electron microscopy study. J. Anat. 108: 23.

CLARKE, I. C. 1971d: A method for the replication of articular cartilage surfaces suitable for the scanning electron microscope. J. Microsc. 93: 67.

CLARKE, I. C. 1972: 'A study of the structure and wear response of hip joint cartilage'. Ph.D. Thesis, University of Strathclyde, Glasgow.

GARDNER, D. L., and McGILLIVRAY, D. C. 1971a: Surface structure of articular cartilage: Historical review. Ann. Rheum. Dis. 30: 10.

GARDNER, D. L., and McGILLIVRAY, D. C. 1971b: Living articular cartilage is not smooth. Ann Rheum. Dis. 30: 3.

GRAHAM, J. 1969: Research report to McLaughlin Foundation. University of Strathclyde, Glasgow.

GREENWALD, A. S. 1970: 'Transmission of forces through animal joints'. D.Phil. Thesis, University of Oxford.

GREENWALD, A. S., and HAYES, D. W. 1969: A pathway for nutrients from medullary cavity to the articular cartilage of the human femoral head. J. Bone and Joint Surg. 51(B): 747.

HAMMAR, J. A. 1894: Ueber den feineren Bau der Gelenke. Abth. II Der Gelenkknorpel. Arch. Mikros. Anat. 43: 813.

HAMMOND, B. T., and CHARNLEY, J. 1967: The sphericity of the femoral head. Med. and Biol. Eng. 5: 445.

LITTLE, K., PIMM, L. H., and TRUETTA, J. 1958: Osteoarthritis of the hip. J. Bone and Joint Surg. 40(B): 123.

MAROUDAS, A., BULLOUGH, P., SWANSON, S. A. V., and FREEMAN, M. A. R. 1968: The permeability of articular cartilage. J. Bone and Joint Surg. 50(B): 166.

MACCONAILL, M. A. 1951: The movements of bones and joints. J. Bone and Joint Surg. 33(B): 251.

McCALL, J. 1968: 'The microarchitecture of articular cartilage and its load deformation response'. M.Sc. Thesis, University of Strathclyde, Glasgow.

McCALL, J. 1969: Load deformation studies of articular cartilage, J. Anat. 105, 212.

MILLINGTON, P. F., GIBSON, T., EVANS, J. H., and BARBENEL, J. C. 1971a: Structural and mechanical aspects of connective tissue. Pp. 189-248 in 'Advances in Biomedical Engineering', Volume 1 (Ed. R. M. Kenedi). Academic Press, London.

MILLINGTON, P. F., MITAL, M. A., and CLARKE, I. C. 1971b: Articular cartilage—Recent studies. CHEMECA 70. Butterworths Australia Session 4, p. 74.

MITAL, M. A., and MILLINGTON, P. F. 1970: Osseous pathway of nutrition to articular cartilage in the human femoral head. Lancet 7651: 842.

MITAL, M. A., and MILLINGTON, P. F. 1971: Surface characteristics of articular cartilage. Micron 2: 236.

MITAL, M. A. 1970: 'Biomechanical characteristics of the human hip joint and a technique of homotransplantation of its articular cartilage'. M.Sc. Thesis, University of Strathclyde, Glasgow.

MUIR, H., BULLOUGH, P., and MAROUDAS, A. 1970: The distribution of collagen in human articular cartilage with some of its physiological implications. J. Bone and Joint Surg. 52(B): 554.

PARKER, A. 1972: 'In-vitro studies on the biomechanics of articular cartilage'. Thesis (in preparation), University of Strathclyde, Glasgow.

RADIN, E. L., and PAUL, I. L. 1969: Failure of synovial fluid to cushion. Nature 222: 999.

RADIN, E. L., PAUL, I. L., and POLLOCK, D. 1970: Animal joint behaviour under excessive loading. Nature 226: 554.

WALKER, R. S., SIKORSKI, J., DOWSON, D., LONGFIELD, M. D., WRIGHT, V., and BUCKLEY, T. 1969: Behaviour of synovial fluid on the surfaces of articular cartilage. Ann. Rheum. Dis. 28: 1.

WEISS, G., ROSENTHAL, L., and HELFET, A. J. 1968: An ultrastructural study of normal young adult human articular cartilage. J. Bone and Joint Surg. 50(A): 663.

WRIGHT, V. 1969: p. 15 in 'Lubrication and Wear in Joints', (Ed. V. Wright). Sector Publications, London.

VARIATIONS IN THE PHYSICO-CHEMICAL AND MECHANICAL PROPERTIES OF HUMAN ARTICULAR CARTILAGE

1: PHYSICO-CHEMICAL PROPERTIES

ALICE MAROUDAS[1]

INTRODUCTION

The purpose of part 1 of this paper is to describe briefly the structure of articular cartilage as revealed by physico-chemical studies, and to show how this structure affects cartilage nutrition and load bearing.

The non-cellular component of cartilage, the so-called matrix, consists of collagen fibres embedded in a gel of proteoglycans and water, the latter containing a number of low-molecular-weight solutes (see figure 1). Since cells occupy a small fraction of the total volume of articular cartilage, the physico-chemical and mechanical properties of cartilage are defined chiefly by the properties of the matrix.

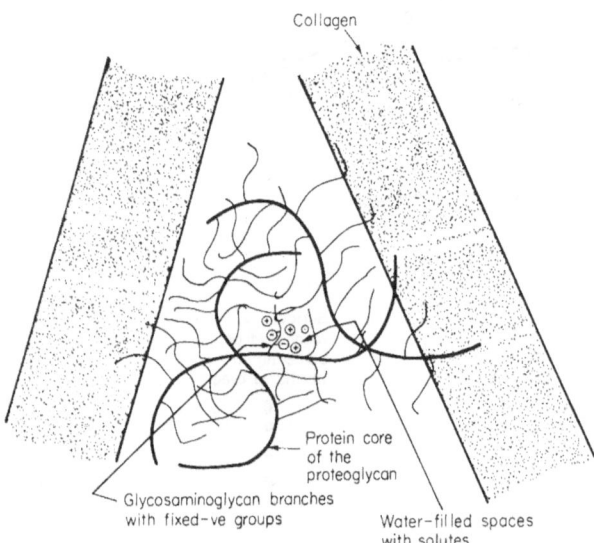

Figure 1. Schematic diagram of the network structure of the proteoglycan – water gel.

While the collagen network is responsible for the integrity of the tissue and its mechanical strength, it is the proteoglycan – water gel with its very fine pores (10–40 Å in diameter) which controls the diffusion of solutes and the movement of water through cartilage. The negatively charged carboxylate and sulphate groups of the glycosaminoglycans determine the ionic equilibria between

cartilage and synovial fluid; they also determine the osmotic pressure existing within cartilage.

Water, collagen, and proteoglycan content vary in cartilage within fairly wide limits. Variations in the amount of collagen and its orientation are important from the point of view of the tensile properties of cartilage and will be discussed in part 2 of this paper. The proportion of proteoglycans to water in the matrix determines the number and size of pores and is therefore the chief parameter controlling the transport properties of cartilage.

VARIATIONS IN GLYCOSAMINOGLYCAN CONTENT

A detailed study has recently been made (Maroudas et al., in press) of the topographical variations in the glycosaminoglycan content over the area of a joint both in normal and in fibrillated specimens. The area chosen for this study was the femoral head, but experiments were also carried out on the acetabulum and the femoral condyles, and these gave similar results. The glycosaminoglycan content is quantitatively related to the concentration of negatively charged fixed groups, and these can be determined by physico-chemical methods (Maroudas et al., 1969; Maroudas and Thomas, 1970). Hence fixed-charge density was used as the experimental parameter, because its determination is very rapid, and can be carried out on very small quantities of tissue. The results are shown in table 1 and figure 2. All results are expressed as milliequivalents per wet weight of cartilage, since in this way both the variations in glycosaminoglycan content and in water are taken into account. (Figures 4, 5 and 6, however, are taken from other sources and give results in milliequivalents per cubic centimetre.) The following were our main conclusions.

(i) Apparently normal cartilage (i.e. cartilage in which the surface layer was intact) showed no statistically significant variations ($P > 0.1$) in fixed-charge density over the area of the femoral head.

(ii) Areas of cartilage which showed signs of fibrillation, however slight, had a lower fixed-charge density than normal specimens. They also had a higher water content.

(iii) If no distinction is made between apparently completely intact cartilage and cartilage showing

---

[1] Biomechanics Unit, Imperial College, London, England.

Alice Maroudas

varying degrees of fibrillation, fixed-charge density does exhibit considerable variations from area to area, the mean being highest on the superior surface (site 1, figure 2) and lowest below the fovea (site 3). This is consistent with the frequency of fibrillation observed at these sites. Out of the 33 adult femoral heads examined only one showed fibrillation at site 1, while 22 showed surface roughening or more serious fibrillation at site 3.

heads which were either completely normal throughout or showed small areas of surface roughness only. The two groups were compared with respect to the mean fixed-charge density at site 1. The cartilage at the given site was normal in 32 out of 33 heads tested. No significant difference was found between the experimental sites of the two groups. It may therefore be concluded that local changes, even if severe, do not affect the concentration of glyco-

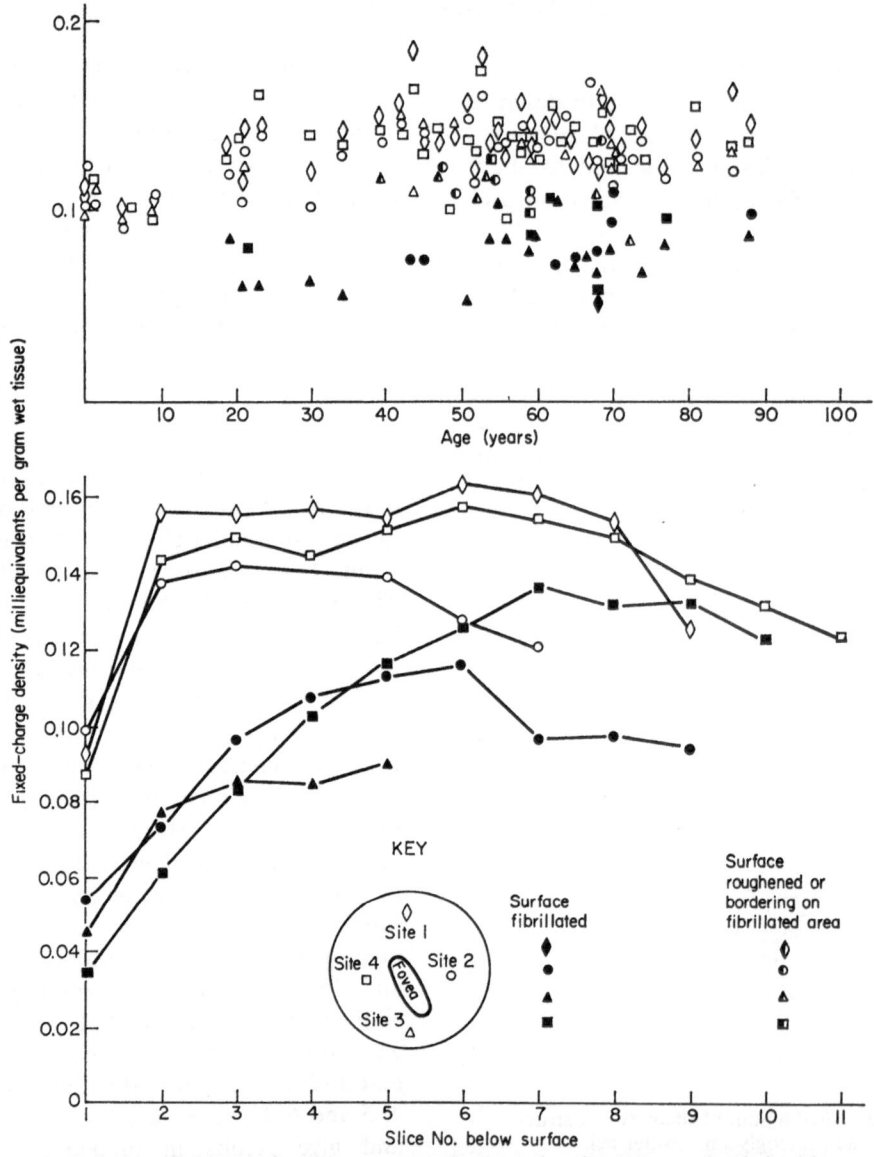

Figure 2. Variation of glycosaminoglycan content (expressed as fixed-charge density) with age, site, and degree of fibrillation for human articular cartilage from the femoral head.

Figure 3. Variation of fixed-charge density with depth from the articular surface, in 72-year-old subject.

(iv) In order to determine whether localised degeneration at a given site leads to changes in fixed-charge density at other sites, where cartilage remains visibly normal, the mature femoral heads tested were divided into two groups. Group I consisted of those which showed, somewhere on their surfaces, fairly extensive fibrillation, ulceration or cartilage loss, and group II consisted of

saminoglycans at other sites of the same joint. Degeneration therefore appears to be initially a local phenomenon and does not seem to be accompanied by changes in the concentration of negatively charged groups over the whole joint.

(v) In normal cartilage from adult subjects neither fixed-charge density nor water content showed any variation with age.

	Site 1	Site 2	Site 3	Site 4	Pooled specimens all sites
All samples					
Total number of samples	33	35	38	38	146
Mean fixed-charge density	0·1374	0·1174	0·0989	0·1220	0·113
(milliequivalents per gram of whole tissue)					
Standard deviation	0·0178	0·0231	0·0282	0·0253	0·021
Normal samples					
Number of normal samples	32	22	13	29	96
Mean fixed-charge density	0·1405	0·132	0·136	0·132	0·135
(milliequivalents per gram of whole tissue)					
Standard deviation	0·0178	0·0176	0·0175	0·021	0·021
Fibrillated samples					
Number of fibrillated samples	1	13	25	9	48
Mean fixed-charge density	0·050	0·095	0·084	0·0895	0·088
(milliequivalents per gram of whole tissue)					
Standard deviation	—	0·0155	0·0234	0·0225	0·021

TABLE 1: Mean values of fixed-charge density of adult human cartilage at different sites of femoral head

Figure 3 shows typical graphs of fixed-charge density *versus* depth below the articular surface for both normal and fibrillated specimens. Fixed-charge density is lowest near the surface, rises to a maximum in the intermediate zone and decreases again in the deep uncalcified zone, without, however, reaching as low a level as in the superficial zone. Also, it can be seen that all fibrillated specimens have a lower fixed-charge density throughout their depth than normal cartilage.

The extent of glycosaminoglycan depletion in the fibrillated specimens appeared relatively more pronounced in very thin cartilage specimens and near the articular surface in thick specimens. This might be interpreted as being due to a diffusive leakage of glycosaminoglycans across the articular surface in fibrillated areas.

RELATION BETWEEN GLYCOSAMINOGLYCAN CONTENT AND THE RATE OF FLUID MOVEMENT IN CARTILAGE

The concentration of proteoglycans in the proteoglycan – water gel determines the number of water-filled pores and their diameter. It is therefore not surprising to find that hydraulic permeability varies inversely with fixed-charge density, as is shown in figure 4.

The rate of fluid expression from cartilage subjected to an applied load, apart from being dependent on the hydraulic permeability, is also a function of the osmotic pressure existing in the tissue. The osmotic pressure, in turn, is directly related to the concentration of negatively charged groups (Ogston, 1970). One would therefore expect a strong inverse correlation between the deformability of cartilage and fixed-charge density. Such relation has been experimentally demonstrated (see part 2, figure 10).

TRANSPORT OF SOLUTES IN CARTILAGE

The diffusion of solutes from a uniformly mixed synovial fluid into cartilage is governed by two main factors (Maroudas, 1968):

(i) the equilibrium distribution of the solute between cartilage and synovial fluid;
(ii) diffusion coefficient of the solute within cartilage.

Distribution coefficients. In the absence of specific interactions small non-electrolyte molecules such as

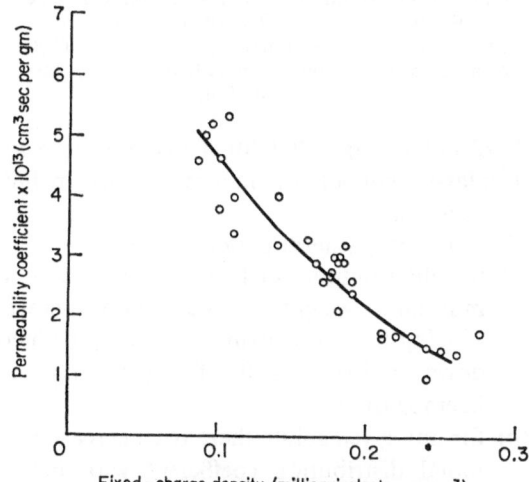

Figure 4. Correlation between hydraulic permeability and fixed-charge density. (Reprinted from *Biophys. J.* **8** (1968): 575, by permission of the Editors.)

urea distribute themselves equally between the interstitial fluid within the cartilage and the outside solution. This indicates that nearly all the water in cartilage behaves as solvent water.

However, where higher-molecular-weight solutes are concerned some of the water becomes inaccessible because of the steric exclusion exerted by the proteoglycan molecules (Ogston, 1958).

Figure 5 shows plotted values of the molal distribution coefficient *versus* fixed-charge density obtained for solutes of different molecule size, ranging from glucose (molecular weight 180) to dextran

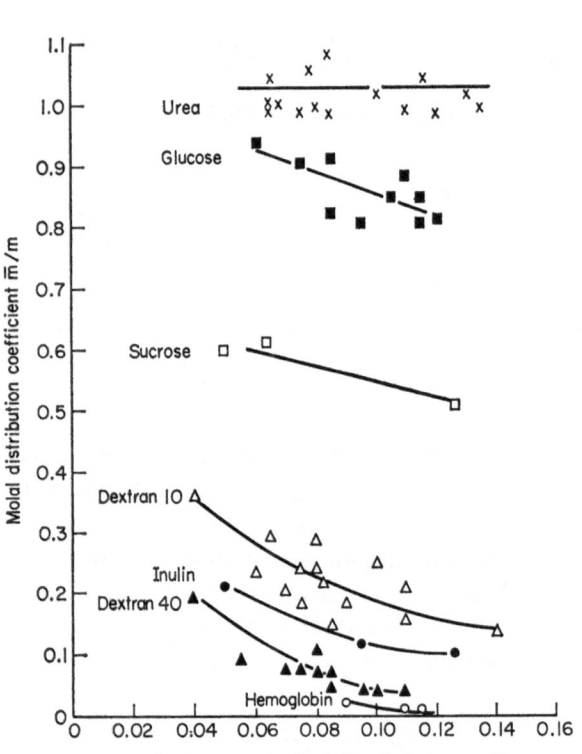

Figure 5. Distribution of solutes between cartilage and Ringer's solution: effect of molecular size of solute and fixed-charge density of cartilage. (Reprinted from *Biophys. J.* **10** (1970): 365, by permission of the Editors.)

40 (molecular weight 40 000). The curve for a low-molecular-weight solute, viz. urea, is also included for comparison.

The following general patterns emerge:

(a) the distribution coefficient decreases as the molecular weight of the solute increases, ranging from approximately 0·85 for glucose down to less than 0·1 for dextran 40 and haemoglobin;

(b) for any given solute there is a decrease in the molal distribution coefficient with increase in fixed-charge density of the cartilage. This decrease is small for glucose but becomes more and more significant as the molecular weight of the solute increases.

The distribution coefficients of the sodium and chloride ions closely obey the Donnan equilibrium law (Maroudas, 1970a and b).

Diffusion coefficients

For molecules which are small in comparison with the cross-sectional area of the pores in the proteoglycan – water gel, the movement of the solute is governed by free diffusion. Therefore for all small-molecule solutes the ratio of the diffusion coefficient in cartilage to that in water, $\bar{D}/D$, depends only on the water content of cartilage and the tortuosity of the pores, and is found to be approximately equal to 0·4.

For molecules whose dimensions are not negligible in relation to pore size, the diffusion coefficients depend not only on the friction between the solute and the solvent but also on the friction between the solute molecules and the matrix.

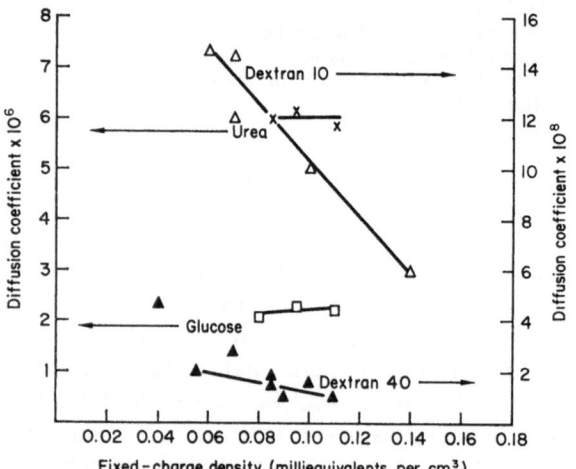

Figure 6. Diffusion coefficients of solutes in cartilage: effect of molecular size of solute and fixed-charge density of cartilage. (Reprinted from 'Chemistry and Molecular Biology of the Intercellular Matrix', **3**: 1400, Academic Press 1970, London and New York.)

Figure 6 shows graphs of the diffusion coefficient *versus* fixed-charge density for typical small and large-molecular-weight solutes. It can be seen that (a) the diffusion coefficient decreases considerably with increase in the molecular weight of the solute and that (b) although the small solutes show no significant variation with fixed-charge density, large solutes $\bar{D}$ show decreases with increase in fixed-charge density.

SUMMARY

The main physiological implication of the distribution and diffusion studies is that while the smaller nutrients and metabolites should be able to diffuse freely in and out of cartilage under all physiological conditions, the transport of the larger molecules is likely to be restricted and closely dependent on fixed-charge density. Thus the passage of substances such as enzymes, antibodies, and proteoglycan fragments may be controlled by local variations in fixed-charge density.

2: MECHANICAL PROPERTIES

M. A. R. FREEMAN, G. E. KEMPSON AND S. A. V. SWANSON[1]

INTRODUCTION

Two of the functions of articular cartilage are (1) to provide a suitable bearing surface and (2) to distribute applied loads over the subchondral bone. The performance of these functions depends on the deformation of cartilage under stress, and therefore the mechanical properties enter into a description of the physiological functioning of the tissue, which is a prerequisite for an adequate account of osteoarthrosis.

This paper describes indentation and tensile tests, and relates the results to the structure and function of articular cartilage.

INDENTATION TESTS

Methods

Human femoral heads were mounted, immersed in Ringer's solution at 37°C, in a rig which enabled any part of the articular surface to be indented. Up to about thirty areas were tested on each head, using either a plane or hemispherical-ended indenter 3 mm in diameter, to which a constant load was applied. This gave average stresses under the indentor of about 28 kgf/cm². The displacement of the indentor into the cartilage was recorded continuously while the load acted and after it had been

[1] Biomechanics Unit, Imperial College, London, England.

removed. Corrections were made for the deformation of the bone, and the thickness of each tested area of cartilage was measured. The two-second Creep Modulus was calculated, adapting the methods of Waters (1965a and b); for fuller details see Kempson et al. (1971).

Results

Deformation-time curves for areas on one femoral head (figure 7), show 'instantaneous' deformation followed by time-dependent deformation (creep) which tends to an equilibrium value. Following the removal of the load, the deformation recovers completely if enough time is allowed.

Figures 8 and 9 show, respectively, maps and histograms of the two-second Creep Modulus for the cartilage on six heads, from subjects in the age range 19 to 82 years; the grading represents the degenerative state of the head according to the categories of Byers et al. (1970). For a total of 45 tested areas on several heads, the two-second Creep Modulus was correlated with the total glycosaminoglycan content measured by Dr Helen Muir on specimens removed from the tested areas after mechanical testing ($r = 0.854$, $P < 0.001$).

Figure 10 shows, for tested areas on seventeen

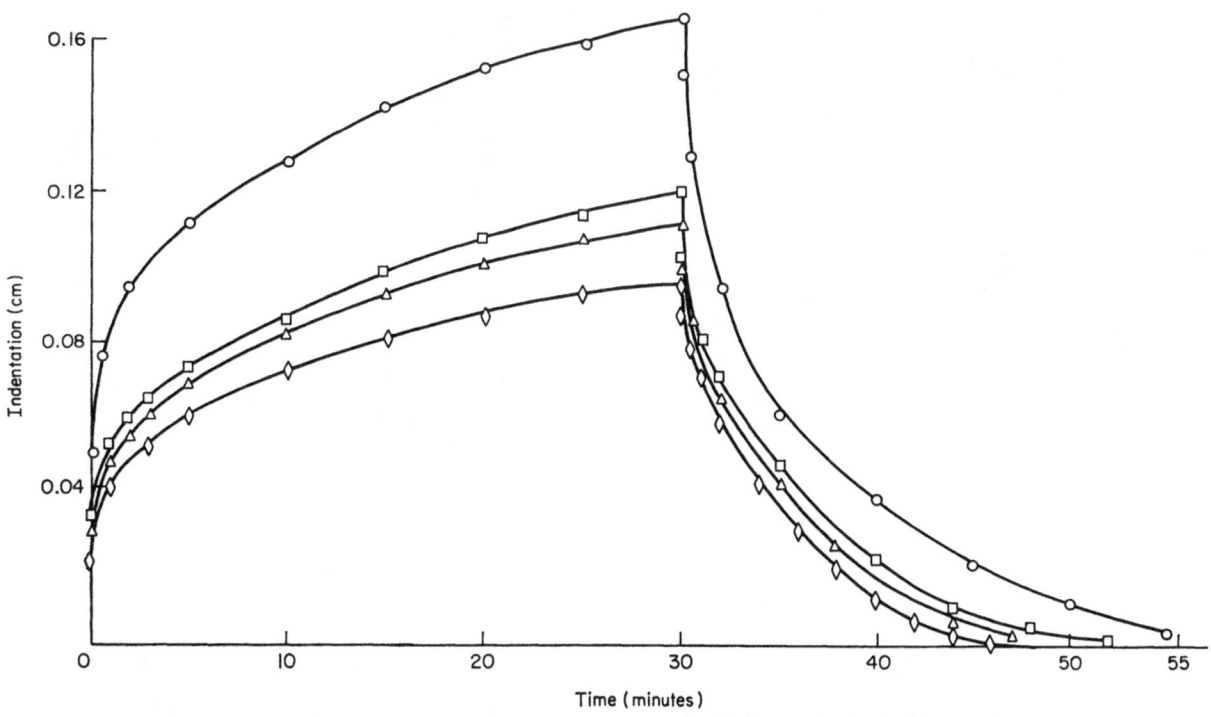

Figure 7. Indentation against time curves for four areas of cartilage. Male aged 68, plane-ended indenter, load 5 lbf (2·28 kgf).

157

other femoral heads, the creep strain at two minutes plotted against the fixed-charge density.

Discussion

Three things are clear:

(1) a reduction in mechanical stiffness is one of the earliest stages in the degenerative sequence;

(2) this accompanies a reduction in the glyco-saminoglycan content;

(3) the stiffness of the material varies regularly, the stiffest material being in a band running anteriorly and posteriorly from the superior surface.

In attempting to define more exactly the deviations from 'normality', some arbitrariness is un-avoidable because of (*a*) the variations within 'normal' material and (*b*) the difficulty of quantifying visual observations. In figure 10 symbols are used to show the grade of departure from the 'normal' glossy surface state of each tested area. There is a strong tendency for glossy cartilage to have crept less than non-glossy, and a less strong tendency for glossy cartilage to have a higher fixed-charge density.

These results suggest that the most sensitive indicator of the smallest changes is the compressive stiffness of the material, which has the added attraction that it is directly relevant to the mechanical functions of the tissue.

TENSILE PROPERTIES

Methods

It was desired to examine the tensile properties of cartilage in planes parallel to the articular surface and in known directions relative to the dominant collagen fibres in the surface layer. In several femoral condyles the surface cleavage pattern (Hultkrantz, 1898), which is assumed to show the dominant direction of the superficial collagen,

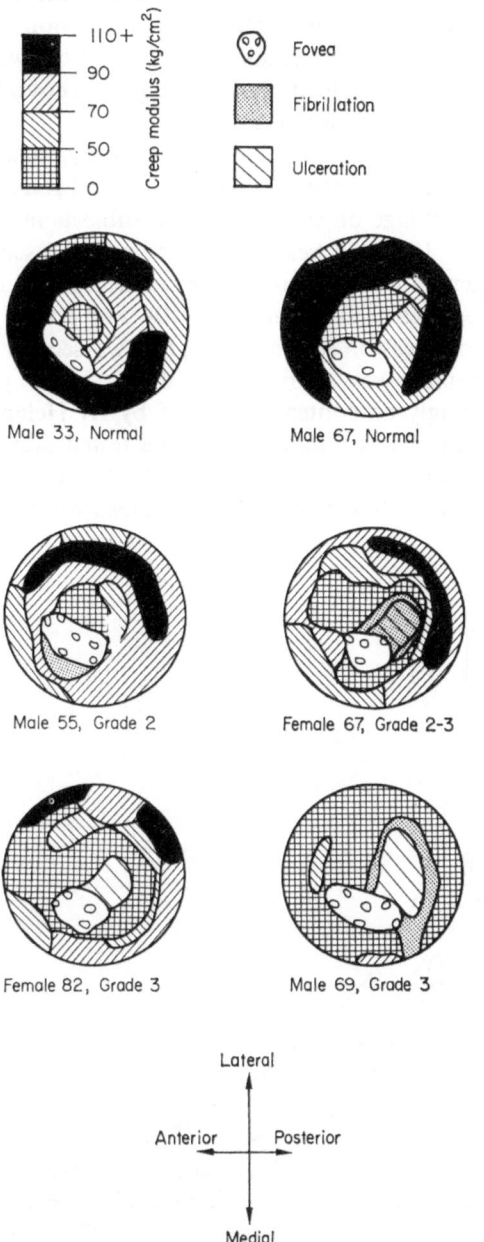

Figure 8. Layered maps showing the variation of two-second Creep Modulus over six right femoral heads.

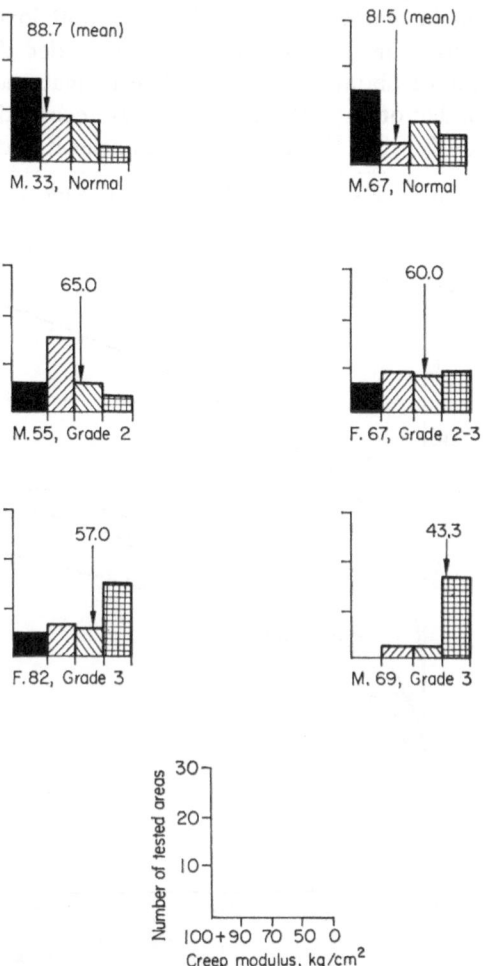

Figure 9. Histograms of the two-second Creep Modulus, corresponding to the maps in figure 8.

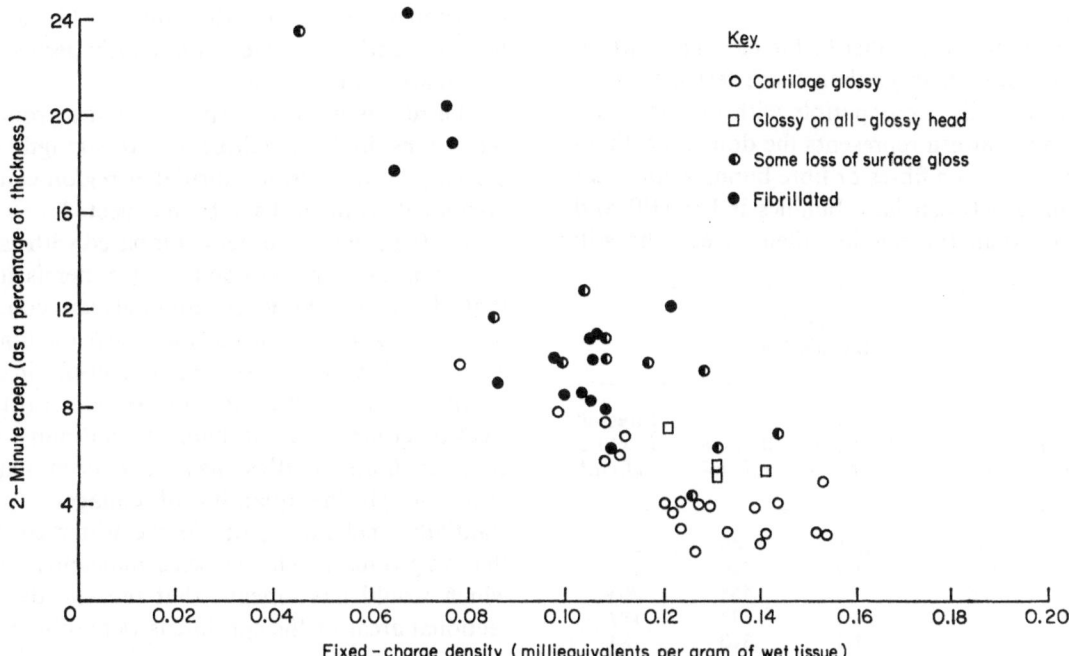

Figure 10. Creep strain at two minutes *v.* fixed-charge density for seventeen femoral heads.

obtained. In unpricked condyles, sites were selected so that tensile specimens 32·0 mm × 10·0 mm could be extracted with their lengths either parallel or perpendicular to the presumed collagen direction. The full thickness of cartilage at each site was sliced to give a set of specimens each 0·2 mm thick; the central parallel portions were 6·35 mm long × 1·6 mm wide, and elongations were measured optically on a 6·35-mm gauge measure. Tests were conducted at 5 mm/min crosshead movement, in Ringer's solution.

Results

Figure 11 shows the nominal stress – engineers' strain curves for four specimens taken from two

sites on the same femoral condyle. The two tested parallel to the surface cleavage pattern were both taken from the same area; the specimens from intermediate depths in the same area gave stress – strain curves between the two extreme ones shown. Similarly, the stress – strain curves for the series of specimens tested perpendicularly to the surface cleavage pattern formed a series of which only the first and last members are shown in figure 11.

Table 2 shows chemical compositions, stiffnesses, and fracture stresses at different depths for two sets of specimens: one set from a 'normal' condyle and the other from an area 5 mm from the edge of a visibly fibrillated area.

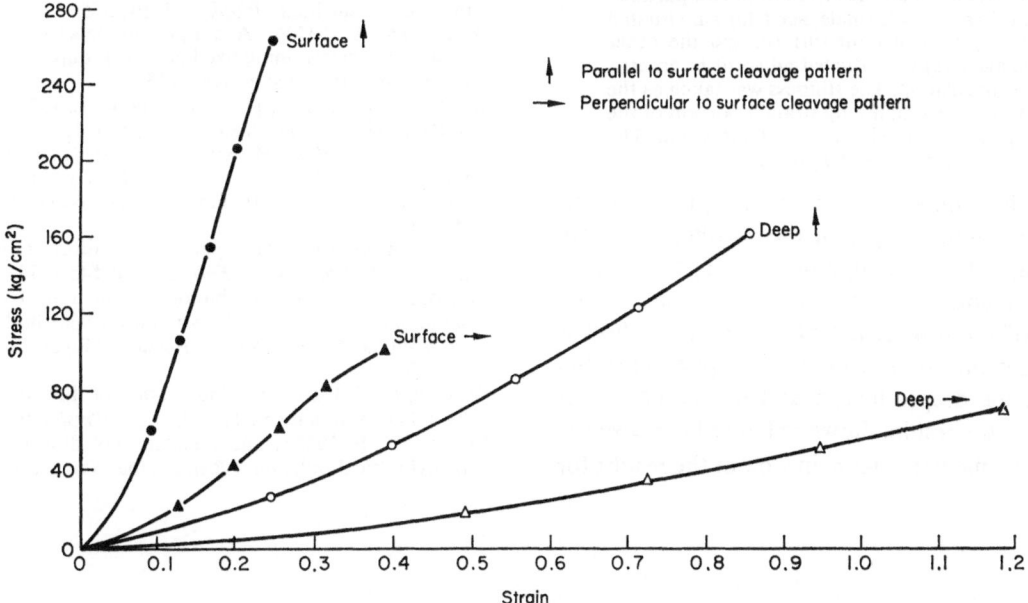

Figure 11. Curves of nominal tensile stress against engineers' strain for the surface layer and the deep layer of two sites on the same femoral condyle in in a 16-year-old male. One site was parallel to the cleavage pattern and the other perpendicular to it.

Discussion

Specimens tested parallel to the cleavage pattern were stiffer and stronger than those tested perpendicular to it. This is compatible with the view that the cleavage pattern represents the dominant direction of the collagen fibres or fibre bundles, and that the bonding between fibre bundles is less stiff and less strong than the bundles themselves. The still

Layer No. (200 μm)	Normal (Male aged 40)			
	Total G.A.G. % dry wt.	Collagen % dry wt.	Stiffness kgf/cm^2	Stress at fracture kgf/cm^2
1	5·7	60·6	1140	224
2	7·7	61·4	640	187
3	8·6	59·0	455	203
4	7·6	58·3	377	187
5	9·5	51·7	310	184
6	8·6	55·1	280	180

Layer No. (200 μm)	Adjacent to fibrillation (Female aged 48)			
	Total G.A.G. % dry wt.	Collagen % dry wt.	Stiffness kgf/cm^2	Stress at fracture kgf/cm^2
1	5·7	63·7	250	38
2	6·6	58·5	300	109
3	12·3	61·8	205	94
4	14·9	68·6	200	85
5	13·1	56·5	120	62
6	14·0	43·4	79	56

TABLE 2: Chemical contents, stiffnesses, and fracture stresses of all the layers from two parallel-oriented sites. One site (male, aged 40) was situated on a visibly normal joint surface, and the other (female, aged 48) was situated adjacent to an area of visible fibrillation. The stiffness was taken as the gradient of the substantially straight portion of the graph, ignoring curved portions at either end. The layers were 200 μm thick

considerable difference between the properties in the two directions deeper in the cartilage suggests that the layout of the superficial collagen persists, to some extent, down to the deepest layer. All accounts of the disposition of the collagen show a decreasing proportion of fibres to be parallel to the surface in the deeper layers, and this is also compatible with the results shown in figure 11. However, it would be difficult to claim, in view of the results for the deepest layer, that the collagen fibres in that layer are entirely or even dominantly perpendicular to the articular surface.

The results in table 2 show that the pronounced reductions in both stiffness and strength of the cartilage near a visibly fibrillated region cannot be attributed, as might have been expected, to a reduction in the collagen content compared with cartilage on a normal femoral condyle. Neither is it likely that the increased glycosaminoglycan content in the deeper zones of the cartilage near the fibrillated area could account for the reduction in tensile properties. Two other possibilities arise: (*a*) in the weaker group, the mechanical continuity of the collagen fibre bundles may have been impaired, even though the quantity of collagen was substantially unchanged; or (*b*) the water content of the two groups may have been significantly different, which would have meant that the effective cross-sectional areas of the specimens were different.

Summary

The tensile results, in common with the results of the indentation tests, suggest that measurements of mechanical properties are amongst the most sensitive early indicators of degenerative changes.

REFERENCES

BYERS, P. D., CONTEPOMI, C. A., and FARKAS, T. A. 1970: A post mortem study of the hip joint. *Ann. Rheum. Dis.* **29**: 15.

HULTKRANTZ, J. W. 1898: Uber die Spaltrichtungen der Gelenkknorpel. *Verhandl. d. anat. Gesellsch. Kiel* (1898).

KEMPSON, G. E., FREEMAN, M. A. R., and SWANSON, S. A. V. 1971: The determination of a creep modulus for articular cartilage from indentation tests on the human femoral head. *J. Biomech.* **4**: 239.

MAROUDAS, A. 1968: Physico-Chemical properties of cartilage in the light of ion exchange theory. *Biophys. J.* **8**: 575.

MAROUDAS, A. 1970*a*: Distribution and diffusion of solutes in articular cartilage. *Biophys. J.* **10**: 365.

MAROUDAS, A. 1970*b*: A simple physicochemical micromethod for determining fixed anionic groups in connective tissue. *Biochim. Biophys. Acta* **215**: 214.

MAROUDAS, A., EVANS, H., and ALMEIDA, L. 1973: Cartilage of the hip joint. *Ann. Rheum. Dis.* **32**(1): 1.

MAROUDAS, A., MUIR H., and WINGHAM, J. 1969: The correlation of fixed negative charge with glycosaminoglycan content of human articular cartilage. *Biochim. Biophys. Acta* **177**: 492.

OGSTON, A. G. 1958: The spaces in a uniform random suspension of fibres. *Trans. Faraday Soc.* **54**: 1754.

OGSTON, A. G. 1970: The biological function of the glycosaminoglycans. *In* 'Chemistry and Molecular Biology of the Intercellular Matrix', Vol. 3. Academic Press, London and New York.

WATERS, N. E. 1965*a*: The indentation of thin rubber sheets by spherical indenters. *Brit. J. Appl. Physics* **16**: 557.

WATERS, N. E. 1965*b*: The indentation of thin rubber sheets by cylindrical indenters. *Brit. J. Appl. Physics* **16**: 1387.

MOTION IN THE HIP: THE RELATIONSHIP OF SPLIT LINE PATTERNS TO SURFACE VELOCITIES

J. D. GRAHAM[1] AND T. W. WALKER[2]

GENERAL

In 1968, one of the authors (J.D.G.—then a McLaughlin Travelling Fellow at the University of Strathclyde, Glasgow) began a study of the mechanical properties of articular cartilage as they relate to the structure of the material.

The internal structure of cartilage was studied by scanning electron microscopy; some interesting facts concerning the relationship of behaviour to structure being brought to light (Graham, 1970). Perhaps more intriguing, however, was the distinctive pattern of a grain in the cartilage surface. This grain was demonstrated in the same way that Langer's Lines can be shown in skin—a pin dipped in Indian ink was used to make small puncture wounds in the cartilage, inserting the pin perpendicular to the surface.

If a portion of such a specimen is subjected to scanning electron microscopy, it can be seen that the perforation is indeed a split, not unlike those which occur in wood when one attempts to drive home a large spike without first drilling a hole for it. The split seems to pass between bundles of fibres parallel to the direction of the split.

A reproducible grain pattern exists for every synovial joint. For example, the pattern of split lines for the femoral head is as shown in figure 1. Work was begun at the University of Toronto in 1970 to study these patterns and determine their function.

FORCE ANALYSIS

It was decided to apply vector mathematics to the investigation (Walker, 1972). The hip joint was selected as the object of this study because much was already known about its kinematics. A set of orthogonal reference axes was defined with origin at the centre of the hip and the three principal directions oriented so as to fall within the conventional anatomical planes (x-axis in sagittal and transverse planes; y-axis in sagittal and coronal planes; z-axis in coronal and transverse planes—see

[1] Department of Surgery, Toronto General Hospital, Toronto, Ontario, Canada.
[2] Institute of Bio-Medical Electronics and Engineering, University of Toronto, Toronto, Ontario, Canada.

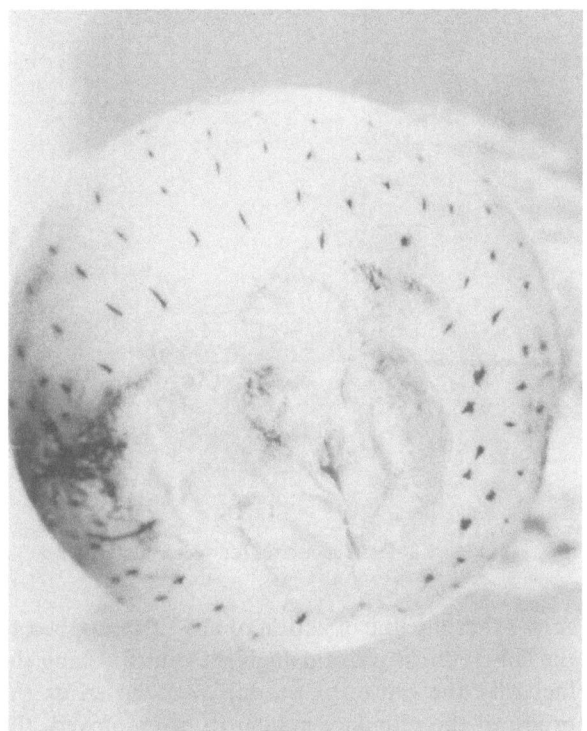

Figure 1. Cartilage split lines on femoral head with osteoarthritis.

figure 2). Data giving the loads carried by the hip, the position of the femur relative to the pelvis, and the speeds of rotation of the femur relative to the pelvis were selected from the literature and referred to these co-ordinate axes.

Load studies published by Rydell (1966) give three dimensionless components of load,

$$P_x, P_y \text{ and } P_z,$$

where

$$P = (\text{load across hip})/(\text{body weight}).$$

Expressed in vector form, this becomes

$$\vec{P} = [P_x P_y P_z] \times \begin{Bmatrix} \hat{\imath} \\ \hat{\jmath} \\ \hat{k} \end{Bmatrix}.$$

This vector, of course, has a magnitude given by

$$|\vec{P}| = \sqrt{(P_x^2 + P_y^2 + P_z^2)}.$$

It can easily be shown for a case such as that encountered for the hip—i.e. a spherical shape whose surface has a coefficient of friction of practically

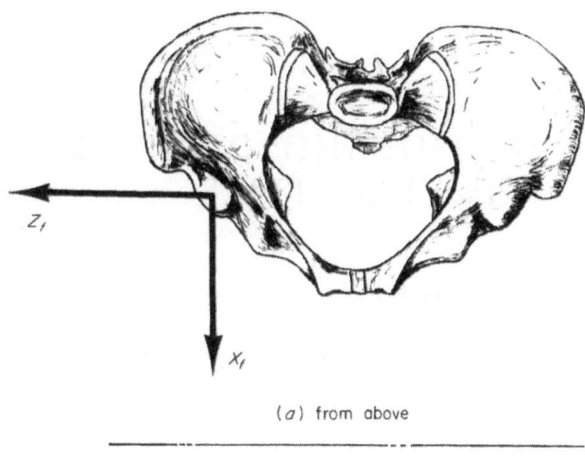

(a) from above

(b) from the front

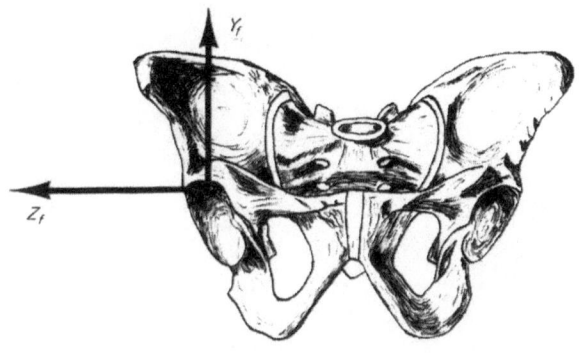

Figure 2. Reference axes for vector analysis.

zero—that the line of action of any force applied to the sphere must pass through the centre. Using the fact that the centre of the hip was chosen as the origin of the reference system to be employed, the path traced by the loaded area within the hip joint can be determined from the equation

$$\vec{L} = -\frac{\vec{P}}{|\vec{P}|}.$$

This was done as part of the investigation by Walker (1972) but the results had no significance except to allow calculation of surface speeds at the points to which load is applied.

Gait studies, of course, give the angular position of the femur relative to the pelvis. For our work, we used data concerning human gait published by Eberhart *et al.* (1968) and by Murray *et al.* (1964) in order to obtain three curves of angular position versus time, referred to the chosen reference axes. These three curves were differentiated with respect to time, yielding curves of

$$\omega_x, \omega_y \quad \text{and} \quad \omega_z,$$

the three components of the angular velocity vector for the hip; i.e.

$$\vec{\omega} = [\omega_x \, \omega_y \, \omega_z] \times \begin{Bmatrix} \hat{i} \\ \hat{j} \\ \hat{k} \end{Bmatrix}.$$

Of course,

$$|\vec{\omega}| = \sqrt{(\omega_x^2 + \omega_y^2 + \omega_z^2)}$$

and, ω is expressed in units of radians/second.

Plotting $|\vec{P}|$ and $|\vec{\omega}|$ versus time (see figure 3) shows that the hip moves very slowly under heavy load at about the 50 % point in the gait cycle, or the point of toe-off.

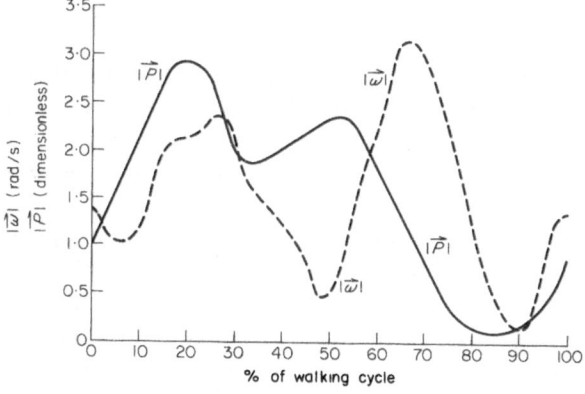

Figure 3. Graph of load and rotational speed versus time for the hip.

The angle λ between the $\vec{P}$ and $\vec{\omega}$ vectors was computed using the formula

$$\lambda = \cos^{-1} [|\vec{P} \cdot \vec{\omega}| / (|\vec{P}| \times |\vec{\omega}|)].$$

The results of this calculation showed that, at times, the hip carries a combination of radial and thrust loads (i.e. $\lambda \neq 90°$) although at high values of $|\vec{P}|$, the thrust loads are very small ($\lambda \approx 90°$).

Comment on the importance of these last two results in understanding the function of synovial joints is beyond the scope of the present paper. It may be stated, however, that they indicate that analysis of the lubrication of such joint such as the hip demands more sophisticated models than have commonly been applied.

DISPLACEMENT AND VELOCITY ANALYSIS

A point on one of the surfaces within the hip possesses a position vector

$$\vec{p} = [p_x \, p_y \, p_z] \times \begin{Bmatrix} \hat{i} \\ \hat{j} \\ \hat{k} \end{Bmatrix}.$$

One property of this vector is that

$$|\vec{p}| = \sqrt{(p_x^2 + p_y^2 + p_z^2)} = \text{radius of femoral head.}$$

In addition, for the point to which load is applied,

$$\vec{p} = \hat{L}.$$

Paths of points on one surface over the mating surface may be determined from

$$\vec{p}_t = \vec{p} \times \{T_t\},$$

where $\vec{p}_t$ is the position vector for the point at the instant t, $\vec{p}$ is the position vector of the point with the hip in its neutral position (subject standing still) and $\{T_t\}$ is a transformation matrix based on the position of the hip at the instant t.

Plotting paths of typical points results in the generation of looped figures as shown in figures 4 and 5. Figure 4 shows paths of points on the femoral

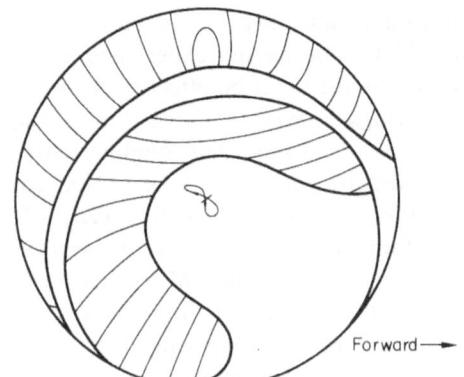

Figure 4. Paths of points on femoral head over acetabulum.

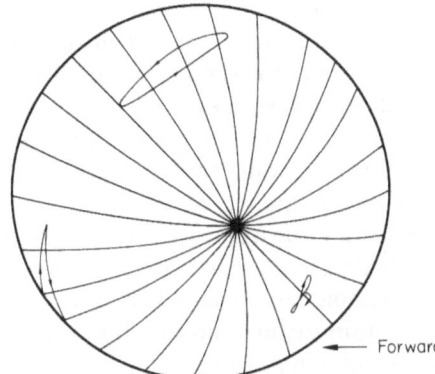

Figure 5. Paths of points on acetabulum over femoral head.

head over the surface of the acetabulum, while figure 5 shows paths of points on the acetabulum over the femoral head. In both cases, the background is intended to show the shape of the cartilage surfaces being studied, the faint lines representing the cartilage grain.

The velocity of the opposing surface at a point on the acetabulum is given by

$$\vec{v}_a = \vec{\omega} \times \vec{p},$$

while the same quantity for a point on the femoral head is

$$\vec{v}_{fh} = -\vec{\omega} \times \vec{p}.$$

In figures 6 and 7, the velocity vectors v_a and v_{fh} respectively for the points traced have been added to figures 4 and 5. The base of the arrow represents the location of the moving point, the length of the

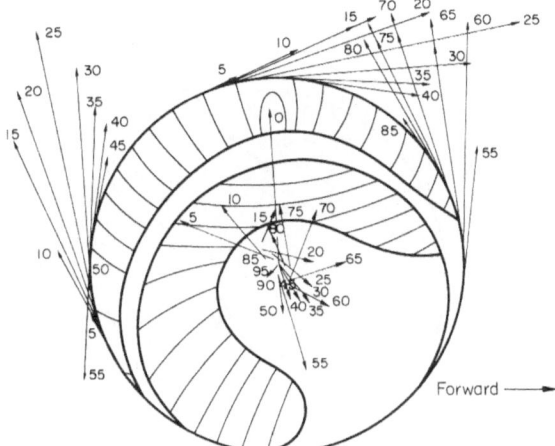

Figure 6. Velocity vectors referred to acetabulum.

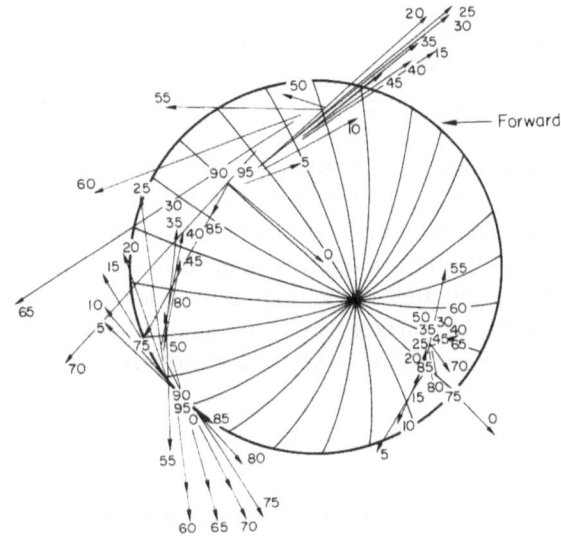

Figure 7. Velocity vectors referred to femoral head.

arrow is proportional to the magnitude of the velocity

$$(|\vec{v}| = \surd(v_x^2 + v_y^2 + v_z^2),$$

and the number beside the head of the arrow gives the phase of the gait cycle, in per cent. The main point of this paper is to point out that the velocity vectors tend to cross the split line patterns at right angles, particularly for the larger vectors (higher speeds).

Figure 8 shows the path traced by the point of application of load over the femoral head and the velocity of the acetabulum relative to the femoral head at this moving load point. At approximately 15–25 % through the gait cycle, when weight bearing is maximum (see figure 3), the speed of the two cartilage surfaces relative to each other is high, while the loaded area shifts very little. This leads one to predict that this particular portion of the femoral head might be the most vulnerable to wear. This is in agreement with clinical observations, as can be seen by comparing the location of the area of

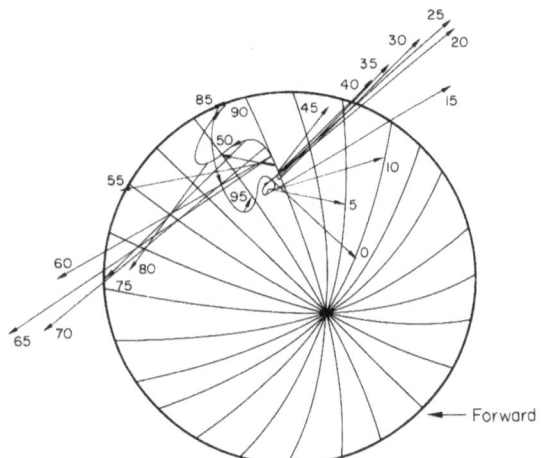

Figure 8. Velocity vectors referred to path of load
point over femoral head.

cartilage degeneration visible in figure 1 with the
location of the load point at maximum load.

Because the magnitude and direction of the
vectors related to the split line pattern is difficult
to appreciate while keeping in mind the phase of the
gait cycle, an animated movie has been made to
clarify the presentation. Regrettably, there is no way
to put the information contained in the film onto
paper, but the movie does show that the general
pattern of surface velocities relative to the split
lines seen in figures 6 and 7 does hold true for other
points.

SUMMARY

We felt at the beginning of our study that the grain
in cartilage might play a role in the lubrication of
synovial joints. After nearly two years of work,
however, the only conclusion we can state is the
aforementioned perpendicularity of the grain with
the relative motion of the cartilage surfaces. Before
we can say any more, we must determine how the
joints are lubricated. We are presently working
toward this end.

It seems at this time that, if a thick fluid film
exists in the joints, it must be of an elastohydrodynamic or hydrostatic nature. For a squeeze
film to form, the cartilage must be more resistant to
deformation than is indicated by the data presently
available. These data also do not allow an accurate
assessment of the fluid film thickness predicted by
elastohydrodynamic theory. The authors have,
therefore, just begun a study of the load-deflection
behaviour of cartilage which, it is hoped, will allow
determination of whether a squeeze film or an
elastohydrodynamic film can exist in the hip.

The probability of the existence of a hydrostatic

film in the joints seems almost nil, unless a fluid-
pumping mechanism within the cartilage layers can
be demonstrated. The 'weeping' of cartilage in
response to load which was described by McCutchen
(1962) cannot provide sufficient pressure to counter-
act load hydrostatically, and various studies have
shown the metabolic activity within articular
cartilage to be insufficient to support any other
manner of 'pump'.

This all leads to the conclusion that lubrication of
synovial joints can only be accomplished by a
boundary film. Engineering experience has shown
that this mechanism can be effective in preventing
wear of bearings carrying high loads at low speeds.
Engineers have also found that roughness similar to
that suggested by the grain in cartilage surfaces
improves the effectiveness of this type of film.
The ways of machine tools show less wear if they are
scraped so as to create minute transverse ridges;
and the cylinder bores of internal combustion
engines are less prone to scoring if a cross-hatch
pattern is machined into their surfaces using a very
fine hone. The authors feel that, unless the existence
within the joints of a thick lubricating film can
somehow be demonstrated, it must be concluded that
boundary films lubricate synovial joints, and this
fact should be taken into account in the development of techniques of orthopaedic practice, and in
the design of endoprostheses.

In conclusion, we would suggest that the data
presented in this paper may be helpful in proceeding
to a better understanding of synovial joint lubrication through knowledge of the action of one
surface against its mate. Furthermore, there appears
to be a reasonable relationship between the pattern
of motion within the hip and the clinical pathology
seen so frequently in practice.

REFERENCES

EBERHART, H. D., INMAN, V. T., and BRESLER, B. 1968:
   The principal elements of human locomotion. *In* 'Human
   Limbs and Their Substitutes. (Eds. Klopsteg and Wilson),
   Hafner, New York.
GRAHAM, J. D. 1970: McLaughlin research paper presented
   at Congress of Anatomists, Leningrad.
McCUTCHEN, C. W. 1962: The frictional properties of animal
   joints. *Wear* **5**: 1.
MURRAY, M. P., DROUGHT, A. B., and KORY, R. C. 1964:
   Walking patterns of normal men. *Journal of Bone and
   Joint Surgery* **46A**: 335.
RYDELL, N. W. 1966: Forces acting on the femoral head—
   prosthesis. *Acta Orthopaedica Scandinavica No.* 37:
   Supp. 88.
WALKER, T. W. 1972: 'A theoretical and experimental study
   of the motion and lubrication of the human hip. MASc.
   Thesis, Department of Civil Engineering and Institute of
   Bio-Medical Electronics and Engineering, University of
   Toronto.

STRESS-STRAIN-TIME RELATIONS FOR SOFT CONNECTIVE TISSUES

J. C. BARBENEL, J. H. EVANS, AND J. B. FINLAY[1]

INTRODUCTION

The application of engineering techniques to biological and medical problems has led to an interest in the mechanical properties of body tissue. Soft connective tissue shows complex-load-extension behaviour which is time-dependent, and any attempt at a full description of the mechanical properties must include all three variables (Fung, 1972).

Attempts at such descriptions using spring and dashpot lumped parameter models have been limited by the complexity of behaviour of the tissues (Frisen et al., 1968). The fullest description of load-deformation behaviour has to some degree ignored time-dependent behaviour and vice versa.

LOAD-DEFORMATION BEHAVIOUR

For reasons of experimental simplicity the majority of available data have been obtained under tensile quasi-static loading and, as a result there are many mathematical descriptions of uniaxial stress/strain relations. In qualitative terms the stress strain curves are invariably concave to the stress axis demonstrating an increase in stiffness with increasing deformation. Quantitatively these uniaxial characteristics of tissue have been described empirically by power laws, simple and modified exponential laws and by combinations of these and other elements.

In addition to being non-linear, many tissues are anisotropic and the principal mechanical difference associated with this anisotropy appears as an increased deformation at low loads (figure 1).

The major differences in behaviour between tissues also appears as differences in the initial low load extension. Thus, although the 'moduli' of tendon and skin at high stress are comparable the rupture strains can differ by an order of magnitude.

Continuum descriptions have met with limited success to date, possibly as a result of the paucity of multiaxial data. Work is principally based on classical continuum mechanics or on elastomeric network theory (Crisp, 1972). Representation of the elastic response of some tissues has been made in

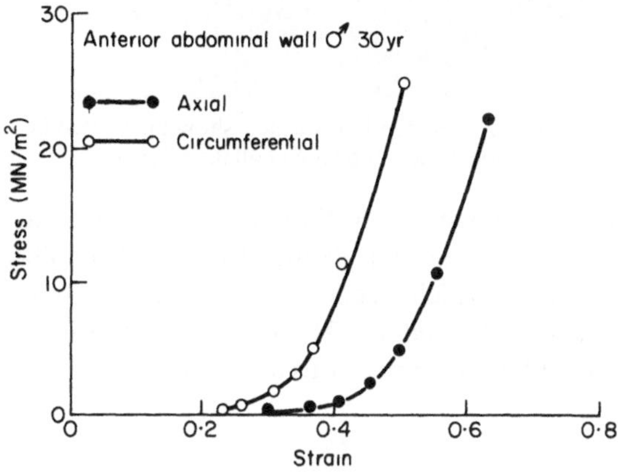

Figure 1. Anisotropic stress strain behaviour of human skin *in vitro*.

terms of stored energy functions. The data for such analysis have usually been obtained in uniaxial tension and the materials often assumed to be isotropic and sometimes incompressible. There appear to have been no attempts to predict (or test) the results of experiments in other modes from the functions obtained in uniaxial loading (Veronda and Westmann, 1968).

Attempts have been made to correlate load-extension behaviour of soft tissues with alterations in the arrangements of the tissue constituents. The adaptation of the collagenous fibre framework to applied load has been observed by light and scanning electron microscopy and detailed qualitative descriptions exist of this phenomenon in skin and tendon. There is little evidence concerning the corresponding deformations in the elastin network and the contribution of this protein to whole tissue mechanics has been inferred by selective component degradation and tests on isolated fibres.

The 'simple' tissues, tendon and ligaments, lend themselves to quantitative assessment because of their highly organised structure. Uniaxial tensile data is appropriate as these tissues function principally as uniaxial force transmitting structures. In tendon the fibrils aggregate in nearly parallel arrays to form fibres. In the unstrained condition these fibres adopt a helicoid form which imparts a wave-like appearance to closely packed fibres of the free tendon surface (figure 2.). The uniformity of fibre direction allows them to be straightened on loading. Approximately 3% extension can be attributed to

---

[1] University of Strathclyde Glasgow, Scotland.

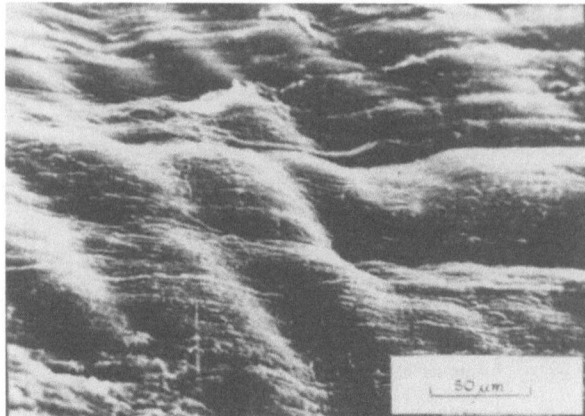

Figure 2. Free tendon surface showing wave-like appearance of packed collagen fibres.

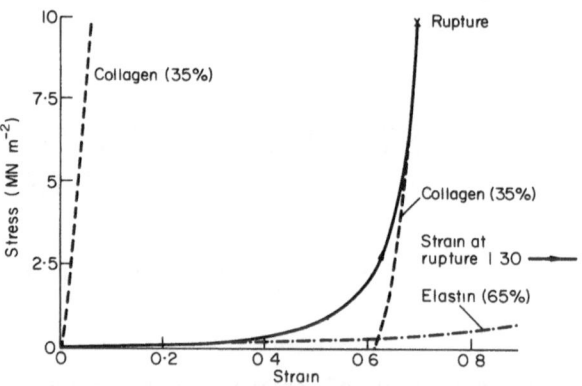

Figure 4. Stress strain behaviour of ligamentum flavum in uniaxial tension compared with collagen and elastin.

this phenomenon so that the maximum strain envelope for tendon is defined by a parallel collagen fibre array which becomes load bearing after the tissue is extended by 3% (figure 3). Stereoscopy has enabled the dimensions and configuration of the

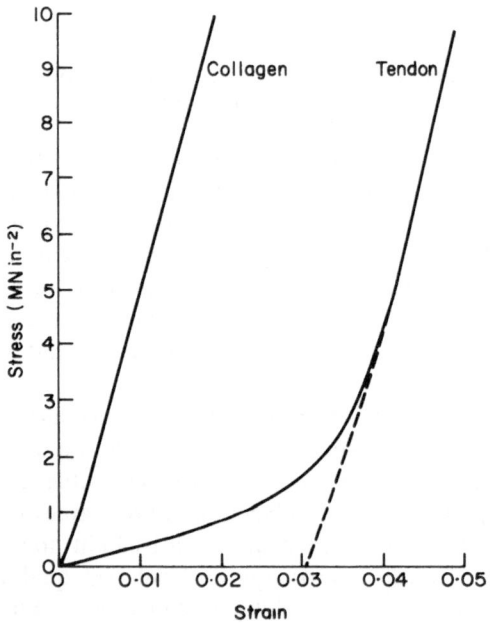

Figure 3. Stress-strain behaviour of tendon in uniaxial tension, and regularly waved collagen fibres exhibiting no resistance to fibre straightening.

fibres to be measured which in turn enabled the strain due to straightening to be predicted. The average periodicity is 70 μm but there is a scatter of wavelength and amplitudes which implies that there is a gradual transition from the fibre straightening to fibre extension phases.

The collagenous structure of ligamentum flavum is similar to that of tendon but the preponderance of elastin significantly influences the tissue's response to loading. Figure 4 depicts the maximum strain limit imposed by the collagen fibres which are assumed to straighten at 60% tissue strain. As collagen represents approximately 35% dry weight of the tissue,

the effective modulus is reduced accordingly to 0·35 of that ascribed to collagen above. Similarly the response of the elastin, which accounts for the bulk of the remaining fibrous material is superposed. This graph assumes that the elastin components do not exhibit an initial lax phase. Thus, in the case of this tissue, both the initial and terminal phases are described in a semi quantitative manner, but there remains the transition which can only be discussed qualitatively.

Transmission electron microscopy of strained tendon has revealed straining at the molecular level. Typically collagen fibrils exhibit a regular cross banding when stained with heavy metals. The average repeat length for unstrained tissue fibrils is 64 nm, but there is some variation. There is an increase in average cross band spacing comparable to the strain experienced by the parent tissue although not all fibrils appear strained. The micro-scale deformation is not observed during the fibre straightening stage (Viidik and Ekholm, 1968).

The role of elastin in the mechanical behaviour of tendon is illustrated clearly by selectively removing this fibrous protein. The initial loading curve is little affected but on unloading the collagen fibre coils do not reform. The tendon then behaves as a parallel array of straight, collagen fibres and a large permanent strain is introduced.

The elastin network has little inherent strength and is probably discontinuous as, without the collagen fibre framework, it ruptures at small strains which are an order of magnitude less than the strain observed in single elastin fibres (Minns, 1972). It seems that it is the interaction of the elastin and collagen which influences the mechanical properties of most soft connective tissues and not the separate behaviour of each network.

TIME DEPENDENT BEHAVIOUR

Biological tissues demonstrate many of the time dependent mechanical properties of viscoelastic materials. Specimens maintained at constant deformation show stress relaxation which is a time dependent decrease in the magnitude of the stress produced by the deformation. Similarly specimens maintained at constant stress exhibit creep. In addition to these quasi-static phenomena viscoelastic materials show complex dynamic behaviour. The application of a sinusoidally varying displacement results in a varying stress out of phase with the applied strain. The phase difference between the two is known as the 'phase angle'. The stress component in phase with the displacement is the *storage component*, and that 90° out of phase the *loss component*.

Soft collagenous tissue when tested either *in vitro* or *in vivo* exhibit rather complex viscoelastic behaviour. Typical of such behaviour are the results obtained by inplane dynamic mechanical tests on human skin *in vivo*. The tests were made using a rotary displacement servo. The equipment facilitates the recording of applied torque for controlled displacements of up to ±20° at frequencies from d.c. up to 20 Hz (Finlay, 1970a and b). Essentially the device consists of a torque motor and power amplifier employing both proportional and derivative position feedback. Figure 5 shows the rotary servo test unit in a typical patient testing situation remote from its associated power supplies and recording equipment.

The area of skin under investigation is defined as a 4 mm wide annulus between a motor-driven disc and a fixed guard ring, as illustrated in figure 6. Both disc and guard ring are attached to the patient by Eastman 910 (a cyano-acrylate adhesive).

The test sequence which proved most useful in practice involved the application of a displacement

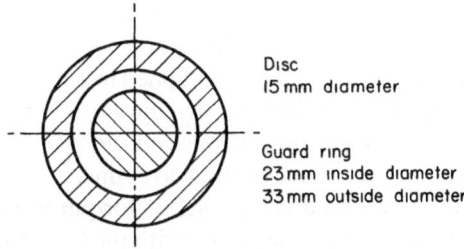

Figure 6. Dimensions of disc and guard ring used for test.

waveform consisting of a series of five identical trapezoids (figure 7).

Each trapezoid consisted of a rotation applied uniformly at a rate of 2 degrees per second followed by a period of one minute, during which time the displacement was held constant and the torque was recorded continuously. Finally, the disc was returned to zero degrees at the same rate of 2 degrees per second. The amplitude was defined as the rotation

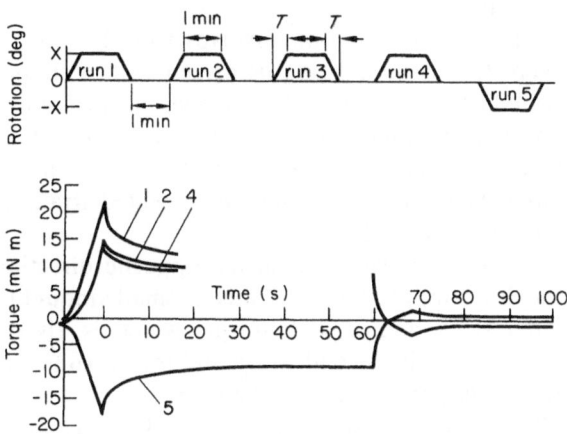

Figure 7. Test sequence of trapezoidal displacements and the torsional responses obtained.

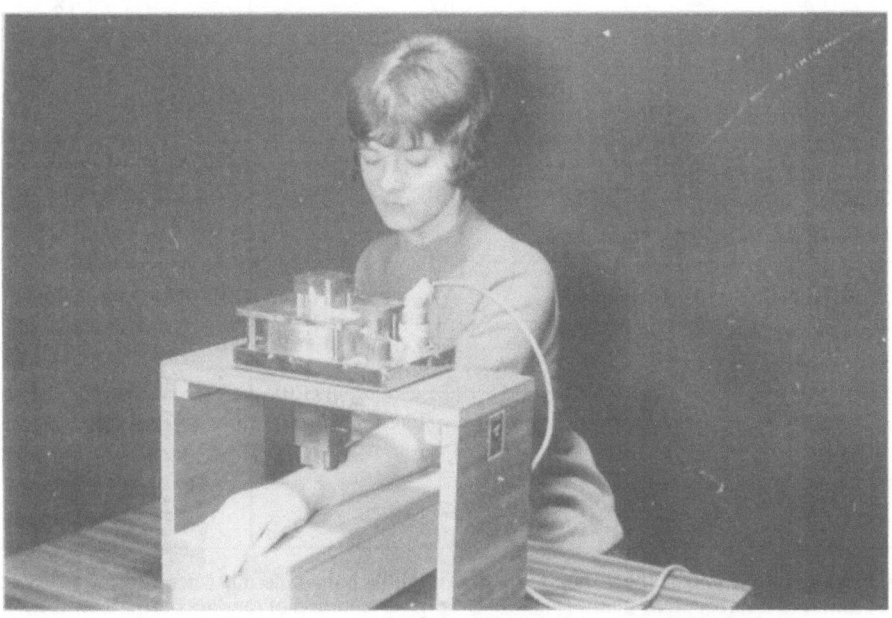

Figure 5. Rotary servo test unit in typical patient-testing situation.

necessary to produce a peak torque in the range 20 to 25 mNm during the first run.

Four such trapezoids were applied in one direction before applying a fifth in the opposite rotational sense. A resting period of one minute was allowed to elapse between each of these five runs.

Figure 7 shows a typical set of torque responses produced by this test sequence with the five response curves plotted from the same origin for the purpose of comparison. This procedure has been used in a survey of the dynamic mechanical characteristics of normal skin (Finlay, 1971) and the skin of patients affected by the Ehlers Danlos Syndrome (Finlay and Hunter, 1972). However, the curves are presented here to illustrate certain of the characteristic features of the mechanical behaviour of human skin.

Three quite separate conditions exist in this sequence of five runs. During the first run the material is being strained for the first time and a small amount of 'initial slack' is evident in the torque-time curve.

The second condition involves runs 2, 3 and 4 where the tissue is being deformed with a similar strain history in each case. As a consequence of this repeated straining the initial slack is increased whilst the peak torque is reduced. Further increases in the initial slack may be obtained by repeated straining to a higher strain level.

Rotation of the tissue in the opposite direction (the fifth run) produces a relatively small amount of initial slack and a peak torque which is always greater than the second, third and fourth runs and sometimes greater than the first run.

To account for this behaviour and the 'residual torque' which persists after the disc has returned to zero, a residual fibre orientation is proposed to exist after all but the most insignificant straining processes of human skin encountered within the normal physiological range. This concept is illustrated in figure 8.

The complex viscoelastic behaviour of skin is well portrayed in the stress relaxation phases of the curves in figure 7. Each run is characterised by an extremely rapid (over a period of 1 to 2 seconds) decay of the torque when the equilibrium rotation has been reached. This is followed by a prolonged decay which ensues over the next fifty or more seconds. There are, also, marked differences in response between the first and subsequent displacements and smaller differences between subsequent tests.

These features are quite typical of skin and are evident even under the smallest of perturbations.

This time dependent behaviour was investigated further by studying the frequency dependent amplitude and phase data of the torque relative to a constant sinusoidal displacement input. A Solartron digital transfer function analyser JM1600 was used to compute the values of amplitude and phase.

Figure 9 illustrates this frequency response data for two quite different amplitudes of rotation. The first at 0·79° r.m.s. produced a torque response of sinusoidal waveform and so appeared to be a *linear* response. At 3·00° rms the second set of data were taken from a torque response that was obviously non-linear—i.e. exhibiting a marked increase in stiffness at the higher amplitude.

The torque amplitude *v.* frequency characteristics

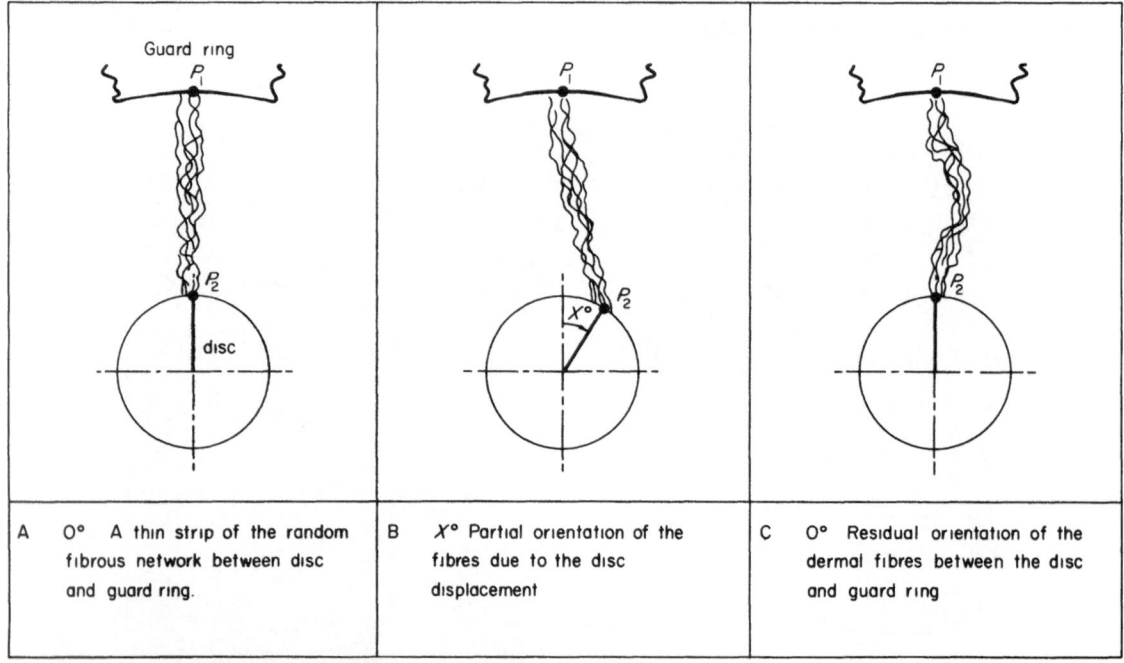

Figure 8. Proposed behaviour of the fibrous network within the human dermis when a disc attached to the skin is moved relative to a surrounding guard ring. A   Original orientation of random fibrous network; B   Partial orientation of the fibres on rotation; C   Residual orientation between disc and guard ring.

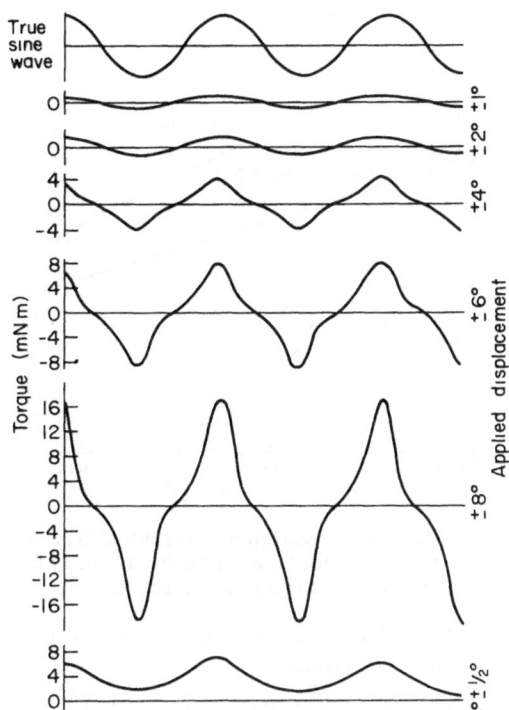

Figure 9. Response of human skin *in situ* to various levels of sinusoidal displacement at 1 Hz.

show a slight increase over the test range of 0·004 Hz to 10 Hz and the difference between the two tests is apparent. However, the phase angle remains constant, having a value about 10° up to 1 Hz whilst a slight increase in phase is evident between 1 Hz and 10 Hz.

Figure 10 is an indication of the variation in the recorded phase-frequency data taken from a group of six individuals. The results are plotted ± one standard deviation and indicate a relatively constant phase up to 1 Hz where there is an apparent 'break point' as the phase starts to increase with further increases in frequency.

The dynamic response of most time dependent materials show marked frequency dependence. However, the results for skin and some other biological materials are markedly different and we can seek an explanation for this frequency independence in the mechanical behaviour of the tissue components.

The component which is often ignored and might

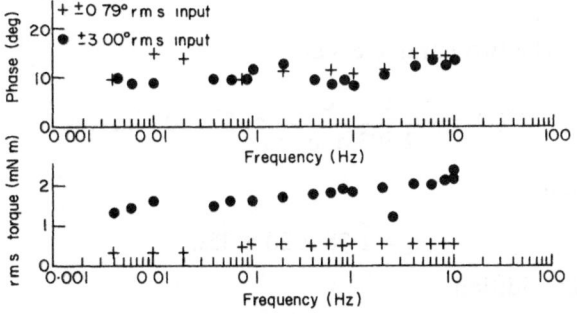

Figure 10. Phase-frequency response of human skin.

well provide the basis of such an explanation is the tissue ground substance. Proteoglycans form the portion of the ground substance that is intimately bound up with the collagen in a biochemical sense. It is known that the skin contains large amounts of hyaluronic acid and a similar long chain molecule, dermatan sulphate, amongst its proteoglycans. In adult pig the amounts of these two components in the proteoglycans of the dermis are 30% and 64% respectively (Loewi and Meyer, 1958).

It is not possible to extract the dermal proteoglycans in quantities sufficient for rheological analysis; but it is known that in synovial fluid hyaluronic acid constitutes about 100% of the proteoglycans (Meyer *et al.*, 1956).

A fall in apparent viscosity with increasing rate of shear has been reported for synovial fluid (Block and Dintenfass, 1963) and hyaluronic acid (Sundblad, 1953). It might be expected that a similar rheological behaviour is displayed by the proteoglycans of the dermis. Both the stress relaxation and frequency response data can be appreciated in a qualitative manner when viewed in terms of such frequency dependent behaviour of the ground substance.

EMPIRICAL DESCRIPTION OF VISCOELASTIC BEHAVIOUR

It is not, at present, possible to obtain quantitative correlation between structure and time dependent behaviour. It is, however, feasible to suggest from the structure of the tissues a mathematical framework in which a phenomenological description of viscoelastic behaviour is possible.

Discrete spring and dashpot models of tissue behaviour have been utilised to describe stress relaxation and dynamic data are, in many cases, and lead to the characterisation using sums of exponential time varying terms. In order for such models to describe the dynamic behaviour of tissues it is necessary that there be specific interrelations between the parameters of the springs and dashpots. These are not generally fulfilled and the stress relaxation and dynamic data are, in many cases, mutually inconsistent. In addition, the technique has the disadvantage of not providing an uniquely defined series of exponentials. The same data can give rise to quite different exponential terms if the method of analysis is changed.

The geometrical variability of the tissues and the complex mechanical properties of the structural components suggest that a continuous relaxation spectrum would be more likely to describe the viscoelastic behaviour than discrete relaxation times. Thus the stress relaxation behaviour is described not by

$$s(t) = s(\infty) + s(0) \sum_i A_i \exp(-t/\tau_i)$$

where A_i and τ_i are constants, $s(\infty)$ the equilibrium stress and $s(0)$ and $s(t)$ the stresses at zero and an arbitrary time t, but by

$$s(t) = s(\infty) + s(o) \int_0^{+\infty} A(\tau) \exp(-t/\tau)\, d\tau$$

or by the analogous logarithmic spectrum (Ferry, 1970)

$$s(t) = s(\infty) + s(0) \int_{-\infty}^{+\infty} H(\tau) \exp(-t/\tau)\, d\ln\tau.$$

The value of $H(\tau)$ can be obtained, approximately, from plots of $s(t)$ against $\log t$. The first approximation is obtained by

$$H(\tau) = -\frac{1}{2 \cdot 303}\left[\frac{d\,s(t)}{d\log t}\right] = \tau.$$

For many tissues, experimental stress relaxation curves plotted in such a manner may be linear for several logarithmic decades. Figure 11 shows such plots for skin and similar results have been reported

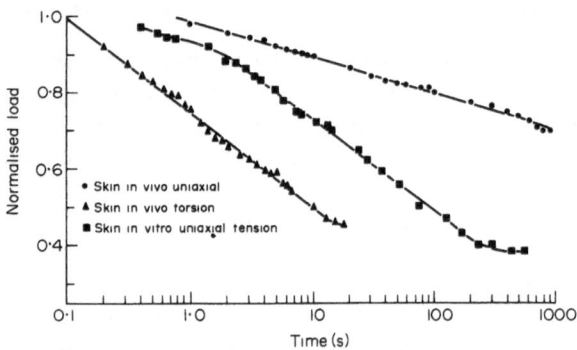

Figure 11. Stress relaxation behaviour of skin. *In vivo* torsion and uniaxial tension and *in vitro* uniaxial tension.

for messentry (Fung, 1972) and tendon (Minns, 1972).

The spectrum for such materials consists of a box of constant height $H(\tau)$ between upper and lower time limits τ_u and τ_l and zero elsewhere. The three parameters characterising the box can be found from the stress relaxation curve (Tobolsky, 1962).

The simplest form of the stress-log (time) plots makes the numerical comparison of stress relaxation curves based on such box spectra relatively easy.

Figure 12 shows the plot of the first and third tests of a set of four *in vivo* trapezoidal torsion tests carried out on intact skin at the same site. The differences between the behaviour of the material is apparent. Table 1 shows the results of a statistical fit of the best straight lines to the data of all four

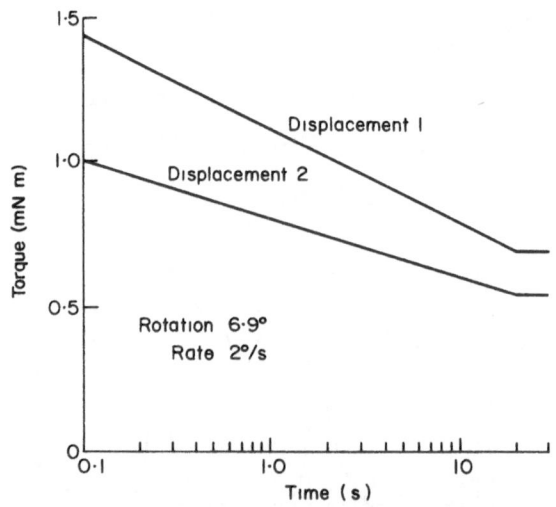

Figure 12. Torque relaxation curves for skin *in vivo*. There is a difference between the behaviour at first deformation and subsequent tests.

runs. The correlation coefficients are unusually high for biological data.

The simplicity and accuracy with which this simple spectrum describes the behaviour of soft collagenous tissues is encouraging. The real test of its usefulness lies in the ability to predict mechanical responses of the material from data obtained in a different testing mode. The stress relaxation and dynamic torsional results obtained on intact skin provide such a test for the self consistency of the analytical method.

For a given rotation the torque (Γ) is proportional to the viscoelastic modulus of the material. Hence interrelations suitable for moduli can be applied to the torque.

	Gradient mNm/ logs	Intercept (at 1s) mNm	Correlation coefficient	Number of points
Displacement 1	−0·367	1·18	0·996	27
Displacement 2	−0·222	0·95	0·996	32
Displacement 3	−0.187	0.80	0·993	32
Displacement 4	−0.162	0.77	0·998	36

TABLE 1

The results of run 3 in table 1 provide an estimate of two of the box parameters $H(\tau)$ and τ_u but not of τ_l which can only be estimated as being less than $0 \cdot 1\ s$.

The two parameters are

$$H(\tau) = -\frac{1}{2 \cdot 303}\frac{d\,\Gamma}{d\log t} \simeq 8 \times 10^{-2}\ \mathrm{mNm}.$$

and

$$\tau_u = 1 \cdot 78 \times 20 \simeq 35s.$$

In addition

$$\Gamma(\infty) = 0 \cdot 55\ \mathrm{mNm}.$$

The in phase torque obtained during dynamic test can be predicted from these box parameters

$$\Gamma'(\omega) = \Gamma(\infty) + \frac{H(\tau)}{2} \ln \left(\frac{1 + \omega^2 \tau_u^2}{1 + \omega^2 \tau_l^2} \right)$$

It is apparent that the predicted torque would be little affected by putting $\tau_l = 0$ at moderate frequencies. In addition at low frequencies the stress approaches an asymptotic value.

Hence it is possible to predict $\Gamma'(\omega)$ for frequencies up to about 10 rad/s—corresponding to the maximum value of τ_l. The results are shown in figure 13 and predict a slow increase of torque with an increase

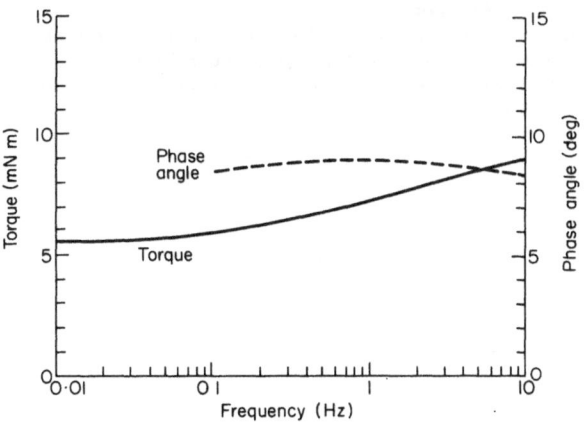

Figure 13. Dynamic response of human skin *in vivo* predicted from stress relaxation behaviour.

in frequency. This result and the stress magnitude correspond reasonably with experimental results displayed in the previous section. The exact magnitude of the stress obtained will depend on the subject's skin thickness and on disc and guard ring sizes.

The out of phase component of stress is given by

$$\Gamma''(\omega) = H(\tau) (\tan^{-1} \omega \tau_u - \tan^{-1} \omega \tau_l)$$

The approximation

$$\Gamma''(\omega) = H(\tau) (\tan^{-1} \omega \tau_u)$$

is satisfactory within the time region of the box but is increasingly inaccurate at both low and high values of ω.

The phase angle δ is given by

$$\text{Tan } \delta = \frac{\Gamma''(\omega)}{\Gamma'(\omega)}$$

This has been calculated and displayed as figure 13 in which the lower frequency limit is approximately equal to $1/\tau_u$. The characteristic feature is the small change in phase angle with frequency. This is in reasonable agreement with experimental results up to 1 Hz displayed in the previous section (see figure 10). Inertia effects might be expected to occur at the higher frequencies and these would produce

experimental results which were higher than predicted (Tobolsky, 1962).

It appears that the viscoelastic interrelations using the box spectra are reasonably accurate. The simplicity by which experimental results are characterised by box spectra also makes it possible to carry out computer analysis of digitised experimental results. The success of such an analysis is shown in figure 14 which is computer plotted. Data was prepared as punched tape which was subjected to

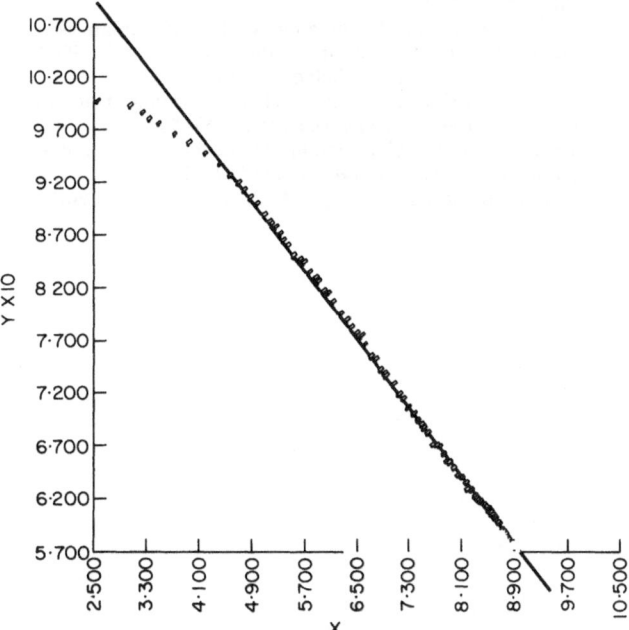

Figure 14. Computer plotted stress relaxation data and line of best fit. The abscissa is ln(time,s) × 3·9, and the ordinate normalised stress.

simple modification followed by statistical analysis to yield the stright line of best fit for selected experimental points.

Further investigation of the utility and disadvantages of the use of the box spectra are presently being undertaken. One disadvantage is the non-linear behaviour of soft tissues. This means that the parameters of the distribution will vary with strain amplitude and that predicted moduli etc. can only be compared with experimental results at similar strains.

It has also become apparent that for some materials and test conditions more complex spectra are required. Thus for stress relaxation tests carried out on articular cartilage of complete rabbit knee joints skewed normal probability distribution spectra are required to characterise the experimental data.

In addition to their practical advantages the use of spectra might discourage the production of some of the more exotic spring and dashpot models and reduce the temptation to identify the components of such imaginary models with real tissue components.

REFERENCES

BARBENEL, J. C., EVANS, J. H., and GIBSON, T. 1971: Quantitative relationships between structure and mechanical properties of tendon. *Digest 9th ICMBE*, Melbourne, p. 150.

BLOCH, B., and DINTENFASS, L. 1963: Rheological study of human synovial fluid. *Australian, New Zealand J. Surg.* 33: 108–113.

CRISP, J. D. C. 1972: Properties of tendon and skin. *In* 'Biomechanics: Its Foundations and Objectives'. (Eds. Y. C. Fung, N. Perrone, and M. Anliker). Prentice Hall, New Jersey. pp. 141–179.

FERRY, J. D. 1970: 'Viscoelastic Properties of Polymers.' Wiley, New York.

FINLAY, J. B. 1970a: Biodynamic studies of human skin. Torsional characteristics in relation to structure. PhD. thesis, University of Strathclyde, Glasgow.

FINLAY, J. B. 1970b: Dynamic mechanical testing of human skin *in vivo*. *Journal of Biomechanics* 3: 557–568.

FINLAY, J. B. 1971: The torsional characteristics of human skin *in vivo*. *J. Biomed. Eng.* 6: 12: 567–573.

FINLAY, J. B. and HUNTER, J. A. A. 1972: Ehlers Danlos syndrome, Part 1. A survey of the *in vivo* torsional mechanical characteristics of skin. (In preparation).

FRISEN, M., MÄGI, M., SONNERUP, L., and VIIDIK, A. 1968: Rheological analysis of soft collagenous tissue. Part I: theoretical considerations. *J. Biomechanics* 2: 13–20.

FUNG, Y. C. 1972: Stress-strain-history relations of soft tissues in simple elongation. *In* 'Biomechanics: Its Foundations and Objectives.' (Eds. Y. C. Fung, N. Perrone, and M. Anliker). Prentice Hall, New Jersey. pp. 181–208.

LOEWI, G., and MEYER, K. 1958: Acid mucopolysaccharides of embryonic skin. *Biochim. biophys. Acta* 27: 453–456.

MEYER, K., DAVIDSON, E. A., LINKER, A., and HOFFMAN, P. 1956: Acid mucopolysaccharides of connective tissue. *Biochim. biophys. Acta* 21: 506-518.

MINNS, M. J. 1972: M.Sc. thesis, UMIST, Manchester, England.

SUNDBLAD, L. 1953: Studies of hyaluronic acid in synovial fluid. *Acta Soc. Med. Upsaliensis* 58: 113–238.

TOBOLSKY, A. V. 1962: 'Properties and Structure of Polymers'. Wiley, New York.

VIIDIK, A., and EKHOLM, R. 1968: Light and electron microscope studies of collagen fibres under strain. *Zeitschrift für Anatomie und Entwicklungsgeschichte* 127: 154–164.

PRELIMINARY STUDIES ON MECHANOCHEMICAL-STRUCTURE RELATIONSHIPS IN CONNECTIVE TISSUES USING ENZYMOLYSIS TECHNIQUES

A. S. HOFFMAN, L. A. GRANDE, P. GIBSON, J. B. PARK,
C. H. DALY, P. BORNSTEIN, AND R. ROSS[1]

INTRODUCTION

In recent studies we have investigated the contribution of the individual components of connective tissues such as artery and ligament to the mechanical behaviour of the whole tissue. Our basic approach is to remove each component of the tissue selectively, using enzymes. The effects of such enzymatic digestions were measured mechanically and were observed histologically by light and electron microscope. Some experiments were performed by applying a sequence of enzymatic treatments to the same specimen; the reversible (low-strain) elastic modulus before and after each step in the sequence was then measured. In some cases only one enzyme was used on a fresh specimen.

In this study we report preliminary data on the sequential and individual enzymolyses of human aorta using the following enzyme sequence:

	Predominant
Enzyme sequence	*component removed*
1a Hyaluronidase (H)	Mucopolysaccharides
or	(ground substance of
	matrix)
1b Hyaluronidase +	Mucopolysaccharides
β-Glucuronidase	(ground substance of
[H + G]	matrix)
2 Collagenase	Collagen
3 α-Chymotrypsin	Microfibril of elastic fibres
4 Elastase	Elastin

The stress-strain curve up to 15% strain was obtained on each specimen before and after each step in the sequence. It is usually both linear and reversible, even after each enzymolysis. We feel that this general technique is especially useful for elucidating the contribution of the different components of connective tissue to its mechanical behaviour.

PROCEDURES, RESULTS AND DISCUSSION

A normal aorta from the left subclavian artery to the diaphragm was obtained at autopsy from a 19-year-old female within 24 hours after death which had resulted from an auto accident. The adventitia was removed and the sample was cut along the long axis, then cut again into 1·5 mm-wide strip specimens, using an aluminium cutter which holds five sharp blades at intervals of 1·5 mm between blades. As soon as the samples were cut they were placed in numbered tubes, washed several times in distilled water and 0·15 N NaCl and stored at 4°C in distilled water.

The sequence of enzyme and buffer solutions used is tabulated in table 1. TNBS (trinitrobenzene sulphonic acid) analysis of the supernatant yielded the content of amino acids released during each enzymolysis. After removal of collagen the supernatant digest was analysed for hydroxyproline (Grant, 1964), a characteristic amino acid of collagen.

The mechanical properties were measured using an Instron TMM tensile test machine with an extension rate of 0·5 cm/minute. The specimens were immersed in distilled water at 37°C during stressing using a specially designed double-walled glass container. The fresh samples were strained initially in the reversible region (15% strain), in order to characterise the mechanical properties before enzymolysis for each sample. To ensure that the 15% strain level remained in the reversible region, repeated strainings were carried out after each enzymolysis on selected samples; these data are tabulated in table 2. It can be seen that the 15% strain level is well within the reversible region for this enzyme sequence.

The typical *complete* stress-strain curves for sequentially treated samples in the order of (H + G), ([H + G] + collagenase), ([H + G] + collagenase + α-chymotrypsin), and ([H + G] + collagenase + α-chymotrypsin + elastase) are shown in figure 1 for five of the specimens. They were obtained to gain some idea of the effect of the sequence of enzymatic treatments on total strength. However, we do not feel that this is a very useful test for understanding the contribution of individual components since it is so subject to sample flaws. As can be seen, the (H + G) treatment has minimal effect on the stress-strain characteristics while the ([H + G] + collagenase) and ([H + G] + collagenase + α-chymotrypsin) treated samples show a lower modulus in the

---

[1] University of Washington, Seattle, Washington, U.S.A.

173

Enzyme[a]	% of substrate (wt)[b]	Buffer	pH	Other additives[c]	Digestion time, temp.
Hyaluronidase '(H)' (Bovine Testes)	5% and 100%	0·01M Na acetate	4·8	0·75 M NaCl 0·01% NaNO₃ (2·5mM N.E.M.) (2·5mM S.T.I.)	100 h at 37 °C
±					
β-Glucuronidase '(G)' (Beef Liver)	5% and 100%	0·01M Na acetate	4·8	0·75 M NaCl 0·01% NaNO₃ (2·5mM N.E.M.) (2·5mM S.T.I.)	100 h at 37 °C
Collagenase (Clostr. Histol.)	5%	0·2M NH₄HCO₃	7·8	0·001 M CaCl₂ 2·5mM N.E.M.	72 h at 37 °C
α-Chymotrypsin (Bovine Pancreas)	5%	0·08M 'Tris'	7·8	0·1 M CaCl₂	79 h at 37 °C
Elastase (Swine Pancreas)	5%	0·05M Na₂CO₃	8·8	0·01% NaNO₃	74 h at 37 °C

[a] All enzymes obtained from Worthington Biochemicals.

[b] The composition of the aorta on the basis of dry weight was assumed to be: 45% collagen, 45% elastin, 5% mucopolysaccharide and 5% microfibril (Ross and Bornstein, 1969). (The muscle cells are assumed to be removed during the initial washing.)

[c] N.E.M. = N-ethyl maleimide.
S.T.I. = Soybean Trypsin Inhibitor.

TABLE 1: Enzyme Protocols

low-strain region (see also figure 2), and no stress development at all in the high-strain region. Thus the collagen in the tissue appears to contribute throughout the entire strain region especially in the high-strain region; the latter observation has been previously established by other workers (Wood, 1954; Kenedi *et al.*, 1965; Milch, 1965).

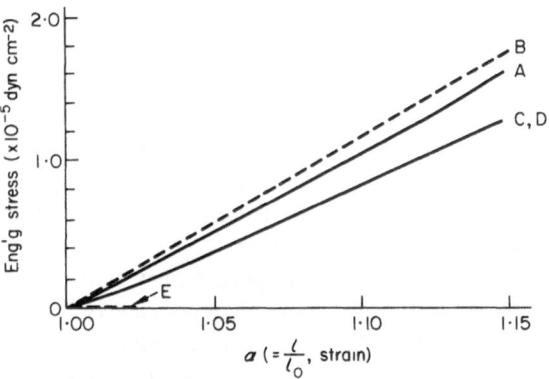

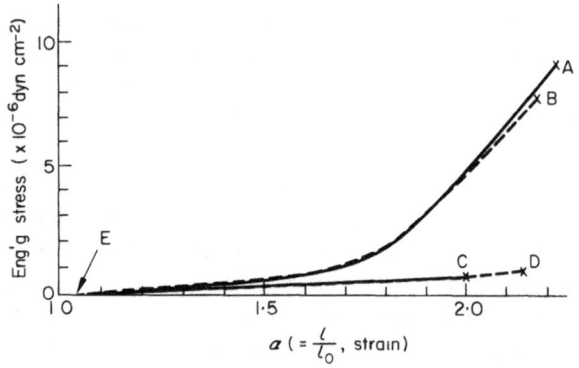

Figure 1. Complete stress-strain curves for sequential enzymolyses. A, Original. B, [H + G]. C, ([H + G] + collagenase). D, ([H + G] + collagenase + α-chymotrypsin). E, ([H + G] + collagenase + α-chymotrypsin + elastase).

Figure 2. Stress-strain curves in the reversible region after sequential enzymolyses. A, Initial. B, [H + G]. C, ([H + G] + collagenase). D, ([H + G] + collagenase + α-chymotrypsin). E, ([H + G] + collagenase + α-chymotrypsin + elastase).

The contribution of collagen in the low strain region can be seen more clearly in figure 2, which shows that the removal of collagen causes a reduction in the modulus in the 0–15% strain range. The additional α-chymotrypsin digestion did not affect the stress-strain behaviour shown by the ([H + G] + collagenase) treated sample. The slight

No. of strainings to 15%	Initial modulus (×10⁻⁶ Dyn/cm²) after			
	Original	[H + G]	[(H + G) + Collagenase]	[(H + G) + Collagenase + α-Chymotrypsin]
1	1·0	1·07	0·91	0·78
2	1·04	1·13	0·91	0·77
3	1·05	1·10	0·92	0·78
4	1·05	1·14	0·91	0·76
5	1·03	1·15	0·93	—
6	—	1·14	0·94	—
7	—	1·13	0·91	—
8	—	1·10	0·92	—
9	—	1·10	0·91	—
10	—	1·12	0·91	—

TABLE 2: Initial modulus changes after each enzymolysis in the sequence *v.* number of strainings

increase of the initial modulus of the (H + G) treated sample (B) is largely due to the high concentration of NaCl used (0·75 M); recent data show little change in modulus when 0·15 M NaCl was used. The ([H + G] + collagenase + α-chymotrypsin + elastase) treated sample disintegrated in the test tube; hence no mechanical test was possible.

Individual (non-sequential) enzymatic treatments on fresh samples of artery show that elastin is the major stress-bearing component in the low-strain region (figure 3). This has also been shown for

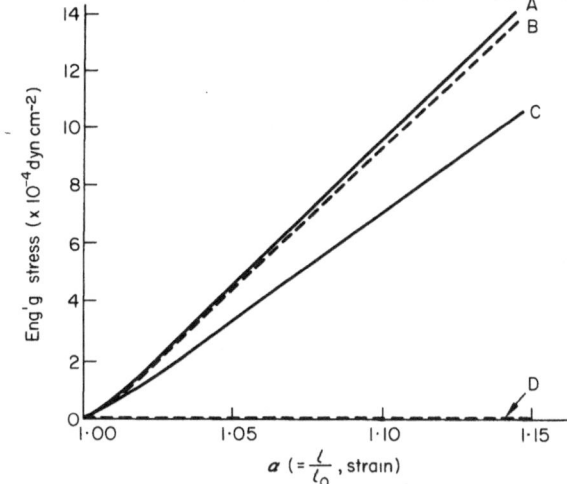

Figure 3. Stress-strain curves in the reversible region before and after collagenase and elastic treatment. A, initial. B, collagenase buffer only. C, collagenase. D, elastase.

ligament (Wood, 1954) and skin (Daly, 1969). These experiments also support those described in figure 2 which suggest there is a small but real contribution of collagen fibres to the stress-strain behaviour in the first 15% strain region. A contribution of this nature by collagen fibres has not been reported before. Such a phenomenon may be related to an 'interweaving' of collagen and elastic fibre networks, forcing the elastic fibres to wind around the crimped collagen fibres and thus to sustain a higher stress (at strains up to 15%) in the presence of these latter fibres than after they have been selectively removed.

When the original artery was cut into samples, the initial modulus of each specimen was measured between 0 and 15% strain. These data are shown in figure 4 for specimens numbered (arbitrarily) between '411' and '446'. It can be seen that the modulus decreases in a 'cyclic' fashion as one proceeds distally along the thoracic aorta. This emphasises the need to characterise each specimen initially, in a reversible region, before enzyme treatments.

In figure 5, the ratios of the initial moduli of specific specimens to (a) the total moles of amino

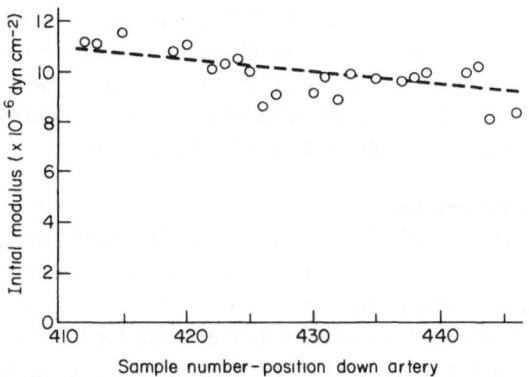

Figure 4. Initial modulus *v.* position distally along the artery.

acids released by the action of elastase or (b) to the weight of hydroxproline released by the action of collagenase on these specimens are plotted as a function of the position of the specimen distally along the artery. If the 'cyclic' variation noted in figure 4 is related to variations in the elastin or collagen content, then one of these ratios should normalise the data, depending on which of these two fibres has a greater influence on the value of the initial modulus. It can be seen that the better correlation is with the elastin data. This supports the conclusion that the elastic fibres are the *main* stress-bearing components up to 15% strain in arterial tissue although the collagen fibre network does have a small influence on the level of stress in this region.

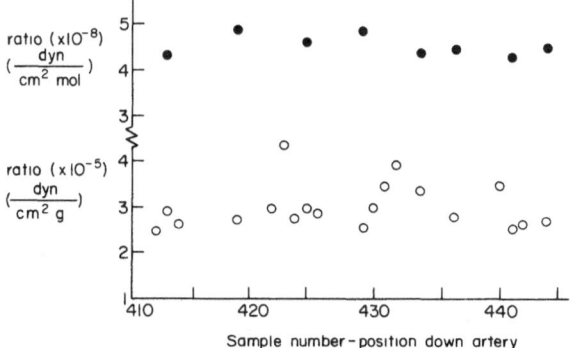

Figure 5. Ratio of the initial elastic modulus to the moles of amino acid from elastin (upper portion) and weight of hydroxyproline from collagen (lower portion) against position distally along the artery.

In this discussion thus far only the mechano-chemical aspects of the connective tissues are considered. However, electron (both transmission and scanning) and light microscopic studies have been performed after each enzymolysis in order to observe the structural changes directly. These experiments are in progress.

SUMMARY

In summary, it is emphasised that one-step, or sequential, removal of the individual components of connective tissues using specific enzymes, in conjunction with *reversible* mechanical tests and microscopic observations can be a powerful tool in clarifying the contribution of these components to the mechanical behaviour of such tissues.

Acknowledgment

Support of this work by NIH GMS grants 16436-03, 16436-04 and the N.I.H. Cardiovascular Training Program (for one of the authors—J.B.P.) is gratefully acknowledged. We thank Dr G. Balian for assistance with the hydroxyproline determinations.

REFERENCES

DALY, C. H. 1969: The role of elastin in the mechanical behaviour of human skin. *Proceedings of 8th Int. Conf. on Medical and Biological Engineering*, Chicago, p. 18–27.

GRANT, R. A. 1964: *Journal of Clin. Path.* **17**: 685.

KENEDI, R. M., GIBSON, T., and DALY, C. H. 1965: Bioengineering studies of the human skin. *In* 'Structure and Function of Connective and Skeletal Tissue,' p. 388. Butterworths, London.

MILCH, R. A. 1965: Matrix properties of the aging arterial wall. *Monographs in Surg. Sci.* **2**: 261.

ROSS, R., and BORNSTEIN, P. 1969: The elastic fiber. *J. Cell. Bio.* **40**: 366.

WOOD, G. C. 1954: Some tensile properties of elastic tissue. *Biochim. et Biophys. Acta* **15**: 311.

PRELIMINARY RESULTS ON IMMOBILISATION-INDUCED STIFFNESS OF MONKEY KNEE JOINTS AND POSTERIOR CAPSULE

A. B. LAVIGNE AND R. P. WATKINS[1]

INTRODUCTION

It is clinically evident that joints become stiff with immobilisation. Upon remobilisation, the joints of younger individuals resolve their stiffness much more readily than do older ones. In particular, the knee joints of some elderly patients, after immobilisation in flexion, fail to regain full range of motion until the posterior joint capsule is surgically released. This experience prompted a study of the course of knee joint stiffness with immobilisation and the effect of immobilisation upon the mechanical properties of a key periarticular tissue, the posterior joint capsule. Monkey knees were selected as the experimental specimens because of their similarity to man's in anatomy, weight-bearing function, prominent posterior joint capsule development, and full extension in range of motion.

Experimentally, joint stiffness due to ageing was measured quantitatively by Wright (1969) who found increasing stiffness with age for an intact limb moved through a limited range of motion. Goddard (1969) measured increased joint stiffness due to short-term immobilisation; he attributed the stiffness to articular gelling. Other biochemical and biomechanical changes due to age, degenerative joint disease and immobilisation of joint components have been described by Mankin (1969), Kempson (1970), Barnett (1969) and others, but these changes have not as yet been correlated with quantitative measurement of the stiffness of both the total joint and individual periarticular tissues. The measurement of these mechanical properties and their change with immobilisation is described in this paper.

PROCEDURE

The left knee joints of 24 female *Macacca nemestrina* monkey were immobilised by pin and cast for varying lengths of time between 0 and 64 days. The right knee joints served as controls. Four of the monkeys were sacrificed at each of the following number of days after surgery: 0, 4, 8, 16, 32, and 64. Immediately after monkey sacrifice, the hind limbs were dissected to remove skin and muscle layers and leave intact the joint capsule and associated ligaments. The dissected limbs were then mounted in a standardised manner on a device called an arthrograph, which moved the joint continuously through flexion and extension. Figure 1 shows the arthrograph with the cover opened to expose the

Figure 1. Monkey knee joint mounted on the arthrograph.

joint. The limb was tested in a liquid and vapour bath of temperature-controlled (37°C) Ringer's solution buffered to pH 7·4. The arthrograph permitted relatively unconstrained motion since the femur was free to slide and rotate in a universal bearing while the joint moved through flexion and extension. The limb was moved either at a constant velocity or with a sinusoidally varying velocity of frequency 0·25 Hz. The force required to move the limb through a given range of flexion-extension and the accompanying amount of rotation about the femoral axis were recorded. The limbs were moved between 100 and 120 cycles. The angle between the femur and tibia changed from 30° to 150° for control limbs and limbs immobilised 0 to 16 days; limbs immobilised 32 and 64 days were moved through a smaller range.

After the joint was tested on the arthrograph, it was dissected further in order to measure the stiffness of the medial posterior capsule alone. This tissue was left intact still attached to the femur and tibia. Physiological positioning of the joint was maintained during dissection by returning the specimen to the jig used to prepare the specimen for

[1] University of Washington, Seattle, Washington, U.S.A.

the arthrograph. Since a mounting clamp remained with the joint, repeatable mounting of the joint either in the jig or on the arthrograph was possible. After the specimen was prepared, it was mounted in the Instron in the same 90° geometry it had in the jig and was tested in a bath of room-temperature Ringer's solution buffered to pH 7·4. Figure 2 shows a side view of a specimen under load in the Instron. Tensile loads were applied to

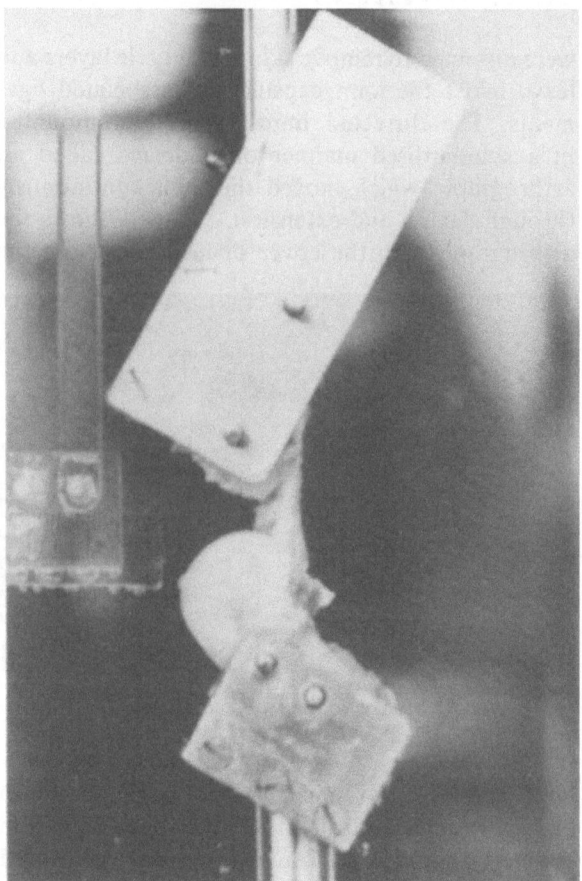

Figure 2. Side view of monkey medial posterior joint capsule mounted on the Instron.

the specimens at a constant rate of extension. The applied force and the corresponding extension were recorded by the Instron; selected extensions were also recorded by photography.

RESULTS

Examples of data from joints tested on the arthrograph are given in figure 3. The force required to move a limb through a given range is shown for limbs immobilised 0, 16, 32 and 64 days. It was found that the force required to move limbs immobilised 4 and 8 days was close to that required for the control limbs. For limbs immobilised 16 days and longer, however, increasing force was required. In figure 3, the relatively flat central portion of the curves corresponds to the range where low applied force readily results in limb motion and tissue

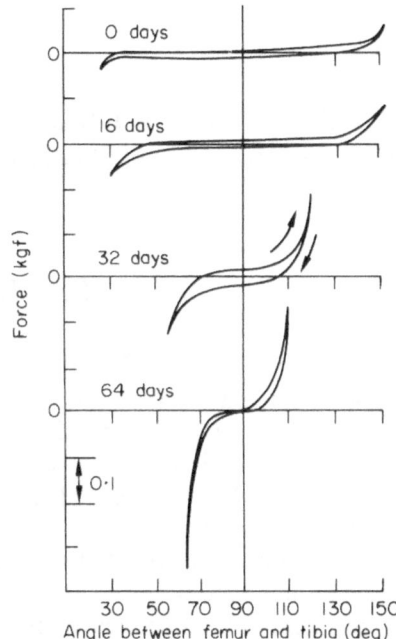

Figure 3. Force required to move monkey knee joints through given ranges of motion. Duration of limb immobilization is given above each curve.

deformation. Since this range decreased markedly with increasing immobilisation, the force required to move a limb through a given arc increased with increasing immobilisation.

Figure 4 gives examples of load-deformation curves for medial posterior capsules from joints

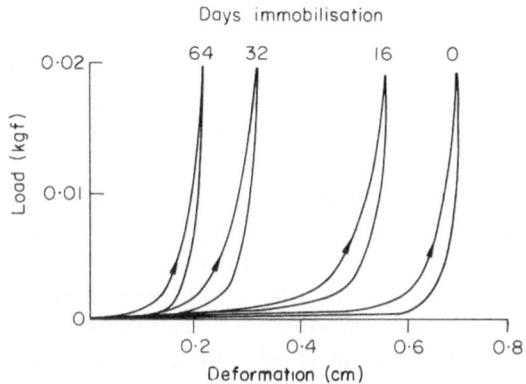

Figure 4. Load-deformation data for monkey medial posterior capsule: low loading values.

immobilised 0, 16, 32 and 64 days. The capsules were loaded to 0·02 kgf at a rate of extension of 0·5 cm/min, and immediately unloaded. In the graphs, zero deformation and load correspond to the capsule mounted in the Instron in the 90° reference position. For the control capsule, deformation between zero and 0·5 cm corresponded approximately to deformation resulting from physiological motion of the knee where the angle between the tibia and femur varied from 90° to 150°. Very small force was required to deform the tissue in this range. When the tissue was loaded above this range and unloaded, a hysteresis loop (representing energy loss) resulted.

Figure 5 gives examples of loading medial posterior capsule to failure at a rate of extension of 0·5 cm/min. The curves for limbs immobilised 16

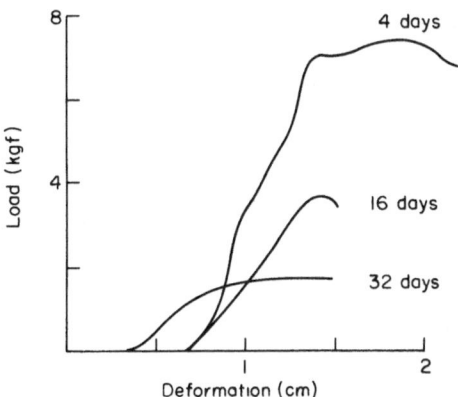

Figure 5. Load-deformation data for monkey medial posterior capsule: high loading values. Duration of limb immobilisation is given adjacent to each curve.

and 32 days are the average of data from 4 specimens, the other curve is the average of 2 specimens. As with the arthrograph data, increasing immobilisation reduced the range where the tissue readily deformed. In addition, immobilisation progressively reduced the tissue strength. When the posterior capsule failed, failure occurred not at the insertion of the capsule into femur or tibia, but at fibres throughout the tissue. In figures 4 and 5, the difference between control limbs and limbs immobilised 4 days was negligible, and the difference between control limbs and limbs immobilised 8 days was small. Visual examination of the specimens revealed abnormal tissues in the posterior capsule from limbs immobilised 32 days; for limbs immobilised 64 days, these changes became major. Increased fatty tissue was present, the capsule was ecchymotic, the meniscus could not be identified, and surrounding soft tissue appeared to encroach upon the articular surface.

DISCUSSION

The principal sources of data scatter were the biological variation of the monkeys and variation in dissection. The monkeys weighed between 3·5 and 8 kg; the average weight was 5·7 kg, standard deviation 1·1 kg. Although most of the monkeys were classified as young adult, 5–6 years of age, the ages of about half were uncertain. Three of

the monkeys were classified as old, 9–12 years of age. Seven left limbs had fractured while the limbs were immobilised.

Although linear corrections to the data on the basis of monkey weight and width of medial posterior capsule were considered, it was evident that the interaction between these factors and the variation caused by monkey age, effect of fractured limb and dissection technique was not simple. Consequently, the data are presented with no corrections.

CONCLUSIONS

Tests performed on monkey knee joints showed that increasing immobilisation progressively reduced the range where either the joint or the medial posterior capsule alone could be readily deformed. In addition, increasing immobilisation progressively reduced the strength of the medial posterior capsule. These effects were not evident after 4 days immobilisation, and only slightly evident after 8 days. At 16 days immobilisation, the changes were appreciable, and at 32 days, not only were the mechanical properties of the immobilised limbs grossly different from those of the controls, but also there were morphological changes visible in the joint tissues.

Acknowledgments

Appreciation is expressed to Professor Jean T. Hodson, Department of Restorative Dentistry, University of Washington, for use of the Instron universal testing machine.

This investigation was supported in part by Grant No. HD-04872 from the National Institute of Health, Washington, D.C.

REFERENCES

BARNETT, C. H. 1969: Factors limiting joint mobility. *J. of Anatomy* **105**: 185.

GODDARD, R., DAWSON, D., LONGFIELD, M., and WRIGHT, V. 1969: A study of articular gelling. *In* 'Lubrication and Wear in Joints' (Ed. V. Wright), p. 134. Lippincott, Philadelphia.

KEMPSON, G. E., MUIR, H., SWANSON, S., and FREEMAN, M. 1970: Correlations between stiffness and the chemical constituents of cartilage on the human femoral head. *Biochimica et Biophysica Acta* **215**: 70.

MANKIN, H. J., and LIPPIELLO, L. 1969: The turnover of adult rabbit articular cartilage. *J. Bone and Joint Surg.* **51A**: 1591.

WRIGHT, V., DAWSON, D., and LONGFIELD, M. D. 1969: Joint stiffness—its characterisation and significance. *Bio-Medical Engineering* **4**: 8.

BIOMECHANICS OF THE ORAL TISSUES

C. H. DALY[1], J. I. NICHOLLS[2], W. L. KYDD[3], AND P. D. NANSEN[1]

INTRODUCTION

Although dentistry seems to be an obvious field for the application of engineering techniques, such studies have been predominantly oriented towards dental materials research and to dental equipment such as the high-speed air drill. However, the effect of stress on the oral tissues has not been the subject of quantitative study. These tissues are known to respond very sensitively to mechanical stress and this effect is made use of very widely in orthodontics. A variety of clinical problems arise from abnormal stresses within the mouth, e.g. ulcers caused by pressure under a denture, periodontal problems.

Two bioengineering studies are being carried out at the Centre for Research in Oral Biology at the University of Washington:

(i) biomechanics of the periodontal ligament *in vivo*, and
(ii) the effects of pressure loading on the oral mucosa.

Progress reports on these studies are presented in this paper.

PERIODONTAL LIGAMENT STUDIES

The periodontal ligament (PDL) is the structure which retains the tooth in its socket. It is a vascular tissue consisting of collagen fibres which penetrate into the cementum of the tooth root and into the alveolar bone of the socket. It has not been clearly demonstrated that these fibres run from cementum to bone in the human although this is known to occur in various animals.

Periodontal disease is the major cause of tooth loss and its control by therapy is a much less certain process than the control of dental caries. The progress of the disease is characterised by a gradual loss of the ligament and the alveolar bone until eventually all tooth support is lost. From the standpoint of biomechanics, periodontal disease is of interest in that one of the few known causes of the disease is the application of abnormal loads to a tooth due to malocclusion or to abnormal oral habits such as bruxism (tooth clenching). One principal diagnostic method used for this disease is the measurement of tooth mobility. This is done in a completely subjective fashion by pressing on the tooth in question and noting its movement relative to its neighbours.

Previous work in this field has largely been concerned with attempts to measure mobility in some quantitative fashion. This work has been reviewed in detail by Mühlemann (1967). No attempt was made in any of these studies to measure the mechanical properties of the PDL. This was not possible because the geometry of the tooth root was not known and the direction of loading with respect to the root was not accurately controlled.

The objectives of the present study are:

(i) to determine various biomechanical characteristics of the PDL itself,
(ii) to relate these properties to pathological changes in the PDL,
(iii) to determine stress distributions on the alveolar bone and relate this to bone resorption,
(iv) to devise clinically useable diagnostic aids.

The maxillary central incisor was selected for this study because of its simple root geometry. Three simple loading methods are possible (figure 1). The

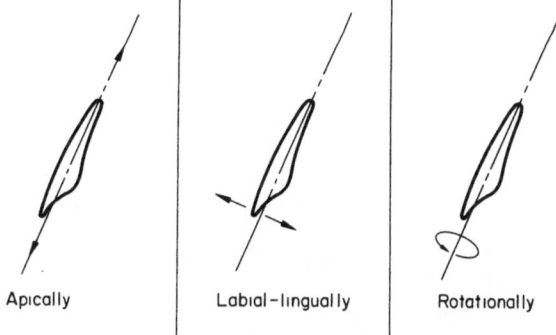

Apically   Labial-lingually   Rotationally

Figure 1. Three convenient configurations for force application to an incisor.

rotation method was selected because of the greatest simplicity in terms of stress analysis, and the possibility of symmetrical loading in either direction. Körber and Körber (1963) used this method of

[1] Department of Mechanical Engineering, University of Washington, Seattle, U.S.A.
[2] Department of Civil Engineering, University of Washington, Seattle, U.S.A.
[3] Department of Oral Biology, University of Washington, Seattle, U.S.A.

tooth loading but the load application method employed did not restrict the load to a pure torque but contained other force components. This in addition to the problems noted above makes the results of doubtful value.

Methods

It is first necessary to determine the position of the axis of symmetry of the root of the selected tooth with reference to the rest of the maxillary dentition. This is done by means of a precision stereo X-ray method (Nicholls *et al.*, 1972). The equipment used is shown in figure 2. The most important feature to note is the cast chrome-cobalt

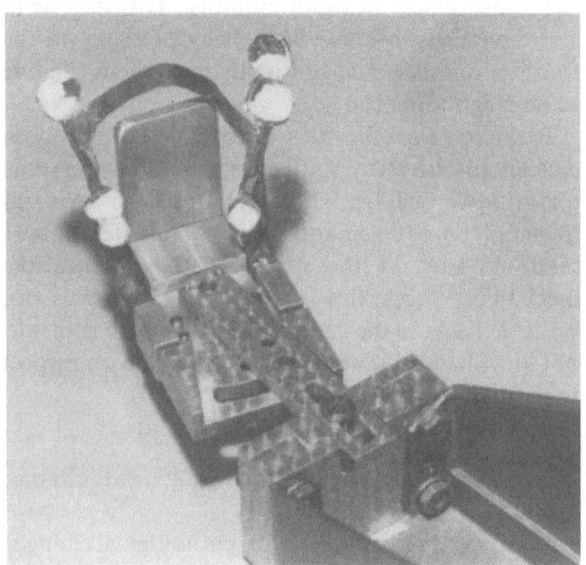

Figure 2. Stereoscopic X-ray device for determining tooth root location. Note the cast clutch used to locate this system to the maxillary dentition.

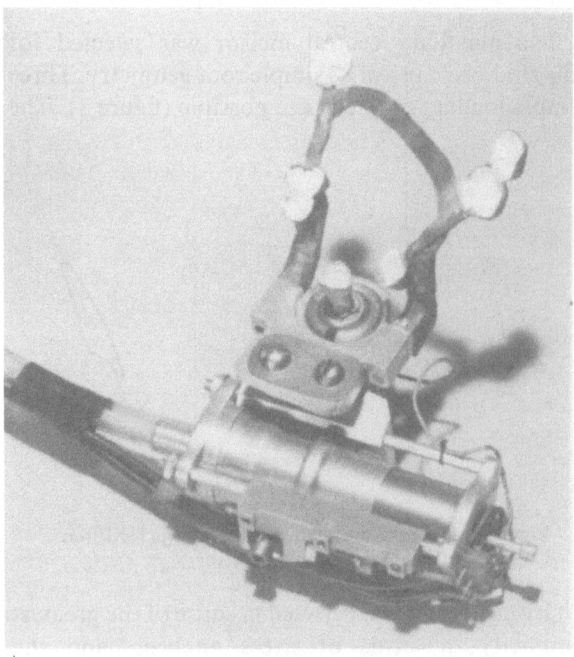

Figure 3. Apparatus used for torque loading of a central incisor.

alloy clutch which provides an accurate and rigid attachment to the dentition. The tooth loading device is attached to this same clutch so as to maintain the same reference (figure 3). The loading mechanism itself can be translated and rotated with respect to the clutch so that its loading axis can be aligned with the predetermined tooth axis. This is done by means of a setting jig (figure 4).

Figure 4. Setting jig used for the alignment of the loading mechanism with the axis of the tooth root.

The micrometer settings for this jig are obtained from the X-ray measurements by means of a simple computer program. Once set, the loading device is attached to the clutch. The loading shaft is attached to the selected tooth by a cast gold coupling piece (figure 5). The loading capability of this device is a maximum torque of ± 0.05 N m with a resolution

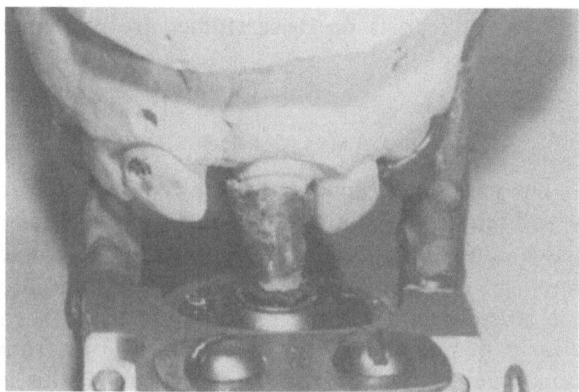

Figure 5. Attachment of the loading mechanism to the incisor.

of <0.1 mN m. Maximum possible rotation is ± 0.04 radian with essentially infinite resolution. The hydraulic operating mechanism shown is an interim system and is being replaced by a servo motor drive system permitting up to 10 Hz cyclic loading and precise control of step function loads or displacements.

Test procedure

Preliminary experiments have been performed on two subjects with the object of determining the order of magnitude of the measured quantities and the qualitative nature of the PDL properties.

The torque and angle measuring systems were switched on for a 30-minute period prior to testing to obtain thermal stability. The data was recorded on a 2-channel chart recorder and on an FM instrumentation tape recorder. The loading device and clutch were inserted in the subject's mouth and the gold coupling piece was attached to the incisor with temporary cement. No anaesthetic was found to be required at the loading levels employed.

Two types of test were performed:

(i) *creep test*—a constant torque was applied in one direction for 60 seconds and was then reversed for 60 seconds. The torque was then removed and the creep recovery was monitored for a further 60 seconds. This test sequence was repeated at 5 minute intervals with increased torque levels being used each time.

(ii) *cyclic tests*—triangular wave or sinusoidal rotations were applied to the tooth using equal peak amplitudes in each direction. Twenty load cycles were applied then the tooth was held at zero rotation for 5 minutes. Further 20 cycle sequences were applied with different recovery times between each sequence.

Results

The results for the first 60-second loading period of a series of creep tests are shown in figure 6. It can be seen that for moderate loads (up to

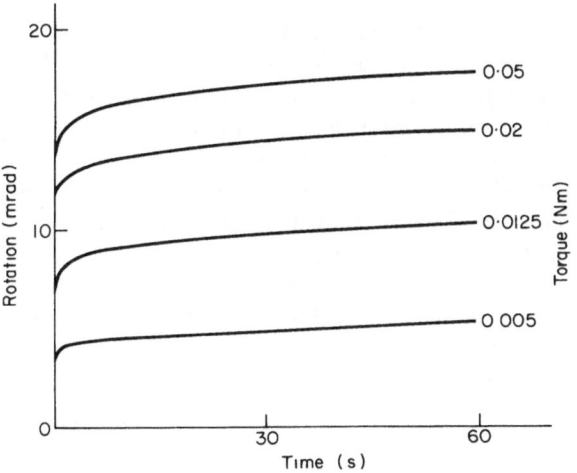

Figure 6. Tooth rotation versus time for various constant torques.

0·01 N m) the PDL behaves in a linear viscoelastic fashion. At higher loads, the non-linear behaviour characteristics of soft tissues becomes apparent.

The cyclic test data are of particular interest because of the possibility afforded by this test technique to apply reversed load cycles to tissue *in vivo* without disturbing the environment of the tissue

in any way. The results shown in figure 7 illustrate the normal effect of successive load cycles on tissues in that the stiffness decreases with each successive cycle and eventually reaches some asymptotic cycle after 10–15 cycles. This behaviour is seen *in vitro* and *in vivo* but in these *in vivo* tests it is found that recovery towards the initial behaviour takes place if the cycling is stopped. This is not found *in vitro*. The time scale of recovery is quite short. This is seen in the third curve in figure 7 which is the first complete cycle applied after a 5-minute recovery period following 20 load cycles. It has often been

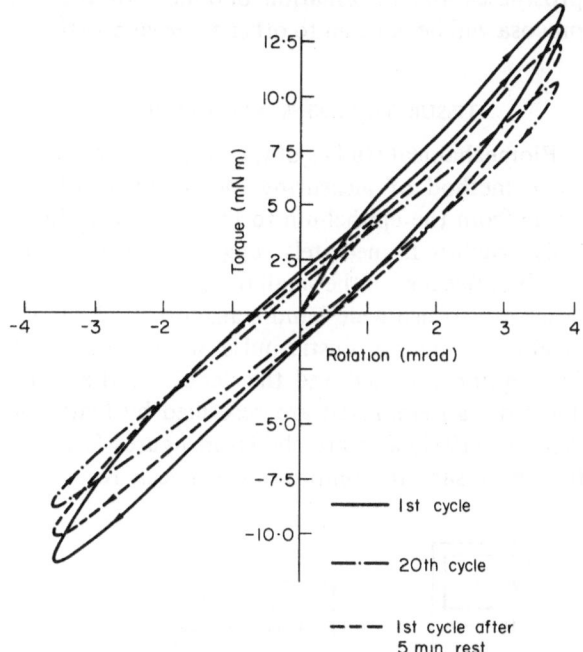

Figure 7. Three load cycles on the incisor using a sinusoidally varying torque with equal positive and negative rotations.

a common practice to assume that, because many tissues are subject to load cycles *in vivo*, *in vitro* test data should be taken on specimens which have been 'preconditioned' by applying several load cycles before making the measurements. The rapid *in vivo* recovery shown here makes this assumption questionable in many situations.

The results presented here are very preliminary in nature. They have allowed the closer definition of the future test programme. At present, the loading device is being modified to permit closer control of creep and cyclic tests and to allow stress relaxation experiments. The planned testing programme includes a series of tests to establish the normal range of behaviour of the PDL and to investigate the changes seen in a variety of mechanical parameters with pathological changes in the PDL.

PRESSURE EFFECTS ON ORAL MUCOSA

The oral mucoperiosteum of the residual ridge and palate is required to withstand pressure loading applied by all complete and some partial dentures. Deformation of the mucosa can cause movement of the prosthesis and interfere with occlusion. Poorly fitting prostheses cause chronic pressure on the mucosa leading to ulcer formation. The process of fitting a prosthesis by taking impressions is subject to errors caused by tissue deformation under the impression compound and by the time dependent nature of the soft tissue. Quantitative study of these effects requires knowledge of the viscoelastic properties of the oral mucosa *in vivo*.

These problems are not restricted to dental prostheses and information obtained for the oral mucosa will be relevant to other external prostheses.

TISSUE THICKNESS MEASUREMENT

Biomechanical studies of the oral mucosa require some method of measuring the thickness of this tissue from the epithelium to the underlying bone. This structure immediately suggests the use of echo ranging. Because of the small thickness involved it is necessary to use wide band transducers which will produce a single acoustic pulse about 50 ns long. The equipment used and the details of the transducer design employed are described by Daly and Wheeler (1971) and are shown in figures 8 and 9. It is necessary to compromise between resolution

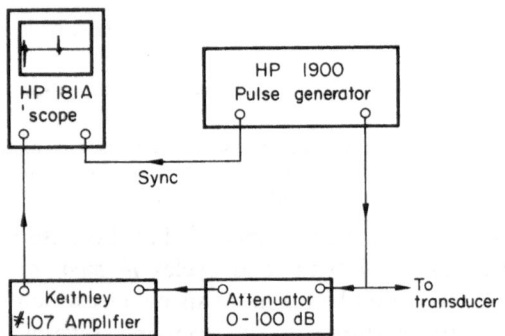

Figure 8. Block diagram of echo-ranging equipment.

and wave attenuation in the tissue. This dictates the present use of a 10 MHz resonant frequency disc of barium titanate as the active element in the transducer. The transducer resonance is damped by the backing piece shown in figure 9. Various materials are being tried but it is not yet possible to obtain the desired combination of high attenuation and a specific acoustic impedance close to that of barium titanate. The first property allows the use of a short backing piece as no echo will be returned from a material with high attenuation. The best material available in this respect is made by centrifuging a fine tungsten powder in an epoxy resin matrix at 100 000 g for 1 hour. A better acoustic impedance match is obtained with lead but the low attenuation

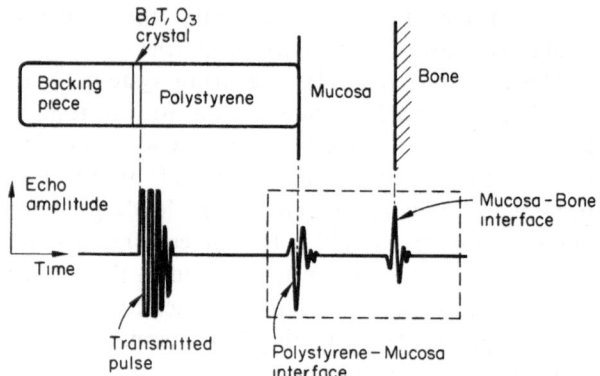

Figure 9. Construction and method of operation of the echo-ranging transducer.

requires a much longer backing piece to delay the echo from the end face of the backing piece.

The present transducer allows the measurement of tissue thickness in the range of 0·5–10 mm. A change in thickness under the transducer of 20 μm can be detected. An analogue read out circuit has been devised and permits recording tissue thickness changes versus time under pressure loading. Calibration of the transducer is accomplished by measuring the acoustic velocity in the tissue. No significant change in acoustic velocity is found even when the tissue is compressed to 40% of its original thickness.

TISSUE INDENTATION STUDIES

Pilot experiments have been performed by indenting a 4 mm diameter transducer into the tissue in a direction normal to the surface (Kydd *et al.*, 1971). The skin overlying the anterior aspect of the tibia was used as a model in these experiments. It has a very similar structure to that of the mucosa and was much more convenient and accessible. Creep loading was used with a constant pressure of 11 × 10^4 N/m² being applied for 10 minutes and then removed. Tissue thickness was monitored during the load and recovery phases.

The results of this procedure for 2 subjects are shown in figure 10. The loading phase shows a

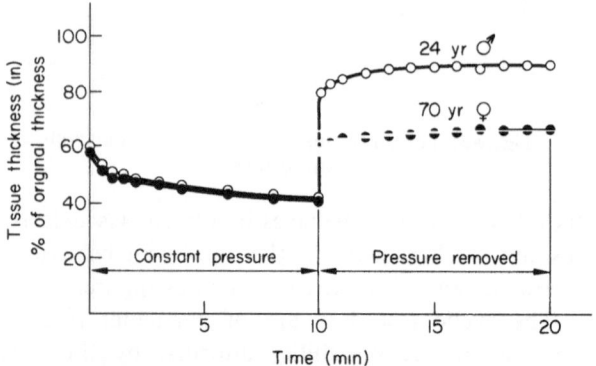

Figure 10. Indentation tests on human skin over the anterior surface of the tibia.

typical creep response and it is seen that the normalised response shown appears to be independent of age. This is true for all of the subjects tested. The recovery phase shows a very marked age dependence and has a much longer retardation time than the loading phase.

The long creep recovery times shown by these experiments have obvious implications for the clinician trying to re-fit a badly fitting denture. A new impression should not be taken immediately after the old one is removed. However, just when it should be taken is not clear as stress-relaxation would probably result in a loose fit for a denture which was an exact fit to the fully recovered tissue. Further study of these time dependent tissue deformations and of those deformations occurring during the impression process is obviously required.

Histological studies of canine oral mucosa subjected to loads as described above for periods up to 6 hours have been described by Kydd *et al.* (1969). Pressure and time-dependent degenerative changes were seen in the epithelium and the collagen fibres of the lamina propria appeared to support the indentor by acting in a net-like fashion. The fibres were seen to be in tension and showed the change in staining reaction to a trichrome stain which has been observed in stressed dermal collagen fibres (Craik and McNeil, 1965).

At present, equipment is being designed to permit tests similar to the above on oral mucosa *in vivo* using techniques derived from the PDL study methods.

The tissue loading device will be servo operated and will be mounted on a clutch to obtain a rigid reference.

The two projects discussed above are providing information on the basic nature of the short term response to stress of the oral tissues. This is an essentially passive response. It is intended to extend these studies to include the long term active response of tissue remodelling.

Acknowledgments

The research was supported by the Centre for Research in Oral Biology at the University of Washington and was funded by U.S. Public Health Service Grant No. DE-02600-04.

REFERENCES

CRAIK, J. E., and McNEIL, I. R. R. 1965: Histological studies of stressed skin. *In* 'Biomechanics and Related Bioengineering Topics' (Ed. R. M. Kenedi). Pergamon, Oxford, p. 159.

DALY, C. H. and WHEELER, J. B. 1971: The use of ultra-sonic thickness measurement in the clinical evaluation of the oral soft tissues. *Int. Dental J.* **21**: 418.

KÖRBER, K. H., and KÖRBER, E. 1963: Die Rotatorische Komponente der Physiologischen Zahnbeweglichkeit. *Zahnärztl Welt* **10**: 309.

KYDD, W. L., DALY, C. H., and WHEELER, J. B. 1971: The thickness measurement of the masticatory mucosa *in vivo*. *Int. Dental J.* **21**: 430.

KYDD, W. L., STROUD, W., MOFFETT, B. C., and TAMARIN, A. 1969: The effect of mechanical stress on oral mucoperiosteum of dogs. *Archs. oral Biol.* **14**: 921.

MÜHLEMANN, H. R. 1967: Tooth mobility—a review of clinical aspects and research findings. *J. Periodont.* **38**: 114.

NICHOLLS, J. I., DALY, C. H., and KYDD, W. L. 1972: A stereoscopic X-ray procedure for locating the centroidal axis of the root of a maxillary central incisor. *J. Biomechanics* **5**: 159.

MULTI-COMPONENT MATERIALS

M. ABRAHAMS[1]

INTRODUCTION

Biological materials are complex natural multi-component or composite materials but the structural elements of the individual components can be simplified to a two or three phase material where one element reinforces the others thus the required balance in properties is achieved. In this respect they are analogous to certain engineering composites, particularly reinforced plastics where a weak polymer matrix is reinforced by strong stiff fibres. Reinforced plastics are already being considered for prosthetic devices but as their theoretical behaviour is now more fully understood they can also serve as a model to assist in the understanding of the more complex behaviour found in biological materials.

In this paper the properties of certain plastic-based composites are first discussed in general terms and compared with the properties of bone, cartilage and tendon. The particular problems of designing with composite materials are next discussed and the effect of the viscoelastic behaviour of the polymeric matrix on composite mechanical properties is outlined. Practical examples of design from basic principles are presented to illustrate new applications using plastic based composites and conventional materials. Finally, the paper suggests that bone replacements by reinforced plastics with similar properties to bone may be a fruitful area for future research.

[1] GKN Group Technological Centre, Wolverhampton, England.

MATERIALS

The majority of biological materials are multi-component materials and the properties of those required to carry load in the human body are similar in many respects to reinforced plastics. The mechanical properties of some well-known reinforced plastics and the comparative properties of bone, cartilage and tendon are given in table 1. In order to understand the properties of these multi-component materials it is necessary to briefly review the function and structure of the components which make them up.

Reinforced plastics consist of a weak polymer matrix incorporating strong stiff fibres. One function of the matrix is to distribute loads among the fibres which cannot generally be used by themselves, but to be of structural value the fibres must be many times stiffer and stronger than the matrix material they reinforce. The ultimate properties of fibrous-based composites depend upon the length and volume of the fibres used and upon their direction and degrees of alignment. The most common fibres in use in the plastics industry are glass fibres but carbon fibres will become commercially important in the future. The most efficient matrices are the thermosetting resins, good examples being epoxy and polyester, but because of limitations in certain properties such as elongation, impact strength and wear characteristics, not to mention difficulties in fabrication, thermoplastic matrices are sometimes preferred. The most common reinforced thermoplastic in the U.K. is glass reinforced nylon where the composite is normally produced by injection

		Fibre angle (deg)	sp. gr.	Tensile modulus (MN/m²)	Tensile strength (MN/m²)	Specific modulus (MN/m²)	Specific strength (MN/m²)
Carbon fibre high mod.	70%/wt	0	1·66	230 000	1140	138 000	690·0
epoxy resin		90		14 000	40	8 450	24·0
Glass fibre type E	58%/wt	0	1·76	29 000	700	16 500	400·0
epoxy resin		90		8 000	40	4 550	23·0
Glass fibre/nylon	33%/wt	random	1·39	6 960	120	5 000	86·0
Figures given below are approximate values							
Spruce		0	0·40	11 000	70	28 000	175·0
Human bone		—	2·3	15 000	100	6 500	43·0
*Tendon (toe extensor)		0	1·10	200	35	180	32·0
*Cartilage (rib)		—	1·20	17	—	14	—

* The figures given are taken from unpublished work by the author carried out some years ago.

TABLE 1: Properties of composites

moulding using short random glass fibres. Other well-known reinforced thermoplastics are poly-carbonate, polypropylene, acetal, etc. In table 1 the tensile modulus and tensile strength of carbon and glass fibre reinforced epoxy resins are compared with glass fibre reinforced nylon. The properties of the first two materials are quoted for 0° and 90° to the fibre axis as the fibres are aligned parallel and hence the materials are highly anisotropic.

The fracture surfaces of a carbon fibre/epoxy resin composite and a glass fibre/polypropylene composite are illustrated in figures 1 and 2. In figure 1 carbon fibres are aligned at 30° to the fracture

Figure 1. Fracture surface of carbon fibre/epoxy composite. Fibres aligned at 30° (diameter, 8 microns) (magnification × 3620).

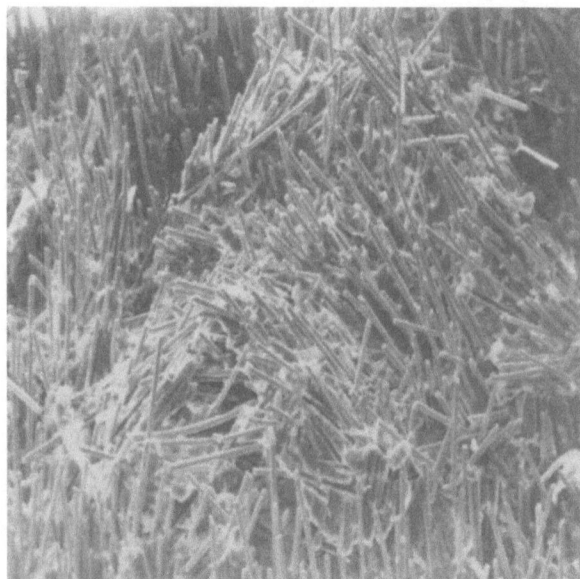

Figure 2. Fracture surface of glass fibre/polypropy-lene composite. Short glass fibres (diameter, 9 microns) aligned in random manner (magnification × 57).

surface, while in figure 2 the random nature of the glass fibres reinforcing the thermoplastic can be observed.

Wood is one of nature's most efficient polymeric composites having as its principal structural element a cellular structure aligned parallel to the axis of the tree trunk. The properties of wood (Dinwoodie, 1971) are closely associated with the cellulose micro-fibrils which make up the basic framework of the cell wall. As the specific gravity of most timbers is <1 they can compete favourably on a specific property basis with many of today's man-made multi-component materials often being considerably cheaper. For comparative purposes the properties of spruce have been included in table 1, and in terms of the specific modulus and strength, i.e. property/sp. gr., it is superior to bone which is generally regarded as being the most efficient composite of them all.

Biological materials are also natural polymeric composites but it is only in the last ten years or so that the mechanical function of the individual elements making up the composite have been clearly understood although the macro properties, particularly bone, have been the subject of study for many years. The author is neither an authority on the physiological function nor on the chemical nature of biological materials therefore the following comments on their individual elements relate strictly to their mechanical function.

Bone is composed of four main elements: collagen fibres impregnated with a complex mineral com-pound similar to naturally occurring hydroxyapatite; bone cells which direct the metabolic process of bone and which also control the relationship between the fibres; and the bone salt and the blood vessels which permeate the substrate providing the nutri-ment and assist in bone repair. The properties given in table 1 are only approximate values as variations in magnitude of several orders have been reported due to age and sex. For the purposes of this paper, however, the structure of dry bone has been simplified to a composite consisting of collagen fibres (49 % by volume), reinforced by hydroxy-apatite crystals (41 % by volume), and other constituents making up the remaining 10%. The strong, stiff hydroxyapatite crystals rather than the weaker collagen fibres provide the reinforcing element.

The functional role of the collagen fibres in tendon, on the other hand, is reversed as the fibres provide the main structural element by reinforcing a con-nective tissue matrix. Tendons consist of parallel bundles of collagen fibres, figure 3, contained in a sheath of dense fibrous connective tissue. Small projections of connective tissue intermingle with the collagen bundles containing fluid and small lymph-atic vessels thus providing the lateral reinforcements

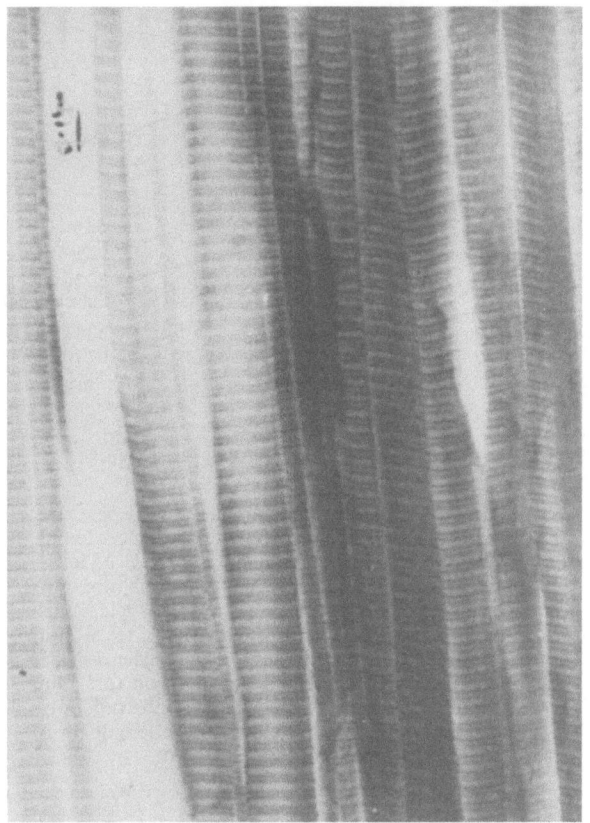

Figure 3. Collagen fibres in horse tendon (diameter, 0·1 microns) (magnification × 58 500).

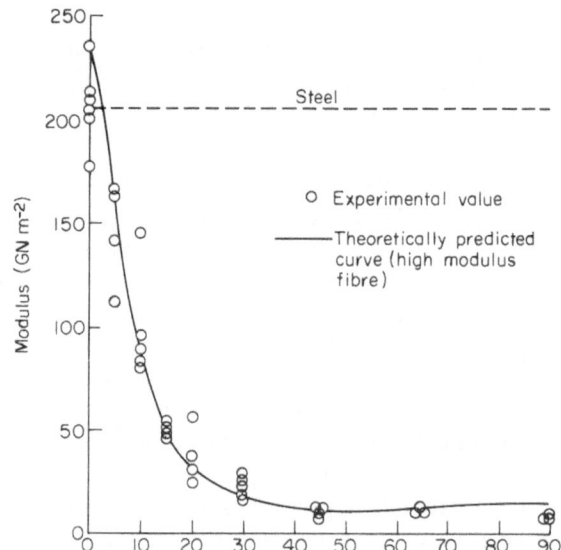

Figure 4. Tensile modulus of a 57%, by volume, high modulus carbon fibre/epoxy resin composite for different angles of fibre alignment.

which bind the structure together. In rib or hyaline cartilage the collagen fibres are arranged in a random manner reinforcing a matrix consisting of a ground substance of mucopolysaccharides which contains the cartilage cells, the chondrocytes, which occupy one-third of the cartilage volume.

DESIGN

Fibres provide the most efficient reinforcement when they are aligned in one direction but the resulting composite will have anisotropic properties. When components are designed using unidirectional composites the designer, in addition to knowing the principal moduli and strengths of the material measured in the direction of the fibres, needs to know the variation of these properties at any other angle relative to the principal axis. This is illustrated in figure 4 where the modulus of a high modulus carbon fibre/epoxy resin composite is plotted for different angles of fibre alignment from 0° to 90°. Although the properties of the composite are equivalent to steel at 0°, at one-fifth the weight, there is a 20:1 reduction at 90°. Isotropic properties in two directions can be obtained by laminating aligned fibres in layers at 0° and ±60° or by using the fibres in a random manner but the properties of such a composite will be limited to one-third of the unidirectional composite for the same volume of

fibres. Fibre-length is of less importance provided a critical aspect ratio, i.e. l/d ratio, is exceeded. For most fibres efficient transfer of load is achieved when the aspect ratio >100.

Of significant importance is the percentage of reinforcement in any one composite. The relationship between the composite elements for elastic unidirectional fibres reinforcing an elastic matrix is expressed by the simple role of mixtures; composite modulus = (fibre modulus × volume fraction of fibres) + (matrix modulus × volume fraction of matrix). Unfortunately, polymeric materials are not elastic but at small strains polyester and epoxy resins have a reasonably linear behaviour for the above relationship to apply. The effect of volume fraction is demonstrated in figure 5 where the modulus of a

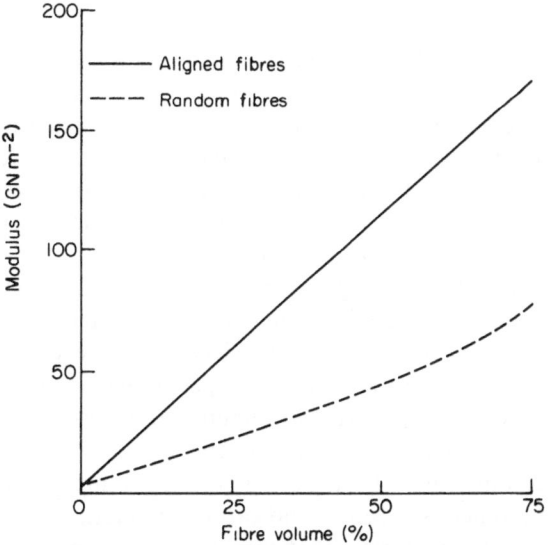

Figure 5. Tensile modulus of a Type A carbon fibre/ epoxy resin composite for different volume fractions of fibre.

Type A carbon fibre/epoxy resin composite is predicted for different volume fractions of fibre. Two curves are plotted, one for aligned fibres and the other for random fibres.

Viscoelastic behaviour

Polymers, like biological materials, are viscoelastic, i.e. their properties are dependent upon time and temperature.

This time dependence can be characterised by creep behaviour which occurs when a material under load extends with time. Creep curves for polypropylene are given in figure 6(a). To make allowances for the time dependent properties of a material the designer can, by making approximation, substitute the time dependent properties in the appropriate engineering formula. The time dependent modulus, $E(t)$ obtained from the slope of an isochronous stress/strain curve is the preferred constant for designing in plastics. The isochronous curve can be plotted from a set of creep curves by taking stress and strain values at ordinates of constant time as illustrated in figure 6(b).

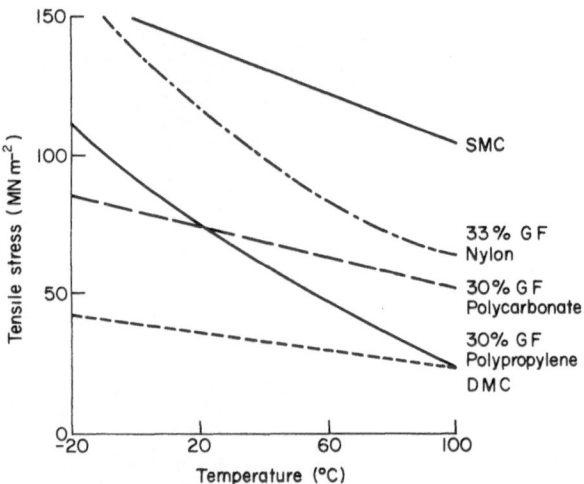

Figure 7. Failure stress as a function of temperature. S.M.C.—glass reinforced polyester resin fibre length 25–50 mm, random alignment. 33 % GF Nylon—glass reinforced nylon fibre length up to 5 mm, random alignment. 30 % GF Polycarbonate—glass reinforced polycarbonate fibre length up to 5 mm, random alignment. 30 % GF Polypropylene—glass reinforced polypropylene fibre length up to 5 mm, random alignment. DMC—glass reinforced polyester resin moulding compound fibre length up to 10 mm, random alignment.

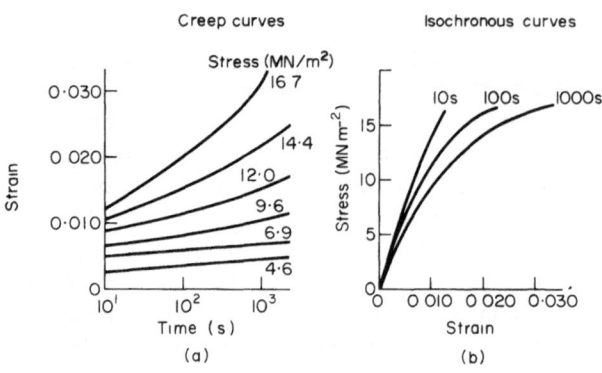

Figure 6. (a) Creep curves for polypropylene copolymer. (b) Derivation of the isochronous stress/strain curve.

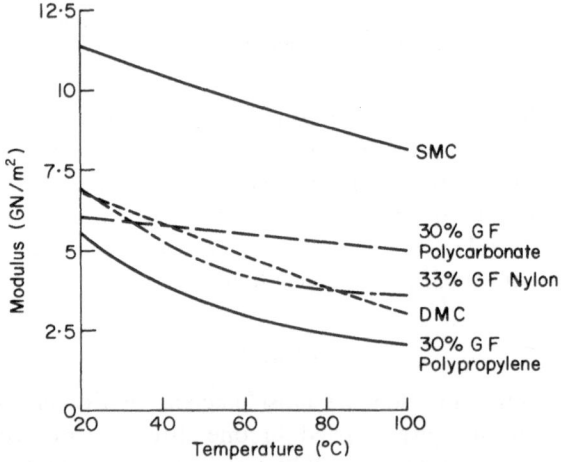

Figure 8. 100 sec, 0·5 % tensile secant modulus as a function of temperature. Key as for figure 7.

The viscoelastic phenomena is more pronounced in thermoplastics than thermoset materials as the former have a linear or branched chain structure and the latter have a cross-linked chain structure. As one would expect, composites based on these materials exhibit similar behaviour but the effects are reduced by fibrous reinforcement.

The effects of temperature on properties must also be considered before a material is selected for a particular application. The decrease in strength, due to temperature, for a range of plastic based composites is illustrated in figure 7. Because of the complex material behaviour the designer should ascertain the effects of time and temperature simultaneously before indicating preference for a particular material. In figure 8 a time dependent secant modulus is plotted against temperature for the materials from figure 7.

Practical examples

Two examples where plastic based composites have been efficiently used to produce advanced components are shown in figures 9 and 10. In figure 9 a carbon fibre/epoxy composite is used to reinforce spruce to produce a light-weight racing oar. The composite is applied in two directions: (a) longitudinally to provide bending resistance; and (b) circumferentially to provide torsional resistance. The net result is that the oar is some 30 % lighter and has a 25 % reduction in cross-section with a 10 % gain in stiffness over a conventional spruce oar.

In the second example, figure 10, a glass reinforced thermoplastic, Noryl, is injection moulded

Figure 9. GKN carbon fibre reinforced racing oars.

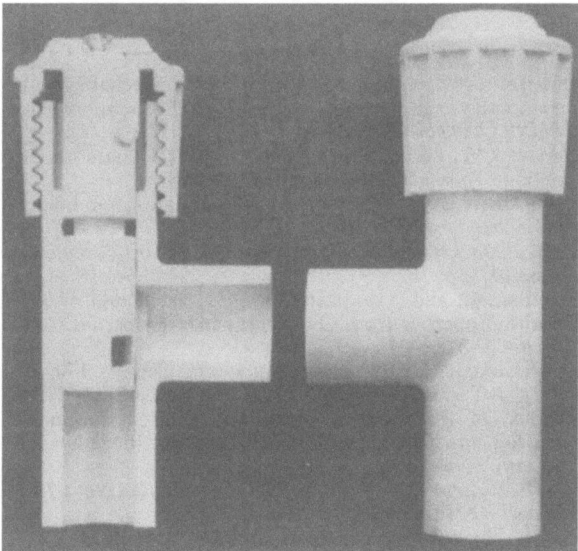

Figure 10. Hot water central heating valve in glass reinforced Noryl.

to produce a hot water central heating valve. The advantages over the brass valve are a reduction in the number of functional parts leading to reduced costs. The valve is designed to operate at 85°C and can be assembled into a steel pipework system by use of adhesives.

FUTURE APPLICATIONS IN BIOENGINEERING

The use of plastic composites in the field of bioengineering is by no means a new concept as both

types of glass reinforced plastics have been used in prostheses for the upper and lower limbs. Carbon fibres composites have also been used, with conflicting opinions in the case of C.F.R.P. jackets for thalidomide children (Composites 1971). Undoubtedly future reductions in carbon fibre prices will result in the development of a wide range of external prosthetic devices which will utilise the high specific properties of the material.

The use of fibrous composites as implants in the body has generally been dismissed because of the doubtful biocompatibility of the matrix polymers and possible dangers of fibre migration from the matrix. There is no reason however why this particular area should not be considered as a subject for future research in the field of bone replacement.

The number of commercially available polymers suitable for long term implantation in the body for orthopaedic purposes is generally recognised as being limited to the medical grades of silicon rubber, pure polypropylene, pure ultra high molecular weight high density polythene and certain acrylic (polymethylmethacrylate) cements, (Scales, and Lowe, 1971). It is unlikely that any fundamentally new polymers will appear on the market in the next 10–15 years so implants will generally be limited to the use of these materials or their derivatives. In fact it is understood that the development of a hydrophilic acrylic (Modern Plastics, 1972) under the trade name of Hydron is meeting with encouraging results from implantation experiments and it is also claimed that this material catalyses bone growth. The serious limitation of all these materials for replacement of the load carrying bones is that by themselves they lack the necessary stiffness and strength to carry the required loads imposed during normal activity. By use of fibrous reinforcement however, these limitations could be overcome.

By use of modern theories developed by Tsai *et al.* (1966) it is possible to predict fairly accurately the stiffness of a composite material from a knowledge of the properties of the individual elements. This is illustrated in figure 4 where the experimental points are shown to agree closely with theory (Dimmock and Abrahams, 1969). By the substitution of time dependent constants in the theory good approximations can also be made for thermoplastic based composites (Abrahams and Dimmock, 1971). If one assumes that the components of bone are reduced to 41 % by volume hydroxyapatite and that the remainder is an isotropic matrix of collagen (which it is not) then by use of the above techniques a model for the theoretical properties of bone can be obtained.

The theoretical curve for stiffness against reinforcement angle for the bone model and for two similar plastic based composite models is given in figure 11. The appropriate values for the modulus of hydroxyapatite crystals and for collagen were obtained from Bonfield (1971). The predicted value

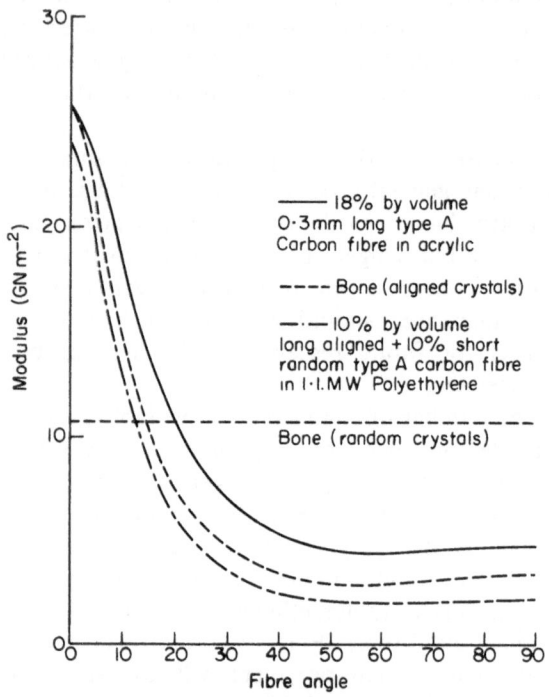

Figure 11. Theoretical prediction of composite modulus against angle of reinforcement.

of the modulus E at 0° fibre angle for the bone model is higher than most reported values but it is the same as the modulus measured by micro strain techniques (Bonfield and Li, 1966). On the other hand the value of E at 90° is lower than the reported values for transverse stiffness indicating that in practice not all the hydroxyapatite crystals are aligned in the longitudinal direction. The predicted value for the random fibre model is also incorrect but in increasing the accuracy (and complexity) of the initial assumptions the computer programme should predict the stiffness of bone fairly accurately. From this point it is a relatively simple matter to match these properties with a carbon fibre reinforced plastic as demonstrated by the two curves in figure 11. The carbon fibre/acrylic curve has been predicted for a composite containing 18%, by volume (33% by weight) short aligned carbon fibres, while the carbon fibre/ultra high molecular weight polyethylene curve

is for a composite containing 10%, by volume, long aligned fibres plus 10%, by volume, short random fibres (35% by weight). In addition, this particular composite would benefit from the good wear characteristics of the high molecular weight polyethylene matrix.

Assuming that a biocompatible plastic based composite could be produced, would it find a place in orthopaedics? Unfortunately, the author is not in a position to answer this question but as there appears to be a serious mismatch in properties between bone and the steels used to reinforce or replace bone in terms of stiffness, strength, weight and corrosion resistance, the indications are that a more accurate replica of bone might have a promising future. Furthermore, the instances of failures reported (Wilson, 1971) for bone replacements by metals could be reduced as the high stress concentrations always present when two dissimilar materials are joined together would be eliminated.

Acknowledgments

The author wishes to thank the GKN Group Technological Centre for permission to publish this paper and would like to record his appreciation for the help provided by his colleague Mr J. Dimmock.

REFERENCES

ABRAHAMS, M., and DIMMOCK, J. 1971: Mechanical and economic comparisons of reinforced thermoplastics. *Plastics and polymers, June,* p. 187.

BONFIELD, W., and LI, C. H. 1966: Deformation and fracture of bone. *Journal of Applied Physics* 37: 869.

BONFIELD, W. 1971: Mechanisms of deformation and fracture in bone. *Composites.* Sept., p. 173.

C.F.R.P. Jackets—a bioengineering application. Composites, June 1971, p. 72.

DIMMOCK, J., and ABRAHAMS, M. 1969: Prediction of composite properties from fibre and matrix properties. *Composites* 1: 87.

DINWOODIE, J. M. 1971: Wood. *Composites,* Sept., p. 170.

SCALES, J. T., and LOWE, S. A. 1971: Factors influencing the choice of materials for bone and joint replacements. *Plastics Institute Symposium* 'Plastics in Medicine and Surgery', Newcastle.

TSAI, S. W., ADAMS, D. F., and DONNER, D. R. 1966: *Tech. Report AFML-TR-66-190.*

TSAI, S. W. 1966: *Tech. Report AFML-TR-66-149.*

Water absorption can be a good thing. *Modern Plastics* Jan. 1972.

WILSON, J. N. 1971: Loosening of cement fixation in prosthetic replacement of the hip without infection. *Plastics Institute Symposium* 'Plastics in Medicine and Surgery', Newcastle.

DYNAMIC CHARACTERISTICS OF THE HUMAN BODY

H. E. von GIERKE[1]

INTRODUCTION

Most studies of the mechanical properties of the human body have been conducted on isolated organs, body segments, limbs or tissue specimens. For these the static and dynamic functions under internal muscle loads or external forces, the geometry and structure are relatively well defined. Results of this type have reached a high degree of sophistication using the whole framework of modern engineering dynamics and stress analysis in the course of the studies. However, if we are interested in the behaviour of the whole body or larger segments of it and are not willing to restrict ourselves to specific force inputs or loads, the number of studies conducted and the information available become less and less. The reasons for this are at least threefold: (1) the overall system becomes extremely complex and its mechanical functions and abuses very manifold. Although engineering systems of high complexity; as for example aircraft, are analysed by dynamic models with respect to their dynamic responses and stress loads down to their individual subsystems and components, they have the advantage of being built of materials with known and understood material properties. (2) Testing of the overall system is almost impossible with simultaneous detailed measurements of the responses of subsystems and components. This is particularly true for the human body *in vivo* and to a large extent even for cadaver materials. (3) Of the many factors determining the dynamic characteristics of the human body (table 1) at least half of them are not constant but vary with time. Several of these parameters determining the body's response are interrelated; for example, the time course of the force input function and the body's geometry determine jointly how the mechanical energy is propagated through the tissue (figure 1). Many of the difficulties listed in table 1 could be overcome by a more genuine interdisciplinary approach and, somewhat connected with this, by posing the overall problem in broader terms leading to results of more general validity. There is no doubt that in the long run this broader attack would be the more rewarding and more economical one. It is hoped that the following brief, but nevertheless critical, review of our knowledge will help to outline some of these proposed more general attacks on the problem.

THE PURPOSE OF INVESTIGATIONS IN HUMAN BODY DYNAMICS

Most investigations in human body dynamics (table 2) were conducted to explain the injuries, performance decrements or sensations caused by periodic or impulsive force environments, or to develop protective equipment against these environments. The body's response in the reversible linear and nonlinear loading range was of primary interest in most cases. However, the injury mechanisms in what are termed 'environmental exposures' in the table involve determining where various force loadings produce primary tissue damage and at what input levels to the body such damage occurs. Although knowledge of the basic tissue strength properties of isolated specimens is helpful in this task, they cannot—or at least not yet—replace experimental studies or accident data analysis on the composite structure or its analogues. It is important to note that always several of the methodologies reviewed in the following must be applied to shed light on body dynamics of interest for one of the purposes listed. When in table 2 the purpose of some investigations with periodic or impact forces is classified as 'methodology', it is meant that the excitation force function as such was of no direct interest but that the data gained with this test added information on body dynamics of interest in connection with other purposes or force excitations.

[1] Aerospace Medical Research Laboratory, Wright-Patterson Air Force Base, Ohio, U.S.A.

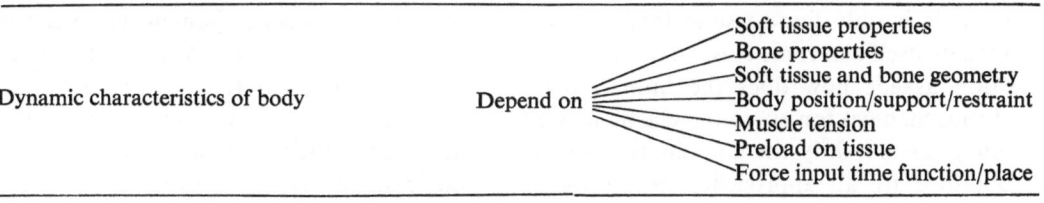

TABLE 1: Factors influencing dynamic characteristics of the human body

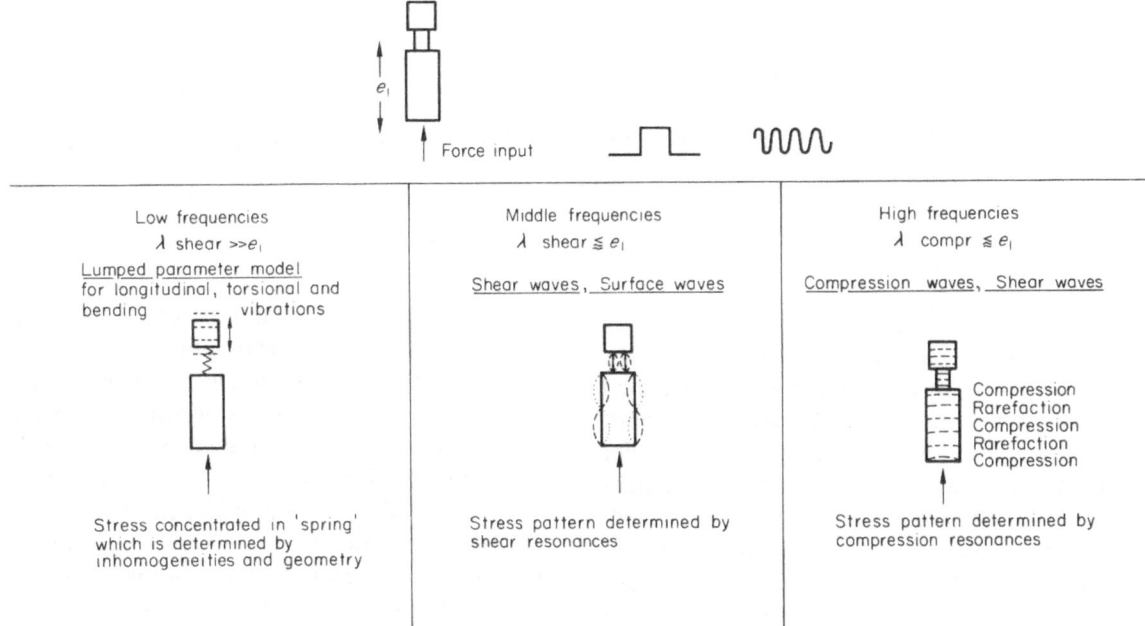

Figure 1. Different modes of propagation of mechanical energy in body tissue. Depending on the time function (frequency content) of the force input the energy is propagated in the form of transverse shear waves or longitudinal compression waves. Stress concentrations depend also on inhomogenities and geometry (λ = wavelength) (from von Gierke,1964).

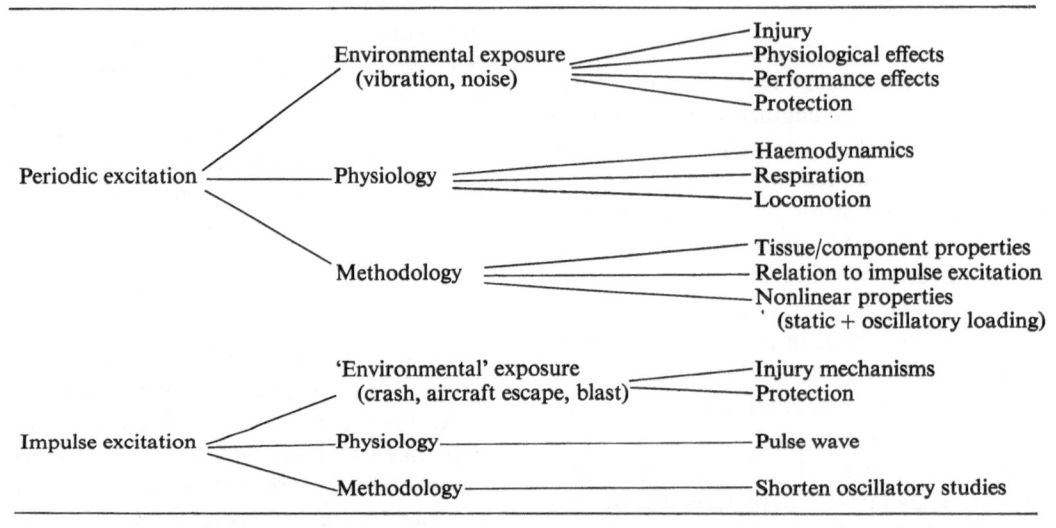

TABLE 2: Purpose of investigations on dynamic characteristics of the human body

METHODOLOGIES

The various approaches to analysing and understanding the human body's dynamic response are listed in table 3. Most of the methodologies have been applied with impulsive as well as oscillatory forces and have been used either to solve specific applied problems or to contribute to the basic overall goal. It is only fair to state that none of the methods by itself can come close to a solution of the question asked: how does the human body respond to mechanical forces, in the elastic as well as the injury range? Only the combination of the results obtained by all approaches can bring us closer to this goal. And it is here, where perhaps the

greatest challenges and problems are, where the last methodology listed, biodynamic modelling, takes on a special position and importance with respect to the others. Not that it could give us a realistic answer without all the other methods—on the contrary, it is entirely dependent on the experimental input from all the other methodologies. But it is the only one which contains in itself the hope of providing the proper framework to combine, correlate, understand and use the total body of information available in this field. In my opinion, this is the area in which the major progress has been achieved in the last decade and which should guide our endeavours in the future (von Gierke, 1971a). This does not mean that it is the area where the major

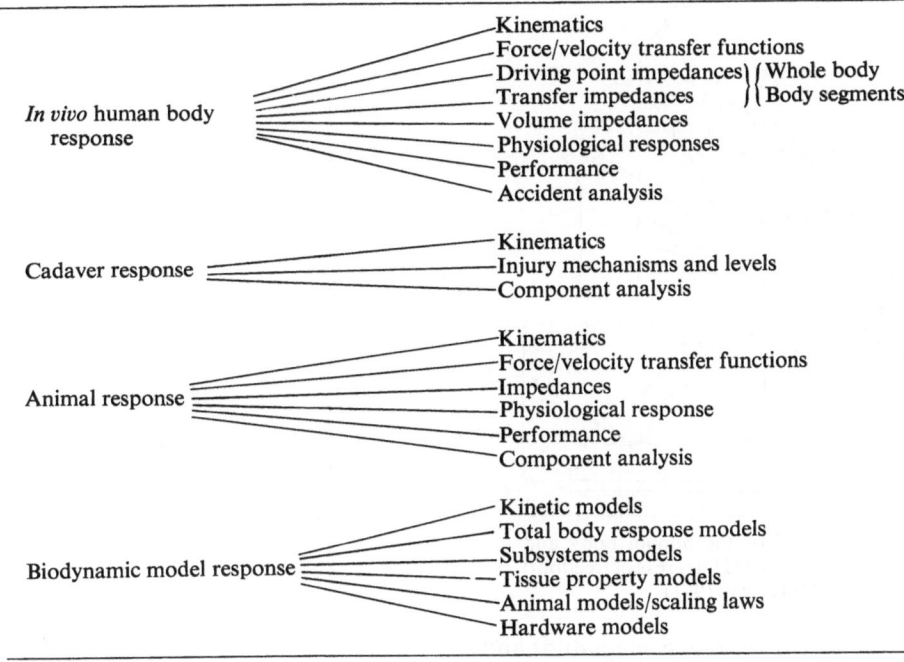

TABLE 3: Methodologies for analysing dynamic whole body characteristics

work or resources are required, but that it must form the foundation, without which the other investigations remain fragments. For this reason the various types of models which resulted from the experimental investigations will be taken as the framework for the following discussion of the present status of the field.

BRIEF REVIEW OF THE BODY'S DYNAMIC RESPONSE

In most cases the effects of mechanical energy on man are the direct result of relative displacement of body parts or segments, organ deformation or local tissue strain. This is certainly true for vibratory as well as impulsive force inputs as far as acute as well as chronic tissue damage and injury are concerned. It is also the case for most known interference effects of vibration with human performance such as manual control (Allen *et al.*, 1972), speech (Nixon *et al.*, 1963), or visual acuity (O'Briant *et al.*, 1970) and also must be assumed to account in a more diffuse and complicated way for such integrated effects as fatigue, discomfort or those effects on performance we are not able to relate directly to head, arm or hand motion (von Gierke, 1971*b*). Correlations of the latter effects with causes other than the motions of single body parts have not yet been tried, with the exception of the probably oversimplifying assumption to relate them to the total mechanical power dissipated in the whole body. The measurement and description of tissue strain suffers from the inaccessibility of living tissue to these types of measurements, the lack of appropriate landmarks and the difficulty of quantitating all those factors listed above as influencing the body's dynamic

response. As a consequence the body's response is best described and understood in terms of biodynamic models which are compatible with available measurements and, at the same time, elucidate which information is still missing (von Gierke, 1971*c*; Sandover, 1971).

Kinetic body response

The gross motions of individual body segments or parts have been studied in internal biomechanics; i.e. as a result of internal energy capabilities (motion, force and torque, gait, etc., as a result of muscular strength, work, power) and in external biomechanics, i.e. under the influence of environmental forces (translation or rotation of the body under blast forces, flailing of head or limbs in the airstream following emergency escape from aircraft and primarily kinetics of aircraft or automobile occupants under crash conditions). Based on kinematic observations kinetic models of the stick-and-joint type or with the proper anthropometric parameters of the individual body segments are constructed. An example of such a model is shown in figure 2. Frictional constraints and limitations of joint articulations are based on empirical fits; however, no attempts are made to relate joint forces with any actual forces acting inside the body or causing injury to body structures. In these models, the elastic behaviour of joints is also neglected, so that oscillatory body responses cannot be realistically calculated. In spite of these limitations, which make these descriptions appear primitive from a biological point of view, these models are extremely helpful for the prediction of body motions as a result of forces such as occurring in automobile

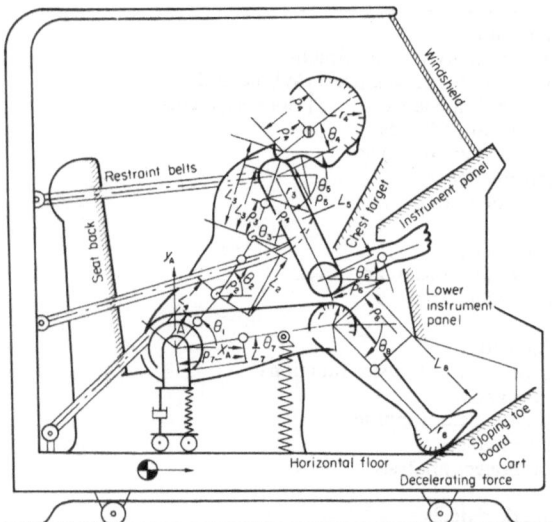

Figure 2. Kinetic model of the human body and restraint system on a test cart (11 degrees of freedom) (from McHenry, 1971).

crashes and to calculate time course, magnitude and location of impact forces to individual body parts such as head, chest or knee. In this context they are of great value in the evaluation of restraints and other protective systems, and in parametric studies of inflatable safety restraints (air bags), etc. In internal biomechanics, kinetic models of this type are used to study walking mechanisms or to calculate positions of body members, of force, lift and torque capabilities incorporating data on isometric muscle strength or other energy capability assumptions. A combination of external and internal kinetic models has been attempted only by including to a partial degree the environmental effects of g-fields into internal models of motion capabilities (Huston *et al.*, 1971). Future developments in this field are foreseen to include time-varying muscle forces and some sort of active closed loop motion control into models of the type shown in figure 2.

Oscillatory body response

Under vibratory force inputs the body exhibits resonances in various frequency ranges which lead to specific physiological manifestations, performance decrements, or injuries. To approximate these complex body deformations and wave propagations, ratios of vibration amplitudes at various body positions have been measured, as for example, the ratio of hip to table, shoulder to table, or head to table amplitude of a sitting subject vibrated on a shake table (figure 3). The second type of measurement relates to volume changes in abdominal circumference, chest circumference and air compressed and forced out of the lungs (figure 4). The third type of measurement frequently reported is the input impedance of the human body at the area of force transmission (the buttocks for the sitting

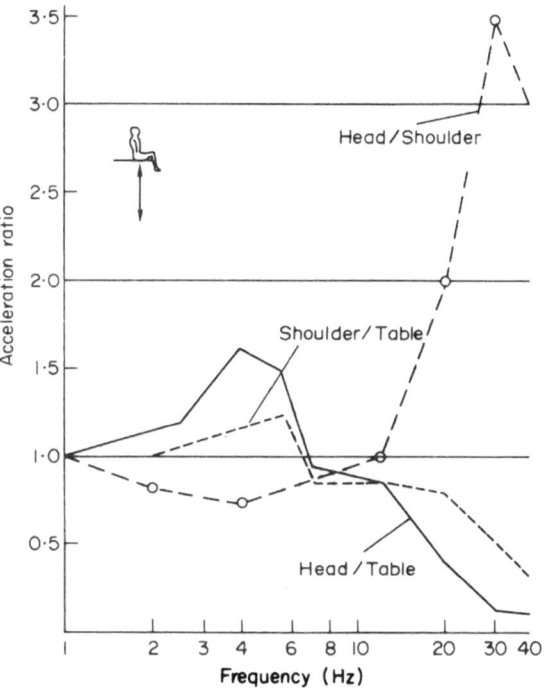

Figure 3. Vertical (Z-axis) vibration transmitted to the sitting human subject (from Dieckmann, 1957).

subject) (figure 5). The combined results of these measurements can be understood and described in terms of total body response models of the lumped parameter type shown in figure 6. Representative model parameters used with this model are given in table 4. These parameters must agree not only with the results of the three types of measurements mentioned above but also with the data resulting from static anthropometry and the data available from detailed subsystems analysis. This is particularly true, for example, in the case of the spinal system with the abundant theoretical and experimental work on the isolated spine, or in the case of the lung-thorax system with the available body of physiological information on respiratory

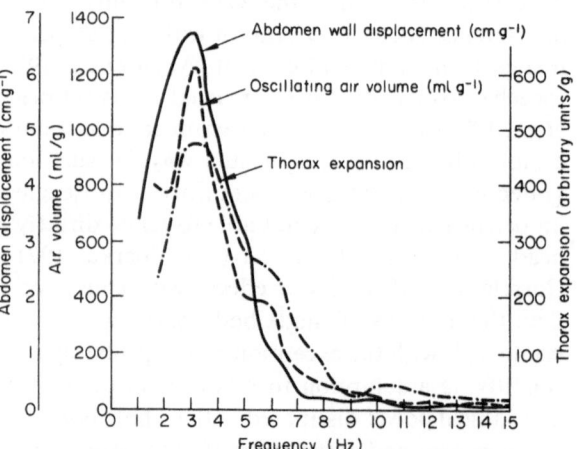

Figure 4. Abdominal wall displacement, thorax expansion and air volume oscillating through the mouth per g longitudinal vibratory acceleration (subject in supine position) (from Coermann *et al.*, 1960).

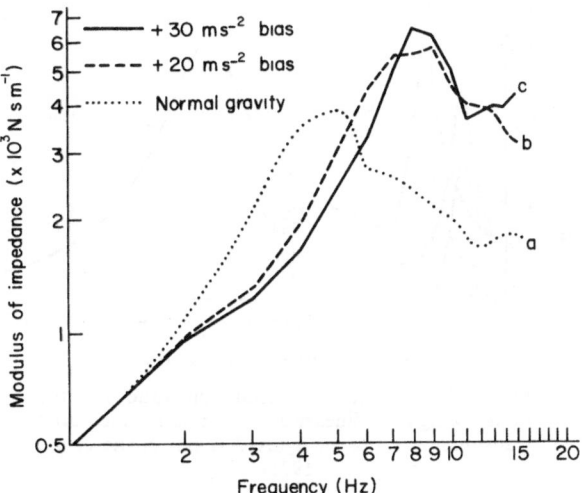

Figure 5. Mechanical driving point impedance of the sitting human subject subjected to Z-axis vibration at various static sustained acceleration levels on a centrifuge (From Vogt *et al.*, 1968); (a) under normal gravity, (b) under +2G$_z$, (c) under +3 G$_z$ acceleration.

V$_o$ (cm³)	4×10^3
A$_r$ (cm²)	2
M$_r$ (gm)	1×10^{-1}
D$_r$ (dyn s/cm)	$1·6 \times 10$
K$_t$ (dyn/cm)	0
A$_w$ (cm²)	2×10^2
M$_w$ (gm)	1×10^3
D$_w$ (dyn s/cm)	6×10^5
K$_w$ (dyn/cm)	1×10^8
A$_a$ (cm²)	2×10^2
M$_a$ (gm)	4×10^3
D$_a$ (dyn s/cm)	1×10^4
K$_a$ (dyn/cm)	8×10^6
M$_t$ (gm)	4×10^4
D$_t$ (dyn s/cm)	4×10^6
K$_t$ (dyn/cm)	1×10^9
M$_p$ (gm)	8×10^3
D$_b$ (dyn s/cm)	$6·5 \times 10^5$
K$_b$ (dyn/cm)	6×10^7

TABLE 4: Typical model parameters for system shown in figure 6.

mechanics. Therefore the fitting of the parameters to all available data is an extremely difficult job. It must be kept in mind that the model will never tell us more than the type of measurements on which it is based. It can explain these measurements, it can make many additional measurements of the same type unnecessary, and can be all around an extremely useful tool; however the assumptions on which it is based should never be left out of sight. For example, it is quite obvious from an analysis of the system in figure 6 that primarily the buttocks system is reflected in driving point impedance measurements at the buttocks. Although such measurements are

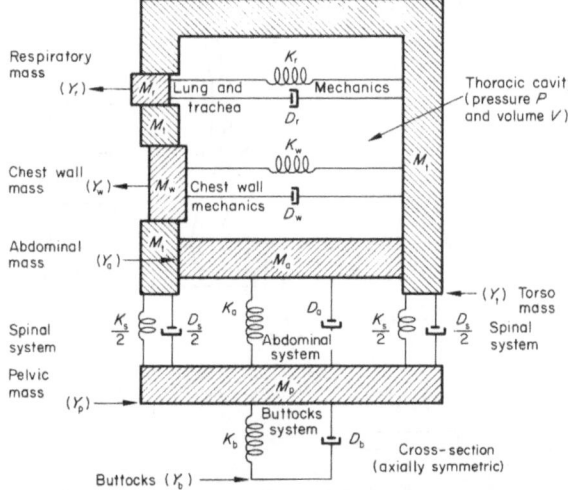

Figure 6. Example of a total body response model for the frequency range 1 to 100 Hz. The model is used to calculate body deformation (spinal compression, pressure in the lungs, etc.) as a function of external longitudinal dynamic forces (vibration or G$_z$ impact) and pressure loads (blast, acoustic pressure, decompression loads) (from Kaleps *et al.*, 1971).

very useful to obtain the total mechanical energy transmitted from the seat to the man, they cannot tell us in such a series system where the energy goes and what responses it elicits there (Payne *et al.*, 1971). If the impedance measured at the buttocks can be represented fairly well by a one or two degree of freedom system, it is erroneous to assume— as it has unfortunately been done—that there are not important separate additional degrees of freedom such as the spinal subsystem or the lung-thorax system, which are not reflected in the buttocks input impedance. Of the measurements possible on the outside (at the surface) of such a system, it appears as if one type of measurement has been neglected or hardly been tried to our knowledge: the input-output impedance or transfer impedance. Treating the system—and also its various subsystems—as four pole, additional information could be obtained by either loading (for the buttocks input case) the torso mass or the chest wall output with known impedances (mass loading), or by trying to reverse input and output; for example, by driving the torso mass at the shoulders and by measuring the input impedance there and the output impedance at the buttocks. It is somewhat unfortunate that the impedance measurements on the sitting human subjects done for the first time more then 30 years ago have been extensively repeated and misinterpreted, but that not too many new and imaginative methods have been added.

Human impedance measurements as well as force-deflection measurements on the buttocks and on isolated spinal segments show a marked nonlinearity. With the exception of the driving point impedance functions, for which measurements on a centrifuge at various static preloads gave valuable information (Vogt *et al.*, 1968) (figure 5), these nonlinearities are largely uninvestigated and deserve further study and definition. A detailed analysis of available data led

Payne *et al.* (1971) to propose the nonlinear spring constants shown in figure 7 for buttocks and spinal spring, which are in agreement with all available measurements on nonlinearities. The linearised four mass system of figure 7 results in the amplitude ratio curves in figure 8, which are similar to the measured ones in figure 3. The lung-thorax system is

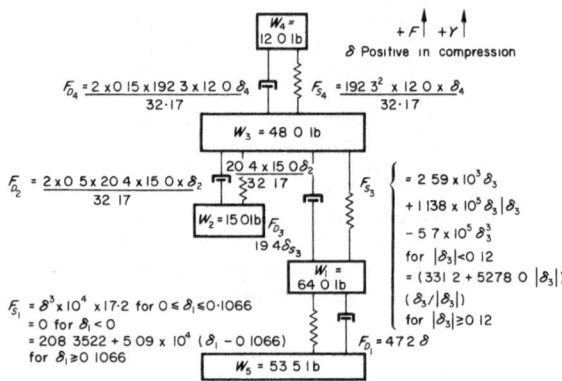

Figure 7. Four degree of freedom lumped parameter model with nonlinear spinal and buttocks characteristics. The model describes longitudinal impact and vibration response (from Payne *et al.*, 1971).

omitted from this model which is being used primarily to study spinal response. How much, however, the absolute magnitude of the transmission factor and of its frequency response depend on the body posture and how much such changes in the frequency response affect the model representation needed to approximate these frequency responses are illustrated in figure 9 for the example of the table to head transmission (Potemkin *et al.*, 1971). If pressure in the lungs and lung damage are of

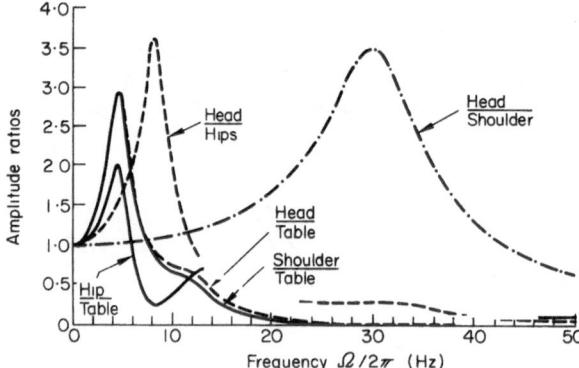

Figure 8. Amplitude transmission ratio for the model of figure 7 (linearised). Compare to measured transmission ratios of figure 3 (from Payne *et al.*, 1971).

primary interest, as in the case of blast or infrasound effects, the pressure-volume relationship for the air in the lungs cannot be linearized as is usually done for approximating the whole body oscillatory case (Kaleps *et al.*, 1971; Bowen *et al.*, 1968).

In addition to the body deformations under oscillating forces so far described, there have been effects observed on the cardiovascular system caused by sinusoidal accelerations (Edwards *et al.*, 1972). In anaesthetised dogs these changes in peak aortic flow were in the 3 to 9 Hz range, and during 3 g vibration at 4 Hz amounted to a maximum aortic peak flow rate of more than twice the control value and a minimum aortic peak flow rate of 10% of the control value. The amplification or reduction of peak flow rate depended on the phase relation between vibration and cardiac cycle. The pulse pressure under these conditions could reach

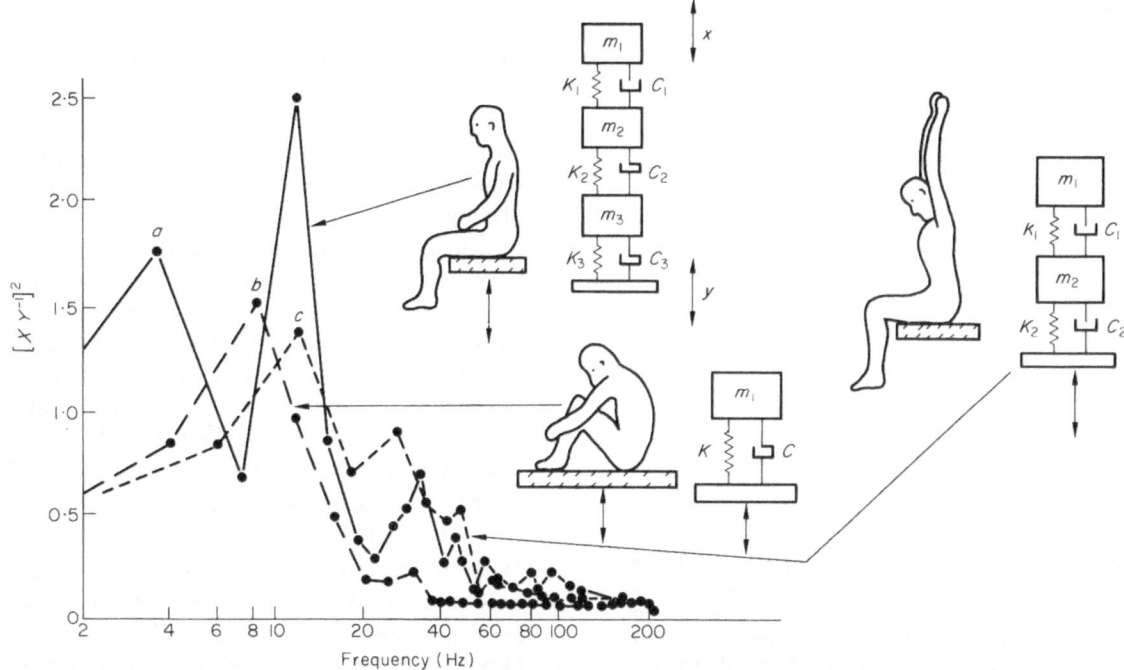

Figure 9. The square of the table to head transmission ratio as a function of frequency and body posture. The networks needed to approximate the functions are also indicated (from Potemkin *et al.*, 1971).

more than five times the control value. Without considering further the physiological implications of these phenomena, which have not yet been studied in humans, it is of biomechanical interest to note that these findings have been explained by modelling the hydraulic aspects of the circulatory system alone; i.e., these effects are apparently not caused by deformation of the vessels because of the mechanical resonance of the abdomen and the chest or by changes in physiological feedback mechanisms. Consequently, it will be necessary to incorporate the model of the cardiovascular system under sustained acceleration and vibratory stress, into the whole body response model discussed above. Such a refined model would have very promising applications to environmental physiology (circulation under hypo- and hypergravic conditions as well as under vibration and impact stress) as well as to problems in clinical medicine (circulatory assist by alternating, heart synchronised forces or pressures).

Impact response

As long as our considerations are restricted to the direct physical effects of the mechanical force environment, there is a clear, well known mathematical relation of the vibration response of the human body to the impact response. This relationship is schematically indicated in figure 10. The left-hand side shows the overall human vibration 'tolerance' curve as the composite of curves of equal tissue strain for various subsystems; i.e. tissue areas. In principle each curve of equal tissue strain under vibration stress on the left-hand side, corresponds to a curve of equal tissue strain under impact stress

on the right-hand side of the figure. The transmission ratios observed in human impact acceleration experiments, as well as the data obtained from accident analyses, support the general trend of these theoretical impact tolerance curves. However, quantitatively the relationship is considerably more difficult due to the fact that the main interest in curves of equal tissue strain for the impact case is at stresses beyond the linear range of the system, namely, in the tissue loading range close to or at the point of irreversible damage (Payne *et al.*, 1971). For this reason the nonlinear lumped parameter model of figure 7 reflects much better experimental data in the impact range of minor probability of spinal or abdominal injury than the linear model of figure 6. Nevertheless, the linear model of figure 6 correctly explains observations, in accident investigations and animal experiments, that for short duration acceleration pulses spinal compression fracture is the most sensitive injury mechanism whereas for longer duration pulses of the same magnitude abdominal injury is more likely to occur (figure 11).

The preceding discussion focused on environmental loads in the direction of the Z-axis of a sitting subject only. The response dynamics are naturally different for the other input directions and different model systems must be designed to fit the impedance, transmission ratios, accident data and subsystem studies. Although a considerable body of information for the other axes is available, the information and its condensation into quantitised models are not yet so complete as for the Z-axis. For X-axis excitation (front to back) impedance

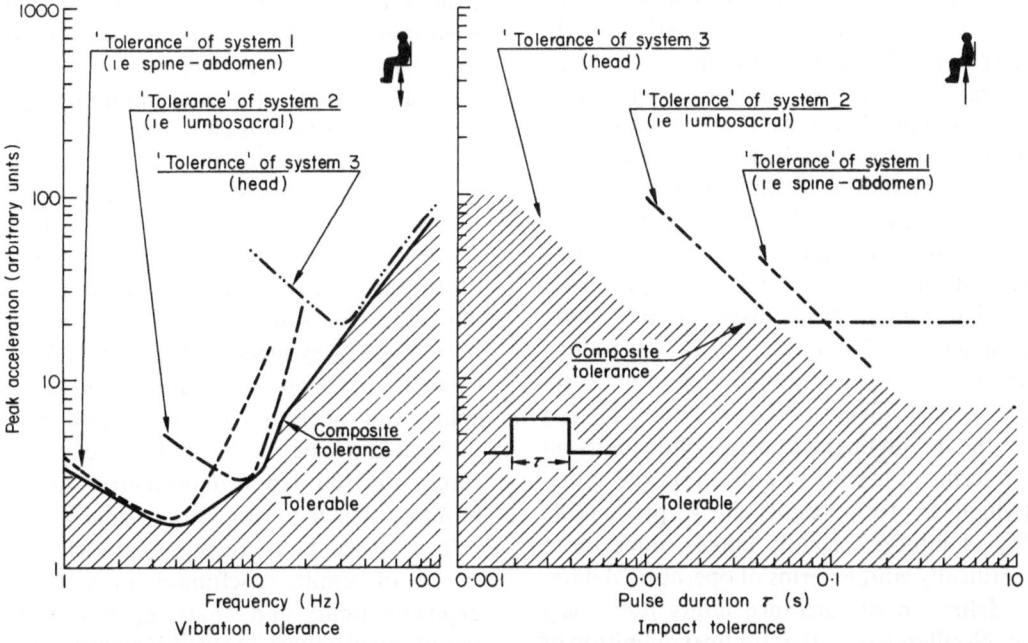

Figure 10. Relationship between vibration tolerance and impact tolerance of a sitting human subject (schematic). The dotted lines indicate lines of equal tissue strain in localised tissue areas; i.e., tolerance lines for the individual subsystems. The composite tolerance curve is the envelope to the individual subsystem curves and represents the overall tolerance curve for the total organism (from von Gierke, 1964).

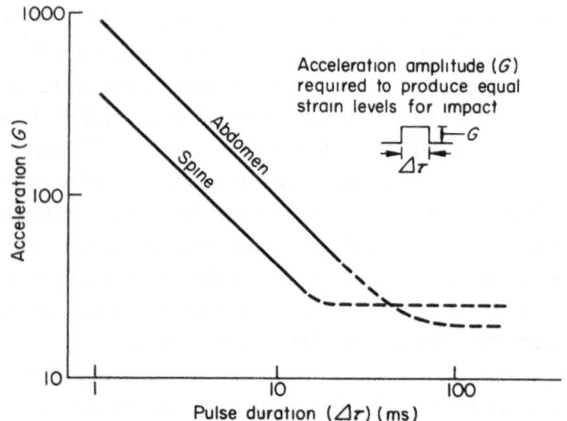

Figure 11. Curves of equal mechanical strain (equal probability of injury) for abdominal displacement and spinal compression in the model shown in figure 6. Note that the cross-over of these curves depends upon the shape of the force input (from Kaleps et al., 1971).

measurements even with static preload are available; however, the response dynamics leading to well defined accident mechanisms are inseparable from the body support (back and headrest) and restraint to permit deducing a generalised model.

APPLICATION OF DATA

The main advantage of presenting human body response data in terms of the biodynamic models discussed in not only the general understanding of the dynamic events elucidated by the model but primarily the application of the model to the solution of practical human tolerance and bioengineering problems. The interaction of man with his environment, for example, with the dynamics of his seat, seat cushions and restraints, is most satisfactorily analysed by means of such body response models. For the case of spinal compression injury as a result of longitudinal (Z-axis) loads as they occur during catapult emergency ejections from military aircraft, the statistical probability of injury is well enough known to correlate it with the strain in the spinal spring of the models in figure 6 or 7. If one limits the considerations to spinal injury only as the most sensitive injury mechanism in the short duration impact range (see figure 11), one can reduce the model to the oversimplified but practically very useful spinal injury model of figure 12. To specify for the design engineer permissible loads on the man in terms of permissible strain on the spinal spring (which in turn results in a predictable probability of injury) certainly makes much more sense scientifically, and in terms of operational data, than any definition of tolerance limits previously attempted. Needless to say, that for finer definition of the type and location of spinal injury a more refined model of the spinal subsystem than the one introduced here is required and desirable. But in general

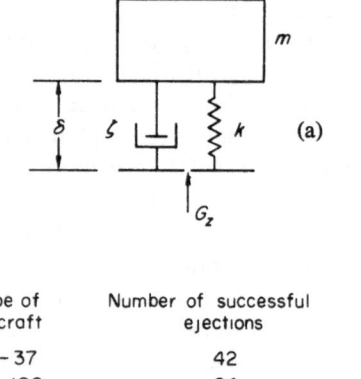

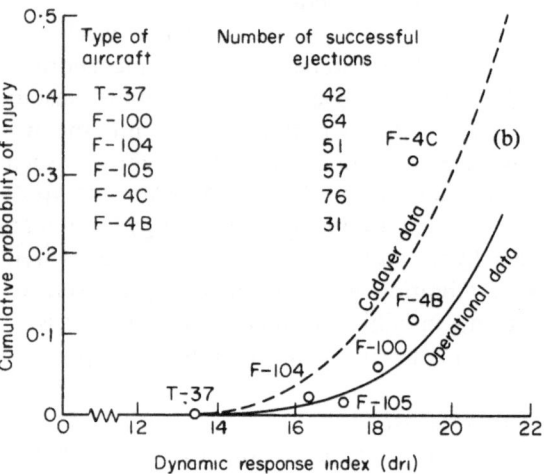

Figure 12. (a) Spinal injury model (from Brinkley, 1968). m is mass (lb s^2/in); δ is deflection (in); ζ is damping ratio; k is stiffness (lb/in); G_z is acceleration input (in/s^2); DRI $= \omega_n 2\delta_{max}/g$, where DRI stands for dynamic response index; $\omega_n = (k/m)^{1/2}$ (rad/s); and g = 386 in/s^2. (b) Probability of spinal injury predicted from cadaver data compared to operational experiences with various U.S. Air Force ejection systems (from Brinkley, 1968).

this way of presenting and using experimental data in an overall probability of injury model—which, if desired, can be backed up by a more refined subsystem injury model—should be looked at as a desirable goal for analysing and treating other injury mechanisms (Payne, 1971).

As an example of how the more complete model of figure 7 has been used to analyse the dynamics of the whole man/ejection seat system, the 'plane of symmetry' model (Band, 1971) derived from the model in figure 7 is presented in figure 13. It allows for calculating fore and aft movement of the man in the seat, includes restraining forces and allows for translational and rotational dynamics in the plane of symmetry (Z–X plane). An example of calculated forces transmitted at the buttocks, spine and neck during a catapult maneuver is illustrated in figure 14 and agrees well with the forces derived from photometric observation of human subjects during ejection loads.

The whole body response models are foreseen to be of similar usefulness in setting vibration exposure limits and analysing vibration isolation requirements once more information on the long term effects and possible chronic injury mechanisms is available and statistically documented. For these purposes the standardisation of 'nominal' human

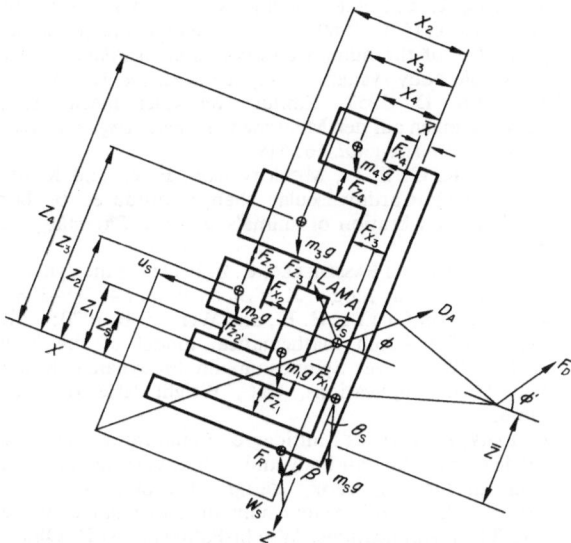

Figure 13. Eleven degree of freedom 'Plane of Symmetry' model of a rocket-powered, free-flight man/seat system. For parameters indicated see reference below. F_R = rocket thrust, F_D = drag of stabilising drogue, D_A = aerodynamic drag of man/seat (from Band, 1971).

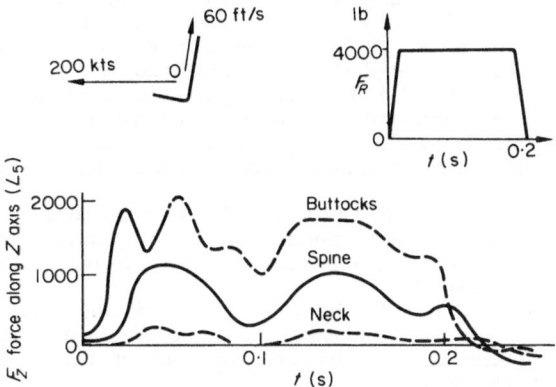

Figure 14. Forces as a function of time during catapult ejection of the model of figure 13. Initial conditions for seat velocities and rocket thrust F_R as indicated. Inclination of rocket thrust to X axis 60° (for details see Band, 1971).

impedance values and transmission ratios is under consideration by various groups. These values are then foreseen to be used in the design of a new generation of anthropomorphic (or better, anthropodynamic) dummies, which incorporate the dynamic response characteristics of man for hardware, vibration and impact tests.

For the planning, interpretation and application of biodynamic data obtained on animal models, it is important to realise that to a first approximation the resonance frequencies increase as the linear body dimensions decrease (von Gierke, 1964, 1972). This fact not only shifts the frequency scale (figure 15) for comparable vibratory responses and the duration scale for comparable impact responses, but results, for the assumption of basically equal tissue strength, in the general conclusion that the

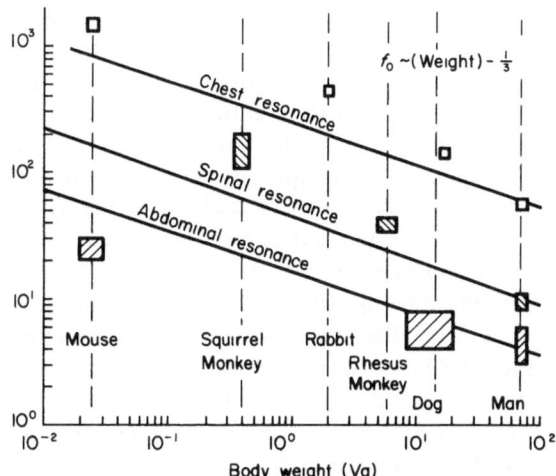

Figure 15. Approximate resonance frequencies of total body response models of the type shown in figure 6 as a function of body size (weight) (from von Gierke, 1971c). The shaded areas indicate ranges for measured data.

smaller the animal the higher the vibration and impact loads leading to similar manifestations. Although experimental data clearly support these theoretical predictions, more detailed work on this subject is desired.

OUTLOOK

This short review could do little more than call attention to the broad scope of its title and sketch some of the typical and important developments in this field. It is hoped that it called attention to the following facts and predictions:

(a) future progress toward a unified, quantitative picture of man's mechanical response to dynamic force environments must come from a refined mathematical description of the biomechanical and biological processes. Experimental data collection should be based on the theoretical model framework available so that no important parameters required for their quantitative integration are overlooked or neglected. Representative statistical samples should be aimed at in describing physical response phenomena as well as physiological or injury mechanisms.

(b) It would be desirable to derive, on the basis of experimental data, a generally valid breakdown of the total body system into subsystems and sub-subsystems. The range of useful and valid application of each subsystem and the coupling between subsystems should be identified together with the biological phenomena hopefully to be described by the particular sub-model approach. For example if a kinetic model analysis of the type shown in figure 2 results in specified force impacts to the skull and the chest, application of a chest sub-model and head-neck sub-model would identify the dynamic responses and biological effects of these specific impacts. After analysis of the head-neck dynamics, it might be necessary to go on to a further, still more refined head injury model, which allows differentiation of the probability of occurrence of the various types of head injury mechanisms. It appears as if a much more systematic approach to the analysis of the system structure must be taken than has been the case in the past. Conceptionally this breakdown should be continuous starting with the models describing the dynamics of the macroscopic body structure, characterised by anthropometric data, down to subsystem models describing basic elastic and strength properties of structural components and finally of the material itself.

(c) 'Active' responses of the body to the mechanical force environments should be incorporated into the passive response models discussed above.

In summary, the preceding discussion of the body's dynamic response to vibration and impact must remain an unsatisfactory accumulation of data until a more coherent effort in data collection, as well as theoretical foundation, allows more direct and uninterrupted correlation between the two. It appears as if the solution does not necessarily rest in an expansion of this research field but in a maturation process and a logical, truly interdisciplinary, step-by-step research programme.

This paper has been reproduced by permission of the U.S. Government. It has been identified by Aerospace Medical Research Laboratory as AMRL-TR-72-53.

REFERENCES

ALLEN, R., WADE, H., JEX, R., and MAGDALENO, R. E. 1972: Vibration effects on manual control performance. Presented at the Eighth Annual NASA-University Conference on Manual Control, University of Michigan, May.

BAND, E. G. U. 1971: Calculation of rocket powered trajectories of a 'plane of symmetry' model of a human subject and ejection seat. *AMRL-TR-7*. AMRL, Wright-Patterson AFB, Ohio.

BOWEN, I. G., FLETCHER, E. R., RICHMOND, D. R., HIRSCH, F. G., and WHITE, C. S. 1968: Biophysical mechanisms and scaling procedures in assessing responses of the thorax energised by air-blast overpressures or by non-penetrating missiles. *Annals of the New York Academy of Sciences* **125**: 122–146.

BRINKLEY, J. W. 1968: Development of aerospace escape systems. *Air University Review*, July/Aug, 34–49.

COERMANN, R. R., ZIEGENRUECKER, G., WITTWER, A. L., and VON GIERKE, H. E. 1960: The passive dynamic mechanical properties of the human thorax-abdomen system and of the whole body system. *Aerospace Medicine* **31**: 443.

DIECKMANN, D. 1957: Einfluss vertikaler mechanischer Schwingungen auf den Menschen. *Intern. Z. angew. Physiol. einschl. Arbeitsphysiol.* **16**: 519.

EDWARDS, R. G. E. P., McCUTCHEON, E. P., and KNAPP, C. F. 1972: Cardiovascular changes produced by brief whole-body vibration of animals. *J. Appl. Physiology* **32**: 386–390.

HUSTON, R. L., and PASSERELLO, C. E. 1971: On the dynamics of a human body model. *J. of Biomechanics* **4**: 369–378.

KALEPS, I., VON GIERKE, H. E., and WEIS, E. B. 1970: A five degree of freedom mathematical model of the body. *AMRL-TR-71-29-8*, Symposium on Biodynamic Models and Their Applications, Oct., Wright-Patterson AFB, Ohio.

McHENRY, R. R. 1970: Multidegree, nonlinear mathematical models of the human body and restraint systems: applications in the engineering design of protective systems. *AMRL-TR-71-29-7*, Symposium on Biodynamic Models and Their Applications, Wright-Patterson AFB, Ohio.

NIXON, C. W., and SOMMER, H. C. 1963: Influence of selected vibrations upon speech-range of 2 cps–20 cps and random. *AMRL-TDR-63-49*, (AD 416 816), Aerospace Medical Research Laboratory, Wright-Patterson AFB, Ohio, June.

O'BRIANT, C. R., and OHLBAUM, M. K. 1970: Visual acuity decrement in whole body ±G_z vibration. *Aerospace Medicine* **41**: 79–82.

PAYNE, P. R., and BAND, I. G. U. 1971: A four-degree-of-freedom lumped parameter model of the seated human body. *AMRL-TR-70-35*, Aerospace Medical Research Laboratory, Wright-Patterson AFB, Ohio, Jan.

PAYNE, P. R. 1970: The human spine—a critical review of existing dynamic data in relation to aircraft escape systems. *AMRL-TR-71-29-9*, Symposium on Biodynamic Models and Their Applications, Oct., Wright-Patterson AFB, Ohio.

POTEMKIN, B. A., and FROLOV, K. V. 1971: Simulated representations of the biomechanical human operator system with random vibration. *DoKlady Akademii Nauk SSSR* **197**: 1284–1287.

SANDOVER, J. 1971: Study of human analogues: part 1, a survey of the literature. Dept. of Ergonomics and Cybernetics, Loughborough University of Technology, England, April.

VON GIERKE, H. E. 1964: Biodynamic response of the human body. *Applied Mechanics Review* **17**: 951–958.

VON GIERKE, H. E. 1971a: In Symposium on biodynamic models and their applications. *AMRL-TR-71-29*, Wight-Patterson AFB, Ohio.

VON GIERKE, H. E. 1971b: Physiological and performance effects on the aircrew during low-altitude, high-speed flight missions. *AMRL-TR-70-67*, Wright-Patterson AFB, Ohio.

VON GIERKE, H. E. 1971c: Biodynamic models and their applications. *J. Acoustical Society of America* **50**: 1397–1413.

VOGT, H. L., COERMANN, R. R., and FUST, H. D. 1968: Impedance of sitting human under sustained acceleration. *Aerospace Medicine* **39**: 675–679.

ESTIMATION OF THE INERTIAL PROPERTY DISTRIBUTION OF THE HUMAN TORSO FROM SEGMENTED CADAVERIC DATA

Y. KING LIU AND JACK K. WICKSTROM[1]

This paper is concerned with the statistical estimation of the inertial property distribution of the human torso, based on an analysis of the data obtained from segmented cadaveric trunks. This estimation is based on a sample of eight cadavers: seven embalmed and one unembalmed. The implications of these biomechanical data on mathematical simulation and manikin (dummy) models of the human body are briefly discussed, especially with respect to dynamic situations.

The study of the inertial properties of the human body is generally acknowledged to have begun with Borelli (1679). He found the centre of gravity of the human body by the following experiment. Nude subjects were stretched out on a rigid platform, which was then moved about a knife edge until it balanced. Thus, he obtained an approximation to the plane containing the centre of gravity of the subject. In the intervening almost 300 years, numerous investigators have added to these studies. These investigations have been reviewed periodically, e.g. the recent survey and expository papers of Contini and Drillis (1954), Drillis and Contini (1966) and Clauser et al. (1969). Since the coverage of these surveys is quite complete, the topics considered therein will not be repeated or further discussed in the present paper except when warranted by the continuity of the subject-matter.

Roughly speaking, the inertial properties of the human body, i.e. the mass, centres of mass, and mass moment of inertia matrix, are found from either cadaveric investigations or in vivo studies using human volunteers. The inertial properties of the major cadaveric body segment, e.g. the head and neck, the torso, and the upper and lower extremities, have been found after appropriate dissection. The intensive study by Dempster (1955) exemplifies this approach. This method is predicted on the assumption that the inertial relationships of a cadaver population are approximately valid for the living. The loss of muscle tone, changes in tissues and body fluids, and the cause of death are some of the possible sources of error in this approach. The preferences for live subjects as opposed to cadavers is obvious; however, it is not without its own inherent limitations. Since dissection in vivo is impossible, the inertial properties are usually determined by using the method of differential change. This method is based on the change in the inertial properties due to a change in body position. For example, to find the weight of the arm in vivo: (1) stretch out the subject on a platform which is supported as a simple beam with known reactions at these supports; (2) move the centre of gravity of the arm from the supine to the vertical position; (3) note the change of the reactions at the points of support and (4) from statics, compute the weight of the arm. Some difficulties have been encountered: (1) the problem is statically indeterminate unless the centre of mass is predetermined accurately; (2) obtaining the difference between scale readings involves subtraction between two almost equal numbers, reducing its significance; (3) the shift of the muscle mass in moving the arm from one position to another also induces error. The determination of the mass moment of inertia matrix of the major body segments in vivo is fraught with even more serious problems than the measurement of its weight.

The reasons for wanting the inertial properties of either the whole body or its major segments vary with the biomechanics problem. For example, the prosthesis and orthotics problem requires primarily the mass, centre of gravity, and mass moment of inertia of the upper and lower extremities, and secondarily, the inertial properties of the torso, head, and neck. In the pilot-ejection problem, however, the trajectory and orientation calculations require the mass, centre of mass, and mass moment of inertia of both the pilot and his ejection seat considered as an ensemble. In an investigation of why certain types of vertebral injuries are sustained by the pilots during ejection, one may need the inertial properties of each segment of the torso associated with each vertebra, i.e. some measure of the distribution of these inertial properties along the torso.

To date, the only work which has been done on the distribution of inertial properties of a given major body segment is the feasibility study of Liu et al. (1971) of the torso. In fact, this paper is an improved and extended version of the previous

[1] Biomechanics Laboratory, Tulane University School of Medicine, New Orleans, Louisiana, U.S.A.

work. A detailed description of the evolution of our
technique and the statistical analysis of the data
obtained are given in the rest of the paper.

IMPROVED METHODS OF PROCEDURE

Seven embalmed cadaver trunks were sectioned
into horizontal segments each containing one vertebra. For each of the segments, the following quantities were determined: (1) the weight; (2) the centre
of gravity; and (3) the mass moment of inertia about

Cadaver number	Cause of death	Age (yr)	Weight (lb)	Height (in)
3029	Suffocation	46	103·0	65·9
2086	Pulmonary abscess	41	147·0	71·2
3107	Pneumonia	56	199·0	69·7
3061	Emphysema	59	189·9	71·1
3026	Myocardial infarction	74	145·3	73·1
3343	Stroke	68	170·7	67·4
3328	Cardiac arrest	61	155·4	61·8
unembalmed	Heart failure	83	155·0	67·0
	Mean	61·0	158·16	68·40
	Standard deviation	13·2	27·82	3·38

TABLE 1: Demographical and anthropometric
data of the cadaver sample

each of the three principal axes. A right-handed
cartesian reference frame is attached to the centre
of gravity of the slice as in the feasibility study,
i.e. the xy and xz planes are the transverse and
midsagittal planes respectively. The demographical
and pertinent anthropometric data of the cadavers
used are shown in table 1.

Preparation of the cadaver

In contrast to the unembalmed cadaver used in
the initial feasibility study of Liu *et al.* (1971), the
seven embalmed cadavers in the present series
arrived with the head and neck severed from the
torso. The plane of the cut extended from the

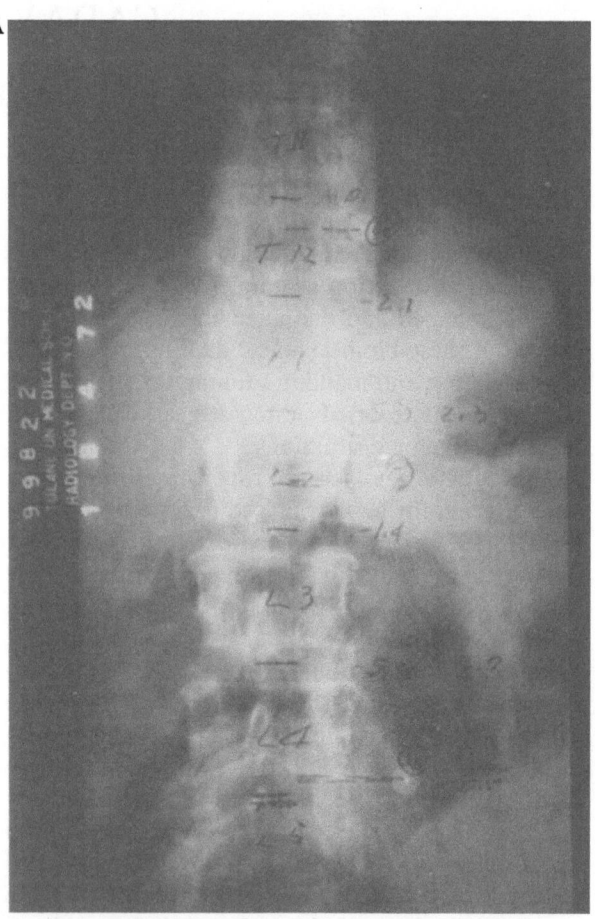

Figure 1. (A) Typical anterior-posterior view of cadaveric spine. *Note:* Bell-Thompson Rule used to help locate
the discs. (B) Typical lateral view of cadaveric spine. *Note:* Colcher-Sussman Rule on sternum to determine
magnification factor.

superior surface of the medial end of the clavicle medially and inferiorly, through the intervertebral disc between the seventh cervical (C7) and the first thoracic vertebra (T1). Upon receipt, the cadavers were quick-frozen in the supine position with dry ice ($-109°$F) in a specially constructed and insulated box. Depending on the somatotype, the time required for the cadaver to be frozen solid varied from 12 to 24 hours. The upper limbs were removed by sawing through the head of the humerus in a sagittal plane, while the lower limbs were removed by an oblique cut 1 inch below and parallel to the inguinal ligament. These cuts were made with a bow-saw and were similar to those made by Dempster (1955), in order to make the data obtained complementary to his major body segment results.

The intervertebral discs were located by multiplane X-ray. Both the anterior-posterior (AP) and lateral views were obtained. A typical set is shown

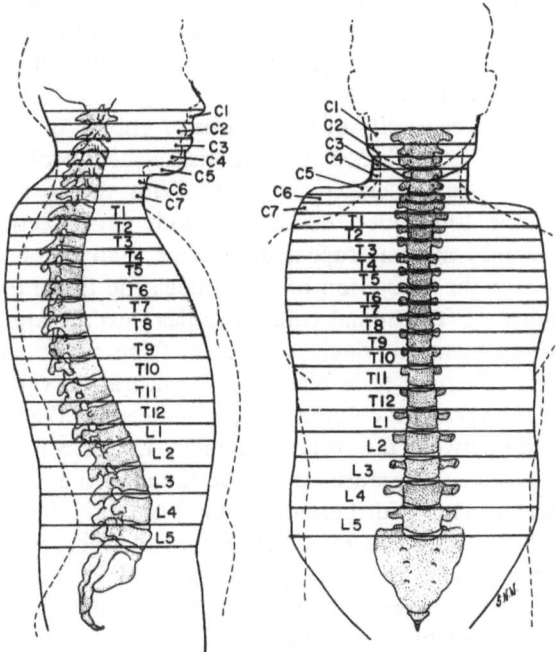

Figure 2. A typical anterior-posterior (PA) and lateral view of the approximate position of the cuts on the cadaver showing approximate tissue shift and location of cuts. The broken outline is the assumed living configuration of the body.

in figure 1. Note that in the AP view (figure 1, A) an opaque Bell-Thompson Rule (100 cm) is shown to help locate the discs. Similarly, the lateral view (figure 1B) is gauged with a Colcher-Sussman Rule on the sternum so that any magnification on the radiograph can be related to actual distances on the cadaver. The approximate position of each disc was marked on the back of the cadaver torso. The transverse cuts were made with a one-horsepower meat-cutting bandsaw (Butcher Boy Model B-12). The first cut transected the disc between L5 and S1.

With the removal of the pelvic mass, the rest of the torso became much more manoeuvrable. With few exceptions, the movable-carriage table plus the vernier-thickness control on the saw allowed us to transect a disc every time. The approximate position of the cuts is illustrated in figure 2. As shown in the figure, the cuts were precise, i.e. each segment had parallel superior and inferior surfaces through the centre of the appropriate disc and perpendicular to the z axis. What appeared to be a very difficult problem in the feasibility study has been satisfactorily resolved.

Experimental procedure

The procedure used for each segment was as follows:

(1) The segment was weighed on a scale having a full scale of three pounds, and divisions of one-hundredth of a pound. Additional dead weights can be attached to change the full-scale capability at will. The mass (m) was found by dividing the weight (w) by the acceleration due to gravity (g).

(2) The thickness of the segment was measured at the anterior, posterior, right, and left edges. These dimensions were averaged to give the average thickness ($\bar{t}$). These measurements were made with a vernier caliper which read to one-thousandth of an inch. The dimensions of the segment along the x and y axes (d_x and d_y) were measured with a ruler which had divisions of $\frac{1}{64}$ inch.

(3) The centre of gravity of each segment was determined in the following manner: the segment was moved around on a knife edge until it balanced; then, with the application of pressure, a line was marked with the knife edge. The process was repeated twice and the intersection of the three lines was taken as the centre of mass. This point was located on the superior surface of each segment. Because of the relative thinness of the segments, it was assumed that the centre of mass was situated midway between the superior and inferior surfaces.

(4) The position of the centre of gravity was then measured with respect to the geometric centre of the vertebral body. The position of the centre of gravity, e_x and e_y, was referred to a set of axes, x' and y', originating from the centre of the vertebral body and parallel to the x and y axes. Positive values for e_x and e_y denote that the centre of gravity is anterior to, and to the left of, the centre of the vertebral body.

(5) With the use of the torsion pendulum described in Liu *et al.* (1971), the moments of inertia were determined about the z axis. In that initial study, the mass moment of inertia values about the x and y axes were also determined with the same pendulum. Bookend-like holders were used to secure the segments while they were torsionally vibrated about the edge-on axes. If the slices are

not positioned so that the axis of rotation bisect the slice, the variability in the period determination can be quite high. For this reason, a special pendulum was constructed to reduce the data variability. Figure 3 shows the mechanical design of this torsional pendulum. The slice is held firmly in place by two parallel circular plates. The centrelines of the two adjustable screws are aligned with the centre of gravity of the piece used in step (3), and the torsion wire essentially bisects the thickness of he slice.

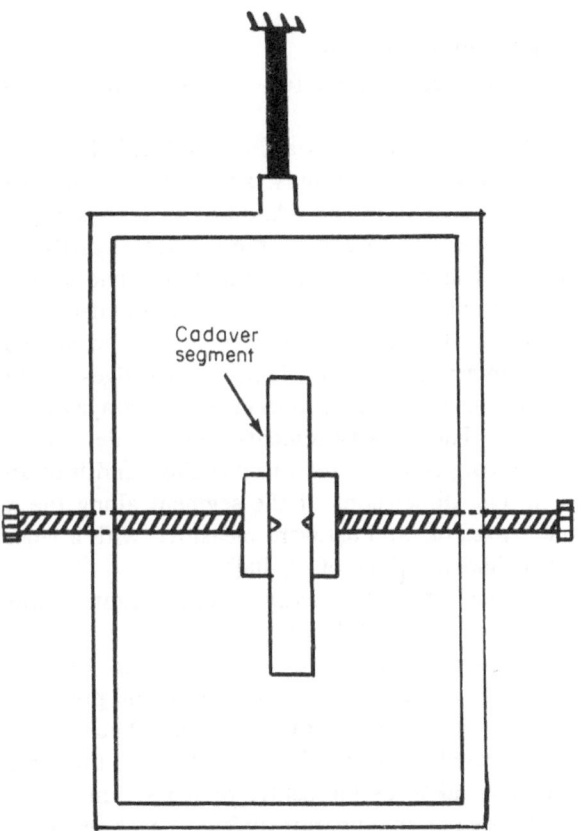

Figure 3. Special torsional pendulum used for the determination of the mass moments of inertia about the edge-on (x and y) axes.

The use of the torsion pendulum to determine the mass moment of inertia is well known from elementary vibration theory. If τ_0 is the period of the torsional oscillation when the platform is empty, and τ_A is the corresponding period when an object of known moment of inertia I_A is placed on the platform, it is easily shown that the unknown moment of inertia I_B of an object B may be found from the relationship

$$I_B = I_A(\tau_B^2 - \tau_0^2)/(\tau_A^2 - \tau_0^2) \qquad (1)$$

where τ_B is the period of oscillation of the platform when an object B is placed on it. The period used for the calculations is from an average of 10 to 20 periods. Time was measured with a stop-watch accurate to one-tenth of a second. An error analysis

involving the use of equation (1) is given in Liu *et al.* (1971).

The mass moments of inertia found with respect to the x, y, and z axes are assumed to be the principal moments of inertia. This, of course, follows from the presumption that the mid-sagittal (xz) plane and transverse (xy) plane of every segment are very nearly planes of symmetry.

DATA ANALYSIS AND RESULTS

The weight of each slice is assumed to be accurate up to one-hundredth of a pound. The total weight of the torso is slightly more than the sum of the individual slices because of the inevitable destruction of tissue as a result of the finite thickness of the saw blade. In the feasibility study a bow saw was used, a body height loss in 'sawdust' of 2 inches was recorded, while in the present series the use of the power bandsaw reduced the number to 1 inch—a 100 per cent improvement. Furthermore, the bandsaw speeded up the cutting process and at the same time very much improved the uniformity of the slice thickness.

The eccentricity of the centres of gravity of each slice is complicated by the tissue-shift problem present in all cadaveric work. With the loss of muscle tone upon death, the body cross-section becomes almost trapezoidal rather than elliptical due to the relaxation of the viscoelastic tissue (see figure 4). The spinal column settles into an equilibrium position which is different from the *in vivo* case. There is a decrease in the natural curvature of the vertebral column, and an increase in the distance between the vertebral bodies and the contacting surface because of the settling of the tissue posteriorly. Furthermore, the normal furrow between the sacrospinalii in the dorsal aspect of the torso *in vivo* is generally flattened out in the embalmed cadaver. These two factors tend to offset each other but it is not clear by what amount. The raw data, without correction, can therefore be quite meaningless. Figure 4 shows the various important measurements. These distances are shown in Greek alphabet for the living and in Roman for the cadaver.

To obtain reasonable values of eccentricity, the raw data were corrected, *in the feasibility study*, on the basis of the following assumptions:

1 As the tissue shifts, the distance from the anterior to posterior edge (see figure 4) remained constant, i.e. $d_x = \delta_x$.

2 In the *in vivo* man, the spinous processes were very close to the skin; i.e. $\gamma = 0$.

3 The distance b from the tip of the spinous process to the centre of the vertebral body on the cadaver was assumed to be the same as that on a laboratory model of a skeleton.

The tissue shift (illustrated in figures 2 and 4) was then assumed to be given by the relation

$$c = d_x - a - b \qquad (2)$$

where a was the distance from the anterior edge of the segment to the centre of the vertebral body. The corrected mass eccentricity, ε_x, was then taken to be

$$\varepsilon_x = e_x + c \qquad (3)$$

It should be noted that assumptions (1) and (2) were slightly incorrect. The distance d_x should decrease somewhat as the body cross-section

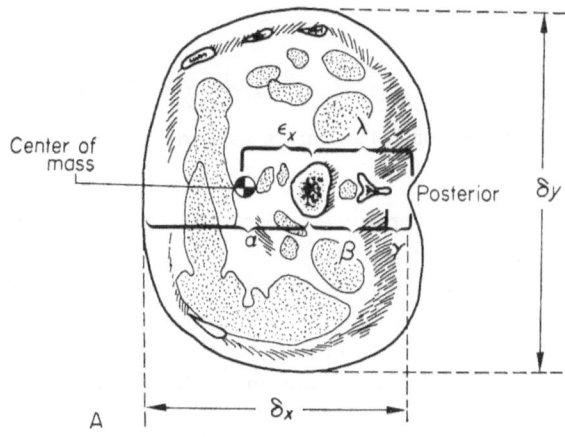

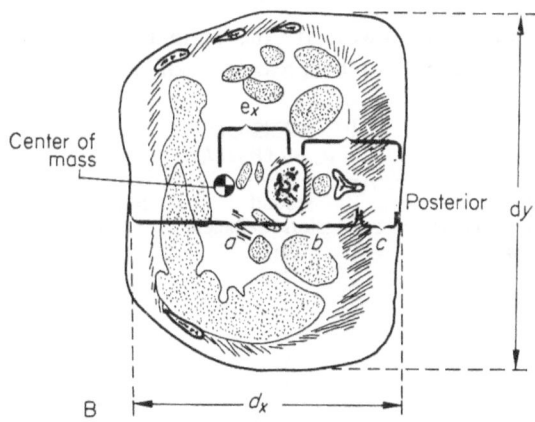

Figure 4. (A) Notation used for describing the important distances in a segment of living body. (B) The corresponding distances in the cadaver.

becomes less elliptical with the shifting of the tissue. On the other hand, there is a certain increase in the amount of tissue between the spinous processes and skin from the viscoelastic relaxation in the absence of muscle tone. Assumption (3) was perhaps the most dubious aspect of the previous study by Liu *et al.* (1971). However, because it was a feasibility study, this sort of rough estimation procedure is allowable.

For the present investigation, it was decided to obtain these critical dimensions from the living

human body. Ten volunteers, whose heights and weights span the 2nd to 99th percentile of height and weight in man, were chosen. The appropriate anthropometric and demographic data are given in table 2. A lead strip was taped to the skin at the lowest point of the furrow along the palpable line of the spinous processes of the vertebral column. The subject lay supine with his arms raised vertically. A Colcher-Sussman Rule was then placed on the sternum along the mid-sagittal plane to obtain the

Volunteer	Age (yr)	Weight (lb)	Height (in)
W. G.	29	205·0	75·3
Y. L.	37	132·0	64·0
R. K.	20	141·0	69·0
S. R.	20	175·0	74·0
U. P.	27	217·0	74·5
J. P.	20	140·0	68·0
K. A.	34	205·0	70·5
S. L.	23	188·0	74·5
W. M.	26	155·0	69·0
W. B.	30	220·0	70·0
Mean	26·6	177·8	70·9
Standard deviation	5·7	32·1	3·5

TABLE 2: Demographical and anthropometric data of the volunteer sample

magnification factor. The volunteer was asked to relax as much as possible in this position. A lateral X-ray was taken with the film cassette located 6 feet away from the focal spot. The 14 inch by 36 inch cassette contained a film (GAF HR 2000) used for the clinical examination of scoliotic patients. A typical radiograph is shown as figure 5. The distance from the skin line to the centroid of each discernible disc, λ, measured on the superior surface of the segment, was recorded.

To obtain the volunteer sample statistics, a preliminary analysis of the basic constituents of the problem was carried out. The individual sliced segments were assumed to be thin discs of elliptic cross-section, with the major and minor axes denoted by d_y and d_x respectively. Their weights were calculated by the simple relationship:

$$w_i = \rho_i A_i h_i = \rho_i (\pi d_x d_y) h_i \qquad (4)$$

where w_i, ρ_i, A_i and h_i are the weight, density, cross-sectional area, and the thickness, respectively, of the ith segment. Assuming that the segment weight and thickness are linearly related to the body weight W and height H, the following basic equation was obtained

$$w = \mu W = AH = (\pi d_x d_y) H \qquad (5)$$

where μ is an appropriate dimensional constant and

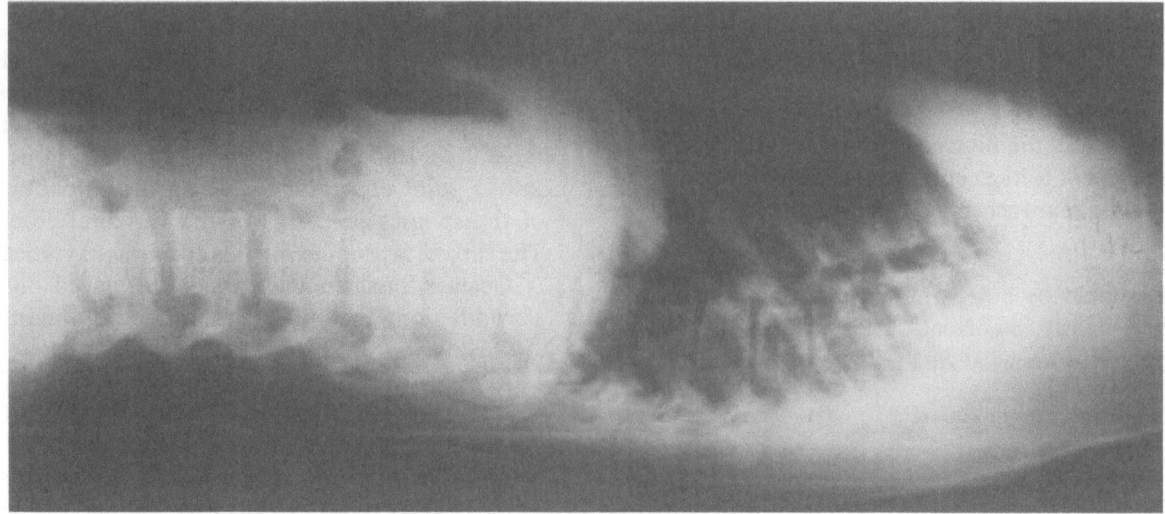

Figure 5. Typical lateral radiograph of a volunteer used in the determination of the distance λ shown in figure 4.

A is the cross-sectional area of the torso. The distance λ is some fraction of d_x and is, therefore, related to the weight and height according to (5), i.e.

$$\lambda \propto d_x \propto (W/Hd_y) \qquad (6)$$

The distance d_y in (6) was measured at three anatomical locations: the chest, waist, and hip breadth (as defined by Hertzberg *et al.*, 1954). The chest breadth, measured at the level of the nipples, was assumed to be the average distance of d_y from T1 to L2. Similarly, the waist breadth was assumed to be between L2 and L4, and the hip breadth between L4 and S1. Additionally, the chest, waist,

and buttock depth were recorded. These data are given in Liu *et al.* (1972).

The sample mean λ, standard deviation σ, and the correlation coefficient with respect to the parameter (W/Hd_y), r, are shown in the first three columns of table 3, and illustrated in figure 6. Linear regression equations to estimate the distance λ in terms of (W/H_y) were calculated for each segment along with its standard error of estimate, S. The regression relationship is:

$$\lambda_{est} = a_0 + a_1(W/Hd_y) \pm S \qquad (7)$$

where λ_{est}, a_0 and S are the estimated value of λ, the first and second regression coefficients, and the

Vertebral level	STATISTIC			REGRESSION $\lambda = a_0 + a_1(W/Hd_y)$		
	Mean value (λ) (cm)	Standard deviation (σ)	Correlation coefficient (r)	First regression coefficient (a_0)	Second regression coefficient (a_1)	Standard error of estimate $(\pm S)$
T1	*	*	*	*	*	*
T2	*	*	*	*	*	*
T3	*	*	*	*	*	*
T4[1]	6·456	0·715	0·93	−4·481	146·847	0·256
T5[2]	6·907	0·818	0·85	0·319	88·378	0·429
T5[3]	6·856	0·864	0·82	0·085	90·835	0·484
T7[4]	6·507	0·771	0·79	0·576	82·168	0·469
T8	6·653	0·795	0·64	1·657	68·497	0·610
T9	6·883	0·792	0·77	0·880	82·289	0·503
T10	7·090	0·866	0·76	0·636	88·459	0·563
T11	7·293	0·979	0·73	0·311	95·703	0·673
T12	7·323	1·016	0·80	−0·643	109·201	0·611
L1	7·418	0·868	0·90	−0·233	104·886	0·382
L2	7·566	0·835	0·82	0·809	92·625	0·472
L3	7·914	0·976	0·89	−6·816	191·526	0·443
L4	8·250	1·277	0·88	−10·716	246·589	0·615
L5	8·853	1·481	0·81	−10·447	250·841	0·870

* Values not available from radiograph.
[1] Number of samples = 3.
[2] Number of samples = 6.
[3] Number of samples = 6.
[4] Number of samples = 9.

TABLE 3: Volunteer sample statistics and regression equations for the distance λ (cm)

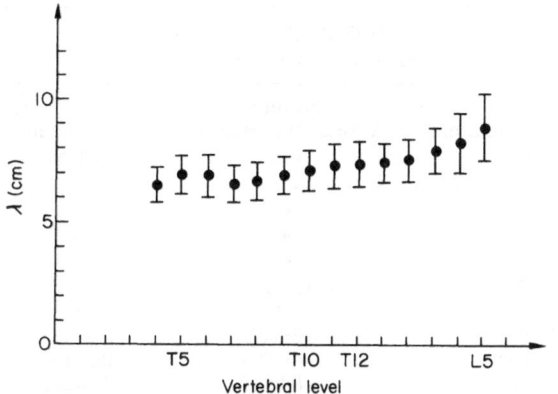

Figure 6. Mean and standard deviation for λ (cm).

standard error of the estimate respectively. The parameters of the regression equation (7) are shown in the last three columns of table 3. Typical computations leading to the results in table 3 are given in the research report of Liu *et al.* (1972).

Knowing the height, weight and d_y of the cadavers, equation (7) allows one to estimate the distribution of its λ values if it had been living. The tissue shift is a complex phenomenon, but it is safe to assume that the bone dimensions are unchanged in the relative rearrangement. Hence, in figure 4 we can equate β and b. Analogous to (3), the formula for the corrected mass eccentricity ε_x becomes

$$\varepsilon_x = e_x + (l - \lambda_{est}) \qquad (8)$$

If one substituted $l = b + c$, $\lambda_{est} = \beta + \gamma$, $\beta = b$, and set $\gamma = 0$, one returns to equation (3) of the initial study.

The weight, corrected mass eccentricity, and principal mass moments of inertia for each vertebral segment are now known. The mean, standard deviation, correlation coefficient, and regression

equations for each of these quantities are given below.

For the weight of the individual segment, equation (5) assumes that it is related directly to the total body weight. Since the transection is carried out with respect to the discs, a short fat cadaver will yield similar weight distribution as a tall thin one, because for a given weight the quantity Ah is relatively constant, i.e. the area A tends to offset the thickness h. However, the body shape, whether a pyramid or an inverted pyramid, is reflected in d_x and d_y. The linear regression equation has the form

$$m_i = a_0 + a_1(d_x d_y H) \pm S. \qquad (9)$$

The sample statistic for the mass of each vertebral segment and its corresponding linear regression equation are given in table 4. These are also graphed in Figure 7.

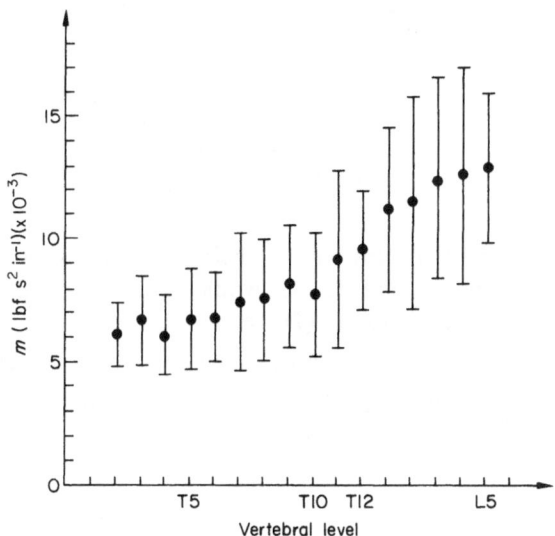

Figure 7. Mean and standard deviation for m (lbf s²/in) ($\times 10^{-3}$).

Vertebral level	STATISTIC			REGRESSION $m_i = a_0 + a_1(Hd_x d_y) \pm S$		
	Mean mass (m_i)	Standard deviation (σ)	Correlation coefficient (r)	First regression coefficient (a_0)	Second regression coefficient (a_1)	Standard error of estimate $(\pm S)$
T1	0·00777	0·00527	−0·52	0·01306	−0·00000172	0·00462
T2	0·00614	0·00146	0·46	0·00372	0·00000050	0·00130
T3	0·00663	0·00187	0·66	0·00038	0·00000100	0·00140
T4	0·00608	0·00163	0·66	−0·00015	0·00000093	0·00122
T5	0·00672	0·00209	0·71	−0·00224	0·00000124	0·00148
T6	0·00682	0·00182	0·79	−0·00371	0·00000142	0·00112
T7	0·00748	0·00297	0·82	−0·00802	0·00000205	0·00169
T8	0·00758	0·00251	0·88	−0·00419	0·00000154	0·00122
T9	0·00810	0·00257	0·80	−0·00225	0·00000135	0·00155
T10	0·00773	0·00256	0·86	−0·00310	0·00000138	0·00130
T11	0·00916	0·00366	0·88	−0·00502	0·00000185	0·00171
T12	0·00958	0·00239	0·76	0·00203	0·00000099	0·00154
L1	0·0112	0·00337	0·83	−0·00010	0·00000146	0·00189
L2	0·0115	0·00433	0·89	−0·00227	0·00000180	0·00200
L3	0·0123	0·00410	0·92	−0·00003	0·00000164	0·00156
L4	0·0125	0·00445	0·93	−0·00075	0·00000179	0·00167
L5	0·0128	0·00311	0·90	0·00371	0·00000128	0·00134

TABLE 4: Sample statistics and regression equations for the segment mass m_i lbfsec²/in

| Vertebral level | STATISTIC | | | REGRESSION $\epsilon_x = a_0 + a_1(W/Hd_y)$ | | |
	Mean value (ϵ_x)	Standard deviation (σ)	Correlation coefficient (r)	First regression coefficient (a_0)	Second regression coefficient (a_1)	Standard error of estimate $(\pm S)$
T1	*	*	*	*	*	*
T2	*	*	*	*	*	*
T3	*	*	*	*	*	*
T4[1]	1·212	0·513	−0·74	3·909	−15·7972	0·34675
T5[2]	0·983	0·477	−0·32	2·075	−6·4164	0·45239
T6[3]	1·132	0·327	−0·28	1·797	−3·8479	0·31435
T7[4]	1·101	0·541	−0·19	1·903	−4·5893	0·53150
T8	1·266	0·478	−0·38	2·856	−8·8540	0·44246
T9	1·500	0·463	−0·19	2·296	−4·3847	0·45411
T10	1·432	0·482	−0·15	2·050	−3·4087	0·47701
T11	1·336	0·539	−0·43	3·107	−9·7070	0·48797
T12	1·365	0·601	−0·48	3·436	−11·4335	0·53735
L1	1·173	0·638	−0·29	2·753	−8·7766	0·61198
L2	1·044	0·735	−0·27	2·667	−8·6519	0·70842
L3	1·168	0·857	−0·22	2·778	−8·7829	0·83597
L4	0·883	0·953	−0·62	6·727	−31·4242	0·74581
L5	0·504	1·076	−0·82	7·658	−37·5359	0·61880

* Values not available from radiograph.
[1] Number of samples = 3.   [3] Number of samples = 6.
[2] Number of samples = 6.   [4] Number of samples = 9.

TABLE 5: Statistics and regression equations for the corrected mass eccentricity ϵ_x.

The mass eccentricity, ε_x being a linear dimension proportional to d_x would have a similar regression equation to (7), i.e.

$$\varepsilon_x = a_0 + a_1(W/Hd_y) \pm S \qquad (10)$$

Table 5 and figure 8 gives all the statistical and regression parameters for ε_x.

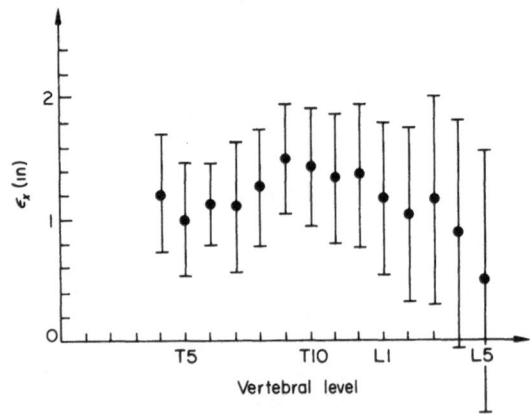

Figure 8. Mean and standard deviation for segment mass eccentricity ε_x (in).

The principal mass moments of inertia for a thin elliptic disc are:

$$I_{xx} = \frac{m}{12}(\tfrac{3}{4}d_y^2 + h^2) \qquad (11)$$

$$I_{yy} = \frac{m}{12}(\tfrac{3}{4}d_x^2 + h^2) \qquad (12)$$

$$I_{zz} = m(d_x^2 + d_y^2)/16 \qquad (13)$$

In (11) and (12), h/d_x and h/d_y are small quantities. Hence, the square of these quantities can be neglected in the primary correlation analysis. Using the same arguments as previously, we get

$$I_{xx} \propto Wd_y^2 \qquad (14)$$

$$I_{yy} \propto Wd_x^2 \qquad (15)$$

$$I_{zz} \propto W(d_x^2 + d_y^2) \qquad (16)$$

Linear regression analyses were performed on the cadaver data using the parameters in (14), (15), and (16). Tables 6, 7, and 8 give the sample statistic and the regression equations for the principal moments of inertia. These are also illustrated in figures 9, 10 and 11. Implicit in the use of these regression equations for mass moments of inertia is the assumption that shift of the centre of mass has little or no effect on the experimentally determined values.

DISCUSSION AND CONCLUSION

The biomechanical data obtained in this experimental series represent an important part of the input parameter needed for the implementation of a new generation of multi-degree-of-freedom discrete parameter or continuous parameter models of the human torso. Since only eight cadavers and ten volunteers were involved, the data have small-sample statistical significance. Generally, the smaller the sample size the more sensitive it is to errors of omission and commission. The following discussion is given to pinpoint the weaknesses in the present study

Vertebral level	STATISTIC			REGRESSION $I_x = a_0 + a_1(Wd_y^2) \pm S$		
	Mean value (I_x)	Standard deviation (σ)	Correlation coefficient (r)	First regression coefficient (a_0)	Second regression coefficient (a_1)	Standard error of estimate $(\pm S)$
T1	0·0459	0·0152	0·86	0·00212	0·00000215	0·00772
T2	0·0912	0·0358	0·96	−0·01769	0·00000380	0·00956
T3	0·104	0·0434	0·85	−0·00490	0·00000365	0·0227
T4	0·0939	0·0339	0·92	−0·01016	0·00000353	0·0131
T5	0·102	0·0401	0·85	−0·01181	0·00000386	0·0214
T6	0·103	0·0409	0·88	−0·03258	0·00000479	0·0196
T7	0·108	0·0577	0·90	−0·07401	0·00000654	0·0253
T8	0·107	0·0555	0·90	−0·04352	0·00000552	0·0239
T9	0·109	0·0548	0·88	−0·03270	0·00000528	0·0259
T10	0·100	0·0531	0·91	−0·04205	0·00000529	0·0220
T11	0·109	0·0731	0·85	−0·06510	0·00000666	0·0383
T12	0·116	0·0511	0·79	−0·00064	0·00000442	0·0311
L1	0·139	0·0656	0·86	−0·01977	0·00000592	0·0340
L2	0·147	0·0848	0·89	−0·04388	0·00000723	0·0380
L3	0·149	0·0789	0·97	−0·04859	0·00000718	0·0178
L4	0·151	0·0814	0·99	−0·04869	0·00000741	0·0138
L5	0·159	0·0629	0·99	0·00014	0·00000620	0·00948

TABLE 6: Sample statistics and regression equations for the mass moments of inertia about the x axis, I_x.

Vertebral level	STATISTIC			REGRESSION $I_y = a_0 + a_1(Wd_x^2) \pm S$		
	Mean value (I_y)	Standard deviation (σ)	Correlation coefficient (r)	First regression coefficient (a_0)	Second regression coefficient (a_1)	Standard error of estimate $(\pm S)$
T1	0·0066	0·0016	0·97	0·00431	0·00000078	0·00038
T2	0·0184	0·0075	0·73	0·00955	0·00000174	0·00513
T3	0·0255	0·0111	0·91	−0·00203	0·00000368	0·00458
T4	0·0278	0·0116	0·88	−0·00142	0·00000339	0·00550
T5	0·0340	0·0160	0·84	−0·00412	0·00000388	0·00872
T6	0·0392	0·0156	0·88	−0·00303	0·00000393	0·00727
T7	0·0476	0·0253	0·97	−0·02790	0·00000665	0·00604
T8	0·0491	0·0233	0·89	−0·01235	0·00000513	0·1048
T9	0·0546	0·0245	0·92	−0·01159	0·00000534	0·00971
T10	0·0534	0·0260	0·96	−0·01602	0·00000539	0·00735
T11	0·0625	0·0346	0·93	−0·02033	0·00000651	0·01235
T12	0·0622	0·0242	0·83	0·01524	0·00000374	0·01361
L1	0·0714	0·0349	0·83	0·00528	0·00000516	0·01952
L2	0·0740	0·0447	0·87	−0·01690	0·00000678	0·02206
L3	0·0732	0·0446	0·99	−0·01949	0·00000740	0·00755
L4	0·0716	0·0437	0·99	−0·01838	0·00000738	0·00738
L5	0·0725	0·0371	0·99	−0·00602	0·00000653	0·00425

TABLE 7: Sample statistics and regression equations for the mass moment of inertia about the y axis, I_y.

in the hope that other investigators will aid in improving and extending these urgently needed data.

The way the head and neck were severed, i.e. the plane of the cut extending from the superior surface of the medial end of the clavicle medially and inferiorly toward the C7 and T1 disc, resulted in a V-notch in the soft-tissue of the T1, and at times the T2, segment. These cuts were made because of a separate study involving the head and neck. Thus, the data is structurally valid only between T2 and L5.

Although our regression analysis is guided by the idealisation that the inertial properties of the torso are similar to those of an elliptic cylinder, the human counterpart is far from being represented by the same. Rather, we hypothesise that the relationships among variables have similar form, as indicated by the mechanics of the problem. Under the usual normality assumptions of regression analysis, we can discuss the significance and confidence intervals of the data obtained. While many parameters can be used, we shall only mention the significance of the sample correlation coefficient r and the associated standard error of estimate S. Tables have been compiled to answer these classes of questions, e.g. Crow *et al.* (1960).

For the population of ten volunteers, we get eight degrees of freedom. If the computed $|r|$

Vertebral level	STATISTIC			REGRESSION $Iz = a_0 + a_1 W(d_x^2 + d_y^2) \pm S$		
	Mean value (I_z)	Standard deviation (σ)	Correlation coefficient (r)	First regression coefficient (a_0)	Second regression coefficient (a_1)	Standard error of estimate $(\pm S)$
T1	0·152	0·124	0·84	−0·07430	0·00013757	0·0666
T2	0·109	0·0406	0·86	0·03801	0·00003942	0·0204
T3	0·126	0·0493	0·91	0·02824	0·00005384	0·0204
T4	0·120	0·0449	0·91	0·03114	0·00004895	0·0188
T5	0·132	0·0512	0·79	0·04421	0·00004827	0·0317
T6	0·137	0·0540	0·78	0·04432	0·00005067	0·0337
T7	0·147	0·0764	0·91	−0·00383	0·00008300	0·0324
T8	0·148	0·0775	0·76	0·01947	0·00007043	0·0507
T9	0·152	0·0749	0·80	0·02074	0·00007192	0·0448
T10	0·146	0·0735	0·84	0·01135	0·00007413	0·0398
T11	0·172	0·0998	0·74	0·01171	0·00008829	0·0674
T12	0·170	0·0711	0·61	0·07491	0·00005238	0·0561
L1	0·196	0·0920	0·72	0·05073	0·00007966	0·0636
L2	0·203	0·113	0·78	0·00200	0·00010479	0·0702
L3	0·202	0·112	0·90	−0·02749	0·00011954	0·0479
L4	0·203	0·112	0·90	−0·02589	0·00011930	0·0476
L5	0·211	0·094	0·93	0·0186	0·00010355	0·0348

TABLE 8: Sample statistics and regression equations for the mass moment of inertia about the z axis, I_z.

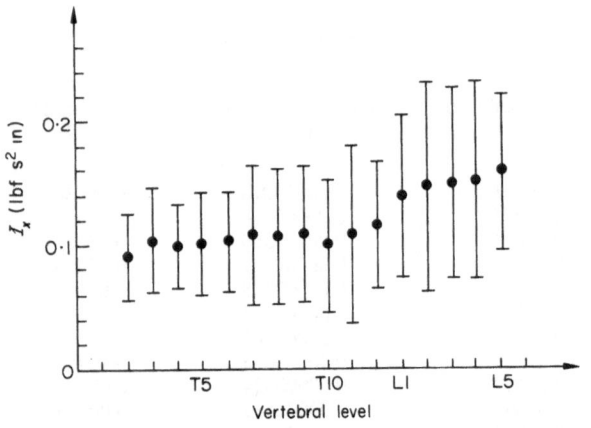

Figure 9. Mean and standard deviation for I_x (lbf s² in).

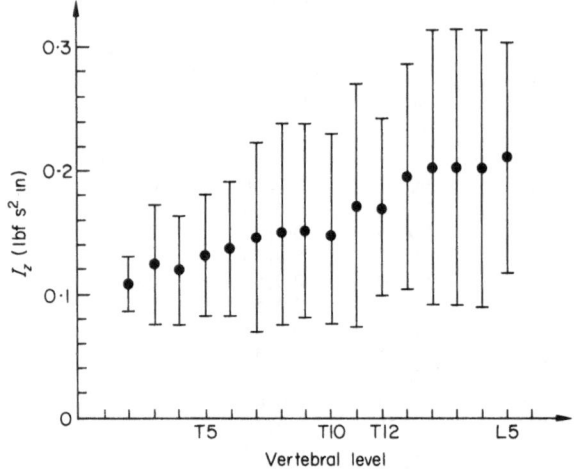

Figure 11. Mean and standard deviation for I_z (lbf s² in).

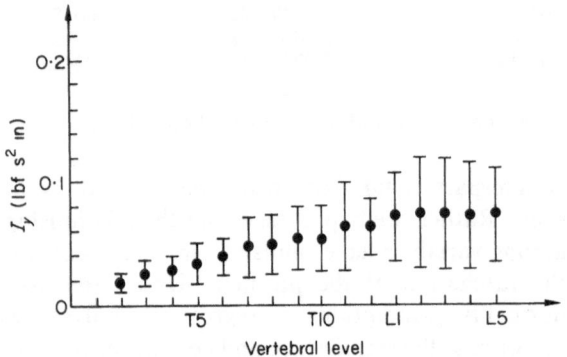

Figure 10. Mean and standard deviation for I_y (lbf s² in).

exceeds either 0·632 or 0·765, we can reject at the 5 per cent and 1 per cent levels of significance respectively the null hypothesis that the distance λ has no correlation with the parameter (W/Hd_y). As we can see from table 3, the majority of our computed correlation coefficients were at the 1 per cent

or better level of confidence. The associated standard errors of estimate for λ is 10 per cent or less of the mean value.

To obtain the distribution of d_x and d_y in the volunteers, the preferred procedure would have been as follows. Lead strips or paint should be applied to four regions: (a) the lowest points of the furrow between the sacrospinalii; (b) the straight line drawn along the midsagittal plane from the midpoint of the manubrium to the navel; (c) two lines along the extreme lateral sides of the torso. The distances associated with each vertebral segment is then taken from the appropriate AP and lateral x-rays.

The estimated value of λ (λ_{est}) is substituted into equation (8) to obtain the corrected mass eccentricity ε_x. This new distance ε_x is then again correlated to the parameter (W/Hd_y). As can be seen from table 5, the resultant correlation coefficient is quite poor. In a way, this is not surprising since equation (8) clearly shows that ε_x incorporated the scatter of two measured quantities, i.e. e_x, l and an estimated value λ_{est}. The dispersion is further exacerbated by the age differences between the volunteer and cadaver population. The mean age of the volunteers is 26·6 years while it is 61 for the cadavers. If data for either older volunteers or younger cadavers were available, one should expect an improvement in correlation coefficient. Additionally, the small sample size results in fewer degrees of freedom. It is well known from regression theory that as the number of degrees of freedom increases, i.e. an increase in sample size, one can reject the null hypothesis at lower and lower values of correlation coefficient. The poor correlation coefficient for ε_x is further reflected in the standard error of estimate.

The estimation for the mass and mass moments of inertia distribution are, in contrast to the poor results for the mass eccentricity ε_x, remarkably good. The very high correlation coefficients suggest the essential self-consistency of the data obtained from the cadaver population. However, one of the unknowns in the regression result is the effect of age on the correctness of our predictions. This question cannot be answered without doing a series using younger cadavers.

The present inertial-property estimation based on weight, height, and certain breadth and width dimensions is believed to be the best available in spite of cautions noted. The schema presented, hopefully, will serve to achieve further knowledge and understanding of the human dynamic system.

An example of a multi-degree-of-freedom configuration model of the human torso under vertical impact where the present data is needed, is the recent work of Orne and Liu (1971). In fact, this previous modelling study was the primary motivation for this work. The use of distributed-parameter models for the study of the structural dynamics of the human torso, e.g. Moffatt (1972), will also need this data after it has been appropriately fitted to continuous functions.

ACKNOWLEDGEMENTS

The work reported was done while the first author was a NIH Research Career Development Awardee (Grant No. GM 40 723-01). This study was supported jointly by Contract No. F-33615-70-C-1565 from the Aerospace Medical Research Laboratory at Wright-Patterson AFB, Grants No. GK 32047 from the NSF and No. FD 00055 from the Food and Drug Administration, U.S. Public Health Service.

H. S. Chan, S. Robinson, and J. Perrien aided substantially in the collection of these data. We also extend our very deep appreciation to Professor Leon Walker of the Department of Anatomy of Tulane University, without whose invaluable assistance in obtaining and handling the cadavers, the entire project would have been impossible.

REFERENCES

BORELLI, G. A. 1679: *De Motu Animalium*, Lugduni Betavoruum.

CLAUSER, C. E., McCONVILLE, J. T., and YOUNG, J. W. 1969: Weight, Volume, and Center of Mass of Segments of the Human Body. *Report No. AMRL-TR-69-70, Wright-Patterson AFB, Ohio.*

CONTINI, R., and DRILLIS, R. 1954: Biomechanics. *Appl. Mech. Rev.* 7: 49-52.

CROW, E. L., DAVIS, F. A., and MAXFIELD, M. W. 1960: 'Statistics Manual'. Dover Pub. Co., New York.

DEMPSTER, W. T. 1955: Space Requirements of the Seated Operator. *Wright Air Development Center Report TR-55-159, Wright-Patterson AFB, Ohio. (AD 10203).*

DRILLIS, R., and CONTINI, R. 1966: Body Segment Parameters. *Office of Vocational Rehabilitation, HEW Publication. Report 1166-03*, N.Y. University School of Engineering and Science, N.Y.

LIU, Y. K., LABORDE, M. J., and VAN BUSKIRK, W. C. 1971: Inertial properties of a segmented cadaver trunk: their implications in acceleration injuries. *Aerospace Med.* **42**(6): 650-657.

LIU, Y. K., WICKSTROM, J. K., and CHAN, H. S. 1972: 'Estimation of the Inertial Property Distribution of the Human Torso From Segmented Cadaver Data'. Research Report, Biomechanics Laboratory, Tulane University Schools of Medicine and Engineering, New Orleans. In press.

MOFFATT, C. A. 1972: Dept. of Theoretical and Applied Mechanics, West Virginia University, Margantown, W. Va., Private Communications.

ORNE, D., and LIU, Y. K. 1971: A mathematical model of spinal response to impact. *J. Biomechanics* 4: 49-72.

DYNAMIC CHARACTERISTICS OF THE TISSUES OF THE HEAD

JAMES H. McELHANEY, JOHN W. MELVIN, VERNE L. ROBERTS[1] AND HAROLD D. PORTNOY[2]

The mechanical causes of head injury have been the subject of much research and controversy. While there is a large amount of literature concerning the overall physiological and pathological effects of head injury, there is considerably less information available on the mechanical characteristics of the tissues of the head. Yet it is these characteristics that determine the mechanism and extent of injury resulting from a blow to the head. As so well put by Ommaya (1968), 'an understanding of the effects of trauma and the development of an exact, rational prophylaxis and therapy for head injury cannot be satisfactorily achieved without a quantitative description of the mechanical properties of the tissues involved'.

Holbourn (1943) proposed on theoretical grounds that injury to the brain is caused by shear strains. These shear strains can be produced in the brain at the point of impact by severe deformation or fracture of the skull resulting in contact with the brain, or they can be produced remote from the point of impact by rotations of the brain within the skull. Holbourn also proposed that concussion is uniquely the result of rotation. Pudenz and Sheldon (1946) and Ommaya (1966) reported on experiments in which *Rhesus* monkeys were fitted with transparent plastic calvaria and subjected to head impact. The motions of the brain during impact were easily visible and tended to confirm Holbourn's predictions of brain rotational movement. Martinez (1963) has shown that brain injury in rabbits can be produced by the rotational motions of severe whiplash alone without impact to the head. A list of damage-producing mechanisms cited by Goldsmith (1966) includes: (*a*) vibration of the entire skull; (*b*) localised large deformations of the skull; (*c*) brain displacement and/or separation at the point opposite to impact; (*d*) establishment of large pressure gradients including negative amplitudes; (*e*) propagation of pressure and strain waves in the cranium; (*f*) rotation of the cerebral mass; and (*g*) neurovascular friction.

The final correlation of head injury theories with head injury experiments has not been forthcoming because of the almost complete lack of knowledge of the mechanism of failure. Goldsmith (1966) has pointed out this problem and has suggested some of the pertinent properties to be determined. Ommaya (1968) has reviewed the scientific literature pertaining to the mechanical properties of the tissues of the nervous system.

The aim of the research described in this paper is specification of the dynamic mechanical properties of the tissues of the head. The mechanical response of biological materials cannot be characterised in the same manner as elastic materials, which obey Hooke's Law. The time dependence of the mechanical responses of these tissues make this description extremely difficult.

MATERIAL PROCUREMENT AND HANDLING

The methods of specimen procurement and preparation (if appropriate) can significantly affect the results of mechanical property tests. A series of experiments on monkey tissues (scalp, skull, brain, and dura mater) indicated that there were no significant changes in their properties over the period of these experiments, i.e. for up to 15 hours.

The majority of the human materials used in this research were obtained at autopsy. Tissue-handling and storage techniques varied with the different materials. The human brain samples were tested between six and ten hours post mortem. Prior to testing, they were kept refrigerated in isotonic saline solution as were the scalp and dura mater samples. Skull bone materials were stored at $-10°C$. Monkey (*Macaca mulatta*) brain and scalp were tested immediately post mortem in the *in vitro* tests. All scalp and brain samples were tested at appropriate *in vivo* body temperatures.

EXPERIMENTAL METHODS

Bulk modulus tests

Monkey and human tissues were tested to deter-

[1] Highway Safety Research Institute, University of Michigan, Ann Arbor, Michigan, U.S.A.
[2] Oakland Neurological Clinic, Pontiac, Michigan, U.S.A.

mine volumetric strain induced by application of uniform stress.

A special stainless-steel piston and cylinder assembly was attached to the driving ram of the closed-loop, high-speed hydraulic testing machine. The piston and cylinder were carefully ground and lapped to ensure a smooth, tight fit. Two neoprene 'O' rings were mounted on the piston to seal it. The cylinder was 2 inches in diameter and the piston $\frac{1}{2}$ inch in diameter, thus providing a very rigid assembly. An acceleration-compensated miniature pressure transducer of the piezoelectric type was built into the head of the piston enabling the pressure in the chamber under the piston to be recorded directly. A second pressure transducer was fitted into the wall near the bottom of the test chamber, to measure the radial pressure in the test chamber.

The tests were carried out in the following manner. The samples were placed in the cylinder which was first filled with mammalian Ringer's solution. The sample displaced the solution through a small valve in the base so that no air was trapped in the cylinder. Cyclic loads were applied to the piston using the programmed loading facility described previously. The displacement of the piston, the piston pressure P_p, and the wall pressure P_w were recorded on a strip-chart recorder. A thermo-couple-controlled heating wire, wrapped around the cylinder, kept the tissue specimen at body temperature.

During these tests, the pressure was increased by increments of 250 lb in^{-2} from 0 to 2500 lb in^{-2}, and the corresponding displacements recorded. Then with a 1500 lb in^{-2} preload applied to the specimen, sinusoidal pressures of 500, 1000, 1500, 2000 lb in^{-2} peak-to-peak amplitude were applied and the corresponding displacements recorded. A dynamic test was made by working the piston in the cylinder at frequencies of 1, 10, 20, ..., 90, 100 Hz.

As a check of the system, a test was conducted on boiled distilled water. The pressure against relative volume changes for water were plotted for the static and dynamic example. The slopes of these graphs gave a bulk modulus of 300 000 lb in^{-2} which agrees well with the value found in the literature.

Tests were made on specimens from 12 human brains, 17 monkey brains, and 11 monkey scalps.

The whole brains were sectioned and cored with circular coring tools that produced right cylindrical specimens $\frac{1}{2}$ inch in diameter and approximately 1 inch long. The test specimens were longitudinal sections from the middle of the left and right cerebral hemispheres. The first section was taken close to the outside of the hemisphere, and the second section close to the mid-line of the brain.

The results from this series of bulk modulus experiments indicate that brain, either human or monkey, has an essentially elastic response when loaded in the manner described above. Since both vertical and radial pressure transducers recorded the same pressure to within 5 per cent this loading is quite close to hydrostatic. The bulk modulus, 300 000 lb in^{-2}, was not significantly different from that for human brain, monkey brain or water. The pressure-volume graph was extremely linear and independent of frequency in the range 0 to 100 Hz. The scalp experiments indicated a bulk modulus of 280 000 lb in^{-2} which was also independent of frequency.

Capillary rheometer tests

A capillary extrusion rheometer test was used in an attempt to determine the viscosity of brain tissue. This type of rheometer continuously pushes brain tissue through a capillary tube by means of a constant-velocity piston. The volumetric flow rate as well as the pressure drop along the capillary was measured. This apparatus and method of data reduction are well described in the literature (Merz, 1958).

The cylinder was identical to that used in the bulk modulus test except for one change. In place of the rigid plug in the bottom of the cylinder, a capillary tube was installed. Various tube lengths (L) and diameters (D) were tried and finally a capillary with an L/D ratio of 95·3 was selected as a standard.

Tissue samples of brain were cored and inserted into the cylinder. The piston was then lowered into the cylinder and triggered to follow one cycle of the ramp function, extruding brain tissue through the capillary tube with a constant velocity. Displacement, piston pressure, and wall pressure were simultaneously recorded.

This process was repeated for a series of piston velocities. The volumetric flow Q was determined from the piston velocity and tube diameter. The average shear rate was estimated from the equation

$$\gamma = \frac{4Q}{\pi r^3}$$

and the shearing stress at the wall

$$\tau = \frac{\Delta P r}{2L}$$

where ΔP = pressure drop through the capillary (lb in^{-2})

r = radius of capillary (in)
L = length of capillary (in).

The results of this experiment are shown in figure 1. No significant differences were found between human and monkey brain in this experiment. The apparent viscosity was found to be 0·46 poise and was constant over the shear strain rate range of 1×10^4 to 5×10^5 s^{-1}. This does not compare well with the value of 43 poise for brain tissue found by

Koeneman (1966). It is clear that this type of test does not provide a direct measurement of the shear properties of brain since the analysis is valid only for fluids. It is hoped, however, that this test will provide a convenient method of comparison of the behaviour of brain tissue with that of candidate substitute materials for the development of a non-biological model of the head.

Free-standing compression tests

Tests of cylindrical specimens of brain and scalp were made by placing the specimens on a piezo-electric load cell and compressing them with the ram of the electro-hydraulic testing machine. The velocity of the ram was constant during these tests, resulting in a constant strain rate. Since the specimens were compressed to approximately 10 per cent of their original volumes, the standard engineering definitions of stress and strain based on the initial

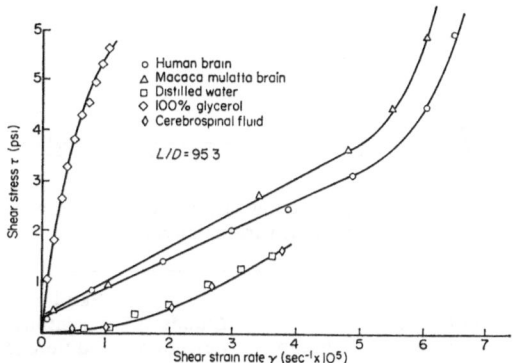

Figure 1. Capillary rheometer test results.

dimensions result in curves that are misleading. Figure 2 shows two presentations of the same stress – strain curve for human brain. The upper curve is determined by computing the stress (σ) as the load (P) divided by the initial cross-sectional area (a) and the strain (ε) as the deformation (d) divided by the initial height (h). The lower curve is obtained by plotting true stress (σ) against natural strain ($\dot{\varepsilon}$) using the instantaneous dimensions. Since the bulk modulus is very large, in the order of 300 000 lb in^{-2}, compared to a compression modulus in the order of 10 lb in^{-2}, it is reasonable to assume a constant-volume process for this test. Thus the instantaneous dimensions can be determined from the initial dimensions and the deformation. The true stress is then

$$\sigma = (1+\varepsilon)\sigma$$

In the compression test the strain is negative and the true stress is less than the engineering stress, because of the increase in area. The true strain is by definition

$$\dot{\varepsilon} = \int_0^H \frac{\mathrm{d}h}{h} = \log_e \frac{h}{h_0} = \log_e(1+\varepsilon)$$

The samples tested were taken from 19 human autopsies and 17 *Macaca mulatta* donors. After the

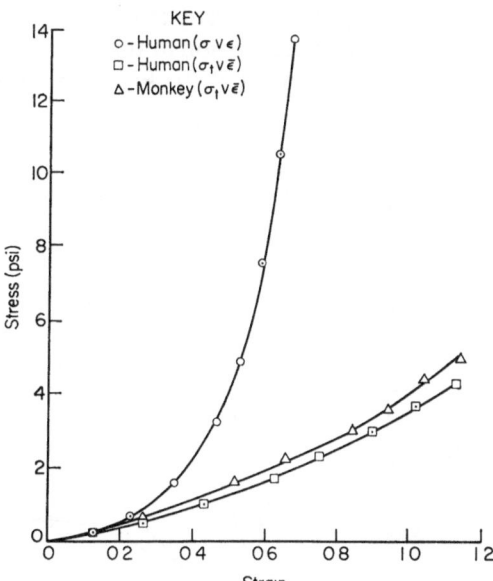

Figure 2. Comparison of human and monkey brain in uniaxial compression: (strain rate 1·0 s^{-1}).

brain was removed, frontal sections were taken every quarter of an inch. Cylinders perpendicular to the plane of section were then removed for test. Most of the cylinders of human brain tissue were taken from white matter with disorganised fibres. The cylinders of monkey brain tissue were either white matter with disorganised fibres or white and grey matter combined.

The response of human brain tissue and *Macaca mulatta* brain tissue in this test showed few significant differences. Figures 2 and 3 illustrate typical examples of the brain tests at various ram velocities. Figure 4 shows typical stress – strain curves for fresh monkey scalp at varying velocities. A strong rate sensitivity is observed in each of these series. There was no significant difference between scalp samples with hair and shaved samples.

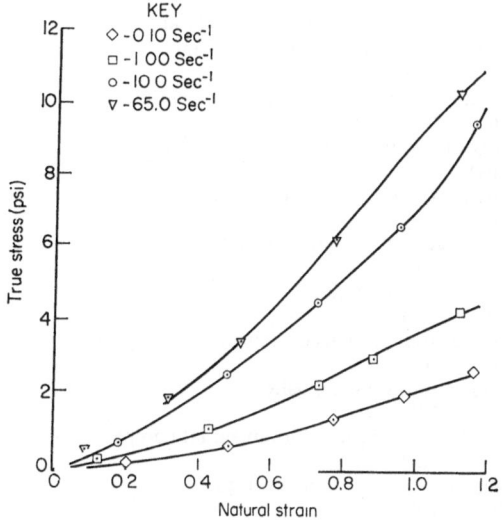

Figure 3. Uniaxial compression tests of human brain at various strain rates.

Skull bone specimens were tested in a similar manner. A complete discussion of this series of tests is given by McElhaney (1970). Figure 5 shows a typical set of stress – strain curves for human skull samples containing the inner and outer tables and the diploë layer or spongy core. In these compression tests there were no significant correlations of any of the measured properties with strain rate. This was due partly to the large variations in the geometries of the various bone layers from one specimen to another. However, this same lack of strain rate influence has been observed in similar tests of porous materials. The results of compression tests on cranial bone at a strain rate of $0.1\ s^{-1}$ are included in table 1.

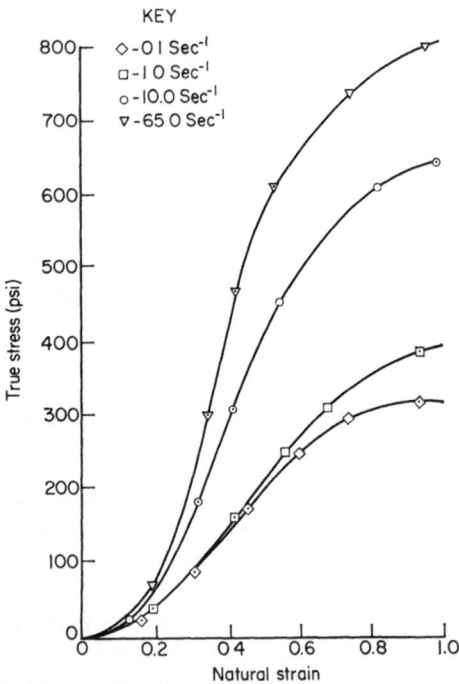

Figure 4. Uniaxial compression tests of monkey scalp at various strain rates.

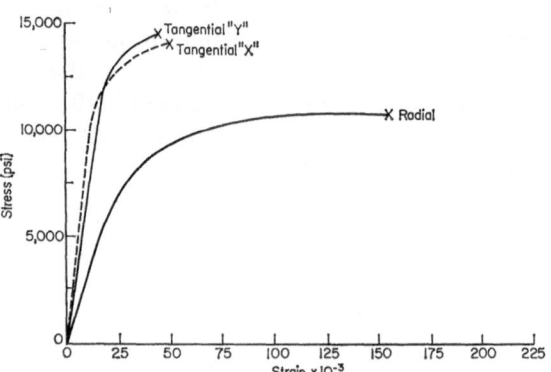

Figure 5. Uniaxial compression tests of human skull bone in various directions.

Property	Number of specimens	Number of donors	Mean	Standard deviation	Skull to skull significant differences
Skull thickness (in)	181	14	0·272	0·047	Yes
Diploë thickness (in)	179	14	0·108	0·042	Yes
Dry weight density (lb/in³)	240	14	0·051	0·019	Yes
Modulus, compression radial (lb in⁻² × 10⁵)	237	26	3·5	2·1	Yes
Modulus, compression tangential (lb in⁻² × 10⁵)	219	14	8·1	4·4	Yes
Poisson's ratio, compression radial	122	14	0·19	0·08	No
Poisson's ratio, compression tangential	327	18	0·22	0·11	No
Ultimate strength, compression radial (lb in⁻² × 10³)	237	26	10·7	5·1	Yes
Ultimate strength, compression tangential (lb in⁻² × 10³)	210	14	14	5·2	No
Energy absorption, compression radial (lb/in³)	237	26	1200	700	Yes
Ultimate strength, diploë direct shear (lb in⁻² × 10³)	348	17	3·1	0·5	No
Ultimate strength, diploë torsion (lb in⁻² × 10³)	90	14	3·2	0·8	No
Modulus torsion diploë (lb in⁻² × 10⁵)	90	14	2·0	1·4	No
Ultimate strength, tension tangential composite (lb in⁻² × 10³)	37	8	6·3	2·7	No
Modulus, tension composite (lb in⁻² × 10⁵)	37	8	7·8	4·2	No
Ultimate strength, tension tangential compact tables (lb in⁻² × 10³)	32	11	11·5	3·8	No
Modulus, tension compact tables (lb in⁻² × 10⁵)	32	11	1·78	0·3	No

TABLE 1: Properties of human cranial bone

Tension tests

Tension tests were performed on specimens of compact cranial bone and dura mater. The cranial bone specimens were made from the compact inner and outer-table bone from fresh human bone plugs. The specimens were machined from the plugs on a Unimat SL combination lathe and milling machine; care was taken to prevent heating during machining. Because of the curvature of the skull and the thinness of the tables the finished specimens were quite small (overall length was 0·625 in, gauge length was 0·10 in, and the cross-section was approximately 0·05 × 0·05 in). Two metal-foil strain gauges with 0·0625 in grid lengths were applied to the specimen with Eastman 910 adhesive. Simple pin grips were used to load the specimen. A seating ball-and-socket slack mechanism was used to allow the testing machine to reach a constant velocity. Test velocities up to 5000 in/min were used routinely in this test. At test speeds up to 200 in/min, the data were recorded as load-strain traces with z-axis modulation providing a time base. At higher strain rates, the load-time and strain-time traces were recorded, and the stress – strain curves were obtained by cross-plotting.

Tests were performed on over 120 specimens from 30 subjects (Wood, 1971). The specimens were taken from the compact layers of parietal, temporal, and frontal bone and the tests were performed at strain rates ranging from 0·005 to 150 s^{-1}. The modulus of elasticity, the breaking stress, and the breaking strain were found to be strain-rate sensitive while the energy absorbed up to failure point did not change with strain rate. The properties showed no important variation with respect to type of bone, side of body, or age of the individual, and there was no apparent variation of properties with respect to direction tangent to the surface of the skull. For the strain rate range studied, the average modulus of elasticity ranged from approximately $1·8 \times 10^6$ to $2·9 \times 10^6$ lb in^{-2}, the average tensile strength ranged from 10 000 to 14 000 lb in^{-2} and the average failure strain ranged from 0·7 to 0·55 per cent. The average value of energy absorbed up to failure point was 42·6 in lb/in^3 with the bulk of the data ranging from 20 to 80 in lb/in^3. A summary graph of the tensile stress – strain curves of human compact cranial bone is shown in figure 6.

Tension tests on fresh dura mater were performed using a test specimen with dumb-bell shape, 2·25 in long with a reduced gauge length section 0·25 in wide and 0·75 in long. This specimen size allowed up to 11 specimens to be cut out of a typical sample of dura mater. The dura mater was obtained at autopsy and placed in a jar of saline solution. If the specimen could not be tested immediately, it was refrigerated in saline solution. An Instron load cell of 1000-lb capacity was used to indicate the load on the specimen; the strain in the specimen was measured by a phototransistorised light extensometer. Two lightweight vanes, attached to the specimen in the test section, blocked a light beam passing from a source to the phototransistor. As the specimen was elongated, the increasing separation of the vanes allowed more of the beam to pass through to the phototransistor. The resulting output of the phototransistor was calibrated against vane separation and permitted measurement of the strain in the specimen independent of any gripping distortions or slippages.

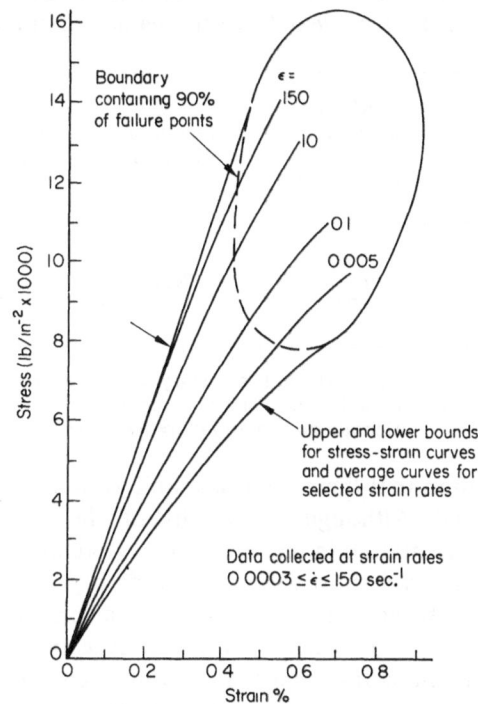

Figure 6. Tensile tests of compact human skull bone at various strain rates.

The macrostructure of dura mater, in the regions relatively free from large blood vessels, appeared to be a membrane with evident directions of fibre reinforcement. Further examination indicated that the direction of apparent orientation viewed on the outside surface of the material was different from that viewed on the inside. Rough measurement of the angle between these two directions showed it to be about 70 degrees. The inner layer, which may be peeled off, is much thinner than the remaining dura mater. Because of this highly oriented structure, specimens were taken from the dura mater in three different planes:

1. Longitudinal—parallel to the sagittal plane.
2. Transverse—perpendicular to the sagittal plane.
3. Diagonal—roughly 45 degrees to both other directions.

The specific mechanical properties determined in the dura mater tension tests were the tensile modulus of elasticity, the ultimate tensile strength and the tensile strain at failure. In all tests the modulus of elasticity was determined, but only where the failure location was in the section between the extensometer vanes, and away from the attachment point of the vanes, were the ultimate strength and strain at failure determined. It is felt that the property of dura mater most pertinent to head injury is the stiffness of the tissue since the dura mater rarely fails in closed head injury. An analysis of variance performed on the data indicated that strain rate effects in dura mater were significant and that specimen orientation effects were not. A statistical summary of the data for modulus of elasticity is shown in table 2. It is evident from the values of standard

Strain rate (s^{-1})	Mean modulus of elasticity $(E\ lb\ in^{-2})$	Standard deviation	Number of tests
0·0666	6027	3138	56
0·666	6430	3238	25
6·66	8799	3720	42

TABLE 2: Statistical summary of dura mater tensile modulus of elasticity for all specimen orientations and all failure locations

deviation that there is considerable variation in the material. Although the results of the study of fibre orientation in the test sections were inconclusive, they did show a trend towards high modulus values being associated with fibre orientations along, or at low angles with, the test section, while low modulus values tended to occur when the major fibre orientation was transverse to the test section. Figure 7 shows the effect of strain rate on the mean stress – strain curves for dura mater.

Free vibration tests

Specimens of scalp and dura mater were suspended in tension with a weight attached and specimens of brain were loaded in compression with a weight. The storage and loss moduli were evaluated by observing the vibration of the system after a sudden disturbance. Figure 8 shows a typical strip chart recording of such an experiment on human dura mater and table 3 provides values for the various tissues tested.

Forced vibration tests

Determination of the dynamic shear properties of both *in vitro* human brain, and *in vivo* and *in vitro* monkey brain was made using forced-vibration test techniques.

Human brain sections, taken at autopsy, were packed in polyethylene bags and placed on ice and water within 10 min of removal from the skull. Initial testing was begun within 2·5 h and subsequent

storage was at 3°C, since early tests on *Rhesus* brain had confirmed that gross change occurs in the modulus when the tissue is frozen.

The complex dynamic-shear modulus of samples of *in vitro* brain was determined by application of a

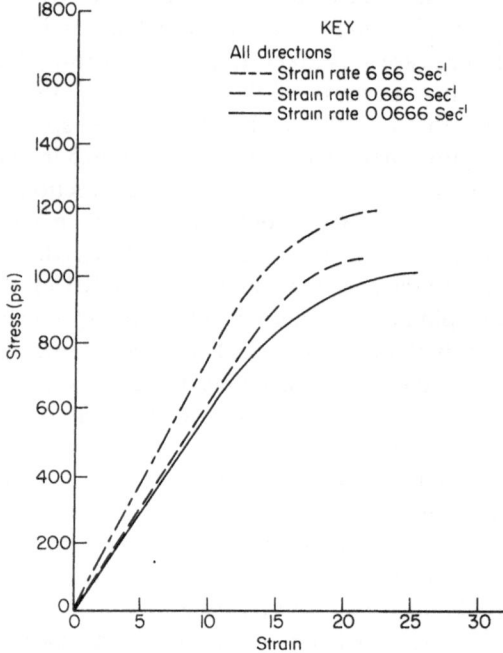

Figure 7. Mean tensile test curves for human *dura mater* at various strain rates.

dynamic shear stress to the test material and measuring the resulting strain. The test apparatus consisted of a sinusoidally actuated mechanism for shearing the sample in simple shear, and electronic equipment to monitor input force, strain level

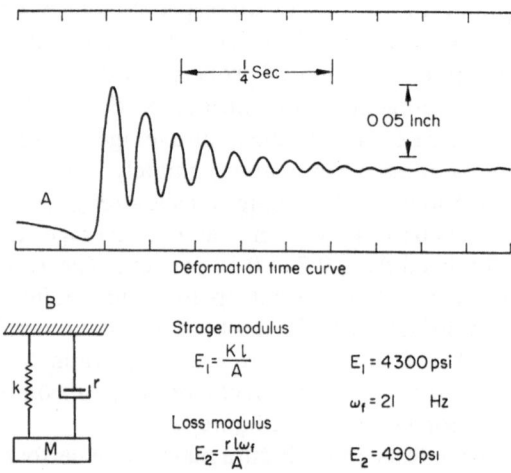

Figure 8. Free vibration test of human *dura mater* in tension.

(output), and the phase angle between them. The sample-shearing mechanism was centrally located on a magnesium-aluminium alloy rod which connected twin electro-mechanical transducers. The driving signal, from a function generator operated in the sine mode, was connected to one transducer. The input force was determined by the current to the driving transducer and was adjustable by means of the function generator signal level control. The other transducer provided an output signal, operating as a velocity transducer. Strain and strain rate were measured by output voltage and amplitude.

The sample-shearing mechanism consisted of a horizontal, aluminium base-plate rigidly attached to the magnesium – aluminium rod and a clear plastic plate which was positioned above, and parallel to, the aluminium base-plate. This plastic

| Material | Brain | | Scalp | Dura | |
	Human	Monkey	Monkey	Human	Monkey
E_1 lb in^{-2}	9·68	13·2	210	4570	2400
E_2 lb in^{-2}	3·80	7·8	74	500	475
ω_f(Hz)	34·0	31·0	20	22	19
E^* lb in^{-2}	10·4	15·3	223	4600	2450

$$E^* = E_1 + iE_2$$

TABLE 3: Typical dynamic modulus values from free vibration experiment

plate was rigidly attached to the main structure of the apparatus and was vertically adjustable. The test sample was sandwiched between these two plates. The sample section, between the twin transducers, was enclosed in a chamber heated by a small electrical heater – fan system. The temperature was maintained at 37°C by means of a temperature potentiometer which utilised an iron – constantan themocouple placed adjacent to the sample.

A total of 57 samples of human brain tissue from 13 individuals was tested *in vitro*, and at 9 to 10 Hz. The results of these tests indicated that the dynamic shear modulus, $G^* = G' + iG''$ could have storage modulus (G') values ranging from 0·062 to 0·138 lb in^{-2}, loss modulus values (G'') ranging from 0·051 to 0·087 lb in^{-2} and an average value of the loss tangent

$$\left(\tan \delta = \frac{G''}{G'} \right)$$

of 0·72.

In order to provide information on the effects of time after death on the properties of brain, and to develop means of extrapolating from *in vitro* test data to *in vivo* test data, a dynamic probe device was developed to test both *in vivo* and *in vitro* monkey brain. The device is essentially a driving-point impedance head with a flat-ended cylindrical probe of 0·1-in diameter. The impedance head and probe were driven by a small electrodynamic shaker. For *in vivo* tests, the probe contacted the brain of the anaesthetised monkey through a burr hole in

the skull which had the dura mater removed from the test site. The tests were performed at vibration frequencies of 80 to 100 Hz. The results of 42 tests on six *Rhesus* monkeys indicate that the *in vivo* storage modulus G' as calculated from the probe data lies in the range 0·181–0·617 lb in^{-2} with an average tan δ value of 0·84. Since *in vitro* monkey brain was found to have similar G^* values as human brain it appears that the probe device produces G^* values approximately twice those given by the flat-plate shearing apparatus, an effect most likely due to the local geometric effects of the probe tip. The probe tests indicated that there was little difference between *in vivo* and post mortem properties, thus the flat-plate shearing properties can be taken as valid indicators of the G^* values of *in vivo* human brain.

MODELLING OF SOFT TISSUE BEHAVIOUR

A variety of modelling methods is available for summarising the results of the free-standing compression tests of brain and scalp. The graphical representation, figure 9, shows the results of over 100 tests of human brain over a range of strain rates from 0·1 to 65 per second. If one allows for the large normal variations inherent in biological

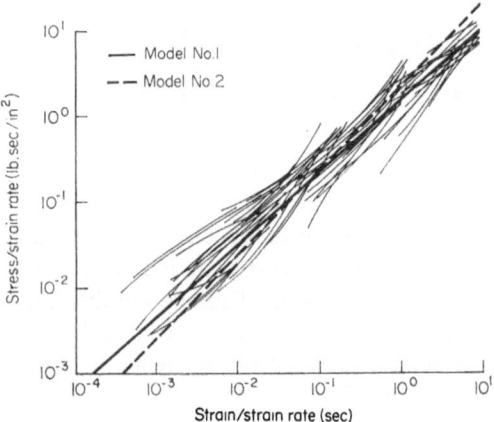

Figure 9. Parametric representation of uniaxial compression of human brain.

materials and the effect of varying sample site, a tendency of results to cluster along a single line may be observed. A similar observation was made for the monkey brain tests and the monkey scalp tests. The equations of these lines are:

$$\sigma = e^a t^b \dot{\varepsilon}$$

where σ = true stress
  $\dot{\varepsilon}$ = true (natural) strain rate
  t = test time
  a, b = empirical constants.

A best fit was obtained for human brain tissue for values of $a = 0·50$ and $b = 0·782$ while monkey tissue yielded $a = 0·69$ and $b = 0·785$. In addition, it is possible to fit several linear viscoelastic models to this curve. A four-parameter fluid (see figure 10) has been chosen because it displays three characteristics of real materials: namely, instantaneous elastic response, viscous flow, and delayed elastic response.

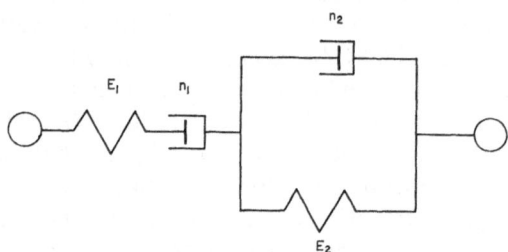

Figure 10. Four-parameter fluid model.

The linear differential equation that describes the response of this model is

$$\sigma + p_1\dot{\sigma} + p_2\ddot{\sigma} = q_1\dot{\varepsilon} + q_2\ddot{\varepsilon}$$

where

$$p_1 = \frac{\eta_1}{E_1} + \frac{\eta_2}{E_2} + \frac{\eta_2}{E_2}; \quad q_1 = \eta_1$$

$$p_2 = \frac{\eta_1\eta_2}{E_1 E_2}; \quad q_2 = \frac{\eta_1\eta_2}{E_2}$$

In the response of this model to a constant rate test, the stress, stress rate, and strain are zero at zero time, and the real unequal roots are

$$\frac{\sigma}{\dot{\varepsilon}} = q_1\{1 + \tfrac{1}{2}[K(K^2-1)^{-\frac{1}{2}} - 1]e^{-(K+K^2-1)\lambda t}$$
$$- \tfrac{1}{2}[K(K^2-1)^{-\frac{1}{2}} + 1]e^{-(K-K^2-1)\lambda t}\}$$

where

$$K = \frac{p_1}{2p_2} = \frac{1}{\sqrt{p_2}}$$

When written as above it is clear that the parameter of stress divided by strain rate is a function of time only. Hence this model displays suitable characteristics to model the compression of brain and scalp. The model values for the dashed line for human brain in figure 9 are

$E_1 = 3\cdot1$ lb in^{-2}
$E_2 = 15$ lb in^{-2}
$\eta_1 = 250$ lb s/in^2
$\eta_1 = 31$ lb s/in^2;

while the parameters that provided a best fit for human scalp are

$E_1 = 1500$ lb in^{-2}
$E_2 = 5300$ lb in^{-2}
$\eta_1 = 7900$ lb s/in^2
$\eta_1 = 310$ lb s/in^2

The results of these experiments indicate that brain is relatively incompressible with a bulk modulus of 300 000 lb in^{-2}. Its response in the capillary rheometer indicates that it is more solid than fluid, and the compression tests have given its compressive response over a wide range of strain rates.

A model of this type is useful in that it summarises a large amount of data. However, it is generally not uniquely defined, for a particular material, by one kind of test. If a model can be found that predicts the response to combined loads, i.e. compression and shear, then it is much more valuable. We are currently seeking such a model by expanding our test program to include torsion and simple shear tests at varying rates. We are also developing several sinusoidal tests over a large frequency range. By combining the results of these experiments it is hoped to arrive at a more general equation for the analysis of the response of the head to trauma.

REFERENCES

GOLDSMITH, W. 1966: The physical processes producing head injury. *Head Injury Conf. Proc.*, Lippincott, Philadelphia.

HOLBOURN, A. H. S. 1943: Mechanics of head injury. *Lancet* 2: 438-441.

KOENEMAN, J. B. 1966: 'Viscoelastic Properties of Brain Tissue'. M.S. thesis, Case Western Reserve University.

MARTINEZ, J. L. 1963: Study of whiplash injuries in animals. *ASME Paper 63-WA-281*.

McELHANEY, J. H., *et al.* 1970: The mechanical properties of the human cranium. *J. Biomechanics* 3-5.

MERZ, E. H., and COLWELL, R. E. 1958: 'A High Shear Rate Capillary Rheometer for Polymer Melts'. *ASTM Bulletin No. 232*.

OMMAYA, A. K. 1966: Experimental head injury in the monkey. Pp. 260-275 *in* 'Head Injury' (Eds W. F. Caveness and A. E. Walker), Lippincott, Philadelphia.

OMMAYA, A. K. 1968: Mechanical properties of tissues of the nervous system. *J. Biomechanics* 1(2): 127-138.

PUDENZ, R. H., and SHELDON, C. H. 1946: The lucite calvarium – a method for direct observation of the brain. II. Cranial trauma and brain movement. *J. Neurosurg.* 3: 487-505.

WOOD, J. L. 1971: Dynamic response of human cranial bone. *J. Biomechanics* 4(1): 1-12.

FURTHER READING

CROSBY, E. C., *et al.* 1962: 'Correlative Anatomy of the Nervous System'. Macmillan, New York.

DALY, C. M. 1966: 'The Biochemical Characteristics of Human Skin'. Ph.D. thesis, University of Strathclyde.

EVANS, J. H., and SIESENNOP, W. W. 1967: Controlled quasi-static testing of human skin *in vivo*. *Dig. 7th Int. Conf. Med. Bio. Engng (Stockholm)*.

DODGSON, M. C. H. 1962: Colloidal structure of the brain. *Biorheology* 1: 21-30.

FRANKE, E. K. 1954: The response of the human skull to mechanical vibrations. *WADC Tech. Report* 54: 24. Wright-Patterson Air Force Base, Ohio.

HICKMAN, K. E. 1965: 'Rheological Behavior of Tissues Subjected to External Pressure'. Ph.D. thesis, Case Western Reserve University.

McELHANEY, J. H. 1966: Dynamic response of bone and muscle tissue. *J.A.P.* 21: 4.

SESSION 'C' DISCUSSION

Dr Radin referred to the calcified cartilage above the osteochondral cement line where it joins the cartilage itself. This is supposed to have a much higher calcium content than bone per unit volume. He considered that this would put cartilage to mechanical disadvantage creating high stress rises at the so-called 'tide-mark' (the boundary between the articular cartilage and the calcified cartilage) and would not seem to be a desirable design characteristic if one wanted to make the articular cartilage stay on the bone for the lifetime of the joint. *Dr Sokoloff*, responding, indicated that the usual explanation based on undecalcified sections is that the calcified layer may reduce the abruptness of the change in stiffness between the cartilage and bone. The fact that the calcified layer in micro-radiographs is as opaque as bone does not tell us whether the two structures in fact have the same stiffness. The latter depends on the physical and chemical interaction of the mineral with the organic components of the matrix. The material properties of the calcified layer cannot be estimated within one or two orders of magnitude simply from its mineral content. *Dr Stockwell* wished to 'draw' Dr Millington out and asked him about the cross-linking of the collagen in the superficial cartilage. It seemed to him to be quite important to know what these cross-links represented. He also asked for any additional information Dr Millington may have regarding their tensile strength. *Dr Millington* stated that there was no mechanical data available relating to the surface properties of cartilage. From morphological studies one can only pre-suppose that the cross-links are fibrils of collagen which branch out and anastomose in the way described in histology textbooks. He had no further information to offer on this. *Mr Growcott* was particularly impressed with the emphasis placed in the various papers on the difference in behaviour under stress of inert materials as opposed to living tissue. He felt there was a great future in investigating more thoroughly the effects of metabolism and the possible regeneration of living tissue under stress. *Dr Edwards* said that at the University of Surrey they are looking at one aspect of the problem postulated by Mr Growcott. In general clinical practice it is seen that cartilage tends to break down under certain abnormal types of mechanical loading if the loading is held for too long. People with certain types of occupation seem to put such abnormal loads on their joints. He believed that this breakdown occurs due to some upset in the nutritional processes, but how this happens he was not quite sure. A group at the University of Surrey have started a series of *in vitro* experiments to investigate the metabolic activity of cartilage under various types of loading. The experiments are of a simple kind such as compressing the cartilage under varying loads for varying periods of time up to 4 hours and investigating the synthesis of mucopolysaccharides. They were able to show that this synthesis is drastically slowed as a result of loading, and the higher the load the lower the rate of metabolic activity and the lower the rate of regeneration. In consequence, he would very positively agree with Mr Growcott and join him in emphasising the importance of looking at the interrelation of the metabolic processes and tissue loading. *Dr Maroudas* commenting on this question of regeneration, pointed out that as collagen has hardly any turnover in cartilage there is obviously very little chance of collagen repair in that particular tissue. In the same tissue, although there is a mucopoly-saccharide turnover, this does not seem to be incorporated. Presumably, the turnover rate is due to increased leakage and as there is no repair as such, fibrillation becomes progressive. *Dr Millington* asked Dr Kempson whether he had any indentor patterns obtained from obliquely applied probes as opposed to the ones quoted in the paper where the indentors were normal to the surface. He also invited Dr Kempson to comment on what the effect of obliquity in the indentor axis might be on the results. *Dr Kempson*, responding, stated that they had no evidence from obliquely applied indentors. All their tests were done with the apparatus arranged normal to the articular surface. He pointed out that the parallelism of the subchondral bone/cartilage interface to the surface mattered more than normality of the indentor. If such parallelism did not obtain, shear stresses were introduced into the mechanism, on the whole lowering the relevant modulus. He thought that Dr Sokoloff found something like this many years ago and there are therefore certain circumstances in which one would have to be very careful in applying this particular technique. *Dr Pawluk* asked Dr Burstein to comment on the modulus of elasticity in human bone. Dr Burstein has tested human bone under sinusoidal and impact loading and found no differences in the modulus of elasticity in compression in the elastic range for human specimens. Dr Pawluk also asked

223

Dr Burstein to indicate whether he had found significant dissimilarities in the elastic range behaviour of bovine and human bone specimens. *Dr Burstein* agreed with Dr Pawluk's suggestion regarding the lack of variability of the modulus of elasticity. In Dr Burstein's tests, in which he used the same specimens for both tension and compression with the same extensometer, he took his specimens up to failure rather than loading them sinusoidally and found that the modulus, while not strictly identical, was not different. He also tested his specimens in the transverse directions and here again found the modulus of the same order. Results that he has presented, although obtained with bovine bone, are fully applicable to human bone also. Preliminary analyses have shown that the ductile behaviour of human bone continues well on into the seventh decade of life. Regarding differences between bovine and human bone: if the specimens are histologically comparable their behaviour, whether they be of human, canine or bovine origin, is essentially identical in the elastic range right up to the commencement of plastic behaviour and eventual failure. Dr Burstein emphasised that differences that do obtain are essentially those resulting from differences in histological structure; there do not appear to be differences in the physical characteristics of the tissues themselves, the only exceptions being density changes or mineral composition. Some of these changes have been examined by Currey (quoted in the reference list of Burstein *et al.*) and the published literature generally tends to substantiate the fact that results are similar provided also that the histology is similar. *Dr Palfrey* asked Dr Maroudas whether she has studied the distribution of charged density radially from the fovea to the margin of the cartilage. It seemed to him that histologically speaking much more variation should be obtained that way than circumferentially. *Dr Maroudas* replied that Dr Kempson and she studied some 30 sites on the femoral head of several specimens and they have found consistently that if there is no surface fibrillation shown by the indian ink method, fixed charged density is found to be extremely reproducible at around 0·14–0·16 with very little variation.

Dr Kempson commented on the tensile properties of cartilage in relation to 'prick' lines. They have taken a series of sections 200 microns thick through the entire thickness of articular cartilage. These sections of cartilage were referenced in relation to the direction of the 'prick' lines on the human femoral condyles. It was found that, particularly in the superficial and tangential zones, the cartilage properties in tension were far superior in the direction parallel to the 'prick' lines to what they

were in the transverse direction. He would support the idea that the 'prick' lines do in fact show predominant orientation. He then asked Dr Millington to comment whether deeper to the surface there is any orientated relationship of structure to the 'prick' lines. *Dr Millington* emphasised that every cartilage specimen appears to be unique, although he agreed that there seems to be a direct relationship between 'prick' lines, fibre orientation and anisotropy near the surface. The deeper zones are much more difficult to analyse at higher magnifications as the most dominant characteristic is an apparent randomness of fibres. A statistical analysis of fibre orientation is required in order to determine any orientative preference. Attempts to do this are in progress. Further evidence is needed before constructive suggestions can be made.

Mr Bennett presented pilot theoretical analyses concerned with soft tissue trauma; in particular, damage to 'flesh' caused by mechanical loading. He proposed to speak of shear stress in flesh, produced by 'prosaic' loading, such as the pressure effects of a prosthetic socket on stumps.

In an aside, he commented that his tools were theoretical rather than experimental, reflecting, not wisdom or aesthetics, but poverty! Theory is cheap, while experimental work in this area is expensive.

The accuracy of the work is completely dependent on the validity of the assumptions shown in figure 1.

Flesh is considered:
1   To include skin, blood, fat, muscle.
2   To exclude bone.
3   To be uniform.
4   To possess a fixed modulus of elasticity.
5   To possess a fixed Poisson's ratio.
6   *Not* to be viscoelastic.

Figure 1. Flesh model—assumptions.

Items (1), (2) and (3) amount to considering flesh to be a homogeneous mélange, which obviously it is not! Therefore at best he is generating a partial truth. The rest of these assumptions (particularly the absence of viscoelasticity) are also gross. The only rationale that can be offered is that the problem is difficult and one must start somewhere. He then went on:

'We are concerned (figure 2) with the stress at a point M contained within a slab of flesh of thickness H, subject to loading $q(x)$. Note that the loading is two-dimensional, that is every slice along the z axis is identical. If we take such a slice of width δ and label the stresses, they would look like the lower half of figure 2 where σ_y represents the stress normal to the upper face and σ_x that perpendicular to the vertical face. The shear stresses are given by τ_{xy} and τ_{yx}.

'The derivation procedure (figure 3) uses the

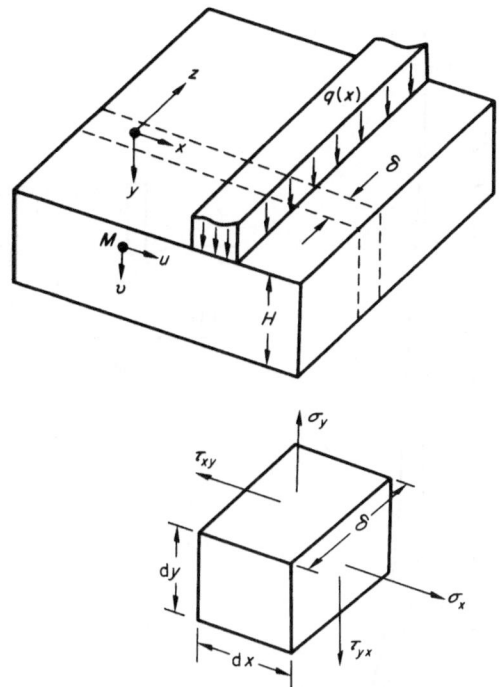

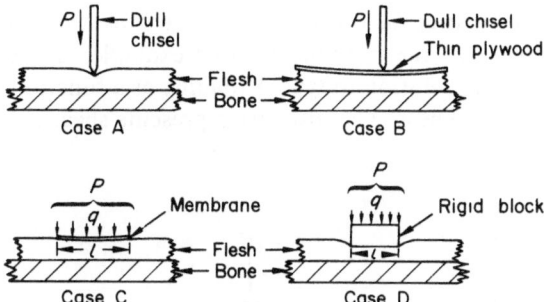

Figure 4. Types of loading studied.

Figure 2. Co-ordinate system employed in this work.

deflection approach, in which the principal unknown is the displacement $v(x, y)$ of the point of interest M. For each load we can write the deflection contribution at M as a combined function of x and y dependent terms. The technique is the separation of variables.

'We have solutions for the following loading types (figure 4): In case A we are pressing a dull 'chisel' edge against a thickness of flesh supported by rigid bone. In case B we have the same situation,

except that a thin flexible sheet (such as plywood for example) has been inserted between the edge and the flesh. In case C the load is applied through a membrane (or bandage) to the flesh. In case D the load is applied through a wide block that is absolutely rigid. In each case the load is the same, P.

'The solution for the shear stress under case B loading is shown in figure 5. It is seen that the expression is messy.

$$\tau_{max} = \frac{(1 - v_0)Pe^{-\bar{\alpha}\eta}\sin\bar{\beta}\eta}{4\bar{\alpha}\bar{\beta}L}$$

where

$$\bar{\alpha} = \sqrt{\frac{s^2 + r^2}{2}}$$

$$s^4 = 2\frac{L}{H}\psi_k$$

$$\psi_k = \frac{H}{2L}\left[\frac{\sinh(H/L)\cosh(H/L) + (H/L)}{\sinh^2(H/L)}\right]$$

$$r^2 = \frac{2(1 - v_0)}{12}\frac{H}{L}\psi_t$$

$$\psi_t = \frac{3L}{2H}\left[\frac{\sinh(H/L)\cosh(H/L) - (H/L)}{\sinh^2(H/L)}\right]$$

$$\bar{\beta} = \sqrt{\frac{s^2 - r^2}{2}}$$

$$\eta = \frac{X}{L}$$

$$L = \sqrt[3]{\frac{2EJ(1 - v_0^2)}{E_0\delta}}$$

$$J = \frac{\delta h^3}{12(1 - \mu^2)}$$

μ = Poissons ratio for plywood
h = Thickness of plywood
E = Modulus of elasticity for plywood

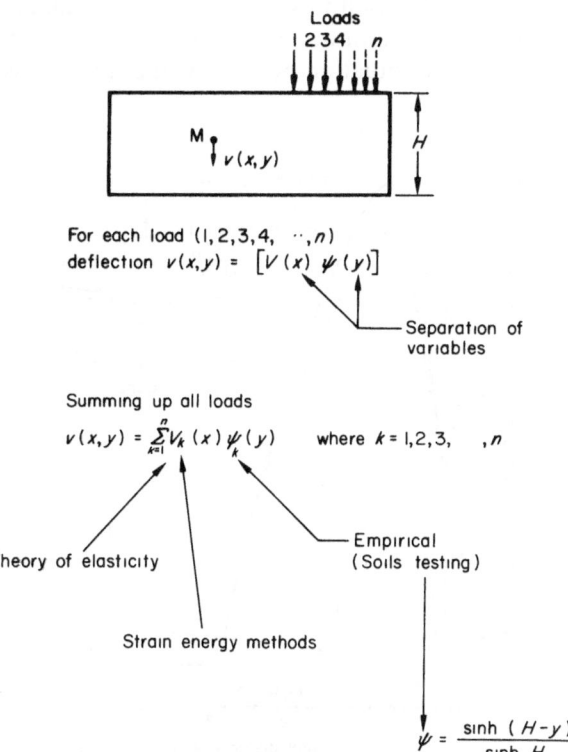

For each load (1, 2, 3, 4, ⋯, n)
deflection $v(x, y) = [V(x)\,\psi(y)]$

Separation of variables

Summing up all loads

$v(x, y) = \sum_{k=1}^{n} V_k(x)\,\psi_k(y)$ where $k = 1, 2, 3, \ldots, n$

Theory of elasticity

Empirical (Soils testing)

Strain energy methods

$\psi = \dfrac{\sinh(H - y)}{\sinh H}$

Figure 3. Derivation procedure.

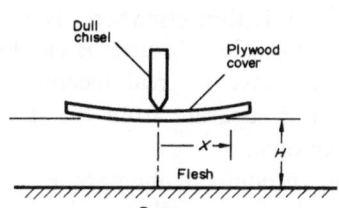

Figure 5. Case B results—Shear stress.

'To clarify the presentation we've substituted real values for the parameters. For example, figure 6 gives the solution to type A loading as a function of flesh thickness. The ordinate represents shear stress

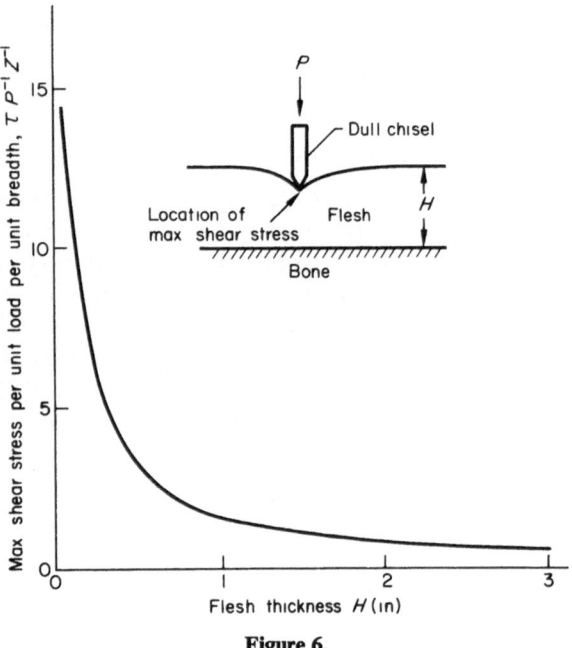

Figure 6.

and the abcissa, flesh thickness. The maximum shear stress is immediately under the loading edge. Note that as the thickness of flesh decreases, the shear stress rapidly increases. In practical terms, this means that shear loading should be avoided where the flesh is thin. The stress values obtained here are the highest of all the cases examined.

'Figure 7 presents the distribution of maximum shear stress with respect to load case A for two flesh thicknesses, $H = 1$ and $H = 3$. Note that the decay with x is large, but that appreciable stresses exist an inch from the loading edge.

'Inserting a thin flexible sheet under the chisel (figure 8) greatly reduces the magnitude of stress. This peak is an order of magnitude *smaller* than the corresponding values without the sheet cover. Using a thicker sheet (bottom of figure 8) reduces the stress level by another order of magnitude, and shifts the peak stress about an inch from the edge load point. It is conceivable that a designer, employing such a thin flexible cover to reduce flesh shear stress at a given location, will shift the location of trauma elsewhere. This is especially true if the flesh is thin elsewhere. Negative values indicating that the flexible cover is breaking contact with the flesh have no real meaning. A thicker cover does not separate from the flesh in the x direction explored.

'Loading through a membrane (a bandage) (figure 9) produces generally low shear stress except at the membrane edges, where the values in practical

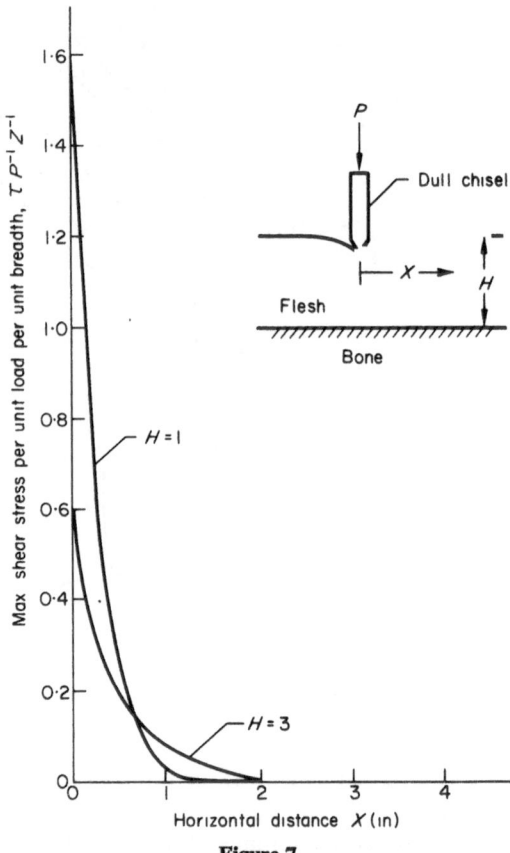

Figure 7.

terms suggest discomfort. Note that the stresses extended beyond the membrane as a mirror image of the stresses under the membrane.

'Loading through a simple block (figure 10) produces large shear stresses, three times as great as those peaks resulting from the membrane loading. As the block width decreases we move towards the loaded edge situation (case A) and the stresses

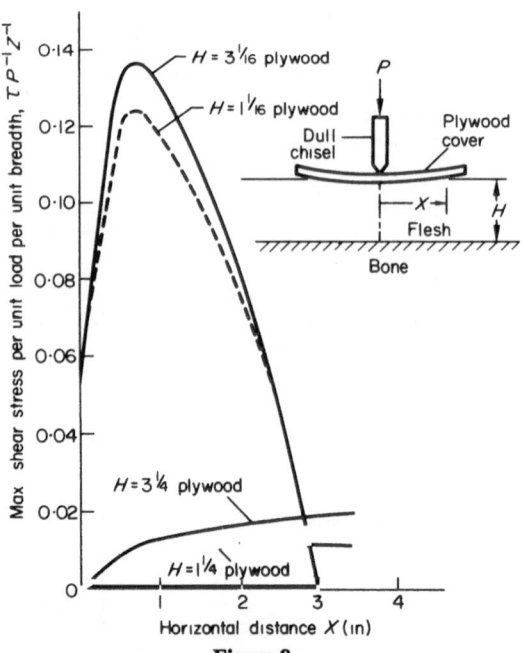

Figure 8.

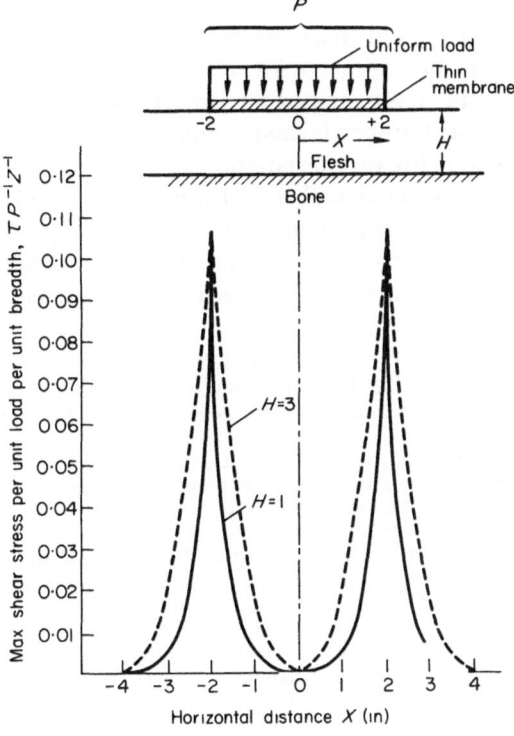

Figure 9.

begin rapidly to increase. It is of interest to note that these stresses are a maximum at the block edge or outboard of the block edge. There is very little shear under the block. This implies that trauma connected with shear stress, resulting from this block type loading, should be found in the zone of free flesh beyond the block. In the particular case of artificial legs, where we can consider the socket to be a rigid block, flesh trauma has been found frequently to extend about $1\frac{1}{2}$ centimetres into the

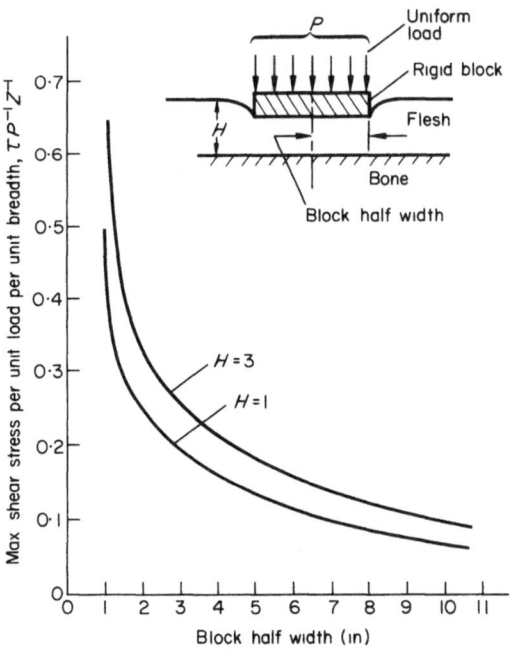

Figure 10.

free zone above ths socket rim. This location may correspond to peak shear stress under mechanical loading.

'It will be of particular interest to fashion socket brims yielding lower shear stresses in the free zone and then to observe the incidence of stump trauma, with low shear sockets, as compared to conventional sockets. I am now engaged in such an effort'.

(*Editorial comment:* The flesh trauma in stumps described by Mr Bennett could arise from several other causes whose influence may equal or exceed that of the shear stress effect.)

The following written contributions were received from Professor A. Viidik.

On the paper by Barbenel *et al.*:

The question whether the 'waviness' of the collagen fibre bundles is planar undulation or a spatial helical arrangement has for a long time intrigued investigators. While Rigby, Hirae, Spikes and Eyring (1959) described a waviness, a helical arrangement appeared to be demonstrated very convincingly in the photomicrographs of Verzar (1965). Both investigations used incident light polarizing systems and rat tail tendons.

The waviness pattern concept was supported by Viidik and Ekholm's (1968) micrographs taken from sections of rabbit hind limb tendons cut parallel to the surface. Viidik (1972a) also demonstrated this 'live' in simultaneous mechanical testing and incident light photomicrography. He showed that the waviness disappeared during the later section of the 'toe' part of the specimen's stress-strain curve, and on inspection gave no indication of a helical arrangement, the pitch of which should have decreased during straining. Other experiments performed by Viidik (1972b) on rat tail tendon fibres demonstrated the same pattern. A thorough analysis of this problem with various polarising microscopy techniques was recently reported by Diamant, Keller, Baer, Litt and Arridge (1972). Their results support the concept that the subunits of the tendons are undulating or are crimped in correspondence with the bands seen in the intact tendon in polarising microscopy. Their calculations seem to exclude the presence of a helical arrangement and they also confirmed that the periodicity gradually vanished when the specimen was strained.

REFERENCES

DIAMANT, J., KELLER, A., BAER, E., LITT, M., and ARRIDGE, R. G. C. 1972: Collagen; ultrastructure and its relation to mechanical properties as a function of ageing. *Proc. R. Soc. Lond.* B. **180**: 293–315.

RIGBY, B. J., HIRAE, N., SPIKES, J. D., and EYRING, H. 1959: The mechanical properties of rat tail tendon. *J. Gen. Physiol.* **43**: 265–283.

VERZAR, F. 1965: 'Experimentelle Gerontologie'. Ferdinand Enke, Stuttgart.

VIIDIK, A. 1972*a*: Simultaneous mechanical and light microscopic studies of collagen fibres. *Z. Anat. Entwickl.-Gesch.*
136: 204–212.

VIIDIK, A. 1972*b*: Functional properties of collagenous tissues.
International Review of Connective Tissue Research (Eds.
D. A. Hall and D. S. Jackson). Vol. VI, Academic Press,
New York (in press).

VIIDIK, A., and EKHOLM, R. 1968: Light and electron microscopic studies of collagen fibres under strain. *Z. Anat.
Entwickl.-Gesch.* **127**: 154–164.

On the paper by Abrahams:

I would like to congratulate Dr Abrahams on a
very interesting paper and, being a biologist primarily in the field of morphology and tissue
mechanics, add a few remarks from this field. Until
recently collagen was considered to be the only
material of mechanical importance in tendons
(Partington and Wood, 1963). Now the importance
also of other components is being investigated
(Jackson, 1972; Minns, 1972) and is found to be of
significance for certain parameters. From a biochemical point of view, interaction between collagen
and glycoproteins (Anderson, 1972; Anderson and
Jackson, 1972) and collagen and glycosaminoglycans
(Öbrink, 1972) have been demonstrated, although
nothing is yet known about their functional significance. It would therefore be extremely interesting
from a biological point of view to investigate
synthetic multicomponent materials of known
component properties and compare their composite
behaviour for various component proportions with
the known physico-chemical properties of isolated
collagen, glycoprotein and glycosaminoglycans.
This would provide guidance on the influence of
component interactions in the biological systems.

REFERENCES

ANDERSON, J. C. 1972: Tendon glycoprotein and interaction
with collagen. *Scand. J. Clin. Lab. Invest.* **29**, *Suppl.* **123**:
p. 4.

ANDERSON, J. C., and JACKSON, D. S. 1972: Tendon glycoprotein and its interaction with collagen. *Scand. J. Clin.
Lab. Invest.* **29**, *Suppl.* **123**: p. 4.

JACKSON, D. S. 1972: Personal communication.

MINNS, M. J. 1972: The role of the fibrous components and
ground substance on the mechanical properties of biological
tissues. *M.Sc. Thesis*, UMIST, Manchester, England.

PARTINGTON, F. R., and WOOD, G. C. 1963: The role of non-
collagen components in the Mechanical Behaviour of
tendon fibres. *Biochim. Biophys. Acta* **69**: 485–495.

ÖBRINK, B. 1972: Interactions between tropocollagen and
glycosaminoglycans. *Scand. J. Clin. Lab. Invest.* **29**, *Suppl.*
123: p. 25.

SESSION 'C' SUMMARY

L. SOKOLOFF

This morning's session, entitled Tissue Mechanics, has dealt with material properties of the body. It has largely been an exercise in anatomy; anatomy, that is, conceived as structure at gross, microscopic, fine and molecular levels. One of the perspectives of bioengineering in its burgeoning is that it poses penetrating questions to professional anatomists and provides tools for answering them.

The gross anatomical studies we listened to somehow seemed a little more grizzly than the others. Nevertheless knowledge of the dynamic properties of the body and its parts will hopefully find humane application in protection from physical injury. In three of the gross studies, partial mathematical models for describing the physical behaviour were developed. *Drs Liu and Wickstrom* told us about the distribution of inertial properties of the human trunk. They sawed frozen cadavers into horizontal pieces and made measurements on each segment individually. *Dr McElhaney* and his co-workers were concerned with the elastic properties and strength of the tissues of the head—from the outside in, the scalp, skull, dura mater and the brain. A limited four-parameter fluid model was found most suited for the elastic, viscous and delayed elastic response of the brain and scalp. Future directions for exploring the basis of brain trauma were proposed. *Dr von Gierke* described his complex general biodynamic model of the whole living body. This takes into account the impulsive and oscillatory responses under various types and magnitudes of loading. *Mr Bennet* presented some theoretical considerations which may be applied to reduction of shear stresses which cause ulceration of amputation stumps.

At a microscopic level, *Drs Barbenel, Evans and Finlay* spoke of the viscoelastic properties of soft collagenous tissues. Tendon and ligaments, because of their predominantly uniaxial arrangement of their collagen, were tested in tension; dermis, more randomly arranged, was measured by a torsional technique. This was applied to the skin of living subjects. A continuous spectrum analysis was found superior to discrete time constants for describing dynamic stress relaxation behaviour. *Dr Daly* summarised biomechanical studies of oral tissues that all of us appreciate as we grow older and lose our teeth. His laboratory made torsional measurements of mobility of the upper incisors in living subjects. Viscoelastic behaviour, presumably arising in the periodontal membrane, was found. There has been a crying need for objective means of following the course of periodontal disease, and this may now be on the way. *Dr Burstein* and his colleagues presented a nice account of plastic behaviour in bone. They pointed out that there are important physical and failure characteristics related to differences in the structure of bone in various species. One must therefore be cautious in extrapolating data from one experimental circumstance to another, and in over interpreting the osteone as the mechanical unit of bone. *Dr Sokoloff* reviewed our state of ignorance of the structure and function of the cement lines that surround osteones. This was considered useful in determining whether cement lines cement osteones together or allow them to slip by each other and so contribute to non-elastic behaviour of bone. Almost as through collusion, *Dr Palfrey* contributed enlightenment on this score. In an elegant electron microscopic examination of the osteochondral junction, no collagenous bridge was found between the two tissues. I found of special interest his observations of differences in the transverse appearance of collagen in the two tissues because it is known by chemical means that they are of different molecular species. Perhaps it should be noted too that species differences exist in this osteochondral structure too. We have not found it to be particularly radiopaque or stained by Bodian's method in mice as it is in man.

Dr Graham's attractive film indicated that the direction of motion of the human hip is perpendicular to the collagenous grain in the tangential layer of the articular cartilage. From the group at the Imperial College in London (*Drs Maroudas and Kempson*), we heard about the basis for the viscoelastic behaviour of joint cartilage. It is a two-phase material in which the highly charged glycosaminoglycans of the ground substance contribute osmotic pressure and stiffness, and collagen provides the tensile strength. It is in this context that the meticulous scanning electron microscopic examination of the collagen of cartilage by *Drs Clarke and Millington* has found useful application.

This brings us up to the nature of composite materials addressed by *Dr Abrahams*. The connective tissues—and perhaps all tissues by definition—are composite materials anatomically. Two different physical ideas are comprehended by this term. In one the properties of the material are understood

SESSION D

Clinical Measurement and Data Analysis

Chairman: Professor J. E. Jacobs, Ph.D., D.Sc.

Associate: Dr D. W. Hill, M.Sc., Ph.D., F.Inst.P., C.Eng., F.I.E.E.

as the sum of the individual components, one contributing the effect that the other lacks. In the other concept, a physical or chemical inter-action between the 2 components provides new properties greater than the sum of the parts. This is an example of what I referred to as the macro-molecular anatomy of the fibres and ground sub-stance of connective tissue. Abrahams made an interesting point in suggesting that this might be applied to development of prosthetic materials which more closely simulate the elastic properties of bone than does steel. It was in the same sense that I replied to *Dr Radin* that one could not make inferences about the physical properties of the calcified layer of articular cartilage simply on the basis of it having mineral and soft components unless we knew how they related to each other.

Finally I would be remiss if I did not extend congratulations to Professor Kenedi from all of us on his splendid new facilities and this symposium. He has over the years given much of himself to the Bioengineering Unit at Strathclyde and the Unit would not exist as we know it without him. He has always been a warm and gracious host and I would like to express my appreciation to his staff, particularly to Drs Paul and Evans, for their unprovoked hospitality.

COMMENTS ON CLINICAL MEASUREMENTS

D. H. BEKKERING[1]

As long as there have been physicians and patients, the physicians have been observing the patients. Observation is an art and a skill. It is a subjective, qualitative happening, and clinical observation is no exception. It is learned by experience during education and it has to be obtained anew by each new student in medical practice.

Measurements are quantitative and are not a subjective experience of the observer. Measurements can be repeated by others and can be standardised. They form an objective playground for scientific discussions. There is a general feeling that measurements deserve more scientific merit than observations. The same holds for clinical measurements.

In the last decades we have seen a firm shift in their relative importance from clinical observations to clinical measurements. The patient orientation of clinical measurements leads to the constraint that clinical measurements need to be non-destructive and, if possible, non-invasive. This constraint leads to a discrepancy between quantities than can be measured on a patient and quantities about which the physician wants to be informed.

Thanks to modern techniques of data handling and simulation-techniques, there is a trend to offer the physician directly the quantity about which he wants to be informed instead of the quantity measured on the patient. As an example the cardiac output computer may be introduced.

For a long time to come a specific group of post-operative patients will continue to be monitored in the intensive care ward by measuring the central blood pressure via an indwelling catheter. Although the aortic blood pressure pulse gives valuable information about the state of the patient, it is cardiac output that is the quantity that the physician eagerly wants to know for this class of patients. Ir. K. H. Wesseling[2] and his co-workers developed a cardiac output computer (figure 1) that calculates from the central pressure pulse, the stroke volume of the patient, but for a constant; this is obtained by calibrating the instrument by a dye-dilution determination. The physician is thus able to see the cardiac output of the patient in digital form.

There is another development going on in clinical measurements: it is the effort to measure everything

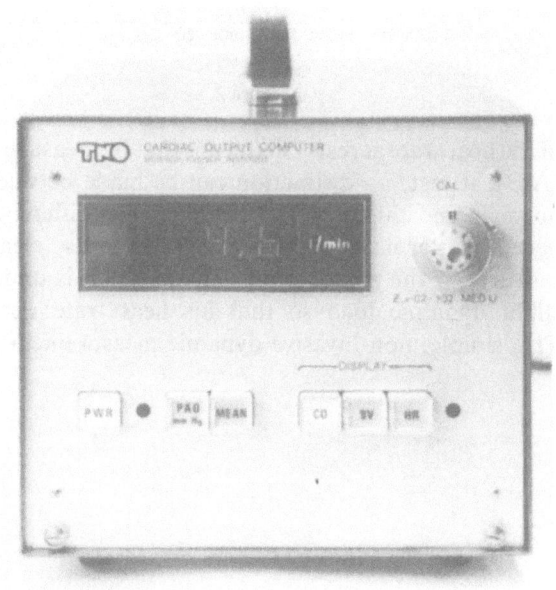

Figure 1.

that is measurable on a patient, with the expectation that by doing so, no relevant information will be left unknown. This attitude may lay a heavy physical and psychological burden on the patient, not knowing where all those activities in and on his body may lead to. There is a positive point in all this activity; many non-invasive measurement techniques that may be of use in population-health screening are obtained as 'spin-offs'. There is a great need for easy-to-perform, non-invasive measurements.

A third tendency, worth mentioning in a Symposium on Perspectives in Biomedical Engineering is the attitude, more and more widespread now in clinics; that of making measurements on the patient under dynamic load. For centuries knowledge about patients had been collected while they were lying quietly in bed. The successes of linear systems analysis, however, penetrate now into clinical measurements.

As an example figure 2 shows the results of measuring a timespan from the carotid pulse, which is closely correlated to left ventricular ejection time (LVET). By plotting LVET as a function of heart rate, in this example two regression lines appear: one for patients of a specific group and one for normals. The two lines intersect in the region

[1] Institute of Medical Physics TNO, Utrecht, The Netherlands.

233

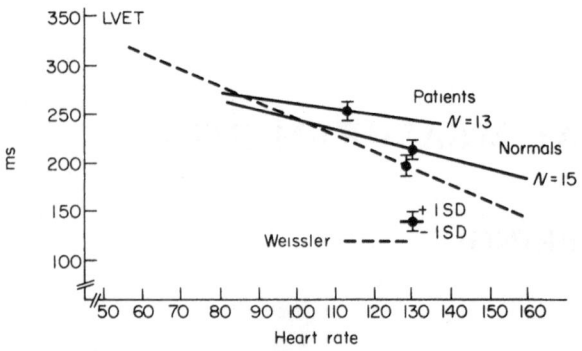

Figure 2.

of the heart rate at rest. This means that by measuring LVET at rest, no distinction can be made between normals and this group of patients. The difference between normals and patients becomes clear, however, if the person under investigation is under slight dynamic load so that his heart rate rises. This simple, non-invasive dynamic measurement is

under development by J. E. W. Beneken[2] and co-workers in collaboration with Dr J. Th. C. Vonk, Head of the Cardiology Department of the Universitg of Nijmegen.

In conclusion, taking the shift from clinical observation to clinical measurement for granted, there are three clearly discernable tendencies:

(i) To offer the physician the quantities he wants to be informed about, instead of quantities that can be measured on patients. Doing so will save the clinician elaborate calculations.

(ii) To measure everything that is possible on a patient—some reservations on this are imperative.

(iii) To make measurements on a patient under dynamic load conditions.

[2] Institute of Medical Physics TNO, 45, Da Costakade, Utrecht, The Netherlands.

SONAR

JOHN MacVICAR[1]

The value of ultrasound in medicine has been recognised only in the past four decades. It was first used as a therapeutic agent, since it can produce localised heat if applied to a specific area at a selected frequency and duration. The heating is due purely to the mechanical effects of the sound-waves. At a frequency of 800 kilohertz, the continuous ultrasonic beam will produce heat and can be focussed to have its effect within the body while not affecting the superficial structures. Tissue damage, however, can be caused and the British Council of Physical Medicine and Rehabilitation in 1952 recommended the use of no more than 800 kilohertz for 30 minutes, with an output of up to 25 watts per square centimetre for this type of ultrasonotherapy.

Actual tissue destruction caused by ultrasonic energy was demonstrated in 1944 on the skull and brains of cats, dogs, and monkeys, using continuous ultrasound at a frequency of 835 kilohertz for 10 to 15 minutes (Lynn and Putnam, 1944). Fry and his colleagues (Fry et al., 1950, 1951) did several experiments on the destruction of tissue by ultrasound. However, they were unable to elicit the precise cause of the cell damage, but thought that it was not entirely due to thermal effects. Though ultrasonotherapy is not now used extensively as a means of heat treatment, selective destruction of tissue, especially at depth, can be utilised for prefrontal leucotomy and in the treatment of Parkinsonism and Ménière's disease. Its use as a means of commissurectomy is currently being investigated.

DEVELOPMENT OF DIAGNOSIS BY SONAR

The use of sonar as an adjuvant to diagnosis was first described by Dussik (1942). He measured the transmission of ultrasound through the skull, from a source on one side, to the other side. He thus attempted to build up a picture of the structures within the skull which the sound had traversed, by recording the amount of ultrasonic energy transmitted through them. The picture obtained was determined by the amount of attenuation, reflection, and refraction to which the ultrasonic beam had been subjected as it passed through the different tissues and structures. The bone forming the vault of the skull absorbed or reflected the beam of sound in such a way as to make the results unsatisfactory, and this method of ultrasonic examination was discontinued.

Utilisation in diagnosis, of the reflected echoes of ultrasound from different tissues, was first suggested by French et al. (1950), who detected tumours deep in the brain by the sonic reflections obtained from them. The advantage that they enjoyed was that the skull vault had been removed prior to the examination, since the patients examined were in the post-mortem room. The reflected echoes were obtained because of the difference in the specific acoustic impedance of the different types of tissue, i.e. the difference between the normal brain tissue and the tumour tissue, and this remains the basis of all the diagnostic systems which have been developed. When a beam of ultrasound is directed through a structure, there is a reflected echo at each tissue plane and the greater the difference in the composition of the tissue the greater the specific acoustic impedance will be, and hence the greater the echo. The echoes will be more apparent if the ultrasonic beam is at right angles to the tissue plane. The extent of attenuation of the ultrasound within the tissue can also be made use of in diagnosis.

It was unfortunate that some of the earliest work was carried out in an attempt to differentiate between simple and cancerous lesions, particularly those in the female breast (Wild and Reid, 1952). This was attempting to make a diagnosis which often can only be made by microscopic examination after excision biopsy. The ultrasonic frequency used was 15 megahertz and echo patterns were poorly defined and only close to the surface. There seemed to be an increase in the number of echoes from cancerous lesions, and this was assumed to be due to the greater number of nuclei in malignant tissue as a result of mitosis. It is impossible to consider that the size of the ultrasonic beam was such that it would be influenced by the individual nuclei. Acoustic coupling between the ultrasonic transducer and the area to be examined was only achieved by immersion of the patient in a bath or water tank, which understandably proved a drawback.

The late firm of Kelvin and Hughes was producing in its Hillington factory in Glasgow an ultrasonic

[1] Department of Obstetrics, Queen Mother's Hospital, Glasgow, Scotland.

flaw detector for metals (figure 1). Professor Donald obtained one of these machines for experiment on human tissue. This machine used a quartz crystal for the generation of the ultrasonic beam. (Barium titanate crystals are now favoured in place of quartz.) The ultrasound produced was pulsed, rather than continuous. Using the piezoelectric effect, the echoes returning to the crystal were studied on a cathode-ray tube. The echoes were represented by

Figure 1. Kelvin-Hughes Flaw Detector, Mark IV, the first machine used in gynaecological diagnosis.

vertical deflections from a base-line and this is known as A-scope presentation. The available crystal frequencies were $1\frac{1}{2}$ and $2\frac{1}{2}$ megahertz. Since air is such a poor conductor of ultrasound, acoustic coupling between the transducer and the patient's skin was necessary. This was achieved by the liberal use of olive oil, which eradicated the need for the immersion of the patient. With no modifications, the flaw detector for metals gave promising results in being able to differentiate between solid and cystic tumours within the abdomen (figures 2 and 3, and Donald *et al.*, 1958). The level of ultrasonic energy being pulsed made the likelihood of tissue damage negligible and this was confirmed

by submitting the non-myelinated nerve tissue of newly born kittens' brains to prolonged exposure. Litters of rats *in utero* were treated similarly. No

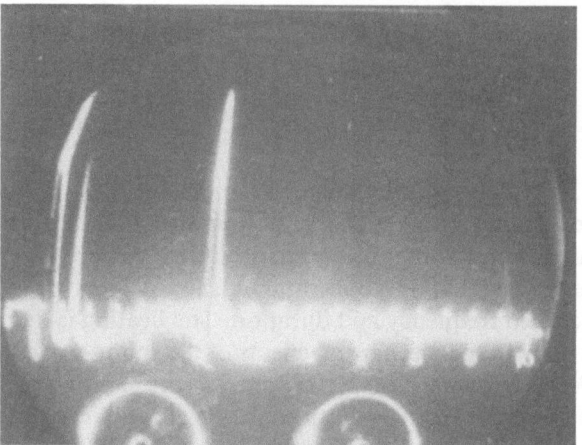

Figure 2. Ovarian cyst detected by Mark IV Flaw Detector ($2\frac{1}{2}$ m/cs). Both front and back wall reflecting strong echoes.

evidence of tissue damage was detected (MacVicar 1958).

It soon became apparent that changes in frequency and transmitter output brought about by adjusting the attenuator of the machine could assist in determining the density of a large mass under review, and that future developments had to embody easy interchange of different frequencies within the range of 1–5 megahertz. Above 5 megahertz, penetration is poor and is useless for anything but the most superficial structures. Below 1 megahertz, the system becomes too coarse, owing to poor resolution.

It was also obvious from the start that the type

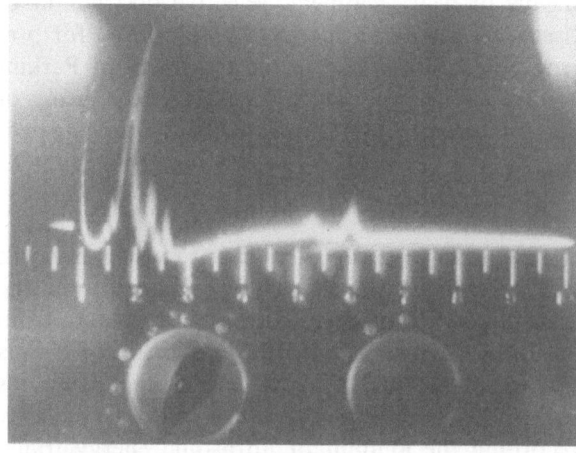

Figure 3. Fibroid of uterus as seen by Mark IV Flaw Detector. Front wall echo obvious but poor back wall echo due to attenuation of sound beam within the fibroid ($2\frac{1}{2}$ m/cs).

of display (A-scan) was inadequate and that comparison between specimens *in vivo* and *in vitro* was of little benefit, since the acoustic properties of tissue change when blood is not coursing through them. For the former, a two-dimensional examination of particular body planes was desirable, and this was accomplished by taking numerous A-scans in succession over the area to be investigated and

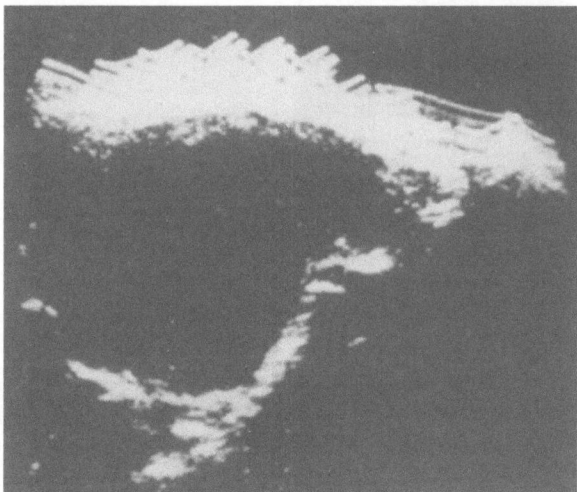

Figure 4. Longitudinal scan from umbilicus to symphysis pubis in a patient with an ovarian cyst. Note easily detected hind wall of cyst and clear space within it (2½ m/cs).

correlating these on one picture. This is what has become known as B-scan and P.P.I. Display. Almost all the available present-day machines give scope for both two-dimensional and A-scan examination. A-scans can be utilised for such work as the measurement of babies' heads *in utero* and the identification of mid-line shifts in the investigation of intra-cranial injury or neoplasm.

CLINICAL APPLICATIONS

With the availability of more machines and the increasing interest of clinicians, the development of ultrasonic techniques has been rapid and it is diffi-

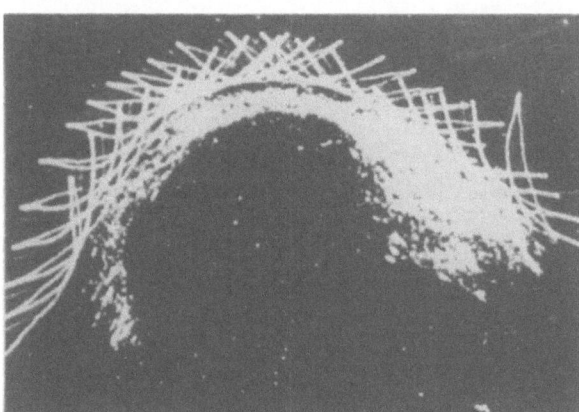

Figure 5. Transverse scan at level of umbilicus in a patient with a large fibroid of the uterus. The ultrasound is being completely attenuated within the fibroid and no rear wall is seen. The multiple echoes on the left of the picture are from gas within the bowel which has been displaced by the fibroid.

cult to credit that it is just in the past ten years that sonar has been widely accepted. The certain diagnosis of tumours within the abdomen, such as ovarian cysts (figure 4), fibroids of the uterus (figure 5), pancreatic cysts, mesenteric cysts, and metastatic deposits within the liver (figure 6), which could not be made by conventional radiological techniques, have proven valuable. Also, in the field of urology, kidney and bladder lesions can be well clarified when studied in conjunction with radiology (Barnett and Morley, 1971, and figures 7 and 8). Information of value in the assessment of the state

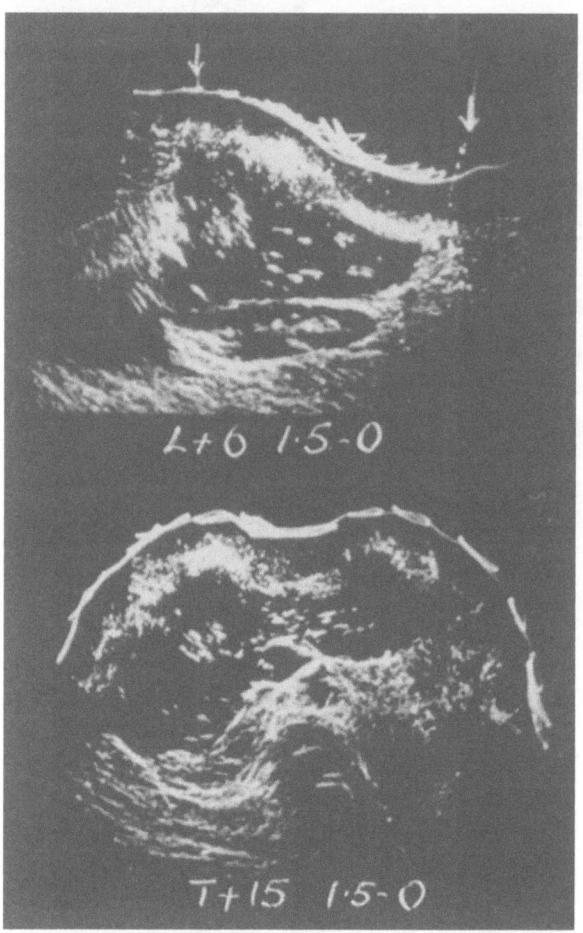

Figure 6. Longitudinal to the right and transverse section taken over the liver which shows multiple intra-hepatic echoes which are due to metastatic cancer deposits.

of the cardiac valves is also available from A-scan presentation and time-position scan. This work was started in 1957 by Effert and his colleagues (Effert *et al.*, 1957), but it is now considered one of the many necessary investigations in patients who may have significant valvular disease.

Diagnosis of intra-cranial lesions, whether neoplasm or injury, has been mainly by A-scope method (Jefferson, 1959; Taylor *et al.*, 1961) and, because of its relative simplicity, upsets the patients very little. This is especially of advantage since the medical condition of these patients may be perilous.

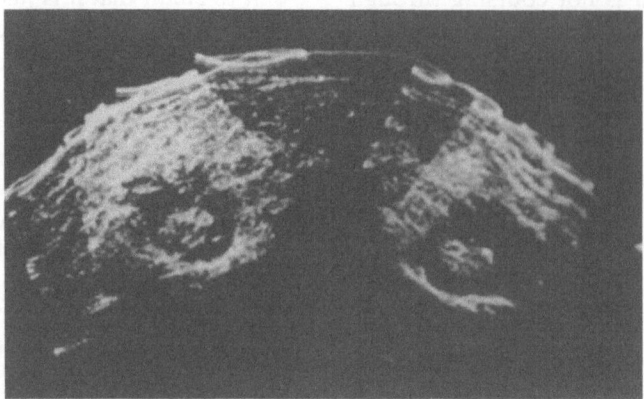

Figure 7. Transverse scan just above umbilicus which shows
both kidneys at least of normal dimensions.

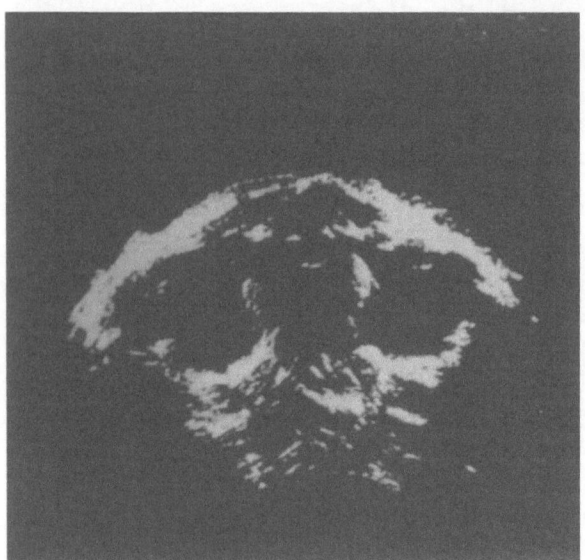

Figure 8. Transverse scan above umbilicus showing two
large 'cystic' areas in the region of the kidneys. This
appearance is due to a bilateral hydronephrosis.

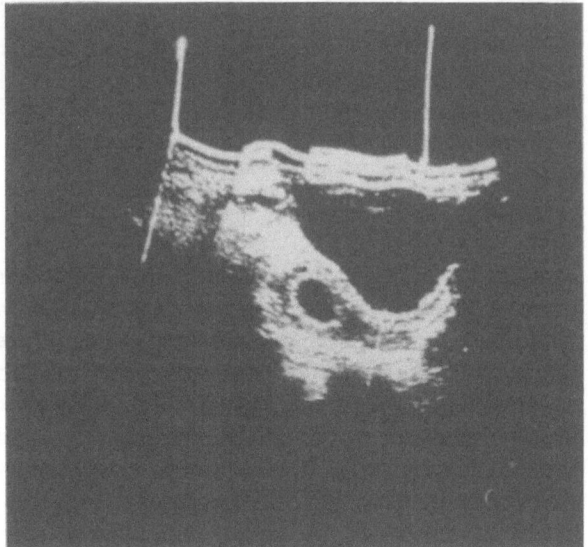

Figure 9. Longitudinal scan from umbilicus to symphysis
pubis showing enlarged uterus containing a pregnancy (7-
weeks' size). The 'cystic' space in front of the uterus is the
bladder filled by urine to displace the bowel and thus
allow the sound to penetrate into the pelvis.

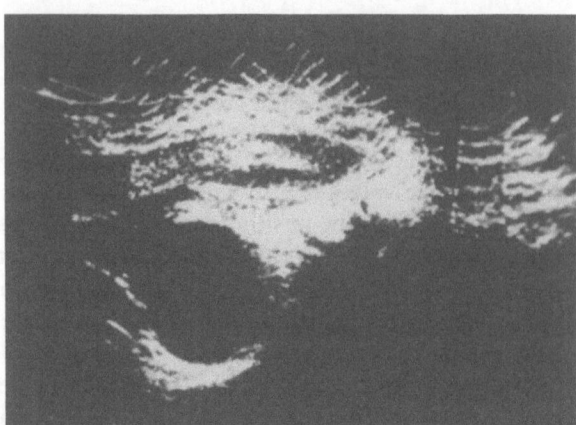

Figure 10. Longitudinal scan from umbilicus to symphysis
pubis showing a pregnant uterus (10 weeks) displaced
forward by a cystic mass which proved to be an ovarian
cyst.

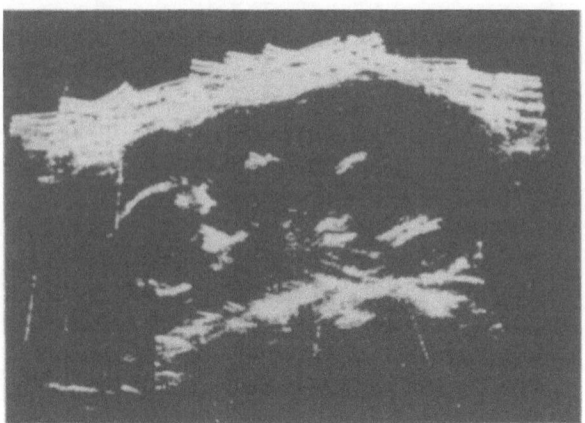

Figure 11. Longitudinal scan from xiphisternum to symphysis pubis showing a 34-week pregnancy with the foetal
head presenting.

In the field of obstetrics, possibly the greatest contribution has been made, primarily owing to the work of Professor Donald. Because of the radiation hazard associated with X-rays, radiology can only play a minor role in obstetrics, but with sonar the growing uterus and foetus can be studied repeatedly from two to three weeks from conception until delivery (figures 9, 10, and 11). Early evaluation of the developing embryo is important in patients whose histories suggest maldevelopment with result-ant unsuccessful outcome. Later in pregnancy, the regular measurement of the size of the baby's head, and its growth, is one of the indications of adequate foetal well-being and if no growth occurs it indicates

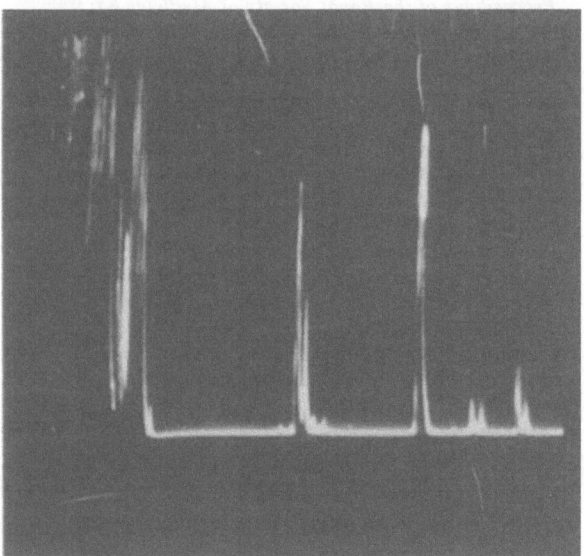

Figure 12. A-scan presentation to measure the bi-parietal diameter. The initial strong echo is from the one side of the skull, the next smaller echo is from mid-line structures and the final strong echo is from the far side of the skull.

that delivery is necessary for foetal survival (Willocks *et al.*, 1967; Willocks and Dunsmore, 1971; and figure 12).

Localisation of the site of the placenta can also be of clinical importance and can be determined as early as 14 weeks, whereas special radiographic techniques are necessary before 34 weeks. The diagnosis of multiple pregnancy can obviously be made by sonar much earlier than the foetal bony skeleton is visible on X-rays. This was carried to excess in the recent case of quins diagnosed in London at 8 weeks (Campbell and Dewhurst, 1970).

Sonar is also used in obstetrics to detect the foetal heart. Alteration in rate can be significant, especially if the foetus is under stress, and several instruments are now available to record the movement of the foetal heart muscle and consequently the foetal heartbeat (Bernstine and Callagan, 1966). The basic principle of these machines is the use of the Doppler effect. Most listeners cannot audibly detect the

foetal heart using the normal stethoscope till about 24 weeks of pregnancy, but with sonar the heart-beat can be picked up as early as 12 weeks. The only drawback of these instruments is that their output of ultrasonic energy is continuous and not pulsed. They therefore deliver to the tissues relatively larger doses of ultrasonic energy than other types of ultra-sonic apparatus, all of which utilise pulsed ultra-sound, though their actual pulses are of higher peak amplitude. Continuous recording of the foetal heart rate in association with uterine contractions is now possible with the FM2 apparatus (Sonicaid) which incorporates a sonar foetal-heart detector and a Smythe-type tocograph. Studies of foetal-heart patterns are possible throughout labour, and can be of value to the clinician on the look-out for foetal distress.

SAFETY OF ULTRASOUND

It has been the practice of most workers to investigate any possible harmful effects which could be attributed to examination by sonar, especially in pregnancy, but no untoward effects have so far been found (Bernstine, 1969). Three centres, in Sweden, the United States, and Glasgow, pooled data concerning the outcome of many early preg-nancies examined by sonar and proved that there was no increase in either the numbers of spontan-eous abortions or in foetal abnormalities (Hellman *et al.*, 1970). A preliminary communication from South Africa in 1970 suggested that certain chromo-somal aberrations could be induced by an ultra-sonic foetal pulse detector (MacIntosh and Davy, 1970), but these have not been substantiated by other workers (Boyd *et al.*, 1971; Bobrow *et al.*, 1971; Watts *et al.*, 1972), and it is stressed that even with high doses of ultrasonic irradiation the chromosomal effect is, if present at all, much less than with X-radiation. Anyone must be mindful of the possible harmful effects of any new apparatus and further investigation into the effects of in-sonation of tissue cultures, especially of placental tissue, is now under way in the Department of Ultrasonics in the Queen Mother's Hospital, University of Glasgow. The benefits, however, which sonar bestows far outweigh any present imaginable risk.

THE FUTURE

Having progressed from uni-dimensional to two-dimensional display, it is obvious that three-dimensional presentation would probably be of more value. Reviews of errors in diagnosis often indicate lack of over-all comprehension, which three-dimensional display might have given. Achiev-ing three-dimensional presentation has, however, been unsuccessful so far, but various lines of investigation are at present under way, with techniques such as the vari-focal oscilloscope and holography (Donald, 1971).

Improvements in the present equipment will gradually take place. Better resolution in azimuth would be of assistance in clarifying the pictures. Methods of presentation may also be improved. With increasing clinical use the potential clinical value will undoubtedly be further realised and much of this must be accomplished by the types of machine at present available. It would be of little value to achieve a method of improved presentation if the cost of such apparatus were to make its use too limited, or if it were to involve discomfort to the patient. In the development of the present machine in use in the Department of Ultrasonics, University of Glasgow (The Diasonograph—Nuclear Enterprises Ltd.) automatic scanning rather than manual use of the probe was tried. However this gave inferior results because a manual operator can see the echoes obtained as he manipulates the transducer across the area under examination, and can thus obtain a result from a build-up of sector scans.

Various projects in progress in the Ultrasonic Department in the Queen Mother's Hospital include improvement in the methods of foetal cephalometry, to make assessment simpler, quicker, and more accurate. Similarly, tomographical or constant-depth scanning may be of value in viewing intra-abdominal structures at different planes. Also, using Doppler systems there is much scope for the measurement of blood flow through blood vessels, which may help with such assessments as cardiac output and the state of the vessels. Combination of Doppler and B-scan systems must also be considered.

Development of sonar has illustrated the advantages to be gained by the combination of the knowledge of the scientist, industry, and the medical clinician. Further development depends on the interest of all of these people.

REFERENCES

BARNETT, E., and MORLEY, P. 1971: Ultrasound in the investigation of space occupying lesions of the urinary tract. *British Journal of Radiology* **44**: 733.

BERNSTINE, R. L., and CALLAGAN, D. A. 1966: Ultrasonic doppler inspection of the fetal heart. *American Journal of Obstetrics and Gynecology* **95**: 1001.

BERNSTINE, R. L. 1969: Safety studies with ultrasonic doppler technique. *Obstetrics and Gynecology* **34**: 707.

BOBROW, M., BLACKWELL, N., UNRAV, A. E., and BLEANEY, B. 1971: Absence of any observed effec of ultrasonic irradiation on human chromosomes. *Journal of Obstetrics and Gynaecology of the British Commonwealth* **78**: 730.

BOYD, E., ABDULLA, U., DONALD, I., FLEMING, J. E. E., HALL, A. J., and FERGUSON-SMITH, M. A. 1971: Chromosome breakage and ultrasound. *British Medical Journal* **2**: 501.

CAMPBELL, S., and DEWHURST, C. J. 1970: Quintuplet pregnancy diagnosed and assessed by ultrasonic compound scanning. *Lancet* **1**: 101.

DONALD, I., MACVICAR, J., and BROWN, T. G. 1958: Investigation of abdominal masses by pulsed ultrasound. *Lancet* **1**: 1188.

DONALD, I. 1971: Sonar—further scope and prospects. *Proceedings of the Royal Society of Medicine* **64**: 991.

DUSSICK, K. T. 1942: Uber die Moglichkeit Lochfrequente Mechanische Schwingungen ab Diagnostiches Hiffsmittel zu Verwierten. *Z. Neurol. Psychiat.* **174**: 153.

EFFERT, S., ERKENS, H., and GROSSE-BROCKHOFFE 1957: Uber die Anwerdung des Ultraschall-Echoverfahrens in der Herzdiagnostik. *Dtsch. Med. Wschr.* **82**: 1253.

FRENCH, L. A., WILD, J. J., and NEAL, D. 1950: Detection of cerebral tumours by ultrasonic pulses: pilot studies in post-mortem material. *Cancer* **3**: 705.

FRY, W. J., WULFF, V. J., TUCKER, D., and FRY, F. J. 1950: *Journal of Acoustical Society of America* **22**: 867.

FRY, W. J., TUCKER, D., FRY, F. J., and WULFF, V. J. 1951: *Journal of Acoustical Society of America* **23**: 364.

HELLMAN, L. M., DUFFUS, G. M., DONALD, I., and SUNDEN, B. 1970: Safety of diagnostic ultrasound in obstetrics. *Lancet* **1**: 1133.

JEFFERSON, A. 1959: Some experiences with echoencephalography. *Journal of Neurology and Psychiatry* **22**: 83.

LYNN, J. G., and PUTNAM, T. J. 1944: *American Journal of Pathology* **20**: 637.

MacINTOSH, I. J. C., and DAVEY, D. A. 1970: Chromosomal aberrations induced by an ultrasonic foetal pulse detector. *British Medical Journal* **4**: 92.

MacVICAR, J. 1958: 'Diagnosis of Intra-Abdominal Masses by Pulsed Ultrasound'. M.D. Thesis, University of Glasgow.

TAYLOR, J. C., NEWELL, J. A., and KARVOUNIS, P. 1961: Ultrasonics in the diagnosis of intra-cranial space-occupying lesions. *Lancet* **1**: 1197.

WATTS, P. L., HALL, A. J., and FLEMING, J. E. E. 1972: Ultrasound and chromosome damage. In press.

WILD, J. J., and REID, J. M. 1952: Further pilot echographic studies on histological structures of tumours of the living intact breast. *Am. J. of Path.* **28**: 839.

WILLOCKS, J., DONALD, I., CAMPBELL, S., and DUNSMORE, I. R. 1967: Intrauterine growth assessed by foetal cephalometry. *Journal of Obstetrics and Gynaecology of the British Commonwealth* **74**: 637.

WILLOCKS, J., and DUNSMORE, I. R. 1971: Assessment of gestational age and prediction of dysmaturity by ultrasonic foetal cephalometry. *Journal of Obstetrics and Gynaecology of the British Commonwealth* **78**: 804.

KINETOGRAPHY AND ITS APPLICATION IN HEART STUDIES

PART 1: REVIEW OF KINETO-CARDIOGRAPHY

W. H. BAIN[1]

For many hundreds of years, physicians have gained information on cardiac function by feeling movements of the heart through the chest wall. They learned to distinguish the barely palpable stirring of the normal heart from the hyperdynamic action which accompanies increased output; the powerful sustained thrust of the hypertrophied heart which is working against a raised resistance; and the labouring heave of the failing myocardium, which has exhausted its powers of compensation and is progressively dilating. Indeed, prior to the development of electrocardiography at the beginning of this century, and catheterisation of the heart thirty years later, palpation of the precordial impulse, inspection of the jugular veins, and auscultation of the heart sounds completed the cardiologist's diagnostic equipment.

In 1861, a means was devised of recording the cardiac impulse (Chaveau, 1861) and subsequent refinements of this method could be regarded as efforts to replace the art of palpation by the science of kinetocardiography (Eddleman et al., 1953). The science has grown slowly, however, because of a lack of absolute measurements of heart action with which to compare the records obtained by the various transducers placed on the chest wall.

When precise measurements of the electrical and barometric events accompanying each cardiac cycle did become easily available, these measurements seemed so much more directly relevant as guides to diagnosis and therapy that analysis of the precordial impulse became once more a bed-side technique.

In recent years however, there has been a resurgence of interest in kinetocardiography. There have been three reasons for this. Firstly, bioengineers have broken through the communication barrier between the engineer and physicist on the one hand, and the clinician on the other, to make available new methods and to suggest clinical applications for old ones. Secondly, there is an increasing awareness among cardiologists of the need to develop direct measurements of myocardial action—as distinct from the consequences of heart action, namely blood flow and pressure. Rheology and haemo-

dynamics have advanced rapidly in the past two decades, but methods and equipment which will give direct information about the integrity of the heart muscle itself have lagged behind and are only now being exploited (Voigt, 1970). Thirdly, there is the natural desire of the clinician to exploit non-invasive diagnostic methods and to search for techniques which will yield precise measurements without the need for intra-venous or intra-arterial intubation, the injection of isotopes, or the flooding of the heart with radio-opaque contrast media.

During the 1960s, the application of angiocardiography to estimation of L.V. volumes (Dodge et al., 1960) and the systolic ejection fraction (Miller et al., 1965) allowed more precise calibration of kinetographic techniques, and more refined methods were developed to assess ventricular function through the intact chest wall (Coulshed and Epstein, 1963; Dimond et al., 1966; Sutton et al., 1970).

Previous workers had shown that the first derivative of the L.V. pressure pulse ($\delta p/\delta t$) bore some resemblance to the external apical records (Agress and Fields, 1959), and in 1968 Gleichmann et al. put these observations on a firmer footing by analysing the first derivative of the apex cardiogram and demonstrating a relationship between $\delta p/\delta t$ and $\delta F/\delta t$ (Gleichmann et al., 1968).

Early workers frequently emphasised that the KCG was of special diagnostic value in assessing cardiac function after a heart attack (Suh and Eddleman, 1959). The characteristic change in the kinetocardiogram as a result of a myocardial infarction is the appearance of a paradoxical systolic outward movement. In a study of 102 patients with both acute and old myocardial infarctions, such paradoxic outward movements (of a duration of 0·2 seconds or longer) were demonstrable in 68 per cent (Eddleman, 1965).

Subsequent investigators have compared the kinetocardiogram with cine-ventriculography for the detection of areas of myocardial asynergy (Lane et al., 1968; McGinn and Lyon, 1968), and in 1971 Gorlin's group in Boston showed that records of the precordial impulse in patients with coronary occlusive disease could give valuable information

[1] Cardiac Surgery Unit, Royal Infirmary, Glasgow, Scotland.

241

concerning L.V. function without resort to invasive techniques (Cohn *et al.*, 1971).

Coronary heart disease is currently engaging a great deal of attention. It has become by far the commonest cause of death in the western world. Among the population at risk (aged 40 to 69 years) the incidence of heart attacks is 15 per 1000 per annum. This means 30 heart attacks per day in the region of Scotland served by the Royal Infirmary, and one third of these are fatal.

Special coronary care units are being established in most large hospitals and many new therapeutic approaches are being explored in an effort to reduce the toll from this disease.

Since coronary occlusive disease produces most of its morbidity and mortality by its effects on the ventricular myocardium, accurate assessment of ventricular function is an essential guide to management and choice of therapy in individual patients.

There is no doubt that the development of a non-invasive technique which would yield information about L.V. function in the patient soon after a heart attack would be of immense value.

Accordingly, a group of bioengineers and clinicians in Glasgow has turned anew to study the low-frequency movements of the precordium. Their particular objective is to develop an instrument which will detect areas of myocardial asynergy both through the intact chest and over the exposed heart in the course of coronary artery surgery. The instrument we have applied to this work is the Displacement Cardiograph (Vas, 1967) which is described in the paper. The paper will also report our basic physiological studies with the instrument in the animal laboratory. We have high hopes that this collaborative study will yield a diagnostic tool of great value in the management of patients with heart disease.

PART 2: THE MAGNETO-KINETOGRAPH AND ITS CHARACTERISTICS

T. G. GRASSIE, T. R. FENTON AND R. VAS[1]

At the University of Strathclyde, studies in the field of kinetography extend into various areas, and the instrument at present used for *in vivo* measurement is one modelled on the lines of the displacement cardio-

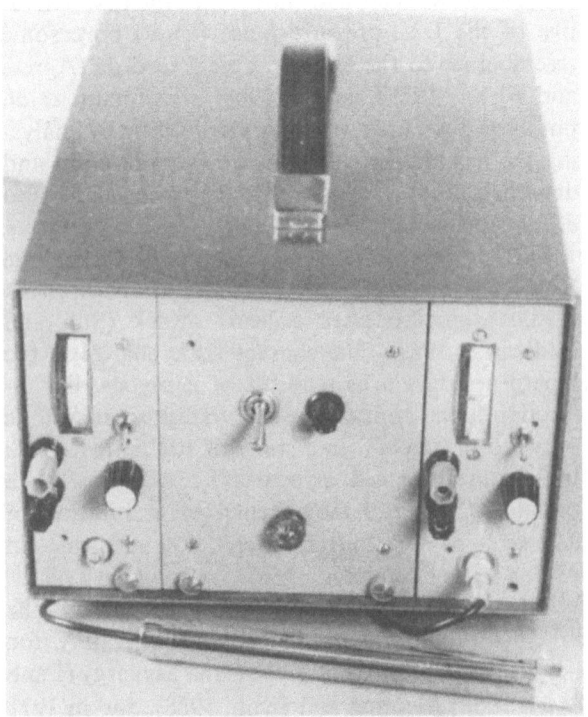

Figure 1. A two-channel magneto-kinetograph.

[1] University of Strathclyde, Glasgow, Scotland.

graph (DCG) first described by Vas (1967). It is an electronic instrument which incorporates a transducer consisting of an inductor wound in the form of a planar spiral coil, and an associated oscillator. The transducer is usually fixed at a convenient distance from the part of the body under examination. Figure 1 illustrates the transducer and its connection to the instrument. Body motion is 'sensed' by the planar coil and is presented at the instrument output as variation in voltage level.

The phenomenon exploited is the dependence of the electrical characteristics of an energised coil on movements of tissue within the field of the coil. The value of the effective complex impedance seen at the coil terminals varies with the electrical properties of the coil environment and with the physical geometry of the environment. At the frequencies of interest it is found that the equivalent inductive reactance increases as the gap between coil and tissue decreases. Since the phenomenon is largely a magnetic field effect, the term magneto-kinetography has been employed. For the dual system dependent upon electric field variations between noncontacting plate conductors, the term electrokinetography would be used.

THE INSTRUMENT

The essence of the signal-recovery system may be explained with reference to the system block diagram (figure 2). The sensing coil is one of the frequency-determining elements of the transduction oscillator, so that the oscillation frequency f_s varies with movement of tissue close to the coil. Typically, with tissue, variations of tens of kilohertz are obtained about a mean frequency of tens of megahertz. The frequency f_s is compared with f_0, the frequency of a reference oscillator inside the instrument.

network of the 'automatic' mode is by-passed and the frequency of the reference oscillator is fixed by manual adjustment of a potentiometer. This mode is commonly used for system calibration.

RESPONSE CHARACTERISTICS

The system has been designed to have a signal bandwidth from d.c. to 150 Hz (3dB). Range and

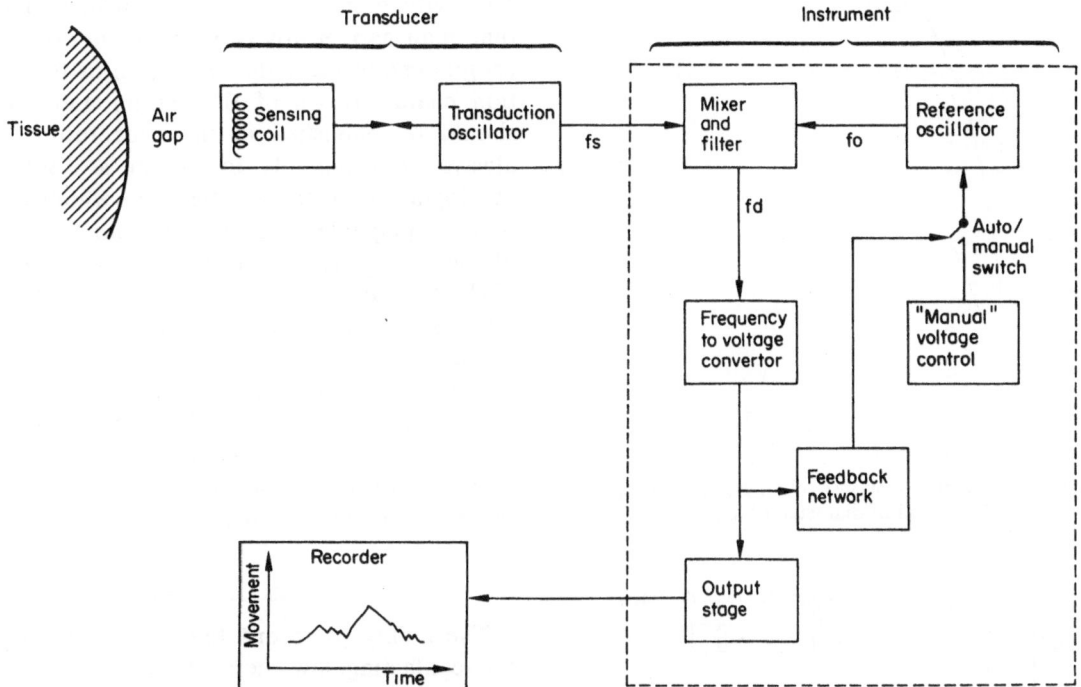

Figure 2. System block diagram.

The output from the frequency comparator (mixer) contains a frequency f_d, being the difference of the two frequencies f_s and f_0 ($f_d = f_s - f_0$). This difference frequency is fed to a frequency-to-voltage converter. The converter output voltage is then amplified and appears as the instrument output. Base-line control of this output is achieved by varying f_0, the frequency of the reference oscillator. In this way the value of f_d is altered and this in turn changes the mean output level of the converter. The reference oscillator is voltage-controlled either automatically through a novel low-pass feedback network or manually.

In 'automatic' mode, the converted output is fed back to the reference oscillator through an active network which has a time-constant (10 s). This effectively compensates for variations which would occur in the mean level of the converter output due to variations in mean position of the tissue relative to the probe. This mode is most useful in eliminating the undesirable frequency shifts induced when the transducer is first applied to the measurement region. For this reason, the instrument is usually used in this mode. In 'manual' mode, the feedback

collimation characteristics follow patterns typified by the curves shown in figures 3 and 4. The 'lateral response characteristic' illustrates the focal nature of the transducer (figure 3). It is seen that for a range of air-gaps, a high degree of area selectivity is achieved. The 'air gap response characteristic' illustrates both the non-linearity and the frequency dependence of the transducer response (figure 4). For small segments on a smooth curve, slope is frequently assumed to be constant, so for movements that are small relative to the sensing range of the transducer, the characteristic may be considered quasilinear. This is usually the case in present clinical applications. The graph shows that sensitivity increases with frequency.

Sensing-coil diameters ranging from a few millimetres to 10 centimetres have been used. The smaller coils respond to movements of small areas, while the larger coils are used to obtain an integrated response over large regions. The wire should have as low a resistance as possible commensurate with practical limitations imposed by coil diameter and the inductance required.

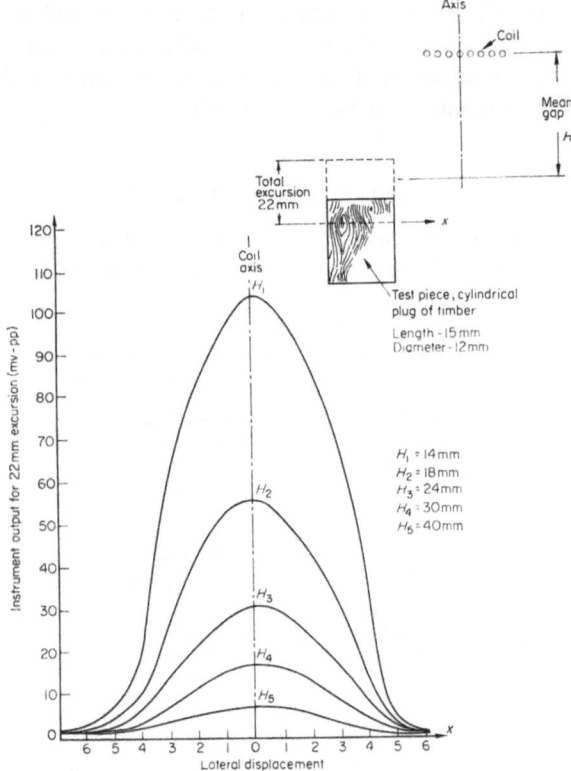

Figure 3. Lateral response characteristic using planar coil of diameter 10 cm.

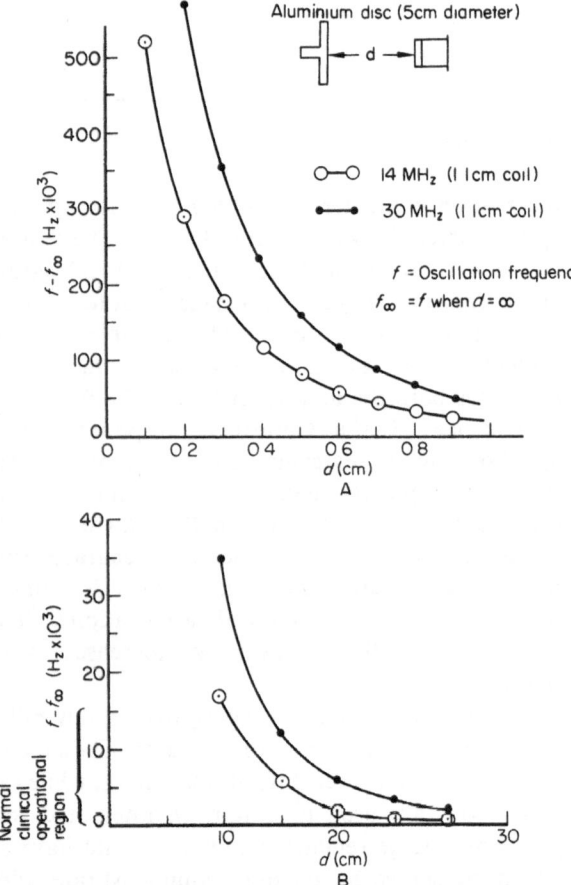

Figure 4. Frequency change of a coil against distance from an aluminium disc. Signal recovery system) sensitivity to frequency change (100mV/kHz).

DISCUSSION

Until recently, the output of the DCG was regarded as being due to movement of the tissue surface. It is certainly true that the instrument is highly sensitive to movement of this surface and it must be accepted that this is the dominant influence. However, there is considerable penetration of the field of the coil into tissue and the instrument does respond to 'in-depth' movements, even though the sensitivity to these movements is substantially below that of air-gap sensitivity. Our present experimental programme includes determination of the penetration characteristics of the system and this data should be published in the near future. It has always been possible to obtain a displacement cardiogram even from patients in whom the apex was not palpable, and on whom no linear microphone apex cardiogram could be obtained. It is likely that this success is due to the 'in-depth' response of the system. However, in clinical investigations described in part 3, the instrument is generally used in a conventional arrangement with the transduction coil separated from the tissue surface by an air-gap. The arguments are developed on the basis of the dominant sensitivity being that to tissue surface movements.

PERSPECTIVES

The ability to record tissue movements with a non-contacting, low-energy device leads to a variety of applications in clinical and physiological research, and to the development of diagnostic tools. Two particular areas of work are described in detail in this symposium: these are heart studies on, firstly, thoracotomised dogs, and secondly, normal human subjects and those affected by myocardial infarction. Some additional research areas in which active interest exists and some potential applications are briefly presented here.

In the cardiological investigations, measurement extends from direct apex movements to movement over the entire torso, the latter being obtained by position scanning of the transducer. Analysis of the kinetograms obtained (Valero, 1967; Valero and Vas, 1969) indicates clear variations due to such factors as bulges, aneurisms, and paradoxical movements. Work on neonates is also proceeding and here the instrument automatic base-line control described earlier is found to be an essential factor in the successful recording of heart movement. The kinetograms are being examined for data on the changes which occur in heart action during the neonatal period, and also for information on the myocardial condition of the newborn infant (Vas et al., in preparation).

Cardiokinetograms have been successfully obtained from inside the body. For this work a miniature transducer has been designed suitable for swallowing. The transducer is positioned in the oesophagus close to the left atrium and it is hoped that analysis of the resulting kinetograms will yield information on atrial action and, in particular, on *a*-wave variations in cases of myocardial ischaemia.

In the field of obstetrics, kinetograms have been obtained from transducers in a range of abdominal sites and from one designed for vaginal insertion.

These kinetograms are being examined for information on placental siting and foetal heart activity. Early work in this area has been reported by Sharf *et al.* (1970).

The sensitivity of the system to movements is probably best demonstrated by its application in vascular investigations in which, for example, movements of such small arteries as the *dorsalis pedis* artery may be monitored. Differences due to occlusion and stenosis may be identified.

PART 3: SOME FACTORS AFFECTING HEART MOTION AND ITS MEASUREMENT

R. VAS, T. R. FENTON[1], F. THEVOZ AND W. H. BAIN[2]

The value of any analysis of heart motion lies in its ability not only to detect but to explain abnormal action. Present techniques and understanding enable clinicians to measure apical movements and to interpret these movements in terms of the mechanical events of the heart (Tavel *et al.*, 1965; Kumar and Spodick, 1970). However, understanding of atypical movements is limited to a few specific conditions (Voigt and Friesinger, 1970; Ingue *et al.*, 1971). This lack of detailed information about heart motion and its measurement, and especially about abnormal motion, is perhaps the main reason for a lack of enthusiasm for kinetocardiography among clinicians.

A principal difficulty in the diagnostic use of heart movements lies in attributing a cause to any degree of abnormality. In general, observed changes from normal recordings can be due to one or both of two causes:

1 physiological influences, for which variable boundary conditions and intrinsic motion are prime factors, and
2 technical influences, including variable measuring conditions and interference by the measuring system itself.

Variable boundary conditions are associated with changes in the physical structure and compliance of the mediastinum, vascular system, etc. Variance in these factors would tend to add constraint to the movements of the heart, thereby altering its absolute motion. The intrinsic factors are those variations in the motion of various parts of the heart relative to themselves, due to different stimulation patterns, adequacy, blood supply, etc.

The object of this study was to investigate whether or not certain factors are able to alter heart motion or its measurement. In this way, if a variance in a factor is shown not to influence heart motion, any atypical heart motion cannot be ascribed to variability in this factor. From such a study, the probable causes of abnormal motion are clarified, which assists in proper 'cause and effect' analyses.

Four potential influences on heart motion were studied, viz.:

1 intraventricular catheterisation,
2 structure and depth of various tissues of the chest wall,
3 thoracotomy and pericardiotomy, and
4 position and orientation of the measuring device relative to the heart.

Generally, it was assumed that a particular factor was able to influence heart motion. If no change in motion was recorded, this hypothesis was rejected. If significant changes were found, it was not always possible to ascribe these changes to any particular source. Additionally, it was assumed that all recordings represent motion of the heart. This was a necessary assumption because of the nature of the transducer used (see part 2 of this paper). Although this question is at present under study, comparison

---

[1] University of Strathclyde, Glasgow, Scotland.
[2] Cardiac Surgery Unit, Royal Infirmary, Glasgow, Scotland.

of human apex recordings made with this transducer, with those made with the apex cardiograph, have shown this assumption to be reasonable.

MATERIALS AND METHODS

Seventeen dogs of various breeds and weighing from 14 to 29 kg were used in these experiments. They were anaesthetised with nitrous oxide and fluorothane, preceded by an injection of pentothal. Ventilation was controlled with a Starling pump and the minute volume was adjusted to maintain normal blood-gas figures. The measuring device used in this work was an electromagnetic transducer (displacement cardiogram DCG) which measures movement in a totally non-contactive manner in either the open or closed chest (see part 2 of this paper).

Electrocardiograms (lead 11 in all recordings) and phasic pressures in left and right ventricles and descending aorta (end aperture, number 9 French gauge catheters) were recorded simultaneously with heart movement on an eight-track recorder (Elema Schonander Mingograph 81). An image intensifier was used to locate and mark the position of the apex on the intact torso.

The dogs were secured supine on a specially

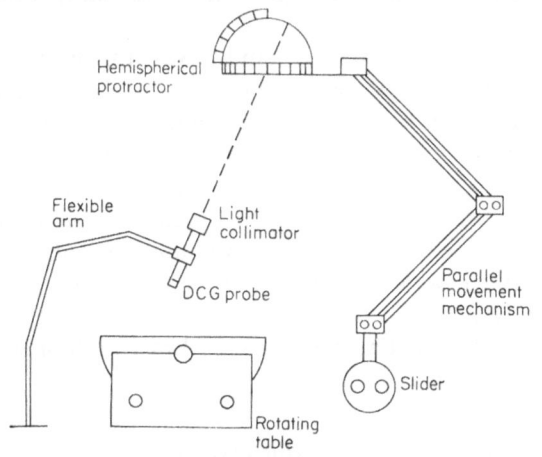

Figure 5. Experimental apparatus.

designed table (see figure 5) which could be rotated through 180 degrees, thereby changing the dog's gravitational axis. The DCG probe was located on a flexible stainless-steel arm which could be fixed in any position. An optical system was developed to measure the angle of the probe in space. A collimator directs a pencil of light along the axis of the probe to the centre of the hemisphere. The brim of the hemisphere is covered with an opaque plate which admits the light beam through a small hole in its centre. The light then illuminates a small area of the translucent dome of the hemisphere. One protractor, which rotates about the central axis of the dome, measures the angle of the probe to the z axis (see figure 6) while the brim protractor measures the angle of the projection of the probe in the xy plane to the y axis. The double parallel and slider mechanism allows the dome to translate in three dimensions, without rotation.

The usual procedure for each situation was to locate the apex region, with an image intensifier in the closed chest, and by direct observation in the open chest. The DCG probe was focused on this region and a number of recordings were made for each situation during maintained expiration, with the probe in various orientations.

A set of initial measurements were taken after anaesthesia. Catheters were then introduced so that their tips were located in the centre of each ventricle and precordial skin, fat, and muscle (exclusive of intercostal muscle) were removed. Then left and right pneumothoraxes were performed, followed by a bilateral thoracotomy to expose the heart within the pericardium. The dog's acid-base balance and blood-gas levels were periodically checked, and were maintained as close to normal as possible. The horizontal position of the table was used for each case.

RESULTS

Despite a significant variation in motion from dog to dog, the same forms of change in heart

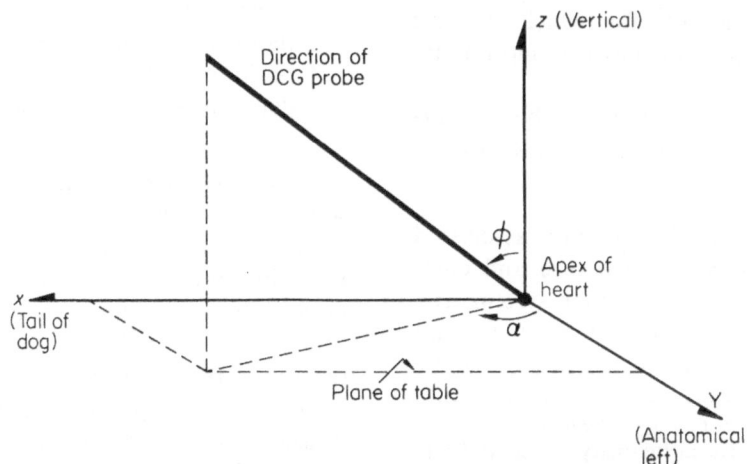

Figure 6. Coordinate system and frame of reference. Orientation of the DCG probe relative to the apex.

motion were found in all dogs. In each example, any change in recordings is expressed relative to that incurred due to repositioning of the probe. No attempt was made to quantify any changes in recordings, so all results remain qualitative in nature and specific to the DCG transducer.

The influence of intraventricular catheterisation of particular interest since catheterisation is a common technique in many animal studies. Figure 7 shows that the presence of catheters in the right and left ventricle had some influence on the motion recorded. This would imply that only in a qualitative nature is it valid to measure the motion of a catheterised heart.

Clearly, intrinsic heart motion did not change during this procedure, yet a family of DCG records was obtained. Since only the orientation of the measuring probe was varied, this indicates that because apex motion is three-dimensional, care must be taken to define which component of this motion is recorded. Otherwise erroneous conclusions may be drawn.

Finally, attention is drawn to the effect of length of an experiment (see figure 12). Significant changes in the recordings with the passage of time are apparent, and because measuring conditions and the boundary conditions of the heart were believed to be constant, these changes could be attributed to physiological

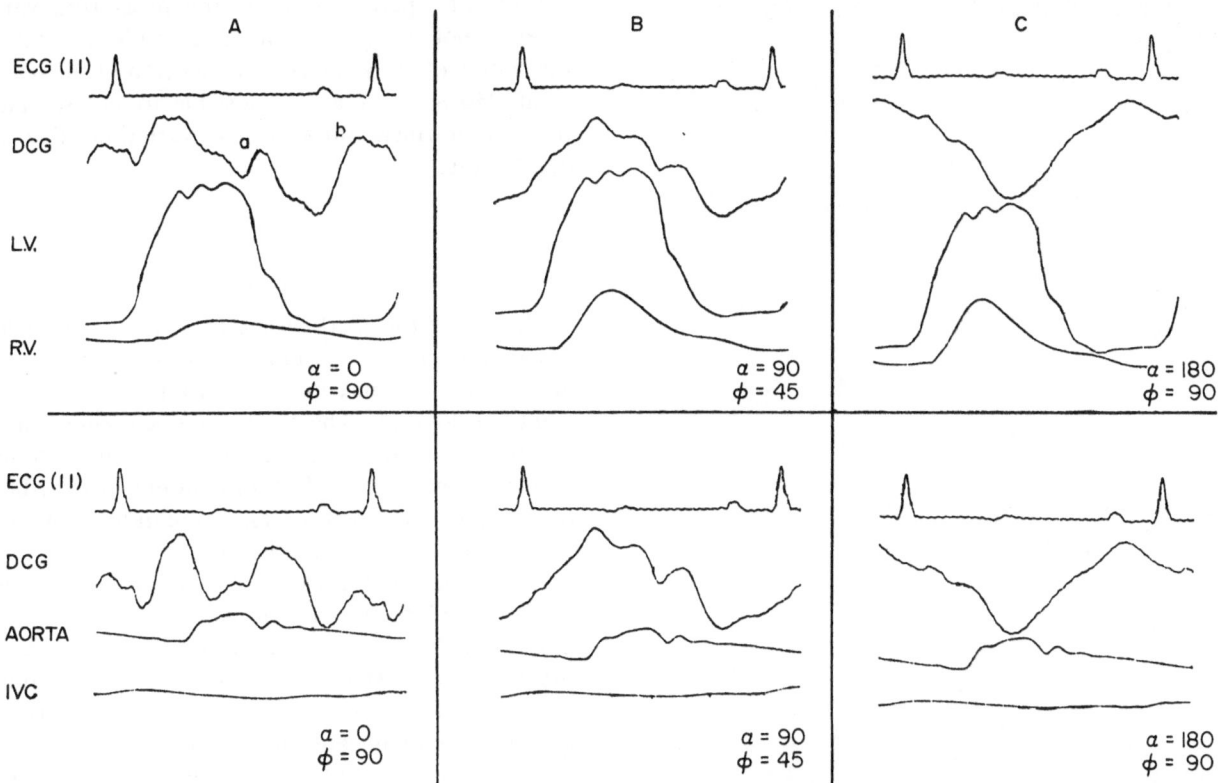

Figure 7. Influence of catheterisation. Upper figures were taken with catheters in the left and right ventricles, lower figures were recorded after withdrawal of these catheters.

As shown in figure 8, absence of precordial skin, fat, and muscle (exclusive of intercostal muscle) have only a slight influence on the motion recorded. This indicates that variations in the structure or depth of these tissues do not affect heart motion or its measurement.

Figure 9 demonstrates the marked changes that follow thoracotomy. A similar degree of variation was found after pericardiotomy (see figure 10). However, these changes were accompanied by gross changes in measuring conditions, as the heart was deprived of its immediate anatomical relations and adopted a new position. The implications of this are left for discussion.

Figure 11 shows a complete spatial history of the apex motion of one dog, taken from 15 orientations.

deterioration resulting from the experimental situation.

DISCUSSION

As previously stated, these experiments were designed to reject certain hypotheses, and to establish that certain factors do not influence heart motion or its measurement. However, in situations where all but one of the considered variables were held constant, it was possible to ascribe a source to any change. This, in fact, was the situation with the precordial tissues studied. The only change was the removal of the tissue, and this was seen to induce small changes in the motion recordings. In a qualitative sense, these influences are of the same order as those incurred by repositioning the probe (i.e. measurement errors). Therefore it would seem that these tissues do not influence heart motion or this method of recording it. The changes produced by

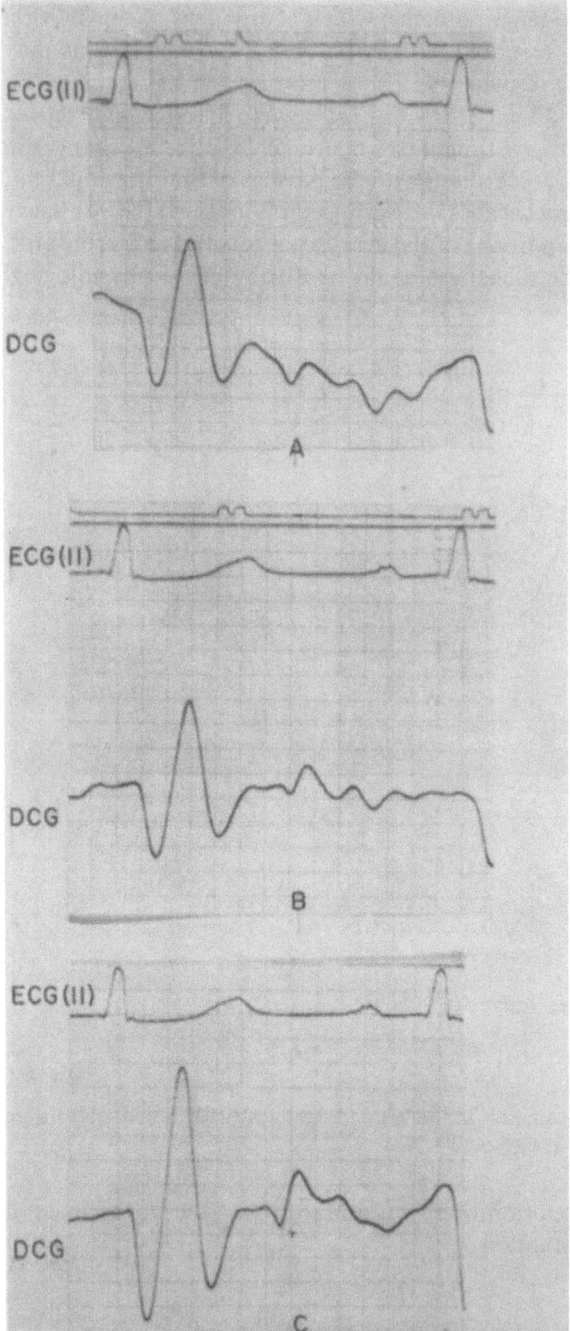

Figure 8. Influence of chest wall tissue (DCG probe in vertical position). (A) Intact dog (just prior to surgery). (B) Without skin and fat. (C) Without precordial skin, fat, and muscle (exclusive of intercostal muscle).

intra-ventricular catheterisation were significant in certain components of motion, and suggest that it may not be justifiable to measure the motion of a catheterised heart, except where the records need only be qualitative in nature.

In the experiments in which orientation of the

probe was changed, and those in which the effect of time lapse was studied, both these factors induced significant changes. Because only one variable was altered in each case, it was possible to decide the source of these variations in recordings.

However, for the surgical procedures (thoraco-tomy and pericardiotomy), in which significant changes in the recordings were obvious, more than one influence was possible. Thoracotomy results in a variation in physical boundary conditions for the heart, as well as altering the measuring conditions of the transducer. Further, the intrinsic motion of the heart could change and, therefore, no conclusions are permissible.

Following pericardiotomy, the measuring variables do not change, so that it is possible to say that the source of variation is in the motion of the heart itself. However, it is not possible to differentiate between intrinsic variation, or variations due to constraints.

SUMMARY

The aim of this work was to investigate the qualitative influence of certain factors on both heart motion and its measurement, using the displacement cardiograph. The factors studied were intraventricular catheterisation, presence of various tissues of the chest wall, thoracotomy, pericardiotomy, and the position and orientation of the transducer relative to the heart.

It was found that precordial skin, fat, and muscle (exclusive of intercostal muscle) do not alter the motion of the heart within the resolving powers of the measuring system. Intra-cardiac catheterisation was shown to influence the motion of the heart to a small, but significant, degree in some cases.

It was not possible to identify the source of changes in recorded motion following thoracotomy. Pericardiotomy was shown to produce significant changes in motion.

Very dissimilar recordings were obtained by merely changing the angle or orientation of measurement relative to the apex. This emphasised the importance of specifying exactly what component of motion is measured.

The pattern of apex motion recordings was found to change with time. In view of such dynamic changes, comparisons for the purpose of establishing change were only valid between situations which existed close to each other in time.

ACKNOWLEDGMENT

The authors wish particularly to acknowledge the support and encouragement received from

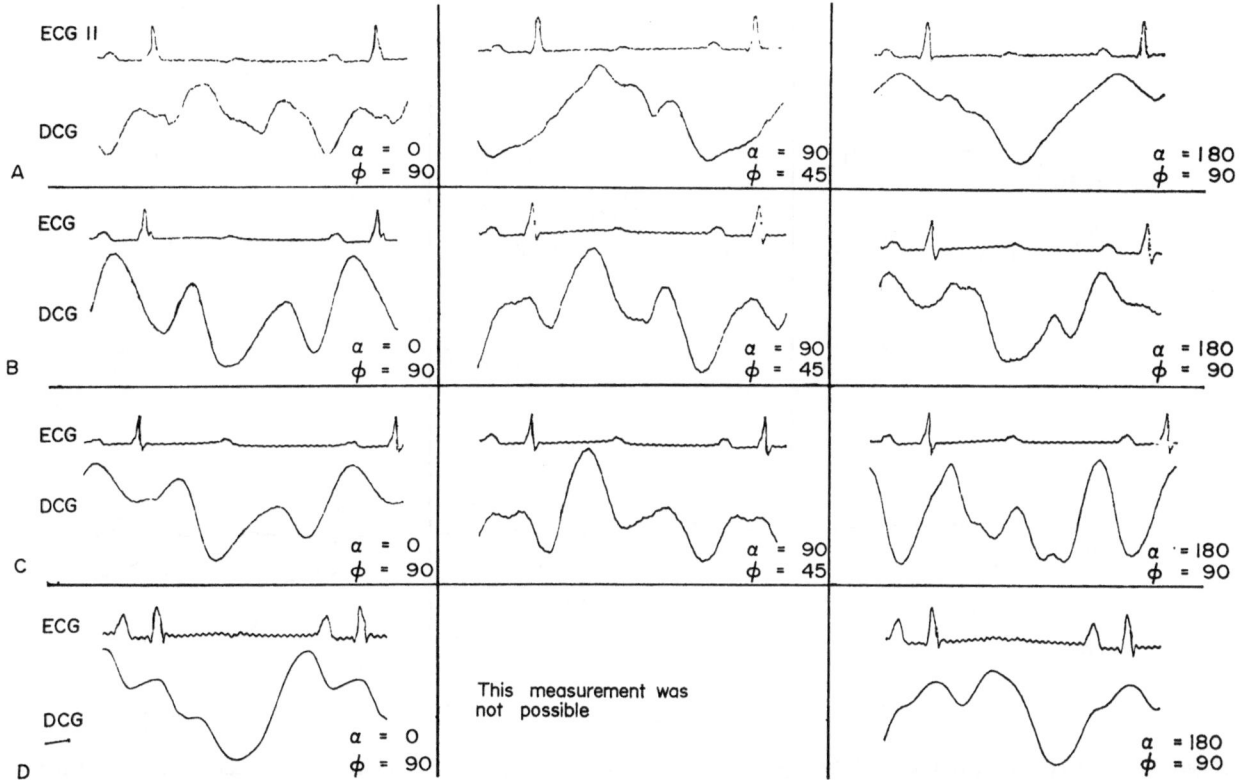

Figure 9. Influence of thoracotomy. (A) Just prior to pneumothorax. (B) Left pneumothorax. (C) Left and right pneumothorax. (D) Bilateral thoracotomy.

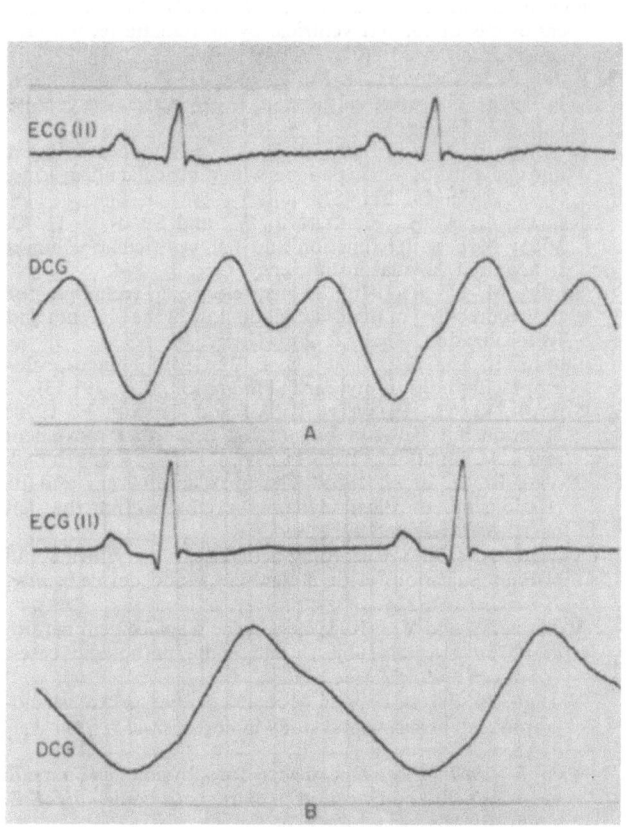

Figure 10. Influence of pericardiotomy. (A) With pericardium intact. (B) Without pericardium.

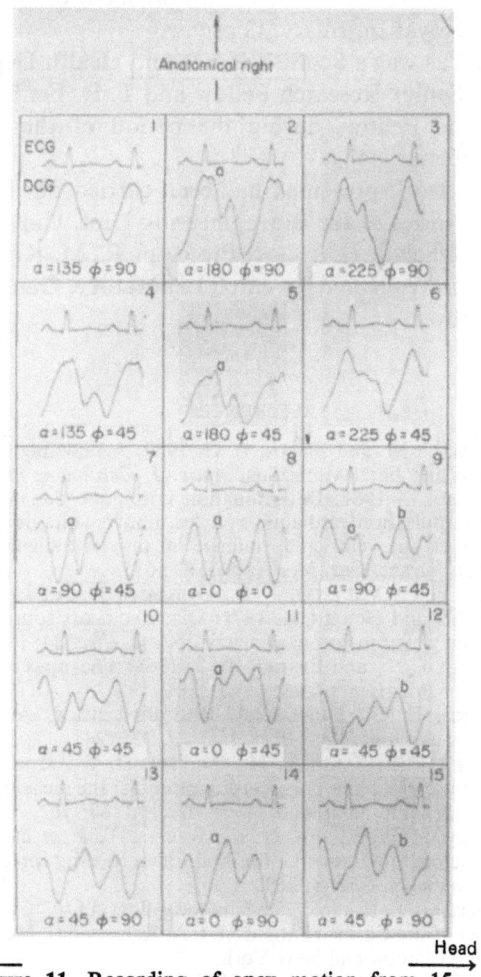

Figure 11. Recording of apex motion from 15 orientations.

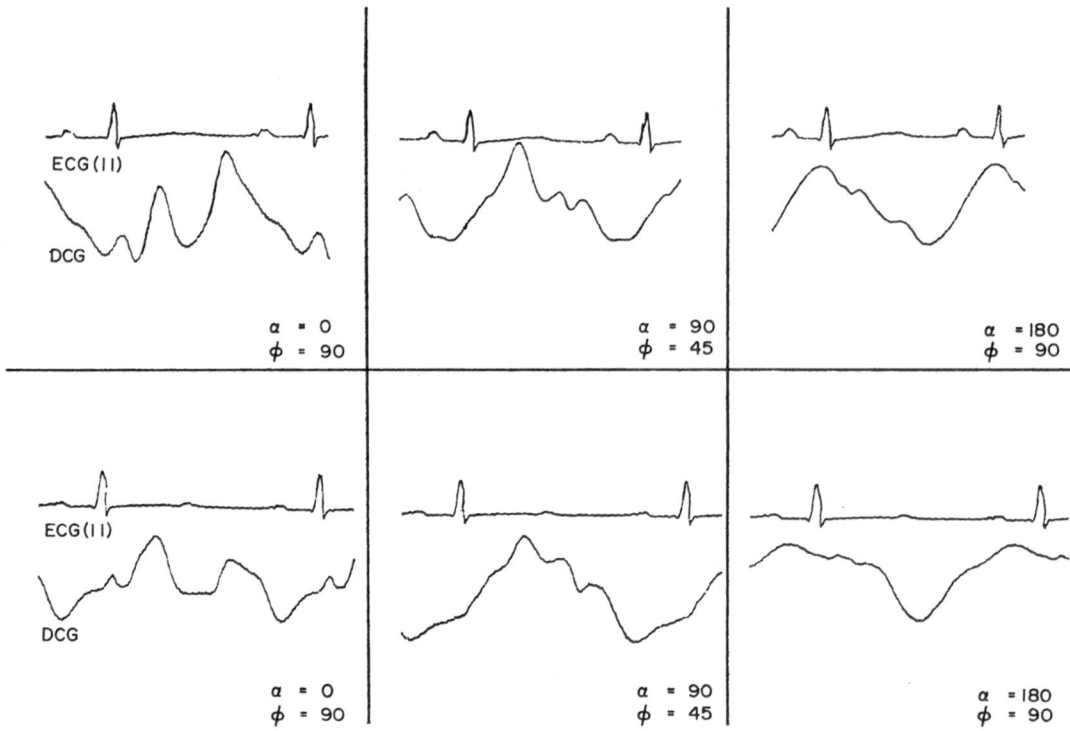

Figure 12. Influence of the duration of the experiment. Upper figures refer to initial conditions (after anaesthesia), lower figures refer to conditions two hours later.

Professor V. Lawrie of the Department of Cardiology, Royal Infirmary, Glasgow.

R. Vas was a Scottish Home and Health Department Senior Research Fellow and T. R. Fenton an Athlone Fellow, during the period of the work described.

The work presented has been carried out in the laboratories of the Bioengineering Unit, University of Strathclyde (Director, Professor R. M. Kenedi), and the Wellcome Research Laboratory, Garscube, Glasgow.

REFERENCES

AGRESS, C. M. and FIELDS, L. G. 1959: A new method for analysing heart vibrations. *Amer. J. Cardiol.* 4: 184.

CHAVEAU, P. 1861: Determination graphique des rapports de la pulsation cardiaque avec les mouvements de l'oriellette et due ventricule, obtenue au moyen d'un appareil en registrant. *Gaz. Med. (Paris)* 31: 675.

COHN, P. F., VOKANAS, P. S., WILLIAMS, R. A., HERMAN, M. V., and GORLIN, R. 1971: Diastolic heart sounds and filling waves in coronary artery disease. *Circ.* 44: 196.

COULSHED, N., and EPSTEIN, E. J. 1963: The apex cardiogram. *Brit. Heart J.* 25: 697.

DIMOND, E. G., DUENAS, A., and BENCHIMOL, A. 1966: Apex cardiography. *Amer. Heart J.* 72: 124.

DODGE, H. T., SANDLER, H., BALLEW, D. W., *et al.* 1960: The use of biplane angiocardiography for the measurement of L.V. volume in man. *Amer. Heart J.* 60: 762.

EDDLEMAN, E. E., WILLIS, K., REEVES, T. J. *et al.* 1953: Kinetocardiogram: I. Method of recording precordial movements. *Circ.* 8: 269.

EDDLEMAN, E. E. 1965: The Kinetocardiogram. *In* 'Examination of the Cardiac Patient' (Ed. A. A. Luisada). McGraw-Hill, London and New York.

GLEICHMANN, U., KREUZER, H., LOOGEN, F. *et al.* 1968: Quantitative measurement of the force and of the rate of

development of force in the apex cardiogram. *Z. Ges. Exp. Med.* 145: 278.

INGUE, K. *et al.* 1971: Ultrasonic measurement of left ventricular wall motion in acute myocardial infarction. *Circ.* 43, 778.

KUMAR, S., and SPODICK, D. H. 1970: Study of the mechanical events of the left ventricle by atraumatic techniques. *Amer. Heart J.* 80: 401.

LANE, F. J., CARROLL, J. M., LEVINE, H. D., and GORLIN, R. 1968: The apex cardiogram in myocardial asynergy. *Circ.* 37: 890.

MCGINN, F. X., and LYON, A. 1968: The phonocardiogram and apexcardiogram in patients with ventricular aneurysms. *Amer. J. Cardiol.* 21: 467.

MILLER, G. A. H., KIRKLIN, J. W., and SWAN, J. H. C. 1965: Myocardial function and left ventricular volumes in acquired valvular insufficiency. *Circ.* 31: 374.

SHARF, M. *et al.* (1970). A new electronic technique for indirect reading of maternal blood flow in the placenta and its localization. *Amer. J. Obst. Gynec.* 106: 292.

SUH, S. K. and EDDLEMAN, E. E. Jr. 1959: Kinetocardiographic findings of myocardial infarction. *Circ.* 19: 531.

SUTTON, G. C., PREWITT, T. A., and CRAIGE, E. 1970: Relationship between quantitated precordial movement and L.V. function. *Circ.* 41: 179.

TAVEL, M. E. *et al.* 1965: The apex cardiogram and its relationship to haemodynamic events within the left heart. *Brit. Heart J.* 27: 829.

VALERO, A. (1967). Recording actual heart movements and arterial pulsations with a new electronic device. *Amer. J. Cardiol.* 19: 224.

VALERO, A., and VAS, R. 1969: On the displacement cardiograph in normal subjects and some pathologic cases. *Haraguah J, Israel Med. Assoc.*

VALERO, A., PELEG, H., and MOLCHO, J. 1969: Focal cardiography. An experimental study in dogs. *Israel J. Med. Sci.* 5: 13.

VAS, R. 1967: Electronic device for physiological kinetic measurement and detection of extraneous bodies. *I.E.E.E. Trans. BioMed. Eng.,* BME 14: 2.

VOIGT, G. C. and FRIESINGER, G. C. 1970: The use of apexcardiography in the assessment of L.V. diastolic pressure. *Circ.* 41: 1015.

DISPLACEMENT CARDIOGRAPHIC STUDIES OF ISCHAEMIC HEART DISEASE

J. B. McGUINNESS, T. SEMPLE, M. van LITH, and R. VAS[1]

INTRODUCTION

When a programme of work on the Displacement Cardiograph[2] was being planned it was thought appropriate to include a small pilot study on patients in order to gain experience of the instrument in the clinical situation, the results of which could be considered along with those from the experimental laboratory.

From previous work (Vas, 1967; Valero, 1967; Valero et al., 1969; Valero, 1970; Scharff et al., 1970; and Ollendorff et al., 1971), it was clear that the instrument measured tissue movement and that when placed over the heart measured mainly cardiac movement. We therefore decided to study patients in whom knowledge of cardiac activity might be an advantage in clinical management.

Patients who have sustained a myocardial infarction have, by definition, left ventricular injury and might be expected to have a disturbance of left ventricular movement and so a study was planned using a group of such patients.

PATIENTS AND METHODS

Two main groups of patients were studied: the first group consisted of male volunteers between the ages of 20 and 65 years with no history or personal suspicion of cardiac disease and, as far as they were aware, in perfectly good health. There were 26 between the ages of 20 and 40 years and nine between the ages of 40 and 65 years.

The second group consisted of 15 male patients in the Coronary Care Unit of the Victoria Infirmary who had suffered a myocardial infarction or an attack of acute myocardial ischaemia within the previous three days. A third small group emerged unexpectedly and this consisted of four men who had volunteered as 'normals'. Two were found on examination prior to the study to have changes on the electrocardiogram which were undoubtedly those of myocardial ischaemia; while in another two, re-examination of the electrocardiogram after an abnormal displacement cardiogram was found

revealed 'borderline' ischaemic changes which, in view of the other normal findings, had not been regarded as absolutely diagnostic at the time of the screening examination.

In each individual, before any instrumentation was carried out, a full clinical history was taken. They were fully examined and this included a chest X-ray and a 12-lead electrocardiogram at rest.

The sensing coil of the displacement cardiograph used in every instance was 5 cm in diameter. It was placed over the point of maximum cardiac apex impulse and if this was not clearly discernible it was placed in the fifth left intercostal space just outside the midclavicular line, this area being explored to obtain the best record. The subject in each instance was supine and the record was taken at the end of a normal expiration with the breath held. For reference, a phonocardiogram was obtained by means of an Elema–Schonander crystal microphone (EMT 25), usually placed over the second left or right intercostal space close to the sternum while lead 2 of the electrocardiogram was recorded from electrodes placed on the trunk.

All of these measurements were displayed simultaneously on an Elema–Schonander Mingograph 81 multichannel recorder which has a frequency response of D.C.—500 Hz. The paper speed was 100 mm per second.

The displacement cardiogram was measured,

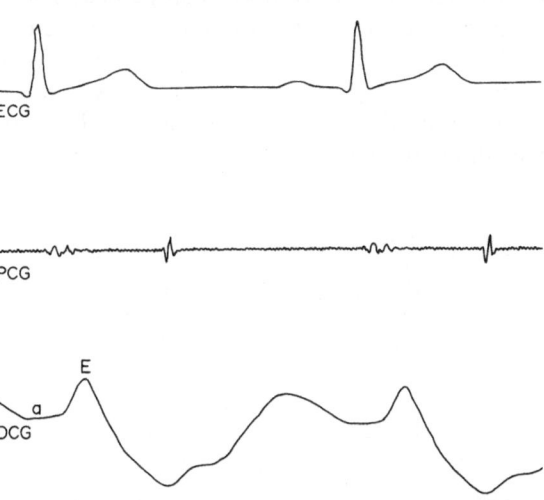

Figure 1. A normal displacement cardiogram showing the 'a' (atrial) and 'E' (ventricular ejection) waves.

---

[1] Victoria Infirmary and University of Strathclyde, Glasgow, Scotland.

[2] *Editorial note:* The 'Displacement Cardiograph' and the 'Kinetocardiograph' (see previous paper) are essentially the same instrument.

Group	Number	T (R–R interval in cm)	E width (cm)	E/T %	a/E₁ %
Normal (20–40 yr)	26	$10 \cdot 25 \pm 1 \cdot 38$	$2 \cdot 12 \pm 0 \cdot 33$	$21 \cdot 1 \pm 4 \cdot 4$	$6 \cdot 1 \pm 5 \cdot 2$
Normal (40–65 yr)	9	$9 \cdot 02 \pm 1 \cdot 78$	$1 \cdot 89 \pm 0 \cdot 24$	$21 \cdot 8 \pm 5 \cdot 3$	$6 \cdot 2 \pm 4 \cdot 8$
Ischaemic heart disease	15	$8 \cdot 38 \pm 1 \cdot 33$	$3 \cdot 55 \pm 0 \cdot 88$	$43 \cdot 9 \pm 13 \cdot 0$	$18 \cdot 3 \pm 12 \cdot 7$

TABLE 1

an average of five consecutive cycles being taken in each instance, and each wave was examined for height and duration, the latter referred to the length of the cardiac cycle. A normal tracing is shown in figure 1.

RESULTS

The significant findings are shown in table 1 where it will be seen that the width of the ejection wave 'E' expressed as a percentage of the duration of one cardiac cycle was significantly prolonged in patients with ischaemic heart disease. On average the 'a' wave also was much more prominent in those who had suffered an ischaemic episode but there was a high standard deviation in both the normal and the ischaemic groups. There was no significant difference between the normal patients below and those above the age of 40 years.

Table 2 shows the individual results in the small group of patients with unsuspected ischaemic heart

Group	E/T %	a/E %
Normal (40–65 yr)	$21 \cdot 8 \pm 13 \cdot 2$	$6 \cdot 2 \pm 4 \cdot 8$
Unsuspected ischaemic heart disease	$53 \cdot 4$	$8 \cdot 1$
	$51 \cdot 4$	$2 \cdot 2$
	$27 \cdot 3$	$17 \cdot 4$
	$21 \cdot 6$	$10 \cdot 9$

TABLE 2

disease. They varied from 49 to 63 years and so are compared with the normal group of patients of 40 to 65 years. It is interesting to note that the first two show a greatly prolonged ejection wave with 'a' waves within the normal range whereas the second two show normal ejection waves but 'a' waves

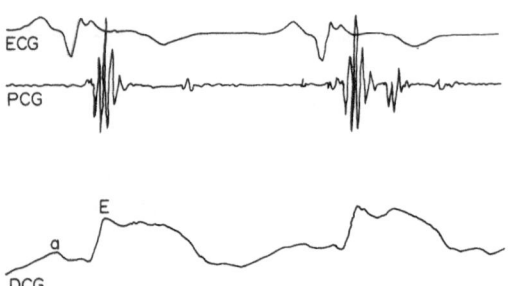

Figure 2. The displacement cardiogram of a patient with ischaemic heart disease showing a prolonged 'E' wave.

higher than the mean normal value. An abnormal ejection wave is shown in figure 2.

DISCUSSION

One of the commonest practical problems met during any study of heart movement is that of finding the true point of maximum impulse and this was our experience. There is no doubt that the position of the probe on the chest affects the shape of the displacement cardiogram but it is interesting to note that despite careful searching a normal pattern could never be obtained in any position in those patients who showed an abnormal E wave. Research is continuing into the use of larger sensing coils and results to date are encouraging. Enlargement of the coil results in a certain loss of resolution of local cardiac movements but promises to give greater and easier reproducibility.

The study has once more shown that the 'a' wave, which represents atrial activity, is prominent in many patients who suffer from left ventricular dysfunction. However this is not a constant finding, which probably accounts for the large standard deviation observed. Individual results obtained in the four patients with unsuspected ischaemic heart disease are interesting in that the 'a' wave is most prominent in those who had no E wave changes and had least marked electrocardiographic changes. It is tempting to postulate that we are in fact observing different degrees or different stages of left ventricular dysfunction.

The character and relative width of the E wave seemed clearly related to left ventricular damage. Indeed broadening and flattening of this wave in ischaemic heart disease was quite striking. The possible significance of this was brought home to us forcibly during the control study; one patient aged 63 years was found to have this pattern on the displacement cardiogram and the electrocardiogram showed posterior ischaemic changes. He took pride in his physical fitness and played golf at least twice a week, in addition to which he was a non-smoker, was normotensive and was not overweight. It was our feeling that he was already carrying out the type of programme which we would have advised such a patient to follow and that no useful purpose could be served in drawing his attention to the abnormalities found although we did of course inform his own doctor (figure 3). He died suddenly on the golf course 10 days later.

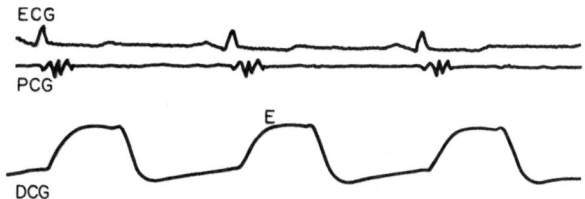

Figure 3. The displacement cardiogram of a volunteer with unsuspected ischaemic heart disease.

CONCLUSIONS

As a result of this introductory study we have begun to realise that the displacement cardiogram is able to detect dysfunction of the left ventricle which shows up as changes in ventricular ejection or changes in left atrial behaviour. There are still practical problems involved in obtaining satisfactory tracings but it is hoped that with further research and further development of the instrument these difficulties will eventually be overcome.

REFERENCES

OLLENDORFF, F., VAS, R., and VALERO, A. 1971: The effect of exercise and acclimatisation on displacement apex cardiogram in normal young subjects. *Brit. Heart J.* **33** (1): 37.

SCHARFF, M., OETTINGER, M., and VAS, R. 1970: Indirect blood flow measurement in the placenta. *Amer. J. of Obstetrics and Gynaecology* **106**: 292.

VALERO, A. 1967: Recording of actual heart movement and arterial pulsation with a new electronic device. *Amer. J. of Cardiology* **19**: 244.

VALERO, A. 1970: Focal displacement cardiography for bedside detection of myocardial dyskinesis. *Amer. J. of Cardiology* **25**: 433.

VALERO, A., PELEG, H., and MOLCHO, J. 1969: Focal cardiography. *Israel J. Med. Science* **5**: 13.

VAS, R. 1967: Electronic device for physiological kinetic measurement and detection of extraneous bodies. *I.E.E.E. Transactions Bio. Med. Eng. B.M.E.* **14**: 2.

BALLISTOCARDIOGRAPHY: PROMISE AND PRACTICE

M. T. MANLEY[1]

BACKGROUND

The recording of the displacement (and its derivatives) of the centre of mass of a subject in space due to mass movements within the cardiovascular system (now known as ballistocardiography) has been practised for many years. In 1877 a barrister named Gordon produced one such record by laying a patient on a bed suspended from long wires and monitoring the motion of the bed by means of a lever arrangement and a smoked drum (Gordon, 1877). The technique appeared again in the literature on a few occasions over the next sixty years (notably Henderson, 1905) but did not receive serious attention until 1939 when Starr built an instrument specifically designed to monitor cardiovascular force actions and named it a ballistocardiograph (Starr et al., 1939).

Starr's instrument consisted of a rigid table suspended on long wires, and movement of the table was opposed by stiff springs giving the loaded table a translational natural frequency of about 15 Hz. Table displacement in the head-foot axis was monitored by an optical method, and was said to be proportional to the force applied to the table at any instant. Because of the high natural frequency of the table when compared to heart rate this instrument is now known as a high frequency (H.F.) ballistocardiograph to distinguish it from other techniques developed later. Starr demonstrated that records obtained from this instrument were, for most young healthy subjects, well defined and relatively repetitive and that respiratory influences upon the record could be ignored.

Although instruments based on the Starr design were used by a number of researchers over the next twenty years it was not until 1952 (Bouhuys, 1952a, b) that the inadequacies of the instrument in producing accurate recordings of body motion were finally established. Experimental and theoretical analysis of the body-instrument system showed that a very light table with a very low translational natural frequency (less than 0·4 Hz) produced a record which is a much more accurate representation of the displacement, velocity and acceleration of the centre of mass of the body during the cardiac cycle (Burger et al., 1953; Von Wittern, 1953).

Early designs achieved this low natural frequency by either suspending the table on long wires or floating it in a bath of mercury but other methods have since been used. Instruments of this type have been designated Ultra Low Frequency (U.L.F.) ballistocardiographs.

One other technique should be mentioned here because of its wide usage (mainly in the U.S.A.) in the early 1950s. This is the so-called 'direct body' or 'shin-bar' method, introduced in 1949, which recorded the motion of the body of a subject on the dorsal tissues by fixing a transducer to a bar which was strapped across the patellae (Dock, 1949). This technique was extremely simple and therefore became very popular but it was not realised at the time that gross distortions were introduced into the record because of the complex dynamic influence of the body on its dorsal tissues, and therefore records obtained by the technique were almost useless as an indicator of cardiovascular dynamics.

Although modern U.L.F. instruments give well-defined repetitive traces for most normal subjects and the limitations of the method are well defined, ballistocardiography has never been well received by most clinicians. The reasons given by most researchers in the field for this clinical scepticism are many and range from the cost of a U.L.F. instrument (although this could be very small when compared with hospital equipment such as X-ray apparatus), to the now-proven limitations of the once popular direct-body method. However, the real reason is possibly a combination of many factors one of which must certainly be the wild claims (often subsequently disproved) which have occasionally been made about the technique over the last thirty years.

SOME CLINICAL POINTERS

Although ballistocardiography is not regarded with favour in many quarters some of the results and inferences which have come from ballistocardiographic studies over the last thirty years are nevertheless well worth considering.

At one time it was claimed that ballistocardiography would replace electrocardiography (ECG), as the BCG was more representative of the mechanical function of the heart. This has not been realised as to date the interpretation of the more complex waveform of the BCG has been mainly subjective,

[1] University of Strathclyde, Glasgow, Scotland.

as with ECG interpretation. Although some researchers have tried to make BCG interpretation an objective exercise (and at present this does seem possible) simultaneous recording of ECG and BCG is more valuable as the time relationships between the ECG and electrical events are well understood.

The development of the U.L.F. BCG has progressed with the aid of cadaver experiments, animal experiments, drug trials and computer and mathematical analysis. Although some of the results have not been conclusive, direct relationships between myocardial contractility, the peak acceleration of blood ejection into the aorta and the amplitude of the early systolic waves of the ballistocardiogram have been fairly well established. (Noble *et al.*, 1966; Winter *et al.*, 1969). Also the information gained from ballistocardiographic studies, when compared with other studies of ventricular function and instantaneous aortic flow, gives a much fuller picture of myocardial function (Starr, 1965). It now appears that the whole of the force from the contractile element of the myocardium is applied very early in systole and causes an extremely high acceleration of blood from the ventricles so that maximum flow velocity is attained early in systole. By mid-systole a slowing of the flow from the heart occurs so that blood is moving from the ventricle only because of its own inertia and thus the acceleration of the blood during this period is back towards the heart.

Figure 1 shows a set of idealised 'normal' ballistocardiograms from an ultra-low frequency instrument with ECG and phonocardiogram (PCG) correlation.

The F–G wave is reported to be presystolic and to be associated with atrial contraction as it bears a

constant time relationship with the P wave of the ECG, whether AV conduction is normal or not (Scarborough *et al.*, 1958). The H–I segment is produced by acceleration of blood into the great vessels by rapid ventricular contraction with the opening of the aortic valve being reported to occur at the H tip (Burchett, 1971) or some time after it with right ventricular ejection beginning at H (Reeves *et al.*, 1957). The I–J wave represents the flow of blood under ·its own inertia from the ventricle into the aorta, with L and M being variously claimed to be the end of systole and the closing of the aortic valve (Scarborough *et al.*, 1958; Reeves *et al.*, 1957). Although these discrepancies do appear in the assumed relationships with some of these 'waves to cardiac events', nevertheless the U.L.F. BCG has been shown to be a reliable external measurement of cardiac forces (Starr and Noordergraaf, 1967).

Also clinical trials have shown that BCGs of patients with ischaemic heart disease (I.H.D.) are usually, but not always of abnormal form. Some of these trials have been conducted with H.F. instruments but even so the changes in form due to age or disease are so striking that they could not be ignored. These findings led to a number of long-term studies on patients with known I.H.D. and also on groups of subjects who showed no clinical evidence of I.H.D. (normal ECG, no chest pains, no circulatory abnormality by history or physical examination, a normal blood pressure, and for some studies a normal cardiovascular X-ray).

Starr and Wood (1961) carried out a remarkable long-term study on 211 initially normal subjects, either up to death or for 23 years, using records obtained from one of Starr's high frequency instruments. It was noted that great differences in amplitude of the original records occurred between subjects (even though in most cases records were of the 'normal' form) and thus the study was conducted using I–J amplitude measurements for each subject.

It was found that a significantly high proportion of those whose initial I–J amplitude was small developed some form of cardiac abnormality. The study also showed that the BCG amplitude diminished with age but by neutralising age influences in the records, by pairing the data, the BCG amplitudes of those who developed heart disease were on average initially only 25 per cent of those who did not.

A later study by Starr (1964) for a period of five years on 221 hospital patients on whom H.F. records were taken is also of great interest here. In this study no attempt was made to classify records quantitatively but rather to use an independent observer (who has no other knowledge of the patient) into one of four categories of normality. The history of each patient was then followed for the next five years or up to death, whichever was the

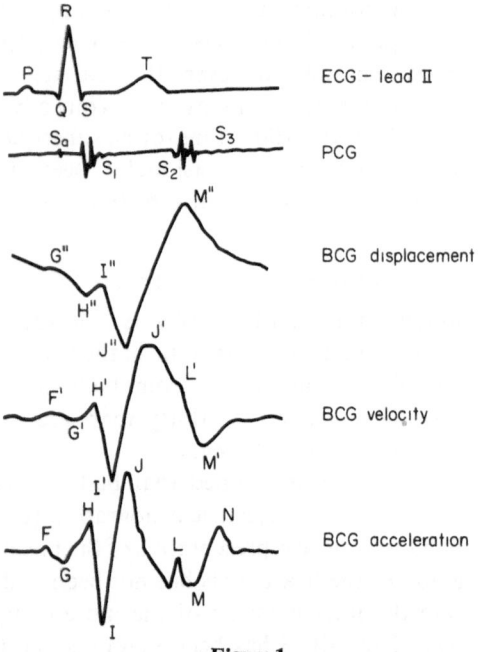

Figure 1.

sooner. The results showed that those patients whose BCG was normal in form had a survival rate over the five year period not significantly different from that predicted from actuarial tables, but that in the group with maximally abnormal records the survival rate was only one-tenth of that predicted.

A later clinical trial reported in 1967 on normal subjects and patients with ischaemic heart disease over a nine-year period showed that an apparently normal subject with an abnormal high frequency ballistocardiogram had twelve times as much a chance of developing a pathological cardiac condition within the time of study than those subjects whose record was normal (Baker *et al.*, 1967). Also, in the group with I.H.D. when the study was started the risk of death or another event occurring in the nine-year period was slightly more than doubled for the abnormal BCG to the normal BCG whatever the age of the subject.

A NOVEL ULTRA/LOW FREQUENCY BALLISTOCARDIOGRAPH

It is results such as these which have persuaded us to investigate the clinical significance of ballistocardiography by building an extremely sensitive U.L.F. instrument which is suitable for use in a clinical environment.

Although a prerequisite of the system was portability and ease of use, the instrument was also required to be extremely sensitive and to give the maximum possible bandwidth of undistorted output.

A dynamic analysis of the body-table system shows that artefacts in the record due to relative motion between body and table are dependent on the mass of the table and therefore this must be reduced to a minimum. The most novel part of this instrument is the table itself which was constructed from carbon fibre composites giving a structure of very high flexural rigidity but very low mass (Manley *et al.*, 1972). The mass of the present table is about 2·5 kg (5·6 lb) but the structure will carry loads in excess of 130 kg (300 lb) when supported over one metre centres.

The table suspension system chosen was an air bearing support with a bearing pressure of 140 KN/m² (20 lbf/in²). The configuration consists of three flat lift bearings to support the body-table mass and four small bearings mounted vertically which bear on guide fins attached to the table and therefore constrain table motion to the head-foot axis. The head-foot acceleration of the table is monitored by an extremely sensitive servo accelerometer (resolution about 1 μg) with a calibrated output and the table has a loaded natural frequency of less than 0·2 Hz. The frequency response of the loaded instrument is flat from 1 Hz to about 40 Hz.

Figure 2 shows the construction of the suspension and table.

The instrument was initially used to record BCGs from young healthy adults simultaneously

Figure 2.

with the ECG which was used as a timing reference. Preliminary statistical analysis of these records shows that greater variation in measured portions of the record, both in amplitude and timing of ballistic events, occurs between one subject and another, than between records taken from the same subject on successive days. The smallest variation occurs in one subject on a given day. Although it may appear self-evident that the physiology of the system would produce this type of result because of the uncontrollable factors involved in monitoring any one subject it is encouraging that the ballistocardiograph records demonstrate this expected result.

Figure 3 shows the type of trace obtained from a normal subject, along with a simultaneously recorded ECG.

A second trial, originally on normal subjects, which has been started, is the simultaneous recording of BCG, ECG, phonocardiogram (PCG) and displacement cardiogram (DCG) (using the displacement cardiograph developed by Vas (Vas, 1967; McGuiness *et al.*, 1972). Again, the results are only in the preliminary stage but a number of interesting finds have arisen. Figure 4 shows the four traces recorded simultaneously.

Initial results have shown that if the H–I segment of the BCG is indicative of early systole, then the peak of the E wave on the DCG, which previous researchers had placed (using the apex cardiogram) at or near the beginning of ejection does not always occur during the same time interval as the H–I segment. In fact, some of our records on 'normal subjects' place the DCG E 'point' up to 0·08 s

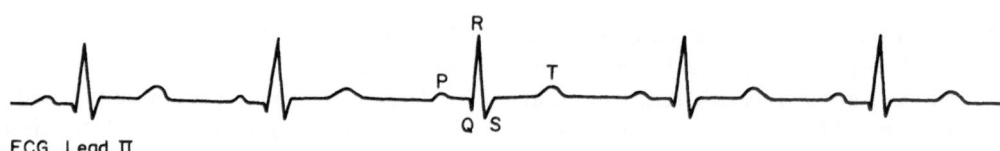

ECG Lead II

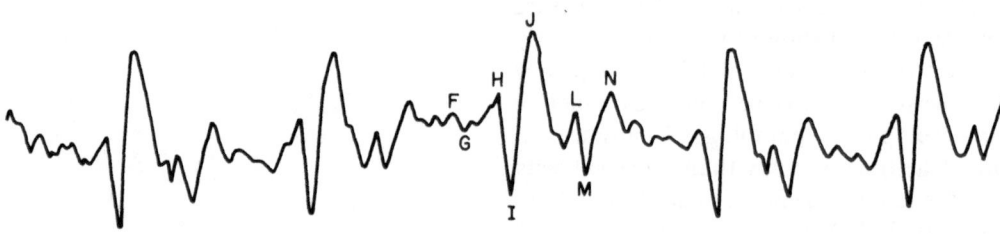

BCG Acceleration record

Figure 3

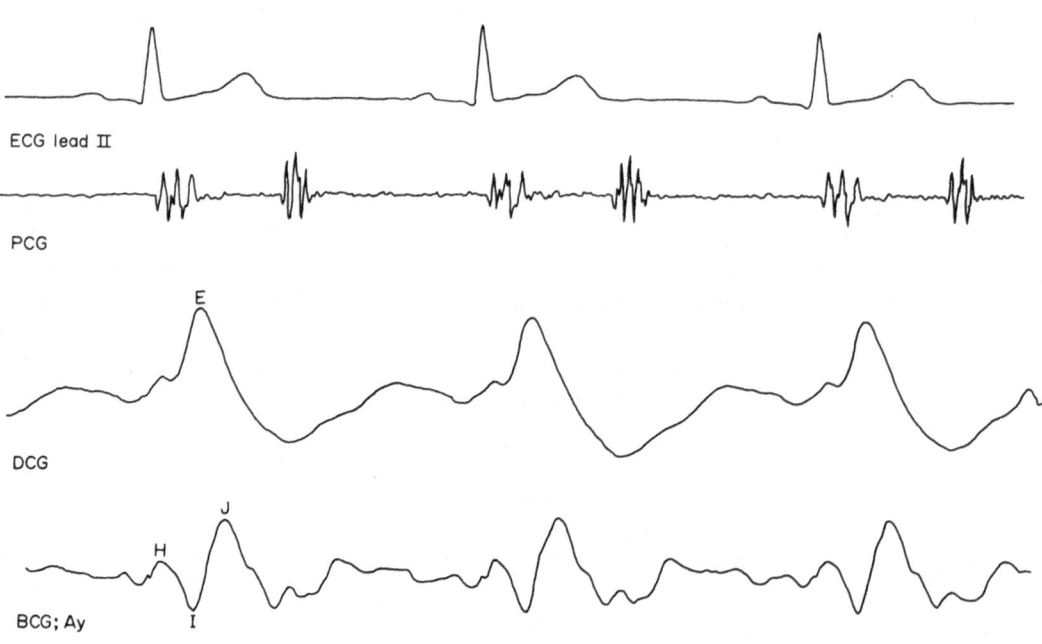

ECG lead II

PCG

DCG

BCG; Ay

Figure 4.

after the I point in the BCG. Many researchers have accepted the theory that the motion of the apex towards the chest wall occurs at the beginning of ejection, but our results to date show that this motion occurs later in the cycle during the inertial flow phase from the ventricle. A recent paper by Kumar and Spodick (1970) also showed that the apex cardiogram E 'point' may be delayed up to 0·1 s from the opening of the aortic valve.

SUMMARY

The preliminary studies reported are continuing. Additionally we are about to commence a series of trials on patients with known pathological conditions so that the clinical significance of the non-intrusive monitoring techniques may be evalu-ated through simultaneous recordings and comparative assessment.

REFERENCES

BAKER, B. M., SCARBOROUGH, W. R., DAVIS, F. W., MASON, R. E., and SINGEWALD, M. L. 1967: Ballistocardiography and ischaemic heart disease: predictive considerations and statistical evaluation. *Proc. R. Soc. Med.* **60**: 30.

BOUHUYS, A. 1952a: High frequency ballistocardiography. *Proc. Koninkl. Ned. Akad. Wetenschap Amsterdam Series C* **55**: 126.

BOUHUYS, A. 1952b: High frequency ballistocardiography *Proc. Koninkl. Ned. Akad. Wetenschap Amsterdam Series C* **55**: 135.

BURCHETT, G. 1971: External evaluation of cardiac function. *J. Am. Osteopath. Assoc.* **70**: 903.

BURGER, H. C., NOORDERGRAAF, A., and VERHAGEN, A. M. W. 1953: Physical basis of the low frequency ballistograph. *Amer. Heart J.* **46**: 71.

DOCK, W. 1949: Recording the motions of the heart and blood for clinical purposes. *Trans. Assoc. Amer. Physicians* **62**: 148.

GORDON, J. W. 1877: On certain molar movements of the human body produced by the circulation of blood. *J. Anat. Physiol.* **11**: 533.

KUMAR, S., and SPODICK, D. H. 1970: Study of the mechanical events of the left ventricle by atraumatic techniques: comparison of methods of measurement and their significance. *Am. Heart J.* **80**: 401.

MANLEY, M. T., PATON, W., and MONTGOMERY, I. 1972: Lightweight carbon fibre structures—their practical use in ballistocardiography. *Composites* (in press).

McGUINESS, J. B., SEMPLE, T., VAN LITH, M., and VAS, R. 1973: Displacement cardiographic studies of ischaemic heart disease. This *Symposium*, p. 251.

NOBLE, M. I. M., TRENCHARD, D., and GUZ, A. 1966: Left ventricular ejection in conscious dogs. *Circ. Res.* **19**: 139.

REEVES, T. J., HEFNER, L. L., JONES, W. B., and SPARKS, J. E. 1957: Wide frequency range force ballistocardiogram. Its correlation with cardiovascular dynamics. *Circulation* **16**: 43.

SCARBOROUGH, W. R., FOLK, E. F., SMITH, P. M., and CONDON, J. H. 1958: The nature of and records from ultra-low frequency ballistocardiographic systems and their relation to circulatory events. *Am. J. Cardiol.* **2**: 613.

STARR, I. 1964: Prognostic value of ballistocardiograms. Studies on evaluation of the doctors experience. *J. Amer. Med. Ass.* **187**: 511.

STARR, I. 1965: Progress towards a physiological cardiology. A second essay on the ballistocardiogram. *Ann. Intern. Med.* **63**: 1079.

STARR, I., and NOORDERGRAAF, A. 1967: 'Ballistocardiography in Cardiovascular Research'. North Holland Publishing Co., Amsterdam.

STARR, I., RAWSON, A. J., SCHROEDER, H. A., and JOSEPH, N. R. 1939: Studies on the estimate of cardiac output in man, and of abnormalities in cardiac function from the heart's recoil and blood's impacts; the ballistocardiogram. *Am. J. Physiol.* **127**: 1.

STARR, I., and WOOD, F. C. 1961: Twenty year study with the ballistocardiogram. The relation between the amplitude of the first record of healthy adults and eventual mortality and morbidity from heart disease. *Circulation* **23**: 714.

VAS, R. 1967: Electronic device for physiological kinetic measurement and detection of extraneous bodies. *Trans. Bio. Med. Eng.* **14**: 2.

VON WITTERN, W. W. 1953: Ballistocardiography with elimination of influence of vibration properties of the body. *Am. Heart J.* **46**: 705.

WINTER, P. J., DEUCHAR, D. C., NOBLE, M. I. M., TRENCHARD, D., and GUZ, A. 1967: The relationship between the ballistocardiogram and the movement of blood from the left ventricle in the dog. *Cardiovasc. Res.* **1**: 194.

MULTIVARIATE ANALYSIS IN MEDICINE AND BIOLOGY

M. J. R. HEALY[1]

Biological research in general, and medical research in particular, is characterised by the expensive nature of much of its experimental material. Once a subject is obtained, it is commonplace to take several different measurements. Multivariate analysis methods are thus particularly appropriate to biological data, and the fact that they are not very widely used is to some extent due to the way in which they are usually presented—as a disjointed collection of recipes in an elaborate mathematical notation. I aim to describe some of the more useful techniques in a way which I hope will illustrate their connections with each other and with better-known statistical methods.

REGRESSION

We may start with the simplest of all techniques, simple regression, with the model

$$E(y) = b_0 + b_1 x \qquad (1)$$

This is, of course, not necessarily a multi-*variate* model; x may not be a variate, in the sense of having a probability distribution attached to it, but equally it may, and the asymmetry, of great importance in some applications, is less important here. The first extension of this model is to a multiple regression model

$$E(y) = b_0 + b_1 x_1 + b_2 x_2 + \ldots b_p x_p \qquad (2)$$

We are at once in possession of an immensely powerful statistical tool. By suitable choice of the x-variables, we can express in these terms an enormous variety of statistical techniques, including all those usually expressed in the form of analyses of variance.

ANALYSIS OF VARIANCE

We now extend this model to a truly multivariate situation. Suppose we have not one but several y-variates. By a pure trick of notation, we can make model (2) still serve our purposes. Putting the different y's and their corresponding b's into vectors, we can write a whole collection of regression models in the form

$$E(\mathbf{y}) = \mathbf{b}_0 + \mathbf{b}_1 x_1 + \mathbf{b}_2 x_2 + \ldots + \mathbf{b}_p x_p \qquad (3)$$

[1] MRC Clinical Research Centre, Northwick Park Hospital, London, England.

Just as model (2) leads to all the complexities of the analysis of variance, so model (3) can give rise to *multivariate analyses of variance*. We can, in fact, calculate not merely the *residual sums of squares* for each y variate after fitting a particular set of x's, but also *residual sums of products* between pairs of y's. The multivariate generalisation of the mean squares of the univariate analysis of variance is thus whole matrices of mean squares and products, readily calculated by the mildest generalisation of well-known formulae. We can compare these matrices by any one of several generalisations of the F distribution—there seems to be no general rule for choosing between these—and thus achieve tests of significance which apply simultaneously to a whole set of dependent variables.

Two cautions are in order. First, the effects of non-Gaussianity on these multivariate F-tests are almost entirely unknown (but see Mardia, 1971). Multivariate analysis is dominated by Normal theory, not least because of the lack of any more general system of distributions such as the Pearson or Edgeworth curves. Multivariate Gaussianity is a tougher restriction even than simple univariate Gaussianity in each separate variate; we have no very satisfactory test for it, and it seems best to place as little reliance as possible on significance levels which may depend critically upon it.

The second caution is more serious. It can be illustrated by a paradox essentially shown by Rao (1966). In its simplest form this states that, given an observation (y_1, y_2) from a bivariate Normal distribution, it is possible for both y_1 and y_2 to differ significantly from 0 at some probability level when the other variate is ignored while, taken jointly, the bivariate observation does not differ from $(0, 0)$ at the same significance level. Put another way, this means that we cannot with impunity pour variates into the analysis; 'noise' variates which do not reflect some effect which we are looking for can dilute genuine effects into non-significance. This is, of course, a result of using a 'blunderbuss' test, but it is one that is not always appreciated.

CANONICAL ANALYSIS

Blunderbuss tests are usually considered (and condemned) in the context of multiple effects, only a few of which are large. Here we have a similar situation with multiple variates, and the remedy is analogous—we consider a limited number of variates, but we select these as *linear combinations* (weighted sums) of the variates originally measured. With our multiple y's, suppose we associate a set of coefficients $c_1, c_2, ..., c_p$ which we assemble into a vector c. We can now construct an ordinary scalar variate $z = c^T y$ and describe this by the univariate model (2). Among other things, we can estimate the correlation coefficient between z and the linear compound $b_0 + b_1 x_1 + ... + b_p x_p$. This correlation is in fact the highest possible for *any* linear compound of the x's – this is one way in which multiple regression can be approached – and is just the multiple correlation between y and the x's. But the coefficients c are still at our disposal. Let us try to choose these to maximise the correlation between z and the x's; we shall then have two linear functions, one of the y's and the other of the x's, with the property that the correlation between them is the highest possible for any pair of such functions.

The mathematics of this double maximisation is not difficult, and it leads to a set of linear equations for the c's which can be written in the matrix form

$$(A + \lambda B)c = 0 \qquad (4)$$

This type of equation crops up all over multivariate analysis. In this case A can be taken to be the dispersion or MSP matrix 'for regression' in a multivariate analysis of variance of the y's while B is the residual MSP matrix. More generally it comes from the problem of maximising a quadratic form $c^T A c$ while holding a second quadratic form $c^T B c$ constant, or equivalently that of maximising the ratio of the two forms—the ratio that generalises the usual F ratio.

The equations (4) are technically *homogeneous* (the right-hand sides are all zero), and non-zero solutions for the c's are only possible for certain values of the scalar λ. These are the *eigenvalues* of the problem; if A and B are $p \times p$ matrices, there are p values of λ (some of which may coincide) and p corresponding *eigenvectors* c. In statistical contexts the λ's are always non-negative and can be interpreted either as variances or squared correlation coefficients.

Reverting to the specific problem, the highest eigenvalue λ_1 is the maximum squared correlation obtainable between any two linear functions. It is called the first *canonical correlation* and the linear functions ($c_1^T y$, the *most predictable criterion*, and its predictor $b_1^T x$) are the corresponding pair of *canonical variates*. We can also interpret the other eigenvalues and their eigenvectors. In fact, if $c_2^T y$ and $b_2^T x$ are the variates obtained from the eigenvectors belonging to the second highest eigenvalue λ_2, then both of them are uncorrelated with the first two canonical variates and λ_2, the correlation

between them, is the highest possible correlation between any two linear compounds obeying this restriction. Subsequent eigenvectors give further constrained maxima.

Canonical analysis appears to provide a major simplification of any multivariate situation when the variates fall into two groups. It is thus a little surprising that it has not found any major applications in practice. A barrier may be the difficulty that seems always to occur in practice in the interpretation of the coefficients c and b. The problem is a well-known one in multiple regression, where the sampling errors of the coefficients are often substantial and not independent. Presumably the same is true in magnified form in the more complex situation. As an example, J. M. Tanner and I in unpublished work have correlated body measurements made at birth with the same measurements made on the same subjects as adults 25 years later. Individually, the correlations are remarkably small, of the order of 0.1 to 0.2; this is because an individual's size at birth is more a function of his mother's size than of his own genetic make-up. The top canonical correlation, on the other hand, is about 0.8, seeming to show that some kind of shape component in the adult is highly predictable at birth. However, the component involved was not readily identifiable from its coefficients and the coefficients at the two ages were quite different in pattern.

The growth example invites a generalisation to more than two groups, since measurements were available at several intermediate ages. There are several possible generalisations of canonical analysis which have very recently been developed by Kettenring (1971). Their usefulness will presumably be limited by the same considerations.

DISCRIMINANT ANALYSIS

An important specialisation of canonical analysis, on the other hand, has found widespread application during the last ten years or so. Suppose we have a set of measurements denoted collectively by y made on the members of several groups of individuals. We now set up as many x-variates as there are groups in such a way that x_i has the value 1 for all units in the ith group and the value 0 for all other units. Then the canonical analysis of x and y amounts to finding the linear function $c^T y$ which has the maximum ratio of the between-group mean-square to the within-group mean-square. This is the *multiple discriminant* problem; geometrically it amounts to choosing the best-fitting straight line, such that the sum of the squared 'distances' of the group means from it, is a minimum (the 'distances' are ordinary Euclidean distances after rotation and axis-stretching have rendered the probability-density contours spherical). Again, a whole set of linear functions drop out of the analysis, providing the best-fitting plane . . ., etc. This is a splendid technique in practice. Inspection of the eigenvalues shows quite clearly how many dimensions are needed to depict the basic data without real loss of information, and in two dimensions at any rate the individuals can be plotted as a check on the theoretical probability-density contours. It appears to be robust to its assumptions of Gaussianity and constant variance – covariance structure. Figure 1 shows a recent medical example (Fraser *et al.*, 1971) in which cases of hypercalcaemia have been divided into five diagnostic groups. These are very well distinguished by combinations of five simple biochemical tests.

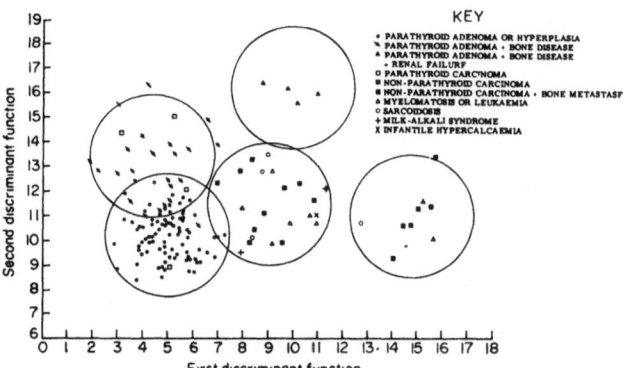

Figure 1. Discrimination between diagnostic groups in hypercalcaemia, based on five biochemical tests (urea, phosphate, alkaline phosphatase, bicarbonate and chloride).

Canonical analysis is developed from equation (2) by the supplying of extra variates on the left-hand side. We may now come full circle by reducing the right-hand side to a single *x*-variable. This produces the classical two-group discriminant problem originally posed and solved by R. A. Fisher. The best-known recent example of this technique in a medical context is the breast cancer discriminant of Bulbrook *et al.* (1962) which has been used in practice for prognostic purposes.

PRINCIPAL COMPONENT ANALYSIS

In all the examples so far there has been a split in the array of variates. All our methods have involved maximising a function, but always relative to another function. Put another way, we have assumed some kind of structure in the data and have asked how this structure is reflected in the measurements. The statistical position is far less satisfactory when the analysis is aimed at discovering structure, even though the methods in common use are in fact the oldest ones around. Multiple discriminant analysis rests upon the equation

$$(B - \lambda W)c = 0 \qquad (5)$$

and it appears to be only a slight change to the basic equation of *principal component analysis*

$$(B - \lambda I)c = 0 \qquad (6)$$

Here B is a dispersion matrix as before, but I is simply a unit matrix. Geometrically, the eigenvectors c now define the *principal axes* of the hyperellepsoid corresponding to B; it is conventional to standardise B, replacing it by the correlation matrix R and to say that, for example, the first principal component accounts for a fraction f of the total variance, where $f = \lambda_1/(\lambda_1 + \lambda_2 + \ldots + \lambda_p)$.

The main weakness of this method is its lack of freedom from scale-independence. Short of arbitrary standardisation (why standardise in terms of

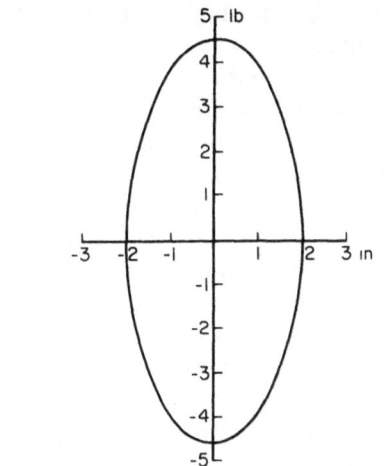

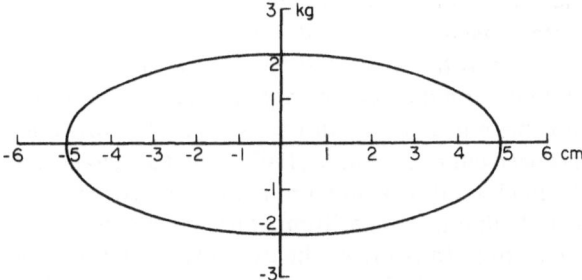

Figure 2. The same correlation ellipse plotted in two alternative sets of units.

the standard deviations, rather than the means, for example?), a change of units of measurement produces a complex change in the principal component coefficients—the components themselves may even change places as figure 2 shows. In one set of units, the first variable provides the first principal component; in another the first component is the second variable. In the same way, the practice of expressing the importance of each component in terms of a fraction of the sum of quantities of different dimensions is one to be questioned.

FACTOR ANALYSIS

It is important to distinguish between principal component analysis and *factor analysis*. Each aims at 'explaining' some part of the variability in the data by means of a few linear functions of the measurements, but what is 'explained' differs in the two cases. Factor analysis ignores the variances and aims only at the covariances. A perfect factor solution is obtained when each variable can be written as a linear function of a few factors plus a unique part which is uncorrelated with the factors and with the unique parts of all the other variables. The unique parts, one to each variable, show that factor analysis does not really reduce the original set of measurements; it 'explains' their covariances by introducing extra variables in the form of the factors.

Until quite recently, the computing methods used for factor analysis were highly simplified and the results were subjective to say the least. The advent of computers caused renewed interest in the maximum likelihood solution originally put forward by Lawley (1940). This solution has been re-derived by different methods by several other workers, notably Rao (1955) who, as something of a theoretical *tour de force*, relates the method to canonical analysis, the factors playing the part of the second set of variates. Unfortunately, the various iterative algorithms proposed all converge extremely badly in practice, and in a rather odd way; the change in the estimates from one iteration to the next is usually very small quite early in the process, but obstinately refuses to become any smaller. This may to some extent be because the coefficients of the linear functions (known in this context as the *factor loadings*) are only determined up to an arbitrary rotation or orthogonal transformation— quite large numerical changes in the estimates may thus have little effect on the fit. Recently, this knot has been cut by Jöreskog (1969), who has maximised the likelihood by a direct numerical approach using a hill-climbing algorithm. The results are very interesting. In many of the examples presented, the maximum of the likelihood occurs on the boundary of the permitted region, where one or more of the variables has no unique part and so becomes essentially identical with the factors. This seems to me to warrant three comments. First, it explains some of the computing difficulties; in most of the suggested techniques the quantities $1/u_i^2$, where u_i^2 is the variance of the unique part of the ith variate, occur as weights, and zero values of the u_i will clearly cause trouble. Secondly, considerable suspicion must be attached to factor solutions obtained by standard methods, which never show this property; indeed, the properties of likelihood maxima at which the derivatives do not vanish themselves require investigation. Finally, if what is needed is a choice of a few single variates to 'explain' the total variability (and this is in fact what many users of principal component and factor analysis

are actually in search of), it may be best to use a method which provides this directly. There is room here for research.

CLUSTER ANALYSIS

Finally, the newest branch of multivariate analysis, and at present the most fashionable, is undoubtedly cluster analysis. Here again we face data which are *a priori* structureless and seek to divide the units into clusters in such a way that units within a cluster resemble each other more than units in different clusters. The enormous literature on this topic indicates that it answers some kind of felt need—there have, for example, been several attempts, none very successful, to cluster medical

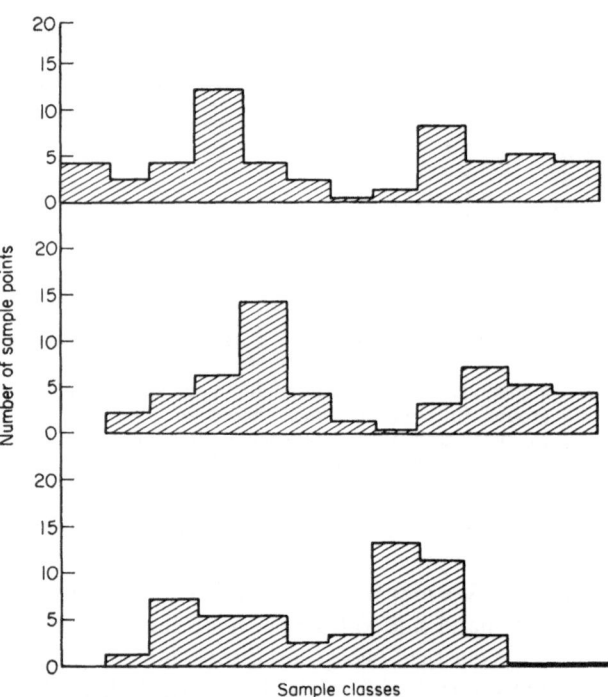

Figure 3. Bimodality in random samples of 50 taken from a 10-dimensional Normal distribution (from Day, 1969).

cases so as to produce new diagnostic groupings. Happily, a recent survey paper (Cormack, 1971) makes it unnecessary to summarise this literature here. What does seem still to be lacking is a treatment of the clustering problem from a fully statistical viewpoint. Generally speaking, many cluster analyses aim to be *predictive*, in that it is intended to allot future data to whatever clusters emerge from the analysis. Yet we have very little information, either theoretical or empirical, about the stability of clusters from a sampling viewpoint. What information we do have is fairly ominous. Day (1969) has treated the very straightforward problem of fitting a mixture of two Normal distributions to a set of data. He points out that the distance between the two means is an essentially positive quantity, and that its estimate must therefore be biased upwards when the true value is zero—when in fact we are attempting to fit a mixture of distributions to data which are actually homogeneous. What is disconcerting is the size of the bias; in 10 dimensions with sample sizes of 50, it amounts to no less than 5 to 6 standard deviations. This means that, in a sample of this size and dimensionality, there will usually be some direction or other in which the sample appears completely bimodal, and figure 3, taken from Day's paper and derived from a single Normal population, shows that this is indeed so. If this result is more generally true, it must cast doubt upon the validity of cluster analysis as a technique until it is provided with the equivalent to a good set of significance tests.

This brief survey has touched only upon the most conventional of multivariate methods. Its omissions may indicate areas where such methods are lacking, or are at least not in general use. One such area is that of multiple 0-1 variates. Cluster analysis operates on such data but very little is known about the statistical properties of the results, to some extent because of the absence of good underlying models. I hope I may have suggested to some statisticians that there is interesting work waiting to be done, and to some biologists that there are already interesting ways of answering some of their questions.

REFERENCES

BULBROOK, R. D., HAYWARD, J. L., SPICER, C. C., and THOMAS, B. S. 1962: A comparison between the urinary steroid excretion of normal women and women with advanced breast cancer. *Lancet* **2**: 1235.

CORMACK, R. M. 1971: A review of classification. *J. Roy. Statist. Soc. A.* **134**: 321-367.

DAY, N. E. 1969: Estimating the components of a mixture of normal distributions. *Biometrika* **56**: 463-474.

FRASER, P., HEALY, M. J. R., ROSE, N., and WATSON, L. 1971: Discriminant functions in differential diagnosis of hypercalcaemia. *Lancet* **1**: 1314-1319.

JÖRESKOG, K. G. 1969: A general approach to confirmatory maximum likelihood factor analysis. *Psychometrika* **34**: 183-202.

KETTENRING, J. R. 1971: Canonical analysis of several sets of variables. *Biometrika* **58**: 433-452.

LAWLEY, D. N. 1940: The estimation of factor loadings by the method of maximum likelihood. *Proc. Roy. Soc. Edinb.* **60**: 64-82.

MARDIA, K. V. 1971: The effect of non-normality on some multivariate tests and robustness to non-normality in the linear model. *Biometrika* **58**: 105-122.

RAO, C. R. 1955: Estimation and tests of significance in factor analysis. *Psychometrika* **20**: 93-111.

RAO, C. R. 1966: Covariance adjustment and related problems in multivariate analysis. *In* 'Multivariate Analysis' (Ed. P. R. Krishnaiah). Academic Press, New York.

COMPUTER PROCESSING OF ELECTROCARDIOGRAMS

P. W. MACFARLANE AND T. D. V. LAWRIE[1]

Electrocardiography is one of the branches of medicine which has benefited considerably from advances in technology in recent years. In the 1950s the patient had to be transported to the electrocardiograph, which was a large immobile piece of apparatus. Nowadays portable recording devices are in common use. To a certain extent the full circle has been turned with the introduction of magnetic-tape recording equipment which in itself is usually larger than the previous generation of equipment, but with the recent availability of cassette tape recorders and integrated circuitry, it will only be a matter of time before this type of electrocardiograph too becomes relatively portable.

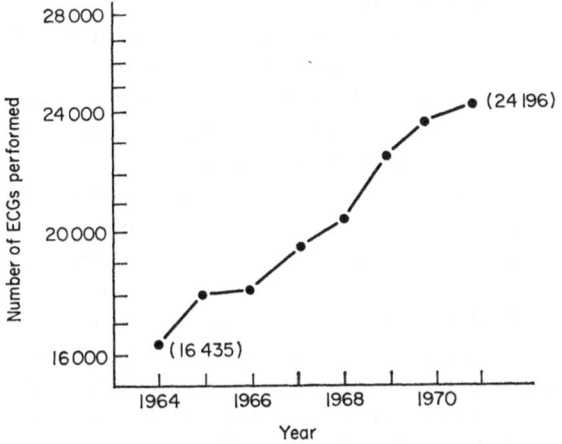

Figure 1. The number of ECGs recorded annually in Glasgow Royal Infirmary and Associated Hospitals, 1964-1971.

With the widespread use of small digital computers in many laboratories, including those in hospitals, the most recent advance in electrocardiography has been towards providing some form of automated interpretation of ECGs. It is fortuitous that this development has occurred at a time when the number of ECGs being routinely requested in hospitals is increasing to a considerable extent (see figure 1).

ECG interpretation is a skilled task which should be undertaken by physicians with considerable experience in this field. It is well recognised that there is an observer error which is inversely related to experience in interpretation. In large hospitals where the reporting commitment may be very heavy, e.g. 500 ECGs per week, it is impossible for experienced cardiologists to set aside the time required for this task and, therefore, more junior personnel have to be used. In some hospitals where there are no cardiologists, general physicians without special training are left to make their own interpretation of ECGs. In view of this and the increasing number of ECGs being recorded, it is to be hoped that automated methods for ECG interpretation will prove acceptable. This paper discusses the introduction of such a system into routine service in the Glasgow Royal Infirmary in January 1971.

METHODS

ECG requests are normally made by medical staff who fill out the appropriate form containing details of the patient's name, address, hospital number, etc. These are imprinted on to one corner with an embossed plate which is provided by the records office for each patient on admission. The hospital number consists of six numerical digits followed by an alphabetic check digit, a unique character calculated from the previous six digits. While every laboratory test is requested on a different form, each has this common identification area.

On the ECG request form, medical staff are required to indicate a provisional diagnosis by ticking off one of sixteen different choices such as myocardial infarction, hypertension, or respiratory disease. In addition they have to indicate whether or not the patient is receiving digitalis therapy. There is also space available for provision of a short written clinical history. In the cardiology department the request cards are sorted according to the various wards, and technicians are assigned a group of patients who are situated in close proximity.

The ECG recording equipment used is shown in operation in figure 2. Each data-acquisition trolley consists of a four-channel tape recorder, ECG amplifiers, and a monitoring oscilloscope. The three-lead ECG is used as derived from the modified axial lead system (Macfarlane, 1969). An example of a three-lead ECG is shown in figure 3. With this type of ECG three signals are recorded simultaneously, representing potential differences in

[1] University Department of Medical Cardiology, Royal Infirmary, Glasgow, Scotland.

three mutually perpendicular directions, normally denoted X, Y, and Z. These leads can be combined in pairs on an oscilloscope to form loops which are known as a vector-cardiogram (figure 4). These loops represent the projections on to three mutually perpendicular planes, of a spatial loop, which depicts the cardiac electrical activity. This latter loop is generated from a knowledge of the amplitude of

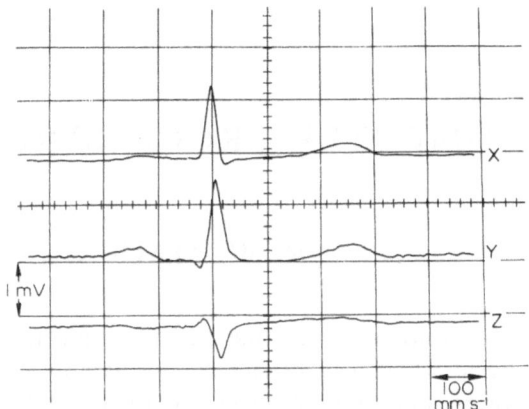

Figure 3. A three-lead ECG recorded at 100 mm/s sweep speed. The leads are called lateral (X), inferior (Y) and anteroseptal (Z).

are no faults. At the end of each recording session, either in the wards or at a clinic, only the completed magnetic tape need be returned to the computer centre within the hospital for interpretation.

Computing methods

The computing system used is based on a PDP8, which has 8K core store, two DEC tapes, extended arithmetic element, analogue to digital conversion facilities, and a small 32K disc which is not essential for ECG interpretation at present.

At the beginning of each interpretation session details of the hospital, date, and technician(s) who recorded the group of ECGs to be analysed are inserted into the computer. In addition, prior to the input of each ECG the details of each patient's name, age, hospital number, clinical classification, sex, digitalis therapy, and ward are input. All of these are obtained from the ECG request form completed by the physician. Two technicians assist at each interpretation session. While one is inserting

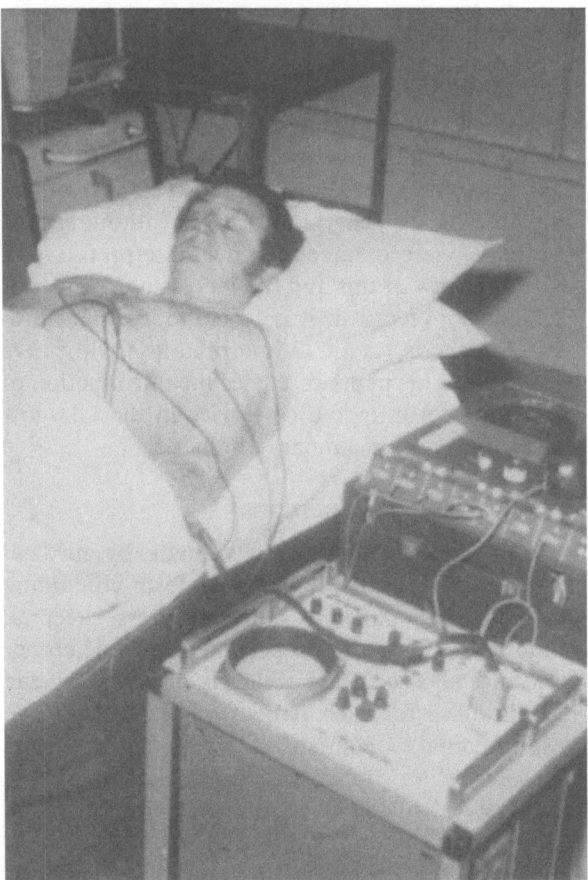

Figure 2. A patient connected to the ECG recording equipment, which consists of ECG amplifiers, a four-channel monitor oscilloscope and tape recorder.

the signals in the X, Y, and Z directions during the cardiac cycle (figure 5). If these signals alone are recorded, the loop and its projections can be calculated mathematically without necessitating an oscilloscope display.

Prior to recording an ECG, a technician records the patient's name, etc., on a voice channel and notes the position of the magnetic tape on a worksheet. Then approximately twenty seconds of ECG are recorded. During this phase the ECG is monitored on the oscilloscope and if for some reason it is unsatisfactory, perhaps due to patient movement, the technician can easily record the tracing again. When a recording has been made it is replayed and monitored again to make sure there

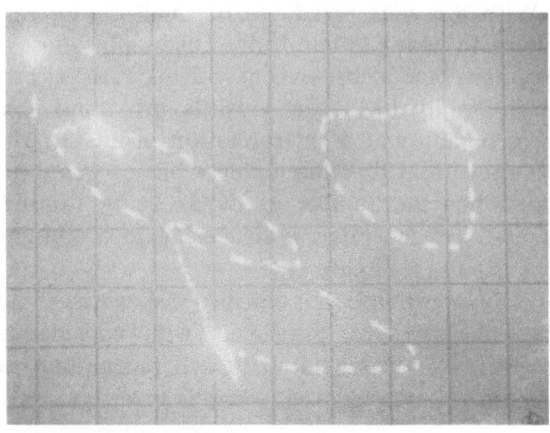

Figure 4. A vector-cardiogram derived from signals similar to those shown in figure 3.

details of the patient into the computer the other is making a copy of the relevant ECG using a multi-channel ink jet recorder. The copy is given to the computer operator who scrutinises the tracing in order to select a technically satisfactory portion for analysis. The ECG is then replayed again but this time for input to the computer. It is monitored using a small oscilloscope, and the operator, having already seen the tracing, is able to indicate to the computer the instant at which to start the sampling of the

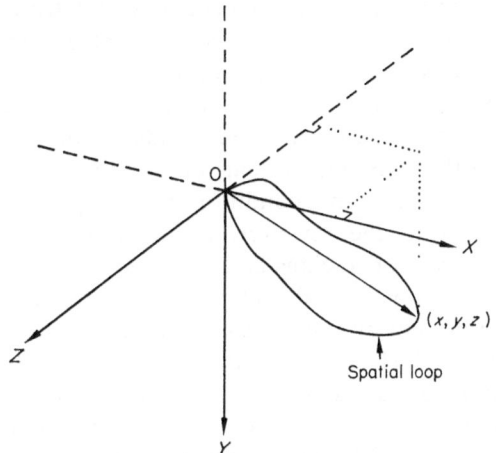

Figure 5. The spatial loop which depicts cardiac electrical activity. The trajectory of the loop is calculated from the amplitudes of the *X*, *Y*, *Z* leads at the same instants in time, as shown for one instant in the figure.

ECG for analysis. This is done by using a 'remote-control' unit, which is a simple level-changing device attached to the analogue-to-digital converter.

On input to the computer the ECG is converted from analogue to digital form at a rate of 500 samples per second. Two seconds of the three leads are input simultaneously and thereafter seven and a half seconds of one lead only are input for analysis of rhythm. This lead is indicated to the computer *via* the teletype and is chosen either because it exhibits the most prominent *P* waves or, on occasions, because it is the lead which is most free of artefact. At the end of ECG input the computer asks whether the portion input is acceptable, and if it is not the operator may request the phase to be repeated. Alternatively the program proceeds to analyse the ECG. During this interpretation, it is feasible to input the details of the next patient, thereby optimising the throughput by overlapping two procedures.

Interpretation methods

The first stage in the interpretation actually takes place during the input of the seven and a half seconds of one lead for analysis of rhythm. At this time the data are smoothed on line using a recursive filter. Thereafter the ECG is differentiated and the wave-recognition procedure for this phase of the

program is carried out mainly on this waveform. All the *QRS* complexes are found, a check on regularity made and an average heart rate computed. Then a search of each *R–R* interval is made to find if any *P* waves are present. These will only be recognised if there is a short zero gradient followed by positive and negative gradients. If this fails, *P* waves are regarded as absent. In other cases a check is then made to find the number of *P* waves in each *R–R* interval. Additional *P* wave routines include a test for flutter waves which show as a series of square waves in the derivative, and a check for *PR* interval variability which would arise for example in cases of heart block or Wenkebach phenomenon. If any abnormality is present, a further analysis is carried out to determine the precise nature of the arrhythmia. This is undertaken using a logical decision tree, of which a small segment is shown in figure 6.

The remainder of the interpretation is common to all ECGs whether or not they exhibit an arrhythmia. The basic methods, summarised below, have been described previously (Macfarlane, 1971). Initially the 'spatial velocity' of the three leads simultaneously input is determined from the following equation:

$$SV = \frac{(a^2 + b^2 + c^2)^{\frac{1}{2}}}{T}$$

where *a* is the change of amplitude of the *X* lead, etc., in the sampling interval *T*. This effectively corresponds to the rate of inscription of the spatial loop. The maximum values of *SV* act as reference points and always occur during the *QRS* complex

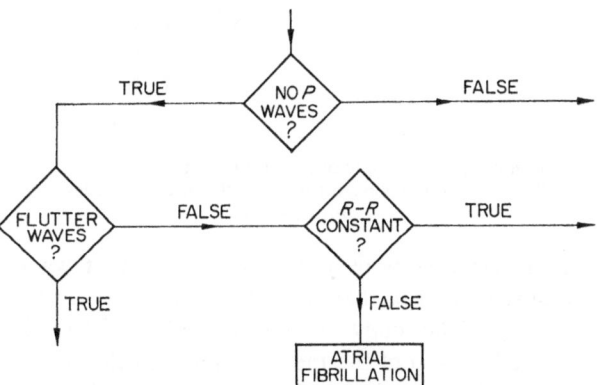

Figure 6. A small segment of the logical-decision tree used in arrhythmia interpretation. There are several different ways of arriving at a conclusion of atrial fibrillation other than that shown.

which can therefore be easily recognised. Data are then scanned outwards from one such reference point until values of the spatial velocity fall below the critical level which determines the onset and termination of the *QRS* complex. Thereafter *P* and *T* waves are found similarly. The amplitude and durations of all these components are then determined and the parameters which characterise the vector-cardiogram loops are calculated from the three simultaneously recorded leads to complete this phase of the program.

The diagnostic interpretation is produced by following a logical decision tree similar to that for arrhythmia analysis. The computer program contains details of normal values of the ECG components (Macfarlane *et al.*, 1971) together with abnormal values found in the various diseased heart states. The newly measured values are then compared with the stored data to check for the presence of any abnormalities.

The duplicated printout from the computer on

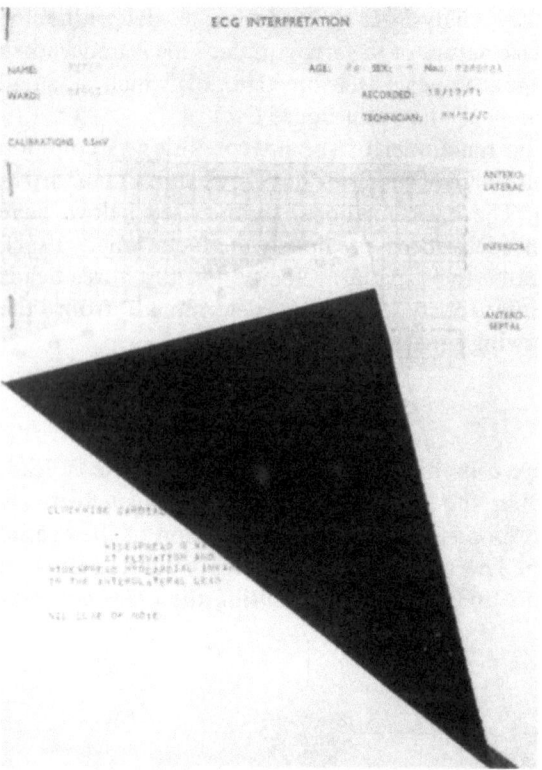

Figure 7. An example of a computer interpretation. The top copy has been folded back for demonstration purposes.

preformatted stationery consists of a set of wave measurements together with their diagnostic interpretation. The copy of the three-lead ECG is mounted on top of the measurements so that these need not be consulted unless desired. This has the effect of making the presentation less formidable and similar to that in use prior to the introduction of computing methods. One copy of the report is sent to the ward or clinic and the other is retained in the cardiology department for filing purposes (figure 7).

DISCUSSION

This service has been offered to Glasgow Royal Infirmary from January 1971. During that year

7 703 ECGs, representing a total of 40 per cent of the commitment of this hospital alone were interpreted by computer. This percentage could not be exceeded for various reasons, the most important being that no more than two hours per day of computer time was available since the PDP8 computing system had to be used for research and development during the remainder of the day.

All of the interpretations were reviewed by experienced personnel before being distributed. This was done to check for the presence of computer errors and so that methods could be updated in the light of experience. In the first six months of operation it was found that 4 per cent of the tracings had serious misinterpretations. For example, the computer occasionally detected bundle branch block where this was not present, or on the other hand failed to detect this condition when it was clearly apparent to the reviewer. In addition a further 4·5 per cent of tracings had less serious errors. For example, myocardial ischaemia was reported as being present on the basis of *T*-wave abnormalities in one lead when, in fact, these were present in two or more. As a result of these errors improvements were made to the methods of interpretation and a new version of the program was released in January 1972. To date this has proved extremely successful. It should be pointed out that the arrhythmia routines have been incorporated into the new version only and the error figures previously quoted do not relate to this part of the analysis.

While these error figures cannot give rise to any complacency, they are regarded as very encouraging. It is thought that if a varied group of physicians, from consultants to senior house officers, reviewed a series of ECGs, an error rate at best comparable with that achieved by automated methods would be obtained. The new program has given rise to the hope that a very low percentage of errors indeed will ultimately result through the use of computer interpretation.

It is thought that the low error rates achieved are due in some measure to the fact that the computer operator is free to choose the technically most satisfactory portion of the tracing for analysis. When upwards of 20 000 ECGs are recorded annually in a large hospital it is unrealistic to expect that each will be technically perfect. Therefore any technique which is proposed for automated ECG interpretation should be capable of dealing with ECGs which are technically relatively unsatisfactory. Of course, if there is no portion of the tracing which has a reasonably stable section of base-line it will be impossible to achieve a sensible answer. In this situation, the physician would make a more accurate interpretation than the computer.

Arguments about the comparative reliability of an automated ECG analysis and a physician's interpretation will continue for years to come. Each mode has its own advantages. On the one hand, the computer can continuously interpret ECGs without tiring, and in each case give an unbiased, objective answer; a physician could not compete in this respect. On the other hand a physician may be able to give a more accurate report on an ECG containing some complex arrhythmia. However, such tracings make up a very small percentage of a routine sample and are mainly confined to coronary care units where ECGs are normally interpreted at the bedside. In addition, the computer can have built into it the experience of many cardiologists, and whereas it may be true to say that on average a consultant cardiologist may make as accurate an interpretation as the computer, such trained personnel are not always available to report ECGs.

Not all the problems associated with the automated interpretation have been overcome. The most notable problem is the absence of techniques dealing with the comparison of several ECGs in order to detect sequential changes following myocardial infarction. This would necessitate storing the results of several tracings so that *ST–T* segment changes, for example, could be followed. To this end we are currently developing techniques in our own department and hope that these will be completed in the very near future.

While telephone transmission of ECGs is commonplace in the United States this is not so in the United Kingdom, because techniques have not been developed and the use of commercially available equipment for this purpose has not been authorised. We are currently carrying out an evaluation of the various methods which could be used for this task.

Our experience to date with automated ECG processing has given rise to sufficient confidence in the method to enable us to proceed with the establishment of a regional centre to provide an ECG interpretation service for several hospitals in the Glasgow area. A PDP8E computing system equipped with 20K core store and two medium-size discs was commissioned in early 1972. In the course of the next few years various hospitals will be linked to this computer for the provision of a routine service. It is to be hoped that this will be the forerunner of several such systems in the United Kingdom for, in our opinion, only through the use of automation can the demand for increasing numbers of ECGs be met. It is felt that in the near future automated ECG interpretation will be the method of choice.

SUMMARY

This paper briefly reviews the methods used in Glasgow Royal Infirmary for ECG interpretation by computer. Details of the recording and computing methods are presented together with a discussion of one year's operational experience. The conclusion drawn is that the method will shortly be the one of choice.

REFERENCES

MACFARLANE, P. W. 1969: A modified axial lead system for orthogonal lead electrocardiography. *Cardiovascular Res.* 3: 510.

MACFARLANE, P. W. 1971: ECG waveform identification by digital computer. *Cardiovascular Res.* 5: 141.

MACFARLANE, P. W., LORIMER, A. R., and LAWRIE, T. D. V. 1971: Normal ranges of modified axial lead system electrocardiogram parameters. *British Heart J.* 33: 258.

SCAN: A DATA-HANDLING SYSTEM FOR USE IN MEDICAL RESEARCH

M. M. JORDAN AND J. MACGREGOR[1]

Since W. Hollerith invented the first punched-card sorter in 1886, researchers have used various types of mechanical, electro-mechanical, and electronic devices to relieve them of the tedious calculations which often accompany research investigations.

As shown by the first and second IBM medical symposia (1959, 1960) the majority of the medical profession only realised the potential of the computer in the early 1960s, although medical researchers had used computers long before this time. This may be due to the suitability of research data for computer processing. The analysis of such data is numerical in nature, for the use of statistical methods has gradually become an established procedure since John Gaunt, in the seventeenth century, attempted to classify mortality data by age and sex. Indeed, most research papers published today contain some 'statistical' results.

Despite the obvious uses of the computer in handling research data, very few systems have been developed specifically for this purpose. The tendency is to develop general record-handling systems which will cope with various forms of input and with the extremely wide range of information which is found on a case-sheet. These systems often need complex coding, and retrieval of data is difficult. There is a real need for a system designed especially for research which provides flexible retrieval and manipulation facilities and so allows the researcher to investigate his data thoroughly. The system described here, known as SCAN (Scheme for Computer Analysis of Numerical data), was designed for this purpose.

SHORTCOMINGS OF EXISTING SYSTEMS FOR RESEARCH PURPOSES

Some of the manufacturers, notably ICL and IBM, do provide statistical software. ICL's statistical routines, based on the Biomedical Data package (BMD) from the University of California (Dixon, 1965) are well known. They provide facilities for multiple regression, analysis of variance, discriminant analysis, principal component analysis, etc. The major disadvantages of these routines are that the input format is complex and they do not provide facilities for file handling.

Within a data-handling package there should be a system permitting data validation and checking by methods such as those described by Healy (1969), in which statistical procedures of distributional examination and the study of transforms are used to pick out discrepancies in data. Once established and checked, there must be further facilities for updating and editing files, and there must be a facility for restructuring the data into various formats suitable for use by statistical routines.

The structure of biological data is usually more complex than industrial or commercial data. Because of ethical and other considerations, it is often quite impossible to ensure that all relevant or interesting data has been acquired and recorded. There may be individual cases where it proved impossible or impractical to perform particular tests of function or to make specific measurements, so that a very large problem is always that of missing data. Flexible data-handling routines are essential to cope with this data and yet it is surprising how very few systems even outside the medical field provide this sort of facility (Cooper, 1969).

An important aspect of computer operations is the financial liability involved. In many instances, individuals are unaware of the true cost of the computer time that they use. It is therefore important that, in designing a system intended for large-scale use, the questions of efficiency and limitation in financial burden be kept foremost. A few large projects for case-sheet handling have been initiated, some in the United States and a few in Britain, e.g. the King's College Hospital project (Knight, 1969). The cost of such systems is very high. Research systems cannot afford this large expenditure and a general system with low running cost is much more practical than a large system specifically designed for a particular investigation.

Within the medical field, very few systems are orientated towards research projects and, of these, most have concentrated on input, neglecting retrieval and analysis. Often such systems are designed for routine case-sheet handling and used in the hope of building a data bank for future research purposes. Examples of this are SWITCH developed at the Western Infirmary, Glasgow (Kennedy et al.,

---

[1] University of Strathclyde, Glasgow, Scotland.

1968), and the system at the Bellevue Hospital Centre, New York (Korein *et al.*, 1966). These systems are useful in improving case-sheet handling, but the results of research investigations conducted on these 'data banks' must be regarded with scepticism for the conditions of sampling and checks on accuracy of results, etc., necessary in statistical work, are not considered.

Researchers at the Hutchinson Memorial Clinic, Tulane University Medical School (Schenthal *et al.*, 1961, 1963), developed quite a powerful system for research based on numerical, fixed field-length records, i.e. all items are coded numerically and each item always appears in the same place in the record. The selection procedures are powerful but there is no statistical output, only simple counts. These researchers tried to extend the system to deal with the case sheet but the fixed-field aspect is too clumsy to cope with all the alternatives needed in a case sheet.

ORGANISATION OF SCAN

The development of the SCAN system can be traced over the last ten years. Initially it consisted of one program on a Ferranti Sirius Computer but since 1965 it has developed into a comprehensive data-handling system based on a large computer. The system was designed initially for a KDF 9 but was later transferred to an ICL 1905 on which it runs at present. The present system is built round the original facilities of the Sirius program, which meant that the analysis part was designed first and the input and other facilities designed to suit it. SCAN is now a useful set of routines for research data and is currently used in many investigations.

SCAN a data-handling system for use in biological research

The organisation of the SCAN system is shown in figure 1. The upper part of the figure illustrates data-input and file-handling routines: the middle part illustrates the analysis requests which follow the file-handling routines; and the lower part shows the type of output from the system. Each of these will now be considered in turn.

Data collection

All data are collected on previously prepared forms which facilitate transfer to paper tape. The paper-tape version is written to magnetic tape and the usual checks for errors carried out. As research data can usually be held in record form with one record per subject, SCAN holds data as fixed-length numerical records.

Updating and file handling

Large volumes of data require comprehensive file-handling routines. The system allows the user to update a file on magnetic tape with new records or items. Any missing items are given a special symbol which the other routines in the system recognise.

File editing

The main editing routine allows the user to carry out calculations on items of his records (e.g. conversions from inches to centimetres) and mathematical transforms (e.g. log, square root). Transforms are often used prior to analysis to normalise the distribution of a parameter. The results of the calculations are stored as part of the record in locations specified by the user.

Analysis

There are several analysis routines, all of which contain record selection-facilities. The system allows the selection of records by series of selectors called hurdles. These hurdles are linked by boolean operators, i.e. 'and's' and 'or's'. Each hurdle refers to one item of a record and any record can be accepted or rejected if this item has a value between specified limits. For example, it is possible to extract all the records of subjects under 30 years of age by specifying that all records with item 3 (age) between 0 and 30 should be selected.

By combining a series of such hurdles with 'and's' and 'or's', quite complex searches can be made. This hurdle routine is an intrinsic part of every output program of the system.

There are several forms of output available, as illustrated by the next few figures. Firstly, lists of selected items from each record, or tables of means, standard deviations, standard errors of items are available. These are shown in figure 2. Each row of this table contains the mean, standard deviation, standard error, and coefficient of variation of a

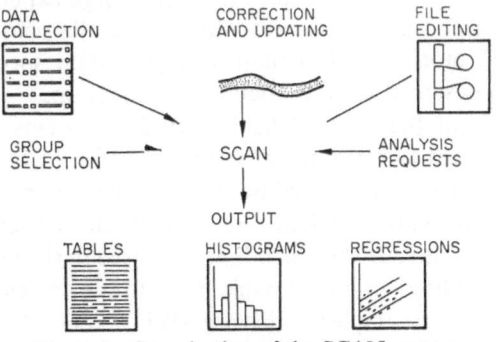

Figure 1. Organisation of the SCAN system.

MASK	1	PASS	76							
	NUMBER	AVERAGE	STD DEV.	STD ERROR	C.V.	M+2STD DEV.	M−2STD.DEV.	M+2S.E.	M−2S.E.	
1	0	0.000	0.000	0.000	0.0	0.000	0.000	0.000	0.000	
2	6	27.000	3.286	1.470	12.2	33.573	20.427	29.939	24.061	
3	15	36.600	2.501	0.669	6.8	41.603	31.597	37.937	35.263	
4	19	44.474	2.796	0.659	6.3	50.066	38.881	45.792	43.156	
5	17	55.765	3.113	0.778	5.6	61.991	49.539	57.321	54.208	
6	14	63.857	2.070	0.574	3.2	67.998	59.717	65.005	62.709	

	NUMBER	CORR.	STD ERROR	T VALUE	GRADIENT	INTERCEPT	STD ERROR
1	68	−0.194	0.118	−1.606	−5.371	1271.460	344.935
2	68	−0.185	0.118	−1.531	−5.245	1509.585	353.294
3	68	−0.267	0.118	−2.249	−0.501	109.163	22.983

Figure 2. Table of results output by SCAN.

particular item. The last four columns, i.e. mean ± 2 standard deviations and mean ± 2 standard errors, are output to make the charting of the data easier. This program also allows the calculation of correlation coefficients and linear regression analysis which are output in tabular form, as shown in figure 2 or in graphical form, as shown in figure 3. Figure 3 shows the line of best fit drawn on the graph plotter with 95 per cent tolerance limits calculated for the line. The original data points are also included.

It is also possible to produce frequency histograms on the lineprinter (figure 4). This routine saves a great deal of time and effort in analysis. Each frequency is calculated as a percentage of the total number of cases included in the histogram.

Other routines in the system include a sub-file creation routine, chi-square, and Student's *t*-test of significance and multiple regression analysis. None of these routines is elaborate but each is easy to use and produces easily understood results.

APPLICATIONS

The SCAN system has now been in use for five years although its development has still continued during this time. The simplicity of operation of the system may be judged by the number of projects already using the system. We have been involved in an investigation into the causes of osteoporosis, in a survey of primigravidae, in a study of diabetes in pregnancy, in a survey of intrauterine contraceptive devices, in a survey of the social aspects of venereal disease, and many other studies. As an example of the type of investigation involved, the investigation described below, although smaller than most trials using SCAN, is fairly typical of the type of problem encountered.

INVESTIGATION OF SKINFOLD THICKNESS
IN NEONATES

It is reported that babies of low birth weight demonstrate a higher incidence of physical and mental impairment in later childhood (MacDonald, 1962; Harper and Weiner, 1965; Drillien, 1969) and that there is an increase in mortality associated with increasing degrees of 'malnutrition' related to susceptibility to infection (FAO/WHO Expert Committee on Nutrition). The babies at greatest risk, therefore, are the 'small-for-dates' (SFD) babies with low birth weights presumably due to deficiencies in the intrauterine environment. To improve the management of these neonates, it is important to develop techniques for the clinical diagnosis of the condition and for defining its severity. At the present time, birth weight related to gestation is considered important in the assessment

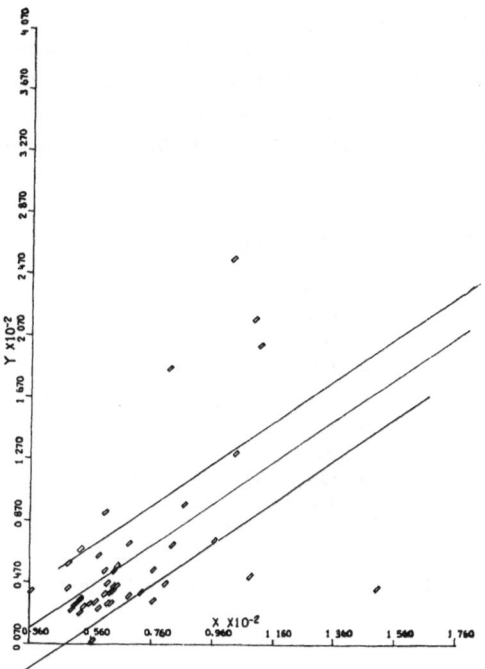

Figure 3. A series of data points with the line of best fit drawn on the graph plotter.

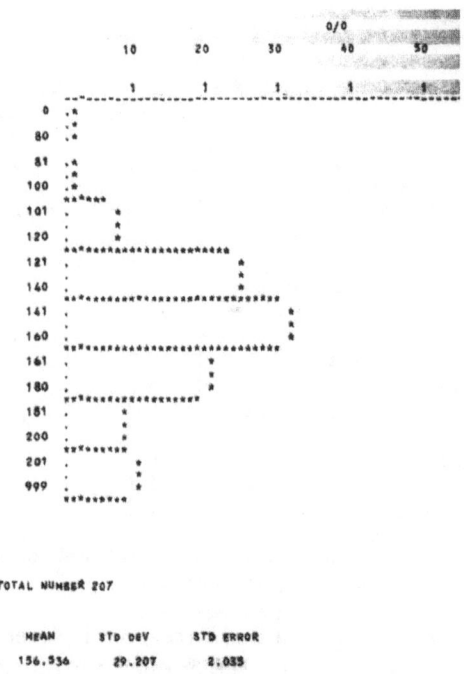

Figure 4. Frequency distribution drawn on the lineprinter.

of the degree of dysmaturity. Birth weight is accompanied by a reduction in tissue subcutaneous fat which is not routinely measured. This investigation was begun, therefore, to develop techniques for measuring subcutaneous fat in neonates, and to assess the usefulness of the methods in classifying the 'abnormal' neonate.

Data collection

The data collected on each neonate is shown in table 1. Five skinfold thicknesses were measured on each infant using a Harpenden caliper. The sites at which the measurements were made are shown in figure 5.

1	Hospital number
2	Sex
3	Birth rank
4	Gestation at delivery
5	Birth weight
6	Length
	Skinfold measurements
7	Biceps
8	Triceps
9	Quadriceps
10	Flank
11	Subscapular

TABLE 1: Items noted on each infant

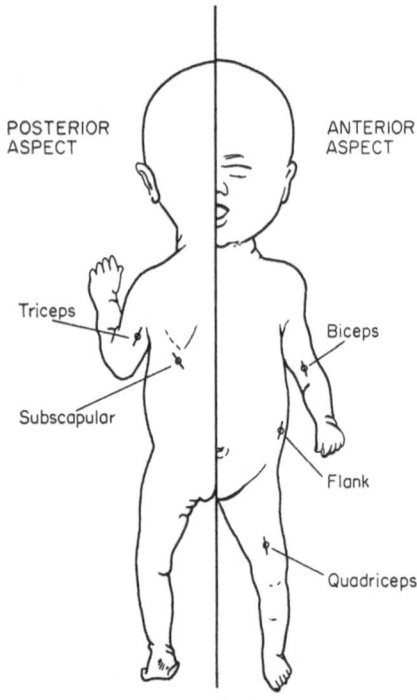

Figure 5. Sites selected for measurement of skinfold thickness.

The investigation was carried out in several parts. Firstly, assessment of error due to variance in site selection and operator error was investigated. Then a pilot study was initiated on a series of 87 'normal' neonates, and, finally, a group of dysmatures was compared with the 'normal' population. Only the second of these investigations is described here with particular reference to the comparison of male and female infants.

Data storage

The information on each baby only amounted to a few items so it was very easy to transfer to magnetic tape. Most investigations using SCAN are much larger than this one and often require several updatings before all data is on tape.

Data editing

It is known that the error attached to such measurements tends to be logarithmic and, therefore, the log transform of each measurement was calculated as was the sum of the limb measurements, sum of trunk measurements, the sum of all five measurements, and the sum of their log transformed values (table 2).

1	Log biceps
2	Log triceps
3	Log quadriceps
4	Log flank
5	Log subscapular
6	Biceps + Triceps + Quadriceps (total limb)
7	Flank + Subscapular (total trunk)
8	Total of all five measurements
9	Log limb measurement
10	Log trunk measurement
11	Log of all five measurements
12	Ponderal index (height/3/weight)

TABLE 2: Calculated items

Analysis

The records were divided into two groups, viz., male and female infants. Initially, the two groups were compared in terms of birth rank and gestation at delivery (table 3). As the table shows, the groups were remarkably similar in both these variables. It seemed reasonable, therefore, to proceed with the comparison of the two groups in terms of skinfold measurements.

	Male		Female	
	Number	%	Number	%
Primigravidae	17	35	14	37
Multiparae	31	65	25	63
Gestation 36 to 37 wks	5	10·4	4	10·3
38 wks	7	14·6	4	10·3
39 wks	13	27·1	9	23·0
40 wks	13	27·1	12	30·8
41 wks	7	14·6	7	17·9
42+ wks	3	6·2	3	7·7
Total	48	100	39	100

TABLE 3: Distribution of birth rank and gestation at delivery by sex in sample of 'normal' neonates

The distribution of each measurement was examined and most were found to be normal. The males, however, tended to have a wider variance in the biceps and subscapular measurements (table 4). This table also shows a comparison of the mean values of each measurement as does table 5 for the transformed data. It can be seen from both these

	Male				Female				Differences
	Mean	Standard deviation	Standard error	Variance	Mean	Standard deviation	Standard error	Variance	Significance level
Length (cm)	50·40	1·90	0·273	4	50·11	2·00	0·321	4	NS
Birth weight (kg)	3·54	0·51	0·070	13	3·41	0·42	0·070	12	NS
Ponderal Index	33·38	0·98	0·143	3	33·39	0·96	0·160	3	NS
Limb									
Biceps (mm)	4·03	0·91	0·131	22	4·26	0·61	0·099	14	NS*
Triceps (mm)	4·62	0·94	0·139	20	4·95	0·76	0·123	15	NS
Quadriceps (mm)	5·91	1·32	0·193	22	6·50	1·29	0·209	20	P < 0·05
Trunk									
Flank (mm)	3·82	0·74	0·108	19	4·23	0·63	0·103	15	P < 0·01
Subscapular (mm)	4·67	1·09	0·159	23	4·96	0·80	0·130	16	NS*
Total limb (mm)	14·56	2·89	0·422	20	15·69	2·30	0·074	15	NS
Total trunk (mm)	8·49	1·72	0·252	20	9·19	1·34	0·220	15	P < 0·05
SUM TOTAL (5 sites)	23·05	4·48	0·654	19	24·88	3·48	0·564	14	P < 0·05

* Significant difference in variance.   NS = Not significant.

TABLE 4: Differences in mean value of length, birth weight, Ponderal Index, and various skinfold measurements of male and female neonates

	Males				Females				Differences
Skinfold Site	Mean	Standard deviation	Standard error	Variance	Mean	Standard deviation	Standard error	Variance	Significance level
Limb									
Biceps	1·37	0·208	0·030	15	1·44	0·152	0·025	11	NS*
Triceps	1·51	0·199	0·029	13	1·58	0·172	0·029	11	NS
Quadriceps	1·76	0·224	0·033	13	1·85	0·209	0·034	11	NS
Trunk									
Flank	1·33	0·188	0·027	14	1·44	0·150	0·024	10	P < 0·005
Subscapular	1·53	0·225	0·033	15	1·59	0·168	0·027	11	NS*
Total Limb	2·66	0·205	0·030	8	2·74	0·153	0·025	6	P < 0·05*
Total Trunk	2·13	0·205	0·030	10	2·21	0·146	0·024	7	P < 0·05*
SUM TOTAL (5 sites)	3·12	0·198	0·029	6	3·21	0·154	0·025	5	P < 0·05

* Significant difference in variance.   NS = Not significant.

TABLE 5: Differences in skinfold measurements of male and female neonates using logarithmically transformed values

tables that differences between males and females do exist.

Table 6 shows the correlation coefficients of each of the measurements and birth weight for both the males and females. As birth weight has already been found to be a reasonable estimate of dysmaturity, it would be hoped that the skinfold measurements would correlate well with it. This was found to be the case in table 6, with most of the correlation coefficients differing significantly from zero at the 0·1 per cent level.

Discussion on the investigation

Although this investigation was not large, the calculations associated with it were considerable. These were as follows:

1 Calculation of transformed values, etc.

2 Calculation of frequency distributions of each variable.

	Birth weight	
Skinfold sites	Male	Female
Limb		
Biceps	0·53†	0·42*
Triceps	0·52†	0·47*
Quadriceps	0·70†	0·51†
Trunk		
Flank	0·47†	0·48*
Subscapular	0·63†	0·35*
Total limb	0·65†	0·55†
Total trunk	0·60†	0·43†
Limb and trunk	0·65†	0·53†

* Significantly different from zero at 1 % level.
† Significantly different from zero at 0·1 % level.

TABLE 6: Correlation coefficients of birth weight against skinfold measurements

3 Calculation of distributions of birth rank and gestation at delivery.

4 Calculation of means, standard deviations, standard errors, and coefficients of variation of both the arithmetic and transformed values for both males and females.

5 Calculation of t-values for differences between means and F-values for differences between variances.

6 Calculation of correlation coefficients of skinfold measurements against birth weight for both males and females.

The system produced all these results from the data stored on magnetic tape with little trouble and saved a great deal of effort on/the part of the investigator. It must also be noted that these results were only a part of the investigation and that the complete trial required much more analysis.

CONCLUSIONS

The numerical nature of many biological investigations makes them suitable for computer processing using statistical techniques. The computer would be a useful tool in these investigations but for lack of software and expertise.

Existing systems have three main faults:

1 The input specifications are too complex. This applies mainly to the manufacturers' routines which are so 'general-purpose' that they lose their simplicity. The designed system should strike a balance between options available and ease of use.

2 Emphasis is on form of input rather than retrieval or analysis. The analysis routines should include graphical output and simple statistical routines which usually cover most of the work needed in a research investigation.

3 Too many systems for use with clinical data have tried to combine case-sheet handling and research. The facilities needed for the two purposes differ quite radically and it is impossible to design a system which copes adequately with both.

Research investigations are not usually large and the cost of developing and using a system should not be out of proportion to the results achieved. Use of coded data facilitates retrieval and cuts costs by reducing computer time.

Finally, a system like SCAN is very useful in a well-designed investigation, but a hazard in a poorly designed one, for it is capable of producing results at an alarming rate. If the user is not certain of his requirements, the volume of data produced by the machine will only add to his confusion. While SCAN is a useful research aid, it is not a substitute for a well-designed trial.

Acknowledgments

The neonate skinfold thickness investigation is a current research project undertaken in collaboration with the Paediatrics Department of the Glasgow Royal Maternity Hospital. The measurements were made by Dr Angela MacGowan, and the results will be published in due course.

REFERENCES

COOPER, B. E. 1969: Statistical computing, past, present and the future. *The Statistician* **19**: 2.

DIXON, W. J., (Ed.) 1965: 'Biomedical Computer Programs-Health Services Computing Facility.' University of California, Los Angeles.

DRILLIEN, C. M. 1969: *Arch. Dis. Child.* **44**: 562.

FAO/WHO EXPERT COMMITTEE ON NUTRITION 1967: *7th report W.H.O. Techn. Dep. Ser. No.* 377.

HARPER, P. A., and WEINER, G. 1965: *Ann. Rev. Med.* **16**: 405.

HEALY, M. J. R. 1968: Disciplining medical data. *Brit. Med. Bull.* **24**: 3.

IBM 1959: *Proc. First Med. Symp.* (1959).

IBM 1960: *Proc. Second Med. Symp.* (1960).

KENNEDY, F., COX, A. G., GLEN, A. I. M., ROY, A. D., and SANDT, C. E. 1968: P. 85 *in* 'Computers in the Service of Medicine'. (Eds) G. McLaughlin and R. A. Skegog.

KNIGHT, J. E. 1969: The problems of implementation, *in* 'Computers in the Service of Medicine'. *Proc. British Computer Soc. Conf.*

KOREIN, J., BENDER, A. L., ROTHENBERG, D., and TICK, L. J. 1966: Computer processing of medical data by variable field length format. (*a*) *JAMA* **196**: 950; (*b*) *JAMA* **196**: 959.

McDONALD, A. D. 1962: *Brit. Med. J.* **1**: 895.

SCHENTHAL, J. E. SWEENEY, J. W., and NETTLETON, W. Clinical application of large-scale electronic data-processing apparatus.
  (i) 1960: New concepts in clinical uses of the electronic digital computer. *JAMA* **173**: 1.
  (ii) 1961: New methodology in clinical record storage. *JAMA* **178**: 3.
  (iii) 1963: System for processing medical records. *JAMA* **182**: 2.

INTERACTIVE COMPUTING

W. J. PERKINS[1]

Numerical data, from measurements taken on a patient or a biomedical experiment, usually require interpretation before any conclusions can be drawn from them. Although numbers are necessary for a precise analysis, our thoughts are normally organised in a qualitative manner, for which pictures and patterns are preferable. Thus the conventional plot of a variable with time can provide an immediate insight into the behaviour of particular aspects of a system. Further information might be obtained by rearranging or processing the data to display other features, whilst future trends might be predicted by analysis of previous patterns. The relationship of two parameters in a system might be deduced by plotting both to a common time scale or they may be plotted directly against each other, either method leading ultimately to rather refined pattern recognition by the observer. A computer allows data to be rapidly processed whilst display of the output on a visual display unit (VDU), in an immediately meaningful form, improves the communication from machine to man, and so acts as a direct aid to the thought processes.

In the study of biomedical systems a biomedical worker needs to be able to interact with the devices for measurement, processing and analysis as with the biomedical system. Communication into a computer can be improved by technology, which may ultimately enable computers to understand plain English, and possible the spoken word; or by education, to encourage computer users to become familiar with computer languages. Though some understanding of computer languages is necessary for biomedical workers wishing to use computers in this interactive mode, few wish to devote the necessary time and effort for complete familiarity, nor do they need to. Interactive programs may be written to allow for adjustment and modification through manual controls or by simple Teletype instructions. Sufficient options may be provided to allow a wide choice of input instructions and so improve the communication from man to machine. The value of an interactive visual display system as an aid to thought is immediately evident to a programmer, who can readily display selected sections of a program, make appropriate modifications and test whether they are acceptable. The relative ease of effecting such changes suggests the possibility of writing a main program in such a way as to allow at least for anticipated modifications. Values of x and y for graphic displays can be transferred from the computer store via digital/analogue converters to the Visual Display Unit (VDU), or read in from a semi-conductor writing tablet, light pen or similar interactive device which allows easy modification of a display.

A common requirement for the analysis of graphical data is a curve-fitting procedure. Within limits of accuracy alternative mathematical equations may fit the same data and a criteria of best fit may also be that one section of the curve is deemed biologically more significant than the remainder. Control of the fitting procedures on a VDU allows for a biomedical fit, using the computer to produce a mathematical fit. A more fundamental method is to simulate one's ideas about the behaviour of a biomedical system on a computer and to adjust parameters, whilst observing the output changes of the model in relation to corresponding experimental data. In biomedical research where it is necessary to determine the design of a system, as well as to understand its behaviour, interactive simulation is rather essential.

Such models require a mathematical definition; for others the data may be available as sets of coordinates that can be displayed on a VDU. For molecular structures, data from X-ray diffraction measurements provide the atom positions in space and the bond lengths in between. A two dimensional screen can only show a projection of any particular orientation of a three dimensional structure. However, a computer can calculate the transformed coordinates for different orientations and display the structure as a rotation about the three orthogonal axes. The orientation, the centre of rotation, the marking of atoms, can all be achieved by manual controls. Possible conformations for a molecular structure may be assessed by rotating sections of the molecule and correlating each position with the energy level associated with the proximity of the atoms. It is thus possible, using computer displays, to observe three dimensional structures on a two dimensional screen, rotation of the projection about the orthogonal

[1] National Institute for Medical Research, Mill Hill, London, England.

axes being equivalent to rotating the structure itself. Depth can also be indicated for stationary pictures by the introduction of perspective, size or brightness discrimination, or the addition of a stereo pair.

In vectorcardiography, only three fixed sections of the complete vector loop are displayed, yet it is conceivable that the effect of cardiac abnormalities might be more significant at other orientations. A relatively small computer with data conversion facilities is capable of displaying and manipulating the three dimensional vectorcardiogram derived from the weighted mean of the electrode lead system chosen. In assessing the effects of three variables upon a system, a family of curves drawn on a two dimensional graph is only suitable for a limited range of the third variable and it becomes necessary to indicate this in some other way, by using symbols, colours, or a contour plot. In some cases, three separate two-dimensional graphs for fixed values of the third parameter may be acceptable but if a computer is used, the complete three dimensional graph can be plotted and viewed either in stereo or with depth perception.

The computer has established its value for the processing, analysis and display of numerical data but where the data is pictorial it needs conversion into a form acceptable by the computer. The value of image enhancement is now evident from the moon pictures and is a technique with obvious application in medicine and biology, for example in radioscopy and microscopy. The reconstruction of three dimensional structures from electron micrographs is of considerable interest, combining modelling, displays and mathematical analysis, for which an interactive computer graphic system is ideally suited.

There are three possible systems at present for biomedical users of interactive graphic displays:

(i) a small computer with appropriate facilities but not so expensive that one person would not be allowed sole use. This has the limitation of small storage capacity and the fact that programming would normally have to be in assembly code.

(ii) A VDU as a time-shared terminal of a large computer. This would require a high speed data link to a remote computer or the VDU could be in the vicinity of a local computer. The need to time share would also introduce a variable delay in the response of the display to manual instructions.

(iii) A local small computer, incorporating interactive graphics and with provision for linking in to a large remote computer, has the merits of both systems without some of their attendant disadvantages.

The current development of large mass stores at an economic price and small size could well influence a decision in favour of the low cost local computer.

STUDIES ON HUMAN OCULAR TREMOR

H. BENGI[1] AND J. G. THOMAS[2]

INTRODUCTION

It has been known for many years that the human eyes are never quite motionless, even when steadily fixed at a point target (Adler and Fliegelmann, 1934). One particular component of the incessant motion is a very rapid tremor of the eyes. The amplitude of this tremor movement is usually less than 30 seconds of arc. This means that the amplitude of vibration of a point on the surface of the cornea is of the order of 0·001 mm. The frequency spectrum is concentrated mainly below 100 Hz, but sometimes extends to approximately 150 Hz. The tremor has horizontal (abduction-adduction), vertical (elevation-depression) and torsional components.

It has been suggested by some authors (Anderson and Weymouth, 1923; Marshall and Talbot, 1942) that tremor may play an essential role in determining visual acuity. Some kind of scanning process is envisaged, and vision is held to be an essentially dynamic process, instead of a static 'snapshot' affair. Other authors (Ditchburn and Ginsborg, 1953) have dismissed ocular tremor as merely a result of incomplete fusion of tetanic stimuli.

Like other types of physiological tremor, the fine tremor of the eyeball is a consequence of involuntary muscular contraction. The tonic impulses which arrive at the extraocular muscles are known to have several sources, and to reach the muscles by different

[1] Department of Electrical Engineering, Middle East Technical University, Ankara, Turkey.
[2] Department of Electrical and Electronic Engineering, University College, Cardiff, Wales.

routes. The various centres of origin include the labyrinth, the proprioceptors and the higher centres in the CNS. A 'block' diagram of the complete oculo-motor system is given in figure 1. In this diagram the Renshaw cell and the spindle are shown as possible sources of tremor, since the stretch reflex is known to be absent from the extraocular muscle system (Hoffmann, 1929; McGauch and Adler, 1932) in spite of the fact that these muscles contain a considerable number of spindles (Cooper et al., 1955).

For more than fifty years, much effort has been devoted to devising methods for recording eye movements. Of the very large number of different methods, only a few are sufficiently sensitive for recording involuntary eye movements, and still fewer are capable of the resolution which is needed for obtaining records of the fine tremor motion. The few tremor records which have appeared in the publications have been found to suffer from various defects, such as excessive noise content, too thick a trace, inadequate resolution in the time or displacement axes.

Three electronic methods which have been developed and used by the authors for recording the human ocular tremor in the form of angular velocity v. time records are described elsewhere (Bengi and Thomas, 1968). The first method involves the use of a piezoelectric strain gauge which is placed in contact with the sclera. The second involves the use of a piezoelectric accelerometer which is fixed to a contact lens. The third method depends on

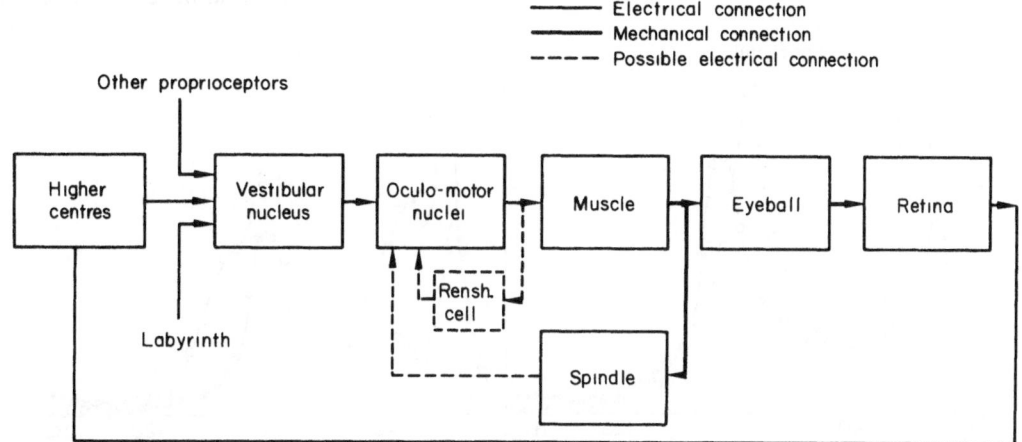

Figure 1. Diagrammatic representation of possible sources of ocular tremor.

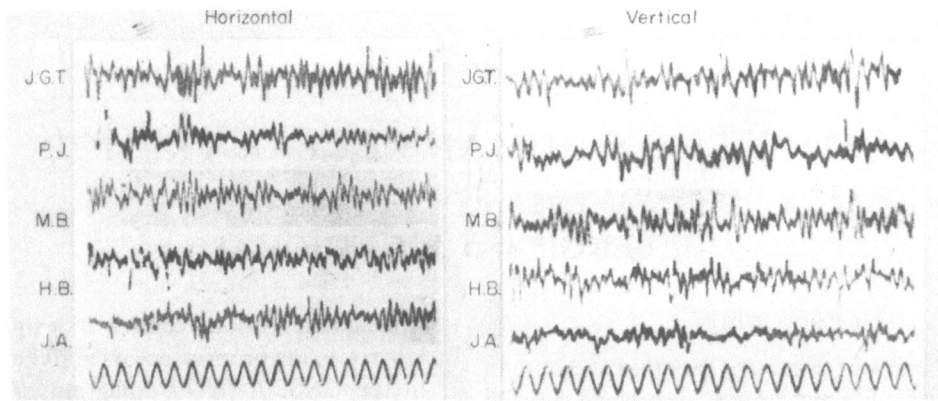

Figure 2. Angular velocity against time records of the tremor of the eyes of five different subjects.

the use of a capacitance gauge for recording the motion of the corneal protuberance in relation to a fixed probe which is placed near the eye. Horizontal and vertical tremor records of five different subjects recorded with the piezoelectric strain gauge, are shown in figure 2.

Visual examination of tremor records is of limited use in specifying the characteristic features of tremor. In order to obtain information about the origin of tremor it is necessary to be able to quantify it. Specifications of the spectral content is a powerful method for achieving this aim. In the case of a random function, such as tremor, the power spectral density is the relevant quantity. In this work the velocity spectral density (VSD) graphs of tremor records (which were available in the form of traces on 35 mm photographic film, and photographed from the screen of a storage oscilloscope) have been determined by projecting the traces on the translucent screen of a specially designed digital trace-reading potentiometer (Bengi and Thomas, 1967), measuring their ordinates at specified intervals of time, and finally punching the data values on paper tapes for a digital computer.

The work here presented covers a comparative study of ocular tremor in several subjects; studies of horizontal and vertical tremor in various positions of gaze, when the eye is loaded; under different

conditions of illumination; and in relation to head posture.

EXPERIMENTAL RESULTS

The results presented are in the form of velocity spectral density (VSD) graphs, computed by means of a digital computer. The ordinates of the graphs are in arbitrary units which are consistent within any set of graphs, so that changes in the amplitude and the shape of the spectrum can be seen when the recording conditions are varied.

Ocular tremor in the primary position

Fixation tremor of five different subjects has been measured. The amplitude and frequency distribution of their tremor spectra have been found to have certain similarities. Horizontal and vertical tremor spectra of two subjects are shown in figure 3. Observations on the spectral density graphs reveal the following: a difference exists between the horizontal and vertical tremors; the frequency spectrum extends up to about 120 Hz, and in none of the graphs is there an appreciable amount of energy in the frequency components above 120 Hz; in all graphs a peak at 70–80 Hz is quite prominent; this peak however varies in amplitude from one subject to another, and its frequency

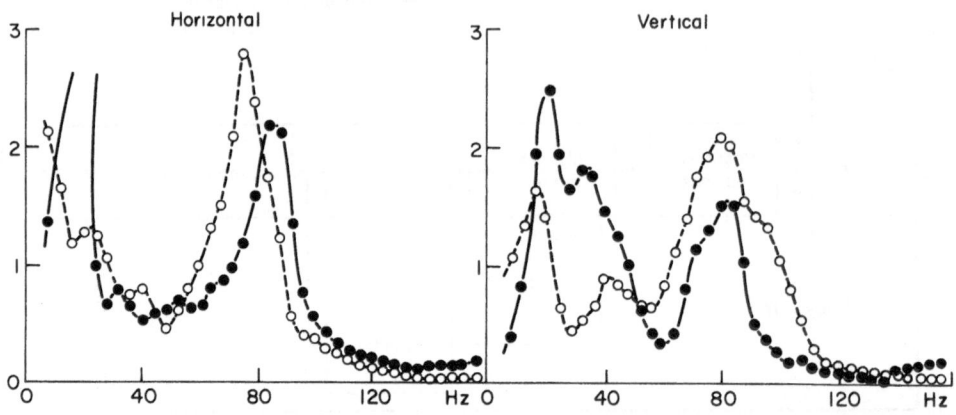

Figure 3. VSD graphs of fixation tremor of two subjects. o–o–o subject J.G.T.; ●–●–● subject H.B.

also shows some variability; there is also evidence of a peak at low frequencies, near 40 Hz, but the amplitude of this peak shows considerable variation. This particular peak is often partially concealed by the variability in the spectra at the low-frequency end, as the graph sometimes rises so sharply that the low frequency peak is not visible. It seems that the amplitude of the high-frequency components in the graph is not closely related to the amplitude of the low-frequency components, because from one graph to another the two main peaks differ widely in amplitude. In both horizontal and vertical tremor, there is often a change in the spectrum from one sample of tremor to the next; i.e. from one 0·625 s interval to the next. In figure 3 each graph represents the average of four individual graphs each computed from 0·625 s record. These average graphs obtained from tremor records made with the piezoelectric strain gauge are much smoother than the constituent graphs, and the two main peaks can be seen fairly clearly.

Comparison of tremor in the primary position when 'fixating' or 'not fixating'

This study was prompted by recent work on the alterations in limb tremor on changing from the resting state to preparing to perform a motor task (Gel'fand *et al.*, 1964; Pal'tsev, 1964; Halliday and Redfearn, 1956). In the present experiments a similar effect has been looked for in the ocular system. Recordings were made with the subject alternately fixating a point target and then ceasing to fixate. The results obtained with one subject are given in figure 4. These graphs were all obtained from tremor records made with the piezoelectric

accelerometer, and a pre-'whitening' with $\alpha = 0·9$ has been applied to the digital data (Blackman and Tukey, 1958). The results show that, when not fixating there is a distinct reduction in the amplitude of the spectrum throughout the frequency range. Some flattening of the spectrum is also seen.

A similar experiment with a second subject, using the piezoelectric strain gauge yielded a similar result. A third subject showed a much smaller difference between the 'fixating' and not 'fixating' conditions.

Fixation tremor in the adducted and elevated positions of the eye

When the eye is steadily 'fixating' some point away from the primary position, the results of figure 5 show that there is a marked change in the spectrum of tremor, compared with that found in the primary position. The graph *b* corresponds to the condition in which the eye is adducted but not elevated, and the graph *d* corresponds to the condition in which the eye is elevated but not adducted. Measurements were made with the eye deviated by 0, 10, 20 and 30 degrees from the primary position.

The principal changes which occur when the eye is deviated from the primary position are, first, the disappearance of the peak which was found at 70–80 Hz in the primary position; and, second, a spreading of the energy of the tremor into the higher and lower components of the spectrum. With increasing deviation of the eye from the primary position, either in adduction or in elevation, the spectrum flattens more and more, and the activity spreads into the higher frequency region of the spectrum. These graphs were obtained from

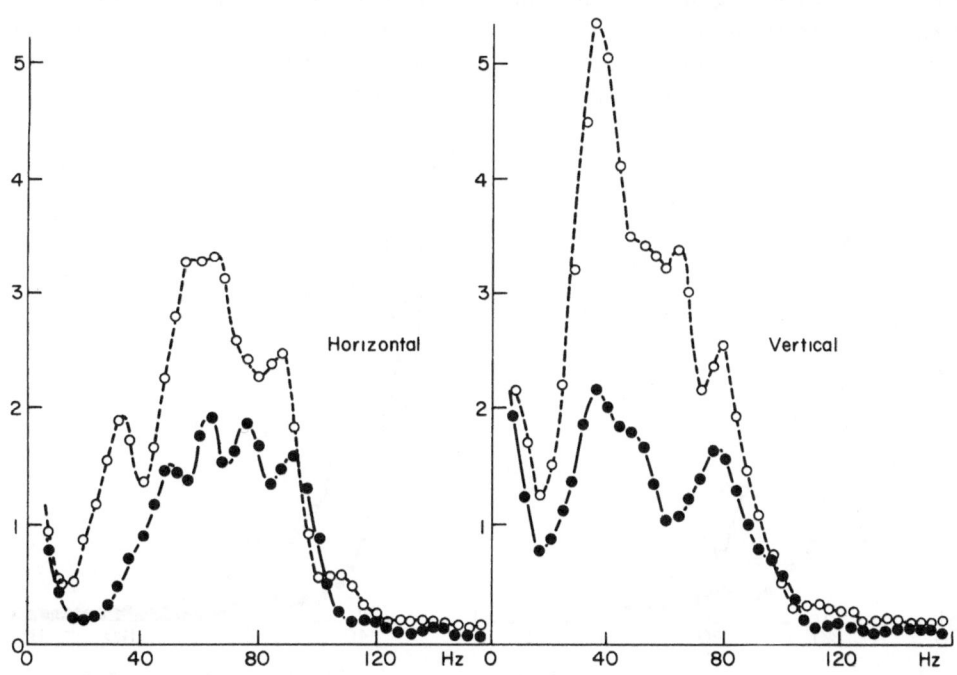

Figure 4. VSD graphs of ocular tremor when fixating and not fixating. o–o–o fixating; •–•–• not fixating.

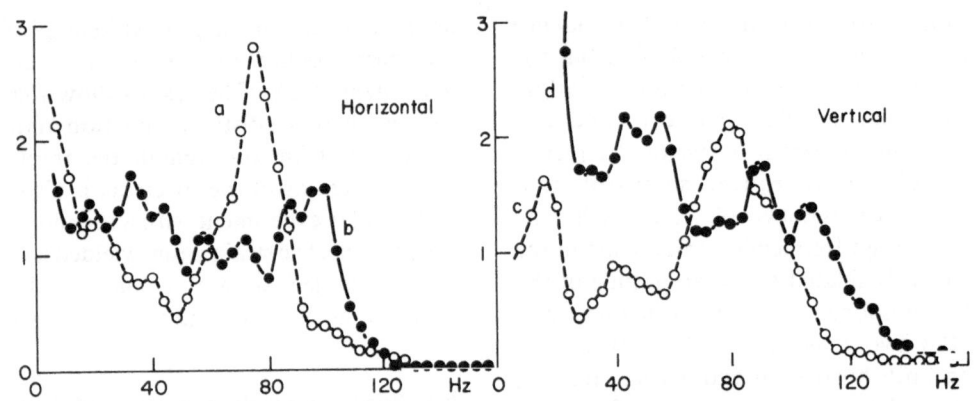

Figure 5. The effect on VSD graphs of moving the eye away from primary position. o–o–o primary position; ●–●–● 20° deviation.

tremor records made with the piezoelectric strain gauge.

Fixation tremor when the eye is loaded

In order to examine the effect on the tremor spectra of changing the mechanical characteristics of the eyeball-muscle system, the effective moment of inertia of the system was increased by adding weights to the contact lens. Results of measurements are given elsewhere (Bengi and Thomas, 1968). As the moment of inertia is steadily increased the following important features are observed: the amplitude of the velocity spectral density is progressively reduced; the low-frequency peak moves steadily downwards in frequency; and the high-frequency peak, at 70–80 Hz, is reduced in amplitude, but is not changed in frequency. Further increase in the added inertia tends to suppress completely the high frequency tremor movements of the eye. Measurements were made with the following set of added inertias: 5, 6·1, 7·3, 10, 12 and 15 g cm² (see figure 6). This experiment was carried out with one subject. His tremor was recorded with the piezoelectric accelerometer when he was fixating in the primary

position of the gaze. He was lying supine so that the weights would not have a tendency to rotate the eye, as they would if the subject were sitting.

Experiments were also made with the electrical model of the eyeball-muscle system shown in figure 7A. This model was based on the mechanical model formulated by Thomas (1967). In the electrical model the components are identified as follows: the capacitors C_p, C_s, C_t and the resistors R correspond to the elastic and viscous damping components of the extraocular muscles respectively; the voltage generators V are the analogues of the force generators in the active components of the muscles; the inductor L_e and the resistor R_e correspond to the mass and viscous damping of the eyeball respectively. The magnitudes of these parameters in mechanical units, both for horizontal and vertical eye rotations, are given by Thomas (1967). The current I in R_e is the analogue of the angular velocity of the eye movements. Transfer characteristics of the model were determined by using an A.C. voltage generator for one of the sources shown in figure 7A, and plotting the ratio I/V to a base of frequency. Transfer characteristics were determined for the following set

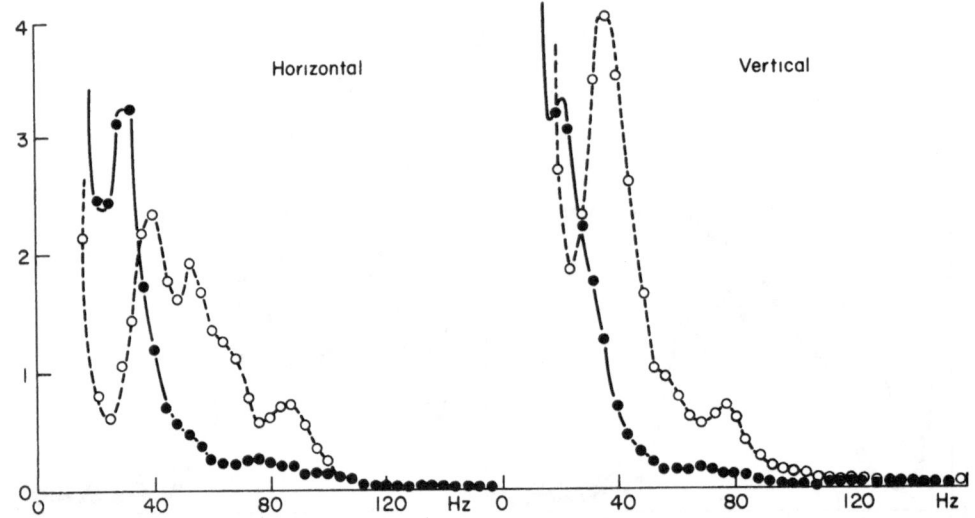

Figure 6. VSD graphs of ocular tremor with added inertia. o–o–o added inertia of 6·1 g cm²; ●–●–● added inertia of 15 g cm²;

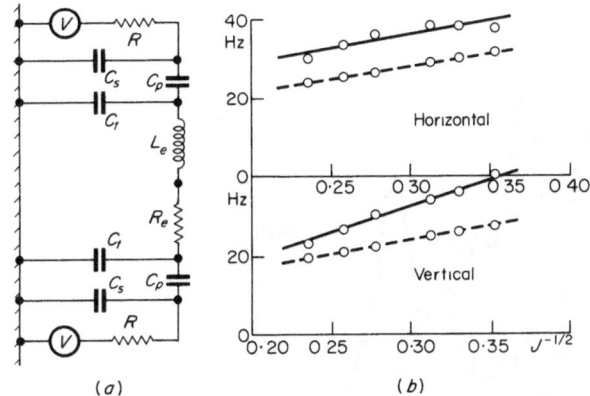

(a) (b)

Figure 7. (*a*). Electrical model of the eyeball muscle system. (*b*), Variation of the position of the low-frequency peak in the VSD graphs as a function of added inertia. O—O—O experimental values; O‑·O‑·O values derived from the model.

of added inertias: 5, 6·1, 7·3, 10, 12 and 15 g cm². There was only one peak in the transfer characteristics (at about 40 Hz for the horizontal eye rotations, and near 30 Hz for the rotations in the vertical direction). As the inertia was increased its amplitude progressively reduced, and the peak moved steadily downwards in frequency, just as did the low-frequency peak in the fixation tremor when the eye was loaded. The variation of the position of the low-frequency peak in the spectral density graphs as a function of $J^{-1/2}$ is shown in figure 7B. Here J denotes the total moment of inertia, i.e., $J = J_e + J_a$, where J_e is the effective inertia of the eyeball ($= 3$ g cm²) and J_a is the added inertia.

The effect on the ocular tremor of changes in posture

The labyrinth is one source of tonic impulses for the extraocular muscles (Cagan, 1956). It was considered that a change in the gravitational stimulation of the labyrinth might cause a change in the tonic impulses in the resting state, and hence cause a change in the ocular tremor. An experiment was therefore made in which tremor was recorded

in two different positions of the head as follows: subject sitting upright, fixating in the primary position of the gaze; and subject lying down, but again fixating in the primary position. The results are shown in figure 8 for horizontal and vertical tremor. The magnitude of the spectral components of the horizontal tremor is seen to be smaller throughout the spectrum when the subject is lying. In vertical tremor the opposite effect is observed. In both types of tremor, the upper limit of the frequency components is not changed by a change in posture.

The effect on ocular tremor of changes in the retinal illumination

The retina has been cited as another of the sources of the tonic impulses to the extraocular muscles (Cagan, 1956). An examination of the effect on the tremor spectra of changes in the retinal illumination was therefore made. Ocular tremor was first recorded under normal ambient lighting conditions when the subject was facing a white screen, without fixating. The lights were then switched off, leaving the subject in complete darkness, and a new recording of the tremor was made. The alternations of light and darkness were repeated several times, so that six or more alternate recordings were made on one film. Experiments were carried out with three subjects, both for horizontal and for vertical tremor. No significant difference between the spectral density graphs, obtained from tremor records made in the light and those graphs obtained from tremor recorded in the dark was observed. This is shown in figure 9.

HEAD TREMOR

For two main reasons, it has been considered relevant to the main study to make recordings of head tremor: First, it enables us to ascertain the extent to which head movement artefact can be

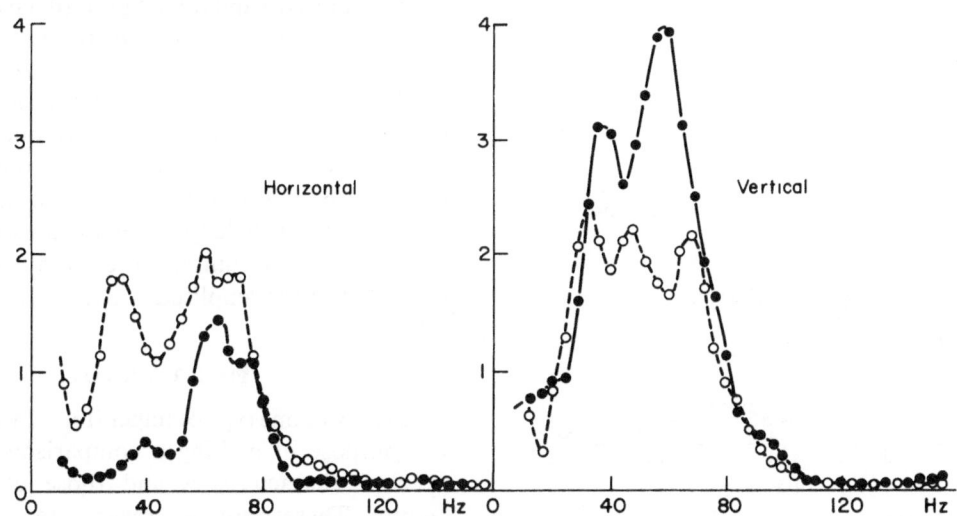

Figure 8. The effect of postural changes on VSD graphs of ocular tremor. o–o–o sitting; ●–●–● lying.

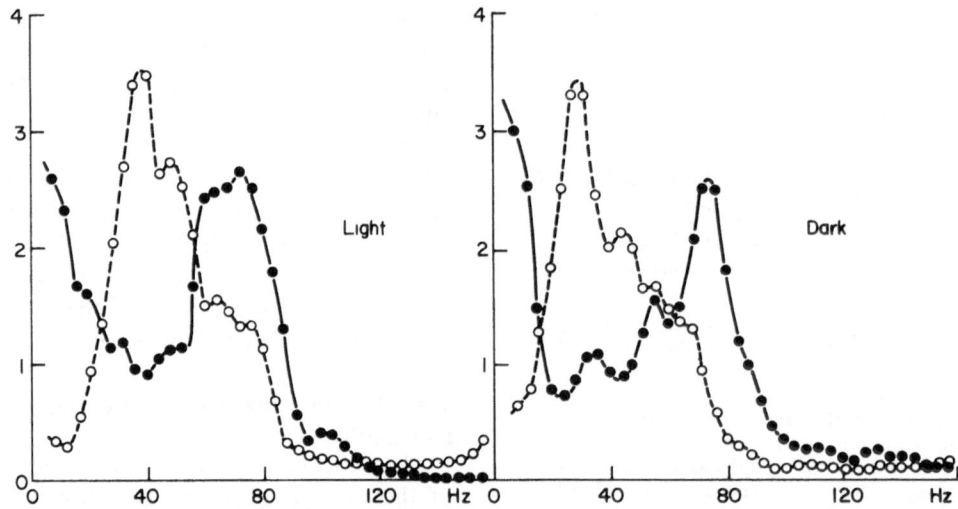

Figure 9. Comparison of VSD graphs of resting tremor when the subject is not fixating in the light or in the dark. o—o—o vertical tremor (subject J.G.T.); ●—●—● vertical tremor (subject H.B.)

expected to contaminate the ocular tremor records which are made with the piezoelectric accelerometer. Second, it allows a comparison to be made between the spectra of ocular and head tremors.

The velocity spectral density graphs of head tremor are given in figure 10. The subject was seated and his head was supported by his biting on a dental impression. Head tremor, in the vertical direction, was recorded by using a piezo accelerometer (of the

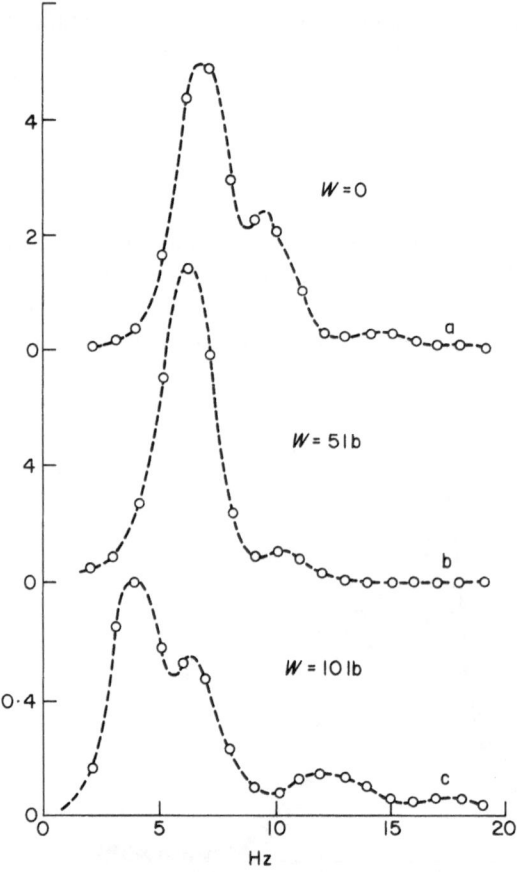

Figure 10. VSD graphs of head tremor with weights fixed on the head.

type used for recording ocular tremor) fixed to a face mask. In figure 10, the graph (a) has been obtained with no load on the head. It has two main peaks, the larger one at approximately 6·5 Hz, and the smaller one at 9·5 Hz. Unlike the peaks in the ocular tremor spectra, these appear to be remarkably steady in frequency from one record to the next. The average spectrum is therefore similar in shape to the constituent curves. The magnitude of the tremor shows some variation from one graph to the next, but this also is smaller than the variability in the ocular tremor. The power in the spectra is seen to be concentrated almost entirely in the frequency range below 12 Hz. The contamination by head tremor artefact of those ocular tremor records which are obtained by using the piezo accelerometer is therefore negligible in the frequency range in which we are interested. (The piezoelectric strain gauge and the capacitance gauge are free from head movement artefacts as they measure the eye movements relative to the skull).

The graphs (b) and (c) in figure 10 have been computed from head tremor records which were made when weights of 5 lb and 10 lb had been fixed to the head respectively. The effect of the smaller weight is seen to cause a shift to lower frequencies of the main concentration of energy in the spectrum, and also to increase the amplitude of these spectral components. The larger weight causes a further shift of the spectrum to lower frequencies, and a reduction in the amplitude of the tremor.

FINGER TREMOR

A study of one type of finger tremor was made, for the purpose of making a comparison between its spectral characteristics and those of the ocular tremor. There were two reasons for choosing the abduction-adduction tremor of the little finger.

Firstly, this motion can be easily isolated from the other tremors of the hand, by binding the other fingers together and clamping them. Secondly, the tremor motion is rapid, and the principal frequencies in the spectrum are intermediate between those of the head tremor and the ocular tremor.

The average velocity spectral density graphs of finger tremor, with 0, 5·3 and 24·7 g added weights are shown in figure 11. The arbitrary units of the ordinates are the same for all graphs.

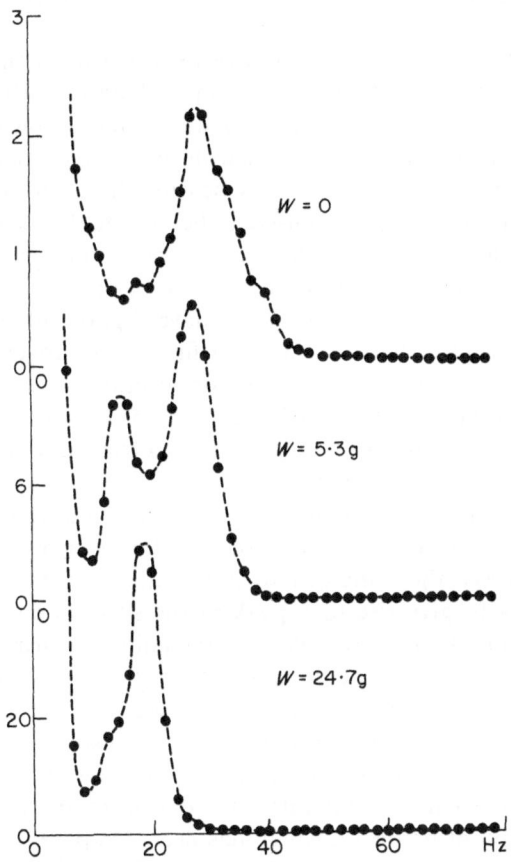

Figure 11. VSD graphs of finger tremor when weights have been fixed to a finger.

When the finger is in the unloaded state, the individual graphs show a main peak at 28–32 Hz, and also smaller peaks. Like the peaks in the head tremor spectra and unlike those in the ocular tremor spectra, the frequency of the peak is almost constant from one record to the next. The average spectrum is therefore similar in shape to the individual curves. The smaller peaks are not entirely averaged out; the one at 18–22 Hz is still visible. With regard to the main peak, it seems that the finger tremor, like head tremor, has a much clearer rhythmicity than is found in ocular tremor. The main peak in the spectrum shifts towards the lower frequencies as the load is increased; however its amplitude increases considerably with loading.

A further experimental result was obtained in connection with finger tremor: When a vibratory force was applied to the finger, so that oscillatory movements of the abduction-adduction type were caused, it was found that the magnitude of the peak angular velocity was strongly dependent on frequency (in a manner resembling that of a resonance curve). Corresponding to the three conditions in which 0, 5·3 and 24·7 g weights were attached to the finger, the frequency of the peak velocity in Hz was found to be at 34, 28 and 19 respectively.

DISCUSSION

In the enormous literature on eye movements, only a small number of observations on the characteristics of the involuntary tremor have been reported. These observations have, with two exceptions, been concerned only with horizontal tremor when the eye is in the primary position.

Workers who have made general statements concerning tremor, have reported the frequency range of tremor as 50–100 Hz (Adler and Fliegelmann, 1934); 30–70 Hz (Ratliff and Riggs, 1950); and 30–80 Hz (Ditchburn and Ginsborg, 1953). The methods by which these figures have been arrived at are not specified. It is likely that they are based on visual examination of the tremor records. A significant feature of these results is that the methods involving the use of a contact lens yield a low value for the upper limit of the frequency spectrum. This supports our finding that the upper frequency limit of the spectrum is determined by the extent of the loading on the eye.

In summarising the experimental results of all previous workers, there is no single feature which is common to all graphs. There are, however, two characteristics which appear in all graphs: the absence of an appreciable amount of power in the frequency range about 100 Hz; and the appearance of a peak in the spectrum in the vicinity of 70–80 Hz.

An analysis of factors which determine the spectral composition of ocular tremor

The experimental results of the present work are examined below in relation to figure 1. (We exclude ballistocardiographic processes from consideration on the grounds of the observations of Habbard and Marg (1960)—that an intravenous injection of curare causes the complete disappearance of tremor.)

(i) *The roles of the eyeball-muscle mechanics, and of the stream of tonic impulses in shaping the frequency spectrum of tremor.* The experimental results obtained when the eye is loaded determine the extent to which the eyeball-muscle mechanics affect the shape of the frequency spectrum of the tremor. The significant feature of these is the displacement to lower frequencies of the low-frequency peak in the spectrum, as the moment of inertia is increased.

Transfer characteristics of the eyeball-muscle system obtained from the electrical model have two noteworthy features: first, the downward movement of the peak as the inertia of the system is increased, and, second, the absence of a second peak in the spectrum at about 70–80 Hz. The downward movement of the peak in the experimental graphs and in the theoretical graphs is shown in figure 7B as a function of added inertia. These results lead us to identify the peak, which occurs at 35–40 Hz in the measured spectrum, with the mechanical resonance of the eyeball-muscle system.

Although the frequency shift of the peak is roughly comparable in the experimental and computed graphs of figure 7B, there is not a good quantitative agreement. The experimental values are systematically higher in frequency than the computed values. This indicates that the parameters in the model which was used to compute the transfer characteristics are slightly in error. The springs are not stiff enough. Nevertheless, the present results lend support to the general correctness of the model. The peak at approximately 70–80 Hz which is seen in our measured spectral density graphs is not associated with the purely mechanical factors in the orbit, because such a peak is not found in the transfer characteristics of the model. This peak must therefore come from a rhythmical component in the tension of the active state of the muscle. It must originate in a 'preferred' firing rate of the individual motor units. This deduction is supported by the reported observations of the electromyograms. Kuboki *et al.* (1958, 1959) have more recently made a Fourier analysis of the electrical activity recorded from the extraocular muscles with 'global' surface electrodes. In their results there is evidence for the presence of a peak of activity at about 70–80 Hz, when the eye is in the primary position. The Fourier components extend upwards to much higher frequencies, but it must be expected from the model of the eyeball-muscle system that the eye tremor will not contain such high frequency components, because of the low-pass filtering action of the mechanical system. The bandwidth of the system is known from the model of figure 7A to be less than 60 Hz. Components with frequencies above this range will suffer progressively increasing attenuations as the frequency is increased. The peak of activity which we have found at 70–80 Hz in the tremor spectra is to be expected, because the response of the system has not fallen off seriously in this range.

A further significant observation made by Kuboki *et al.* (1958*b*) is concerned with the changes in the Fourier components of the EMG when the eye is adducted. For a 20 degree adduction the largest components occur at about 120 Hz. As the eye is adducted further the peak moves to still higher

frequencies. We should expect, in view of the low-pass filtering action of the mechanical system, that such a high frequency rhythmicity in the muscle tension would give rise to a much smaller peak in the ocular tremor spectrum than is present when the eye is in the primary position. In our experiments, when the eye is away from the primary position, the peak which had been present at 70–80 Hz has disappeared. The results of Kuboki and his co-workers are thus in agreement with the predictions of the model which is used in the present work.

All the workers who have recorded from single motor units (see for example Kuboki (1957); Nebiyeridze 1966)) have found regular firing and complete absence of synchronism between neighbouring motor units. It is possible that, although units are not synchronised in their firing, the records of electromyograms in which several units can be seen firing, often have the appearance of synchronism, because the units which have slightly different firing rates will occasionally fire almost together, and will continue to do so for several consecutive intervals before they drift apart again (Taylor, 1962). This spontaneous regular firing is a feature which is relevant to the present work.

An important point which appears to have been overlooked by several authors is that synchronism between the firing of individual units, is not necessary to give rise to a peak in the spectral density graph. If, for example, several units are firing at approximately the same frequency, but each having some dispersion from complete regularity, then a peak will occur in the spectral density graphs, even when there is no synchronisation between units. This is illustrated by the spectral density graph of figure 12, which has been computed from a waveform consisting of a stream of impulses which correspond to the unsynchronized firing of five individual motor units. The dispersion of each unit from completely regular firing is normally distributed, with a dispersion $\sigma = 5$ ms. In constructing

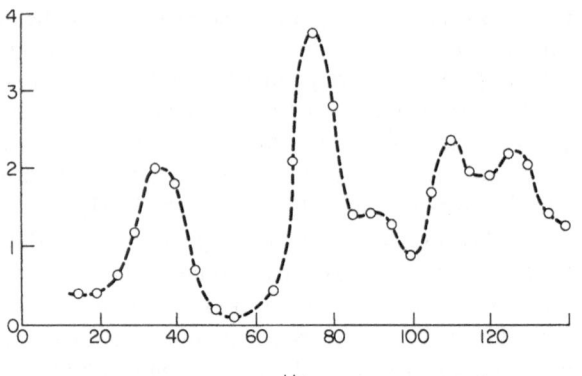

Figure 12. Computed spectral density graph of a waveform consisting of five unsynchronised streams of impulses. Each pulse stream has a basic rhythmicity of 40 pulses/s, with a dispersion of $\sigma = 5$ ms.

the waveform, the impulse streams have been added in a random manner. The main peak, and several harmonics can be seen in the spectrum. If there were a synchronisation between the firing of the individual units, these peaks would be much sharper. It has been shown by several workers that when the firing of several units does, under some circumstances, become synchronised, then a violent tremor or 'shivering' results.

From the results of the present work, and the collected evidence from the results of other workers, it is inferred that the peak which we have found at 70–80 Hz in the velocity spectral density graphs of several subjects is associated with a 'spontaneous', unsynchronised firing of the motor units. The results of Schaefer (1965) indicate that the rhythmicity in the firing is present at the level of the opto-motor nuclei, and may well originate at a higher level.

(ii) *The role of supra-nuclear control in shaping the frequency spectrum of the tremor.* The present experiments have involved two procedures in which supra-nuclear control is exercised over the tonus of the eye muscles. The first of these is in the steady adduction or elevation of the eye. The second is in the voluntary act of fixation.

We have measured the effects of adduction or elevation of the eye for 0, 10, 20 and 30 degrees of deviations from the primary position. The results for 0 and 20 degrees of deviations are given in figure 5. As the extent of the adduction or elevation increases, the energy spreads progressively into the higher and lower frequency components of the spectrum, with the consequent flattening of the spectrum. These results are in accord with the myo-electric power density spectra given by other workers (Kuboki *et al.*, 1958*b*; Scott, 1967); they have shown that the myoelectric activity spreads progressively in both direction of the frequency spectrum as the degree of contraction of muscle is gradually increased.

It is known that adduction or elevation of the eyes occur in response to control from the frontal oculo-motor centre of the cortex. The action of this control over the oculomotor nuclei overrides the rhythmic firing which, in the resting state, we have inferred to be the cause of the peak in the spectrum at 70–80 Hz. If the increased tension in the agonist when deviating the eye from the primary position is caused mainly by recruitment, as has been found by Gel'fand *et al.* (1964) for moderate levels of contraction of muscles of the head, then our results indicate that the additional units have a wide range of firing rates. This causes a spread of energy throughout the spectrum.

As shown in figure 4, the action of fixating causes a different type of change in the tremor spectrum. There is a fairly uniform change in the amplitudes of the components throughout the spectrum.

There is no appreciable change in the shape of the spectrum, i.e., no redistribution of energy among the spectral components, such as has been observed when the eye is deviated from the primary position.

Fixation is known to be in part accomplished by a reflex action. The pathways involved in this reflex are shown in the diagram of figure 1. The frontal oculo-motor centre can exert an overriding control of this reflex. It seems from the present results that this control is much less strong in maintaining fixation than it is in deviating the eyes from the primary position, since the tremor spectrum is much less affected. Whether the reflex or the control from the higher centres is dominant, the act of fixation must involve a feedback of position information from the retina; it is not possible to maintain fixation in the dark.

(iii) *The role of retinal feedback in shaping the frequency spectrum of tremor.* Two of the experimental results given earlier are relevant to this subject. As mentioned above, the action of fixating must involve a feedback from the retina. It has therefore to be considered whether the graphs of figure 4 may, at least in part, be determined by this feedback. The other experiment with a bearing on this subject is that in which the tremor spectra obtained in the light and dark are compared. In figure 9 it can be seen that the spectra obtained in the dark are not significantly different from those which were obtained in normal ambient illumination when not fixating. It is concluded that the feedback from the retina does not play a significant part in shaping the tremor spectrum. The change which occurs between fixating and not fixating is therefore to be attributed to the supranuclear control.

(iv) *The role of other feedbacks in shaping the frequency spectrum of tremor.* The extraocular muscles are known to differ in several respects from other voluntary muscles. Two of these differences involve feedback mechanisms. The first of these is the stretch reflex, which is operative in voluntary muscle systems, except the extraocular muscles. The second is the feedback via the Renshaw cells. There is no evidence for the existence of Renshaw cells in the oculomotor apparatus (Schaefer, 1965). The possibility must be considered that the observed differences in the characteristics of ocular tremor and other types of tremor are to be ascribed to the absence of these feedbacks in the ocular system.

The spectra of head tremor and finger tremor show a clearer rhythmicity than has been found in ocular tremor. Also, unlike ocular tremor, consecutive epochs of tremor record yield very similar spectra. It seems, therefore, that the mechanism which is defining the rhythmicity of head tremor and of finger is absent from the oculomotor system. The stretch reflex, or the action of the Renshaw feedback loop are implicated. Additional evidence to support this view comes from the records of eye and finger tremor when weights have been added to the moving system, and also from the experiments in which sinusoidal forces at various frequencies have been used to set the system in motion. Experimental results show that as the moment of inertia of the eyeball-muscle system is progressively increased the low-frequency peak in the ocular tremor spectrum shifts to lower frequencies and the amplitude of this peak is reduced. This is quite different from the effect observed with the finger tremor: an increase in the inertia causes a very large increase in the size of the peak, in addition to causing the shift to lower frequencies. As reported earlier, the motion of the finger when sinusoidal driving forces are applied at various frequencies shows a frequency dependence in a manner similar to that of a resonance curve. The frequency at which the largest amplitude occurs is found to be approximately the same as the frequency of the peak which is found in the tremor spectrum for each of the loading conditions. The indication that this is not the mechanical resonant frequency of the system comes from the fact that the amplitude of the peak is considerably increased by weighting. A more tenable hypothesis is that the sinusoidal stretching of the stretch receptors causes a reflex response of a similar type to that which has been reported by Lippold (1958).

The results of the present work indicate that a mechanism, which can be plausibly associated with the stretch reflex, is responsible for the differences between the characteristics of ocular tremor and other types of tremor. There is no evidence that a feedback type control is operative in defining the characteristics of ocular tremor.

(v) *The role of the labyrinth in shaping the frequency spectrum of tremor.* As long ago as 1881, the existence of tonic impulses arising from the labyrinth and influencing the extraocular muscles was recognized by Högyes (1881). The function of the labyrinth in the static and dynamic control of reflex eye movements has since been intensively investigated. In the present work, one of the experiments has a bearing on the question of the part played by the labyrinth in determining the characteristics of ocular tremor. The effect on the spectra of changing the position of the head (and hence the orientation of the labyrinth receptors) in the gravitational field has been shown in figure 8. The abduction-adduction tremor when the head is upright is seen to be larger than when the subject is lying down. The converse effect occurs in elevation-depression tremor.

It is known that the response of the succular and utricular maculae to the gravitational stimulus depends on the orientation of the head. There is

unfortunately a difference of opinion in the reference texts about the orientation of the maculae in the head.

In view of the uncertainty about the exact nature of the labyrinth's static responses to the gravitational stimuli, we shall restrict our conclusions from the results of figure 8 to the observation that the labyrinth appears to play an important part in determining the magnitude and shape of the spectrum.

We may note here that Charnwood (1950) has pointed out (in another connection) that in the supine position the eye loses the support of Lockwood's ligament. The indications that the present results of figure 8 are not caused by the loss of support of Lockwood's ligament are: first, the different results obtained in the horizontal and vertical tremor; and, second, the frequency-sensitive nature of the changes. It seems to be more plausible to attribute the changes to the stimulation of the labyrinth.

Thomas (1967) obtained certain unpublished results which appear to be closely related to this subject. He has found that when the head is vibrated sinusoidally at a very small amplitude, the magnitude of the resulting vibrational eye movements depends on the frequency of the vibration. Typical results are shown in figure 13. These correspond to the

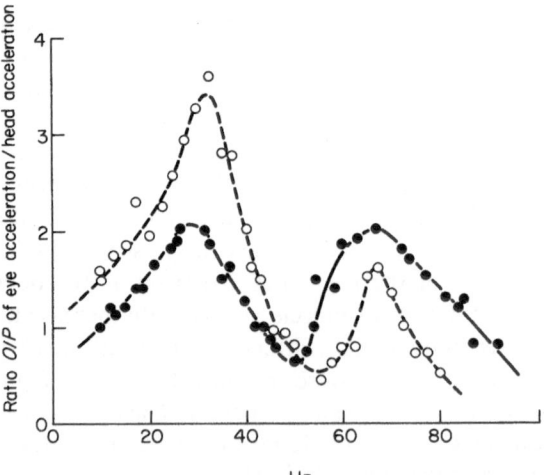

Figure 13. Variation with frequency of the ratio of output of eye accelerometer to bite bar accelerometer when the head is vibrated.

condition in which the eye is loaded with the contact lens and its attachments. It can be seen that the ratio of the magnitudes of the eye rotation and head rotation, when plotted against frequency, takes the form of a graph which has two main peaks. These peaks occur at approximately those frequencies at which we have found the main peaks in the ocular tremor spectra.

The graphs of figure 13 like those of the ocular tremor spectra appear to be made up of two separate parts. The amplitudes of the two peaks are independently variable. This indicates that the two peaks have to be attributed to different causes. The low-frequency peak (at 30–40 Hz) is identifiable with the natural mechanical resonant frequency of the eyeball-muscle system. This has been shown by direct measurements (Thomas, 1967) to lie in this frequency range, when the eye is loaded with the contact lens and its attachments. The fact that the high-frequency peak which is found in the head vibration experiments is at approximately the same frequency (70–80 Hz) as the peak in the ocular tremor spectra (particularly as the vibrational eye movements are approximately two orders of magnitude larger than the ocular tremor) points to the possibility that they have a common origin. It is therefore relevant to our study of the origins of ocular tremor to consider the mechanism whereby head vibrations can cause eye vibrations which are particularly large in a certain frequency band.

It is known that in many animals the labyrinth is very sensitive to vibrations (Kuboki *et al.*, 1959; Aschroft and Hallpike, 1934; Lowenstein and Roberts, 1951). In particular the saccule has been claimed to be the main centre of the vibration sense. The sensitivity has been found to depend on the frequency of the vibration, and a strong synchronization of the action potentials with the cycles of the vibration occurs at some frequencies.

In an attempt by some workers (De Vries and Schierbeck, 1953) to measure the minimum perceptible angular velocity to elicit a response from the labyrinth, it was found that there is no measurable threshold. Very weak stimuli could sometimes elicit a response. It is clear from these results that very small movements of the labyrinth can cause 'spontaneous' firing of the receptors. As the sensitivity of the labyrinth to vibrations is known to be frequency selective, this can explain the fact that head vibrations in the frequency range around 70 Hz are most effective in producing eye rotations. The peak in the tremor spectrum at 70–80 Hz is likewise attributed to the stimulation of the labyrinth by normal random motion. Even with the head stabilised by a bite bar, the haemodynamic processes inside the skull can cause movements.

Acoustic stimulation of the labyrinth with very loud noise has been reported to result in the blurring of vision (Ades, 1951; Davies, 1942; Rosenblith, 1952), and in an increase in the involuntary eye movements (Krauskopf and Coleman, 1955). Unfortunately, the amplitude of the tremor in the published record is far too small to allow an analysis of the type which has been made in the present work. The statement does however go some way towards supporting the inference made from the results of the present work, and from the results of other workers, that the peak in the tremor spectrum at 70–80 Hz is associated with the action of the labyrinth.

Although the experimental results implicate the labyrinth in shaping the frequency spectrum of ocular tremor, it has not been possible to make tremor recordings with subjects having selective damage of the labyrinths. Attempts to stimulate the labyrinth directly by exposing the subject to very loud sounds could, in any case, not be made with subjects having normal hearing, because of the possibility of permanent damage.

SUMMARY AND CONCLUSIONS

Ocular tremor of six human subjects has been recorded by three electronic methods, namely: (*a*) piezoelectric accelerometer; (*b*) piezoelectric strain gauge; and (*c*) capacitance gauge methods.

The velocity spectral density graphs of these tremor records, computed by means of a digital computer in the usual way, have been found to have basic similarities, with two peaks, one near 40 Hz and another at 70–80 Hz. There is a strong indication that each of these graphs represents the composite effect of several factors. Abduction-adduction tremor is found to be different from elevation-depression tremor.

The role of the eyeball-muscle mechanics in determining the tremor characteristics has been determined by increasing the inertia of the system and observing the resulting changes in the tremor spectra. Results gave support to the general validity of Thomas' eyeball-muscle model; and demonstrated that earlier workers who have used a mirror on a contact lens to record eye movements have not obtained a faithful record of the tremor due to the reduction of the bandwidth of the mechanical filter.

Ocular tremor in the resting state, and when the eye is adducted or elevated, has been found to be generally as anticipated from the electromyograms of Kuboki and his co-workers. The present results are believed to be the first to give support to their findings.

The differences in the tremor spectra between the conditions of fixating or not fixating are much less pronounced than those which occur when the eye is deviated from the primary position. The changes are attributed to a weak supranuclear control.

The present results do not lend support to the suggestion that tremor performs a 'scanning' function which is part of the visual process. For, if this were so, the frequency of the tremor would be independent of the direction of the gaze, and other factors, and would be expected to depend on the illumination of the retina.

Ocular tremor has been found to differ from other physiological tremors in several respects: It has components at much higher frequencies; it does not have a clear rhythmicity; and, unlike other types of tremor, the amplitudes of the peaks in the ocular tremor spectrum do not increase when the inertia of the system is increased. It seems reasonable to ascribe these differences to the absence of a stretch reflex.

The peak at 70–80 Hz in the ocular tremor spectrum has been identified with a similar peak which is found to occur in the eye rotations which are caused by vibrating the head. This has been identified with the action of the labyrinth.

Acknowledgment

We acknowledge with thanks the support of the Science Research Council.

REFERENCES

ADES, H. W. 1957: Nystagmus elicited by high intensity sound. *U.S. Naval School of Aviation Med., Project, N.M.* 13 01 99, Report 6.

ADLER, F. H., and FLIEGELMANN, M. 1934: Influence of fixation on visual acuity. *Arch. Ophth. N. Y.* 12: 475–483.

ANDERSEN, E. E., and WEYMOUTH, E. W. 1923: Visual perception and the retinal mosaic. *Amer. J. Physiol.* 64: 561–591.

ASCHROFT, D. W., and HALLPIKE, C. S. 1934: On the function of the saccule. *J. Laryng. and Otol.* 49: 450–458.

BENGI, H., and THOMAS, J. G. 1967: A digital trace reading potentiometer. *J. Sci. Instrum.* 44.

BENGI, H., and THOMAS, J. G. 1968: Three electronic methods for recording ocular tremor. *Med. and Biol. Engng.* 6: 171–179.

BENGI, H., and THOMAS, J. G. 1968: Fixation tremor in relation to eyeball-muscle mechanics. *Nature* 217.

BLACKMAN, R. B., and TUKEY, J. W. 1958: 'The Measurement of Power Spectra', 1st ed., Dover, New York.

CHARNWOOD, J. 1950: Effect of posture on involuntary eye movements. *Nature* 164: 348–349.

COGAN, D. C. 1956: 'Neurology of the Ocular Muscles', 1st ed., Charles C. Thomas, Springfield, Illinois.

COOPER, S., DANIEL, P. W., and WHITTERIDGE, D. 1955: Muscle spindles and other sensory endings in the extrinsic eye muscles. *Brain* 78: 564–583.

DAVIES, H. 1942: Final report on physiological effects of exposure to certain sounds. *OSRD Report No. 889*, NDRC Div. 17.

DE VRIES, H. L., and SCHIERBECK, P. 1953: The minimum perceptible angular velocity under various conditions. *Prac. Oto. Rhino. Laryng.* 15: 66.

DITCHBURN, W. R., and GINSBORG, B. L. 1953: Involuntary eye movements during fixation. *J. Physiol. (Lond.)* 119: 1–17.

GEL'FAND, I. M., GURFINKEL, V. S., KOBS, YA. M., KRIINSKI, V. I., TSETLIN, M. L., and SHIK, M. L. 1964: Investigation of postural activity. *Bisfizika* 9: 710–717.

HALLIDAY, A. M., and REDFEARN, J. W. 1956: An analysis of the frequencies of finger tremor in healthy subjects. *J. Physiol. (London)* 134: 600–611.

HEBBARD, F. W., and MARG, E. 1960: Physiological nystagmus in the cat. *J. Opht. Soc. Amer.* 50: 151–155.

HOFFMANN, P. 1929: Ist es möglich die physiologischer Erfahrungen über die Sehenreflexe mit den Pathologischen in Einklang zu bringen? *Nervenartz* 2: 641–656.

HÖGYES, A. 1881: Uber den Nerven Mechanismus der associerten Augenbewegungen. *Ann. d. Akad. Wissensch. Budapest* 10, 11, 12.

KRAUSKOPF, J., and COLEMAN, P. D. 1955: The effect of noise on eye movements. *U.S. Army Medical Research Lab., Report No. 218*, Project No. 6-95-20-001.

KUBOKI, T. 1957: Studies of discharge intervals of a single motor unit in the human extraocular muscles.

KUBOKI, T., SEKINO, I., and FUKUSHI, F. 1958a: The periodicity in the e.m.g. of human extraocular muscles. *Acta Soc. Ophth. Jap.* 61: 1565–1567.

KUBOKI, T., SEKINO, I., and FUKUSHI, F. 1958b: Periodicity in the e.m.g. of human extraocular muscles. *Acta Soc. Ophth. Jap.* 62: 2361–2366.

KUBOKI, T., SEKINO, I., and FUKUSHI, F. 1959: The periodicity in the e.m.g. of human extraocular muscles. *Jap. J. Ophth.* 3: 66–74.

LIPPOLD, O. J. C. 1958: The effect of sinusoidal stretching upon the activity of stretch receptors in voluntary muscle and their reflex responses. *J. Physiol. (London)* 114: 373–386.

LOWENSTEIN, O., and ROBERTS, T. D. M. 1951: The localization and analysis of the responses to vibration from the isolated elasmobranch labyrinth. A contribution to the problem of the evolution of hearing in vertebrates. *J. Physiol. (London)* 114: 471–489.

McCOUCH, G. P., and ADLER, F. H. 1932: Extraocular reflexes. *Amer. J. Physiol.* 100: 78–88.

MARSHALL, W. H., and TALBOT, S. A. 1942: Recent evidence for neural mechanisms in vision leading to a general theory of sensory acuity. *Biol. Symp.* 7: 117–164.

NEBIYERIDZE, R. B. 1966: Work of the motor units of the eye muscles in the steady state. *Biofizika* 11: 143–146.

PAL'TSEV, YE. I. 1964: Changes in the spectral composition of tremor in relation to the character of the motor task. *Biofizika* 9: 742–745.

RATLIFF, F., and RIGGS, L. A. 1950: Involuntary motions of the eye during monocular fixation. *J. Exp. Physiol.* 40: 687–700.

ROSENBLITH, W. A. 1952: Problems of high-intensity noise: a survey of recommendations. *Psycho-Acoustic Lab., Harvard Univ.*, Report 133.

SCHAEFER, K. P. 1965: Die Erregungsmuster einzelner Neurone des Abducens-Kernes beim Kaninschen. *Pflüger's Archiv.* 284: 31–52.

SCOTT, R. N. 1967: Myoelectric energy spectra. *Med. and Biol. Engng.* 5: 303–305.

TAYLOR, A. 1962: The significance of grouping of motor unit activity. *J. Physiol. (London)* 162: 259–269.

THOMAS, J. G. 1967: The torque-angle transfer function of the human eye. *Kybernetik* 3: 254–263.

INTRA-PARTUM MONITORING: AN OBSTETRICIAN'S VIEW AND A PHYSICIST'S RESPONSE

T. E. TORBET AND D. L. THOMAS[1]

PART I: THE OBSTETRICIANS VIEW (T. E. TORBET)

In 1969 at the Southern General Hospital, an investigation was started to review foetal heart monitoring equipment for the new Maternity Unit. This review was conducted in collaboration with the Department of Clinical Physics and Bio-Engineering of the Western Regional Hospital Board and, following the submission of preliminary reports, it was supported by the new Medical Development Fund of the Scottish Home and Health Department.

Much of the work has concentrated on the interface with the patient but the objective has always been to produce instrumentation for foetal heart and uterine contraction monitoring which would allow for both invasive and non-invasive methods of measurement and would allow the signals to be intercepted at any point from the transducer to the display so that measurements other than beat-to-beat counting of the foetal heart or estimations of intra-uterine pressure could be made. The measurement of the beat-to-beat rate by electrocardiography was pioneered independently by Hon (1960) and Caldeyro-Barcia (1961); later, Hammacher (1962) described beat-to-beat counting by phonocardiography. Commercial instruments based on these principles have been available for some time and more recently the British firm, Sonicaid, have marketed an instrument which uses ultrasound on the Doppler Shift Principle as a method of detecting foetal heart signals for this purpose. These instruments are all satisfactory within their limitations, so long as the currently accepted methods of measurement are employed, but it is unlikely that further progress will be made unless measurements of other parts of the signals obtained, or other parameters, are measured.

Although the application of physics or engineering to obstetric practice has been present for centuries, e.g. the obstetric forceps; there are still those who view the application of the machine, or the terminology of the technical age, to the art of the physician accoucheur with at best good humour, at worst repugnance. Nevertheless, if the midwife is

[1] Southern General Hospital and Department of Clinical Physics and Bioengineering, Western Regional Hospital Board, Glasgow, Scotland.

observed as she makes her observations, it will be seen that she uses simple physical devices, such as a stethoscope and a watch to improve her observations, although her hand is the simple but well-trained sensor through which she perceives uterine contractions (figure 1). By fulfilling her role as an observer and as an administrator of treatment,

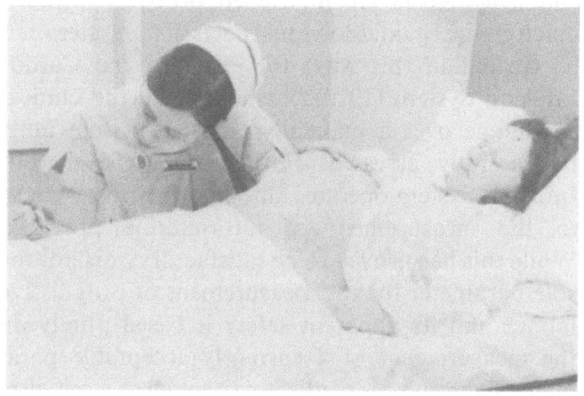

Figure 1.

she becomes part of an observation/treatment cycle which, in modern technical terminology, could be called a closed-loop information system (figure 2). The use of monitors enhances this system by producing a means for continuous observation,

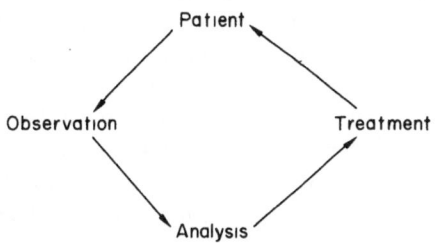

Figure 2. Observation-treatment cycle. Closed loop information system.

by enhancing the detection of clinical changes and by displaying the information for better analysis.

To the scientifically inclined obstetrician, the attraction of completely automating the observation/treatment loop is great, but there are many pitfalls for the clinician whose electronic knowledge is limited; not least a tendency to expect more from a system than it is capable of giving. Conversely, for the physicist or the engineer, there is the problem of not knowing exactly what he is measuring and hence how complex to make the equipment. The end result of such working in isolation is the tendency to over-elaborate on display systems without improving the information obtained, or for that matter the observation/treatment loop and hence the quality of patient care.

Another attractive possibility is to supplant the observer in the loop by a computer. If this is examined in the context of automation terminology, we would say that labour is not a fixed strategy mechanisation system but a multiple strategy system in which the variables are incompletely analysed. Consequently, unless the computer was extremely complex, manual interference is going to be required and hence costs are multiplied probably without much clinical gain. Individual parts of the system can be treated in this way; for example, the Cardiff Infusion System (1970) but here again the clinical advantage over a manually adjusted system must be carefully assessed. Furthermore, the Cardiff Infusion System operates automatically in response to the measurement of intra-uterine pressure. While this has proved to be satisfactory, it is not the sole parameter for the measurement of progressive labour and its apparent safety is based purely on the measurement of a currently acceptable parameter. This point is made, not so much as a criticism of this instrument or accelerated labour, but rather as a reminder that continued exploration of the measurement of uterine action is required and of better assessment of foetal well-being while new techniques are being introduced.

If the therapeutic aspect of the loop is considered, there are further problems. New drugs have allowed greater control over uterine action, pain, anxiety and hypertension. The doses are finely adjusted but the methods of administration remain crude. Simple improvement of syringe pumps and control mechanisms could help in this. Furthermore, these methods of treatment may have effects on the patient or foetus which will require better monitoring at the observation side of the cycle and thus modify once more the observation-treatment cycle. For this reason, we concluded that the cycle should always be open at one point, viz. analysis, to accommodate change.

This preamble may seem to be rather more critical than constructive but it was from this form of thinking that our monitoring trolley was conceived and ultimately delivered, and our system was to follow the pattern shown in figure 3.

The specification given to the physicist was to provide instrumentation at reasonable cost, capable

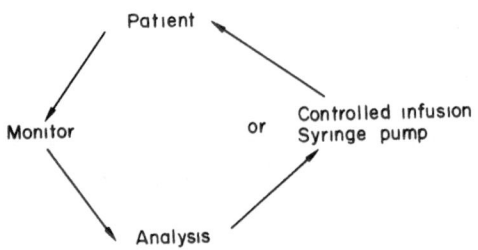

Figure 3. Observation treatment cycle. Electronic measuring and automatic control of drug administration.

of measuring the instantaneous foetal heart rate before and after rupture of the membranes, and also uterine action by both intra- and extra-uterine methods, and to provide a satisfactory display. In addition to this: to provide a suitable method of administering drugs. The ultimate answer was a trolley with an arrangement of amplifiers, meters, chart recorder, and a Palmer Infusion Pump (figure 4). The problems involved in the development will be described later.

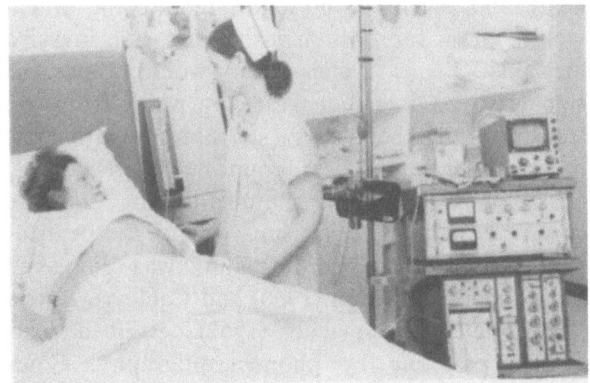

Figure 4.

Two trolleys covering approximately 2700 deliveries were put into use during 1970, and in 1971 the full impact on clinical management was observed. There were 475 patients managed using these trolleys. These were drawn from two sources: 398 were 'high risk' pregnancies and 77 were acute complications in labour (figure 5).

By comparing the gross statistics for 1971 with 1969, the last full year before the introduction of monitoring, some index of the changes brought about by this management can be seen. The year 1970 was a transitional year and the statistics conform to neither pattern. There was an increase in the number of deliveries from 1876 in 1969 to 2697 in 1971. The induction rate increased from 11·4 per cent to 16·8 per cent, as did the forceps rate from

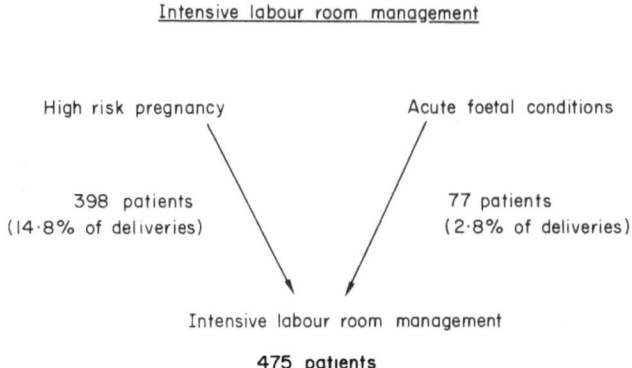

1971

Intensive labour room management

High risk pregnancy   Acute foetal conditions

398 patients   77 patients
(14·8% of deliveries)   (2·8% of deliveries)

Intensive labour room management

475 patients
(17·6% of deliveries)

Figure 5. Intensive labour-room management.

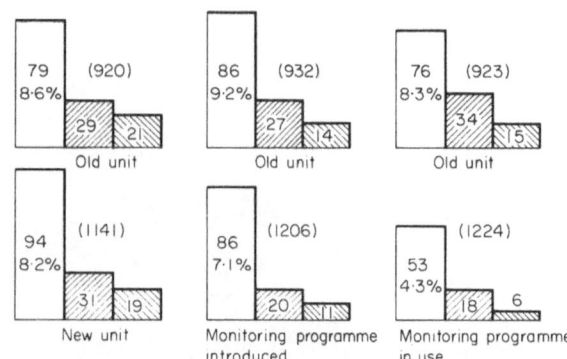

Figure 7. Caesarean sections—6-monthly totals comparing indications 'elective' and 'foetal distress'. (Total deliveries shown in brackets.)

11·04 per cent to 17·8 per cent. The Caesarean section rate declined from 9·1 per cent to 4·3 per cent. Our index for the justification of the policy was a fall in the perinatal mortality from 42·6 per 1000 births to 29·6 per 1000 births (figure 6).

	1969	1971
Total deliveries	1876	2697
Induction of labour	11·4%	16·8%
Caesarean section	9·1%·	4·3%
Forceps delivery	11·04%	17·8%
Perinatal mortality	42·6 per 1000	29·6 per 1000

Figure 6. Comparative statistics.

Further study of the indications for forceps deliveries shows that the increased rate is due to the more frequent diagnosis of foetal distress early in the second stage of labour. This diagnosis is based on foetal heart decelerations, the total dip area being most frequently the basis for diagnosis. The Apgar scores of the babies were in the middle range 5–7. There were no perinatal deaths and the recovery of the babies following delivery was rapid. We concluded that there was an asphyxial element of short duration.

When the indications for Caesarean section are analysed the most remarkable decline is in the numbers in the foetal distress group. There has also been a reduction in the numbers of elective Caesarean sections (figure 7).

Another effect of instrumentation and the use of oxytocin stimulated labour has been in work concentration. By selecting the high-risk patients for planned induction by amniotomy and stimulation of contractions by oxytocin, many of the problems of observation practice can be packed into the working day when most resources are at their best to deal with them. It has been found that labour room work concentrates mostly in the periods 9 a.m. to 10 a.m. when the inductions are carried out, and 5 p.m. to 7 p.m. when many of the patients seem to be delivered. We have found that emergency Caesarean

sections during the night have become rare. To the purist amongst obstetricians this type of practice may be deplored, but in terms of current staffing problems the concentration of deliveries and crises to time when those most skilled to deal with them are available may be an important application of bio-engineering to obstetrics.

One cannot ignore the future. A study of the causes of perinatal deaths in 1969 compared with 1971 shows that the fall in asphyxial deaths is not matched by premature deaths (figure 8). If this is considered with the findings of Drillien (1964) and

	1969	1971
Prematurity	22	27
Congenital abnormality	19	17
Asphyxia	12	8
A.P.H.—placenta praevia and abruptio	14	12
Toxaemia	7	9
Others	3	7
Total	77	80

Figure 8. Causes of perinatal deaths.

others of the sequellae of prematurity, it becomes obvious that this is a problem which merits attention. Further means of study of uterine action are required. Furthermore, it is becoming apparent that the advantages of beat to beat counting are now fully exploited. Surely study of heart action must yield further results.

It is concluded that the bio-engineer has provided the means for significant changes in obstetrics. There are still challenges for further exploration.

PART II: THE PHYSICIST'S RESPONSE (D. L. THOMAS)

General description

In the clinical context a monitoring system is a means of obtaining information about the patient— in this case, the unborn child. We have already illustrated the traditional method used by the

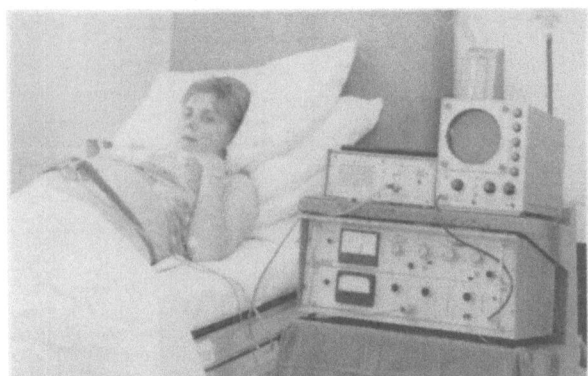

Figure 9. Simple intrapartum monitoring using the dual-function external transducer in conjunction with a standard Sonicaid D205.

midwife for counting the foetal heart rate and for sensing the maternal uterine contractions.

In figure 9 we see the simple and compact electronic monitoring unit which we have designed and constructed doing the same job, but with greater precision and less fatigue than even the most conscientious midwife would find possible. External transducers are employed and the foetal heart rate and uterine contractions are visually displayed on meters calibrated in beats per minute and millimetres of mercury pressure respectively.

Figure 10 shows our comprehensive development of the two-channel monitoring system. There is now the option of using either external or internal methods of obtaining the signals. The meter displays are now paralleled with a permanent chart record which shows immediate past trends and which is also used for retrospective analysis. Additionally, the actual foetal heart signal—EEG or ultrasonic Doppler—is displayed on a long persistence oscilloscope. This signal has intrinsic clinical value in addition to its obvious significance in providing direct assurance that the foetal heart is being properly detected. The various modules of the monitoring system have been suitably matched and interconnected so that purely technical instrument readjustment consequent on changing from external to internal monitoring method (or *vice versa*) on either channel is made as simple as possible. This means that the choice is made, as it should be, on practical clinical grounds.

Foetal heart rate from foetal electrocardiography

The foetal electrocardiogram is measured as the potential difference between the foetal scalp electrode and a maternal electrode—this forms the second electrode on the foetus diffused over that area of its body which is in electrical contact with the uterus. The foetal ECG signal measured in this way has similar significance to that of the vertical leads in conventional electrocardiography although the amplitude of the signal is attenuated to that which would be evident postnatally. In a breech

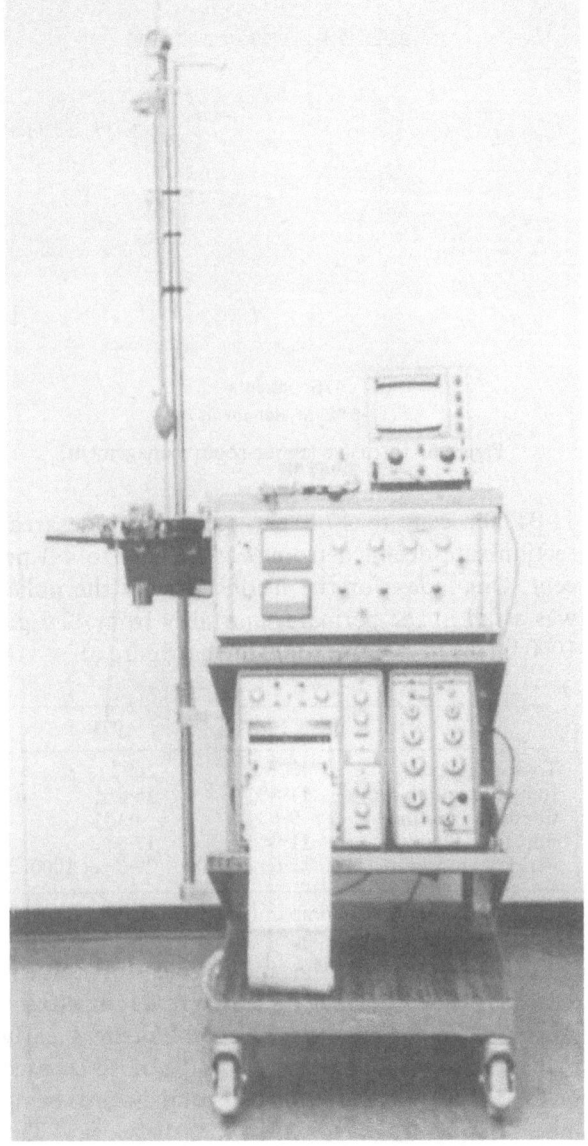

Figure 10. The comprehensive intrapartum monitoring system.

delivery, where the foetal electrode is in fact attached to the foetal buttock, the ECG is comparable with the conventional aVF lead of electrocardiography while in the normal vertex delivery the ECG is inverted with respect to the conventional aVF. This is of practical importance since knowledge of the sign of the QRS complex is required if consistent triggering of the ratemeter is to be achieved.

The scalp electrode used is a length of stainless steel wire 0·18 mm in diameter and enamelled for insulation. The technique was described by Baumgarten (1969) (figure 11), and has the advantages of simplicity, cheapness, disposability, lack of trauma, and it is rarely dislodged by vaginal examination or scalp blood sampling. The wire is inserted under the foetal scalp, using a 23-gauge needle attached to an adapted scalp blood sampling blade holder. The wire passes through the needle and along the hollow blade holder, the ends are bared by burning

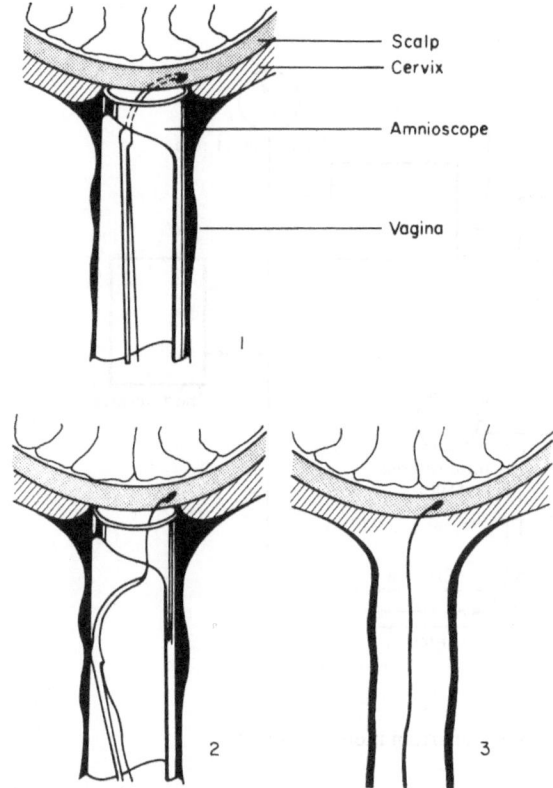

Figure 11. Baumgarten's technique for the foetal scalp electode.

Foetal heart rate from foetal ultrasonic Doppler signal

The externally obtained foetal heart signal is provided by a standard Sonicaid D205 foetal heart beat detector. The pencil beam of this instrument is modified into a fan-shaped beam of included angle 44 degrees by means of a cylindrical perspex lens fitted to the transducer (see figure 12). This has two advantages: the wide angle beam is less sensitive to

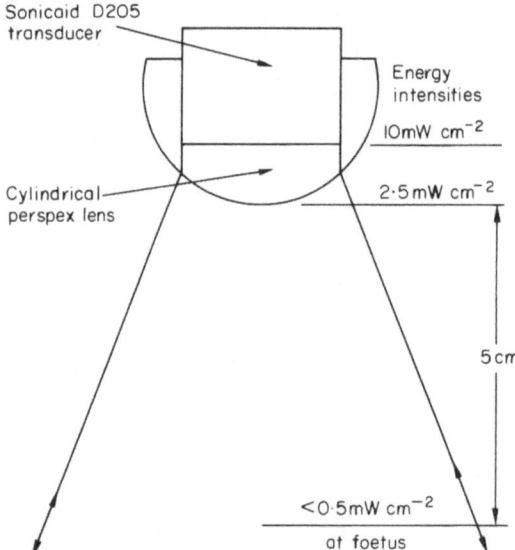

Figure 12. The technique for modifying the pencil beam of the standard Sonicaid D205.

off the insulation and the needle is bent to a gentle curve so that the direction of the point is at right angles to the blade holder. The end of the wire is then bent over the bevel of the needle to form a small hook 2 mm long. This hook can then be inserted under the scalp by gentle manipulation of the needle, care being taken that the whole hook goes under the scalp so that it acts like a barb. The insertion is done through an amnioscope and with a little practice this can be done through the 12 mm size bore. It is removed by plucking the wire sharply.

As already indicated, the second (or 'negative') connection to the foetus is made *via* the mother: this is made with a disposable electrode applied to the maternal abdomen. A third (or 'reference') connection is also required in order to avoid electrical interference; again a disposable electrode is used and it is generally applied to the maternal thigh. The position of these two maternal electrodes is not at all critical.

The foetal ECG is amplified by a high-gain differential amplifier and then displayed on a long-persistence oscilloscope as well as going to an instantaneous ratemeter of the Devices/Neilson type. An instantaneous, rather than an averaging, type of ratemeter is used because only the former correctly indicates beat-to-beat variation of heart rate: in addition, the actual foetal heart rate is more readily recognised in the presence of movement and electromyogram artefacts.

foetal movement and the effect of the lens is to attenuate the ultrasonic energy intensity reaching the foetus. Radiated energy intensities of $10 mW/cm^2$ at the transmitter surface of the Sonicaid D205 transducer and $2.5 mW/cm^2$ at the exit surface of the perspex diverging lens were measured with an ultrasound radiation pressure balance. This agrees with the theoretically calculated attenuation resulting from reflection and absorption by the lens. Because of further attenuation resulting from divergence of the beam and from absorption in the maternal abdominal and uterine walls, the maximum energy intensity reaching the foetus is estimated to be less than $0.5 mW/cm^2$. This energy intensity is very low even in the diagnostic context. An investigation just published by Watts *et al.* (1972) has confirmed the safety of diagnostic ultrasound in general and the safety of low-intensity continuous irradiation in particular. Such lenses have been successfully used both on their own and in the combined foetal heart rate uterine-contraction transducer which will be described later.

Even when the wide-angle lens is used, the output signal from Sonicaid D205 still requires considerable processing before it will reliably drive the instantaneous ratemeter. The block diagram of this

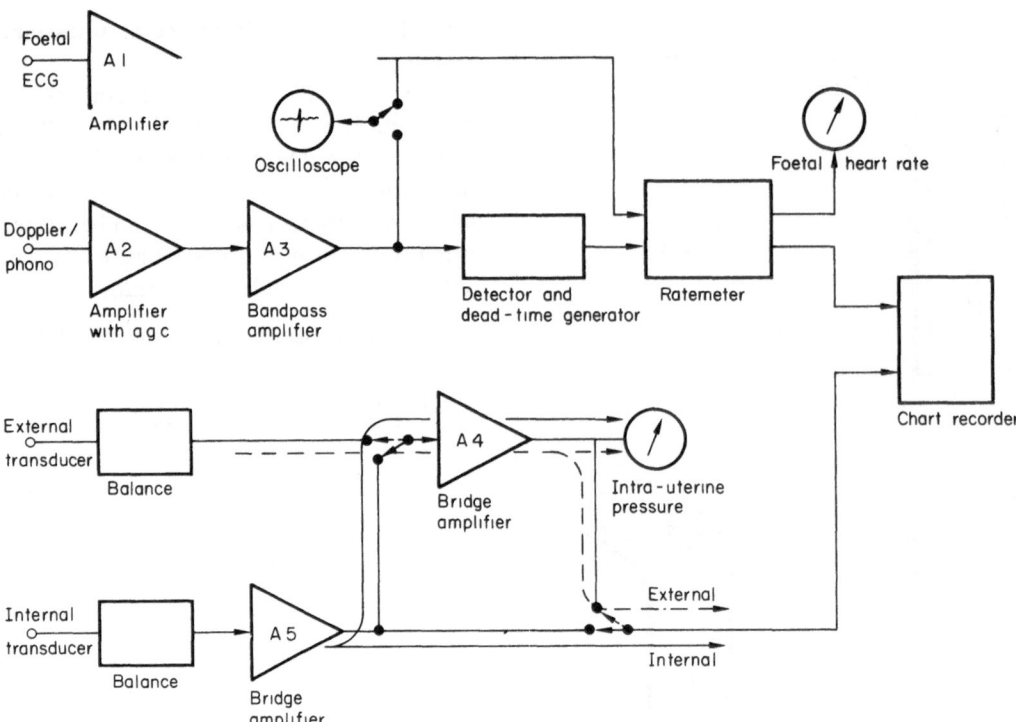

Figure 13. Block diagram of the comprehensive intrapartum monitoring system.

system is included in figure 13. In order to smooth the signal level variations which result from foetal movement and other causes, the first stage of processing is to pass the signal through an amplifier with automatic gain control. This provides a substantially constant output for an input amplitude variation of 100:1 (40 dB). The next stage of processing is a band-pass amplifier of the Wien bridge type which is tunable over the range 70–2000 Hz the bandwidth being 50 per cent of the centre frequency. Analysis of ultrasonic Doppler signals using a speech spectrograph demonstrated that frequency components in the range 100–500 Hz tended to be more distinctly related to the foetal heart cycle than components lying outwith this frequency range. The band-pass filter therefore enhances the signal-to-noise ratio, the noise in this context being, for example, foetal movements at low frequencies and maternal placental blood flow at high frequencies. The tuning facility enables the best frequency to be chosen for each patient and also leaves open the possibility of using foetal phonocardiography where the best signal-to-noise ratio occurs at about 90 Hz (Shelley, 1969).

The final stage of processing is a detector followed by a dead-time generator the need for which may be understood by considering figure 14, which shows a simultaneous record of foetal ECG and foetal ultrasonic Doppler signal. In the former, the QRS complex forms a very well-defined time marker in each cardiac cycle, whereas in the latter there are several amplitude bursts in each cycle. This presents

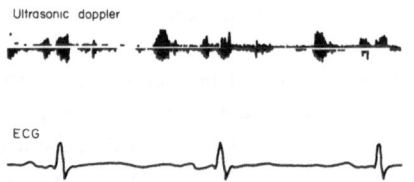

Figure 14. Tracing of a simultaneous record of foetal ECG and foetal ultrasonic Doppler signal.

to the electronic circuitry a recognition problem which is partly taken care of by the gain-controlled amplifier which sets the amplitude of the largest burst just above the detector threshold. However, there remains the possible difficulty of more than one burst in each cardiac cycle being above the threshold level. This difficulty is overcome by the dead-time generator which inactivates the detector for a period of 0·3 s after an amplitude burst has been detected. This dead-time has been chosen on a compromise basis: it defines a maximum detectable heart rate of 200 beats per minute, while at rates which are less than half this maximum, i.e. 100 beats per minute, it is conceivable to 'double-count' the signal. Although the total signal processing which is described in this paper goes far in minimising this 'double-counting' possibility, the effect remains as a fundamental limitation in the use of the foetal ultrasonic Doppler signal for recording heart rate. It is pertinent to note that similar problems with the foetal phonocardiogram signal contribute significantly to the cost and the internal complexity of the Hewlett–Packard Cardiotocograph.

Uterine contractions—direct and indirect measurement of intra-uterine pressure

Direct internal measurement of intra-uterine pressure is achieved using duplicated open-ended catheters which are saline-filled and connected via a three-way tap to a standard physiological pressure transducer. The duplication of the catheters helps to overcome difficulties of blocking and kinking because it is unusual if both catheters become non-patent and fail to respond to flushing. The pressure transducer bridge amplifier is standard (figure 13).

Indirect measurement of pressure is made using a guard-ring pressure transducer (tocodynamometer) working on the principle described by Smyth (1957). Two types of tocodynamometer have been used: one being basically similar to Smyth's original instrument, except that instead of wire strain-gauges the more sensitive semiconductor strain-gauges have been incorporated; the other being the combined transducer mentioned earlier. Both tocodynamometers have an overall diameter of 7·5 cm. The design of the combined transducer is illustrated in figure 15. The ultrasonic lens is concentric with (and in fact part of) a tocodynamometer

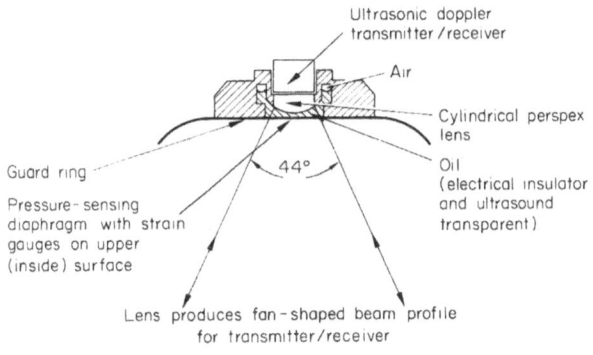

Figure 15. Diagrams of the dual-function external transducer. (U.K. patent application number 49642/71.)

which differs from Smyth's in that the central piston has been replaced by a pressure-sensitive diaphragm on which strain gauges are mounted. The space between lens and diaphragm is filled with an electrically insulating oil having the same acoustic impedance as tissue, so that the ultrasound-diverging property of the lens is preserved; the diaphragm itself is made transparent to ultrasound by virtue of its thickness being an integral number of ultrasonic half wavelengths. In the prototypes the integer was zero, i.e. the diaphragm was 'very thin'. Although this arrangement has been technically successful, it has proved too fragile from a practical

viewpoint and more robust transducers have been designed with a diaphragm thickness of one half wavelength.

A significant advantage of the diaphragm type of tocodynamometer over Smyth's original piston type is the absence of an annular space which can become clogged with debris. This feature can be exploited independently of the ultrasonic lens system and offers an important practical improvement in the design of the guard-ring tocodynamometer.

Possible directions of new development in foetal monitoring

In developing the system for monitoring the foetal heart rate and uterine contractions which has just been described, the guiding consideration has always been to give to the obstetrician or midwife information that is helpful in the management of labour. If this approach is correct, future development of foetal monitoring should seek to increase the amount of useful information available to the obstetric attendant.

With regard to monitoring of the foetal heart and its relation to foetal distress, most interest has been concentrated up to now on patterns of instantaneous foetal heart rate. Thus the custom has evolved of not displaying the foetal heart signal itself, or at least not in a form amenable to clinical evaluation. The actual shape of the foetal ECG would be expected to alter with onset of foetal distress, for example, lengthening of the S–T segment and notching of the QRS complex. We believe that it would be valuable to investigate if such signs can be used as a warning of foetal distress.

It is much more difficult to suggest ways of directly interpreting the foetal ultrasonic Doppler signal because the reflecting surfaces which give rise to this signal are difficult to identify with precision. However a significant element of this difficulty is due to the inability of many standard ultrasonic Doppler instruments to distinguish whether a surface is moving towards or away from the transducer. The reason is not one of fundamental physics, it is one of economics. If it were removed there should be little difficulty in distinguishing the signal due to diastole (myocardium moving towards transducer) from that due to systole (myocardium moving away from transducer).

When considering uterine contractions, the simplest approach is to compare the uterus with an elastic balloon and to take intrauterine pressure as the significant variable. This may not always be an adequate model. Smyth at University College Hospital, London, has described the results of making simultaneous measurements of intrauterine pressure and of force exerted sideways by the dilated cervix on the foetal head. It appears that this force often continues to be exerted for some time after the intrauterine pressure falls away. In such a case, foetal bradycardia corresponding to a type II dip when correlated with the intrauterine pressure, may in reality correspond to a type I dip when correlated with the force on the foetal head. Since the distinction is important when assessing foetal distress, this example illustrates the need for a better understanding of uterine action and how it may be measured.

In conclusion, it is our belief that the most fruitful avenues of new development in foetal monitoring are not to be found in the elaboration of existing techniques into more sophisticated display systems. Rather, they are to be found in modification of existing techniques and adoption of new ones so as to increase the relevance of the monitoring to the control of the actual foetal condition.

REFERENCES

BAUMGARTEN, K. 1969: Private communication.

CALDEYRO-BACIA, R. et al., 1961: Proceedings of the Fourth International Congress on Medical Electronics. Princeton University Press, New Jersey.

DRILLIEN, C. M. 1964: 'The Growth and Development of the Prematurely Born Infant'. Livingstone, Edinburgh.

HAMMACHER, K. 1962: Geburtsh. u Frauenhk. 22: 1542.

HON, E. H., and HESS, O. W. 1960: American Journal of Obstetrics and Gynaecology 79: 1012.

SHELLEY, T. 1969: American Journal of Obstetrics and Gynaecology 105: 597.

SMYTH, C. N. 1957: The Journal of Obstetrics and Gynaecology of the British Empire 64: 59.

TURNBULL, A. C., FRANCIS, J. G., and THOMAS, F. F. 1970: J. Obstet. & Gynaec. Brit. Cwlth. 77: 594.

WATTS, P. LOOBY, HALL, A. J., and FLEMING, J. E. E. 1972: Ultrasound and chromosome damage. British Journal of Radiology 45: 335.

SESSION 'D' DISCUSSION

Mr Longmore commenting on the papers by Mr Bain and Dr McGuinness and their co-authors emphasised the importance, in his view, of the developments described. He pointed out that cardiologists are now faced with approximately 12 per cent of patients who have myocardial infarction of whom approximately half die. Around 25 per cent of those who die do so of pump failure. It is important in intensive care units to be able to distinguish between the patients who are going to die of pump failure in an hour or two and those who are going to get better. The only clues at the moment are provided by methods which measure acceleration of blood or velocity of contraction and non-invasive techniques like the kinetograph, the ballistocardiograph and Doppler shift measurement of blood acceleration in the aorta. It was his hope that these or appropriate variants of these will become available as clinical tools as quickly as possible in their non-invasive form. On a specific point, he wished to make a plea to the research workers concerned not to forget the poor old right ventricle. 'From the time one gets up in the morning until the time one goes to bed the right ventricle will have pumped some ten tons of blood as well as the left ventricle and it is a very important structure.' He concluded by re-emphasising his view that these papers were some of the most exciting ones he has heard for a long time. *Dr McGuinness* concurred with Mr Longmore and stated that it was the possibility of an effective non-invasive technique which excited his clinical group when they were approached by the bioengineers who asked for their co-operation. They have to cope in their coronary care unit with people who have severly damaged left ventricles, and the way in which one determines the competence of a left ventricle outwith these non-invasive techniques are very invasive indeed. It means putting catheters into the left ventricle or into the left atrium or at least into the lungs. They are very encouraged by the results so far and intend to carry on with this. *Mr Bain*, associating himself with Dr McGuinness' comments expressed his thanks to Mr Longmore for his endorsement and indicated that they will most certainly press on with it. *Mr McCutchen* asked about sonar cardiography. *Dr MacVicar* indicated that he believed investigations of this kind are being carried on, to quite an extent as clinical routine, in the Western Infirmary in Glasgow and gave very valuable information. *Professor Jacobs* commented that the main difficulty with sonar techniques is the need of alignment of the sonar path. The beauty of the kinetographic tech-

nique is that physical alignment does not have to be so accurate.

Mr Hanka asked Dr Macfarlane regarding his figures of 4 per cent 'serious misdiagnosis' and $4\frac{1}{2}$ per cent as 'less serious'. These figures in his view meant that every tenth patient has been misdiagnosed. On what basis are figures of this kind judged to be satisfactory or otherwise? *Dr Macfarlane* emphasised that one has to be very clear as to what was meant by the $4\frac{1}{2}$ per cent 'less serious misinterpretations'. As stated in the paper, the computer perhaps reported results consistent with myocardial ischaemia, which was correct, but nevertheless it only detected ischaemic changes in perhaps one lead instead of two or three; this is a less serious error and not an exceptionally significant one. As regards the 4 per cent serious errors, this must be considered by comparing it against the figures which would be obtained from a spectrum of physicians, say from consultant cardiologists to senior house officers, who would report on electrocardiograms as past of the routine service. Examination of such data indicated that a 4 per cent 'error' is quite acceptable. *Mr Hanka* then addressed a question to Dr Healey on cluster analysis. *Dr Healy* indicated that he did not believe in cluster analysis and considered it a failure. *Dr von Gierke* asked Dr Bengi if he made any attempt to separate passive ocular tremor due to the circulation from what he would call active tremor induced by muscle action. Under vibratory conditions, resonances in vision and blurred vision can be observed at about the same frequencies. He could not believe that active muscular action played any role in these resonances, so it looked to him as if the peak in the resonances is determined by the mechanical properties of the eye. If this is the case then the resonance of the very complex eye structure could very well be changed by just changing the gravity factor on the eye. For example, it is known that the eye is deformed under acceleration conditions. He thought it significant and drew Dr Bengi's attention to the fact that if subjects are vibrated, these resonances are observed at around 70 Hz, which is in the range of the second peak of the spectrums shown by Dr Bengi.

Dr Bengi indicated that from the studies detailed in their paper they have concluded that the first peak in the spectrum is a resonance effect related to the eyeball muscle mechanics. The second peak, however, appears to originate from excitation of the otolith in the labyrinth which is very sensitive to movement. In this connection it is relevant to note that when measuring the tremor with the subject lying and with the subject sitting there was a change in the spectrum. *Mr McCutchen* asked Dr Bengi to comment on the fact that tremor of the eye is believed to be necessary in order to see. Was this a form of programmed hunting, because he believed that if the image is fully stabilised relative to the eye, eventually the image is not seen. *Dr Bengi* referred to statements that tremor is a sort of scanning effect and that it plays part in vision. He quoted certain experiments which they carried out to assess whether this was true or not. The first of these was to measure the tremor with the eye illuminated and not illuminated. Both in the dark and light eye situations the tremor was exactly the same. In the second set of experiments they deviated the eye from the primary position by amounts ranging from 0 to 30° in both the horizontal and vertical positions. The spectrum changed with the deviation. Thus as tremor does not change with the intensity of light but changes with position of the eye, the suggestion that tremor has something to do with vision is contradicted.

Dr Kerr Grieve, discussing his methods of obstetric management, referred to a basis of diet control by routine estimation of haemoglobin level (without iron supplement) and weight gain, and the use of a high protein starch free diet (one pound of meat per day) in the prevention of what he called 'gestational failure' (pre-eclamptic toxaemia, abruptic placentae, uterine dysfunction, 'unexplained' intra-uterine death, first stage foetal distress, prematurity and the respiratory distress syndrome). He expressed the conviction that control of gestation (diet and duration) was more important to the health of the mother and survival of the baby than the management of parturition even with the most sophisticated modern bioengineering devices.

Control of gestation was indeed the province of the obstetrician and a more rewarding field of operation for the experimental bioengineer. If this was effective then the management of parturition could be safely left to that expert obstetric physician, the well-trained midwife. There would remain only a few cases which might require special monitoring during parturition and a very few requiring expert surgical intervention by the experienced obstetrician, cases, where failure of gestational control had to be admitted.

In support of this belief Dr Grieve quoted a brief analysis of 10 000 consecutive deliveries (booked and unbooked cases) at Motherwell Maternity Hospital during the years 1960–69. The figures are as follows:

Consecutive deliveries booked and un-booked	10 000
Consecutive surgical inductions booked and unbooked	7 000

Pre-eclampsia (Diastolic blood pressure of 90 on 2 or more occasions)	4%
Pre-eclampsia with albuminuria	0·4%
Caesarean section rate—all cases	0·9%
Caesarean section rate after induction	0·15%

Perinatal death rate consistently below the Scottish average and despite a high incidence of anencephaly and pre-viability; sometimes comparable with the lowest in Britain.

It was a pity that the 1958 British Perinatal Survey had not included data related to dietary performance during gestation. These would have shown a certain relationship between protein deficiency and/or water retention and the various categories of maximum risk parturient such as the grandes multipares, young and old primigravidae, the undersized and the under privileged.

There were two problems for the bioengineers arising out of this preventive approach to childbirth: to devise some method of measuring tissue 'quality' during the process of gestational change so as to distinguish the 'good' from the 'bad' and thus assess the efficacy of diet control; and to produce a monitoring system applicable to the foetus *in utero*, preferably without the use of intra-uterine devices or needles, so as to be able to measure haemoglobin level, weight gain and blood pressure during gestation. It is quite likely that the reaction of the baby to dietary deficiency is similar to that of the mother in toxaemia and it is significant that the risks to the baby seem to be present especially when the mother fails to react. Indeed, at present the risk to the baby can only be surmised where the mother appears to be fully compensated despite an inadequate intake of protein.

If the baby could be monitored as the mother can be at present, timely preventive measures could be instituted including induction of labour.

In conclusion, successful dietary education of the expectant mother especially if widely accepted could well be passed on to the children who would then grow up to a life free from such grave degenerative vascular disorders as coronary artery thrombosis, chronic hypertension and gastro-duodenal ulceration and, of course, gestational failure.

Thus in posing the problems to his bioengineering colleagues Dr Grieve was making a plea for Preventive Bioengineering which in his view would make a positive contribution to obstetric management.

Dr Torbet commented that Dr Kerr Grieve's overall figures for Caesarian section were the same as that of his group for *placenta previa*. This suggested a specific way of treating this particular condition and concluded by saying that all that they were trying to do was to emulate Dr Kerr Grieve's results in the best way they know how by using the bioengineering methods and techniques produced by his colleague.

SESSION 'D' SUMMARY

D. W. HILL[1]

The Session was opened with some very relevant observations by Professor Bekkering of Utrecht. He made the point that it is necessary for the biomedical engineer to be able to extract as much of the information as possible required by the clinician from the limited number of signals which may be available from the patient. He illustrated his theme with the analogue processing unit designed by Ir. Wesseling and his colleagues. This produces from a central aortic pressure signal a number of outputs including the heart rate, systolic, diastolic and mean pressures, and an indication of the stroke volume and cardiac output.

Dr McVicar followed with a revealing account of the use in obstetrical practice over a number of years in Glasgow of ultra-sound. Applications had included the confirmation of *placenta previa*, the diagnosis of multiple pregnancies and the indication of the growth of the foetal skull. It is hoped that ultrasound will be able to confirm at an early stage that a pregnancy will be successful. The technique has much to recommend it, in that it is both non-invasive and does not have the radiation risk to the foetus associated with X-radiation. In some cases, the beating foetal heart has been detected as early as seven to eight weeks.

Mr Bain described the non-invasive technique of kinetography for the evaluation of myocardial movements and the basic work being undertaken in animals to correlate the kinetography signals with physiological events. The clinical prospects for kinetography were well illustrated by Dr McGuinness who showed that it appeared to have a considerable application for the routine screening of patients for possible myocardial infarction. Although the quantitative aspects of non-invasive methods for estimating cardiac output and contractility are still in doubt, there is an urgent need to investigate further kinetography and other techniques such as ultrasound and electrical impedance techniques.

Mr Manley confirmed that Strathclyde is still much interested in ballistocardiography and described the new light-weight apparatus based on a carbon fibre material table top and air bearings. Ballistocardiography should supply valuable supporting evidence in the evaluation of non-invasive methods for the monitoring of cardiac function.

Dr Macfarlane made it obvious that Glasgow is ahead in the U.K. in the computer processing of electrocardiograms and the sharing of the equipment as a service to a number of hospitals.

Mr Healey underlined the need to have a firm statistical treatment of the large amounts of data which could be obtained by the use of current measuring techniques.

Miss Jordan reinforced this with a description of the data processing system developed at Strathclyde and which put adequate statistical processing within the reach of all the research staff on a routine basis.

Mr Perkins described the work of his team on displaying techniques for use with computers and underlined that it should be possible for the user to be able readily to change the display format to suit his particular needs.

Dr Bengi's account of the measurement of human ocular tremor gave an insight into the development of instrumentation which can be used to measure and record the motion of delicate organs such as the eye whilst allowing the subject a reasonable freedom of movement. In many ways this type of study brings out the eventual skills required of a bio-engineer.

The last contributions by Drs Torbet and Thomas emphasised to the full how an obstetrician and a physicist can work closely together to develop means of foetal monitoring which have not only reduced the risk of foetal mortality but proved feasible to use in the arduous conditions of the labour room.

The wide variety of subjects covered in this last session serve to reinforce the fact that biomedical engineering has really come of age and that investigations of many different disciplines have learned how to work effectively together.

---

[1] Institute of Basic Medical Sciences, University of London, London, England.

SUMMARY REPORT OF THE SYMPOSIUM ADDRESS[1]

'Bioengineering—a many splendoured thing—but for whom?'

H. S. WOLFF[2]

After expressing his pleasure at being with the Symposium participants and guests, Mr Wolff commenced his lecture by exhibiting himself (figure 1) looking perplexed about why in the first instance he had agreed to give a lecture at all with the title quoted. He then went on to deplore the way

engineering is taking the human touch out of medicine. His figure 2 (of a scantily dressed female with adoring male in classical garb) illustrated how, in his view, a civilised outpatient department should be conducted, providing that immediacy of touch between patient and operator which gives such

Figure 1.

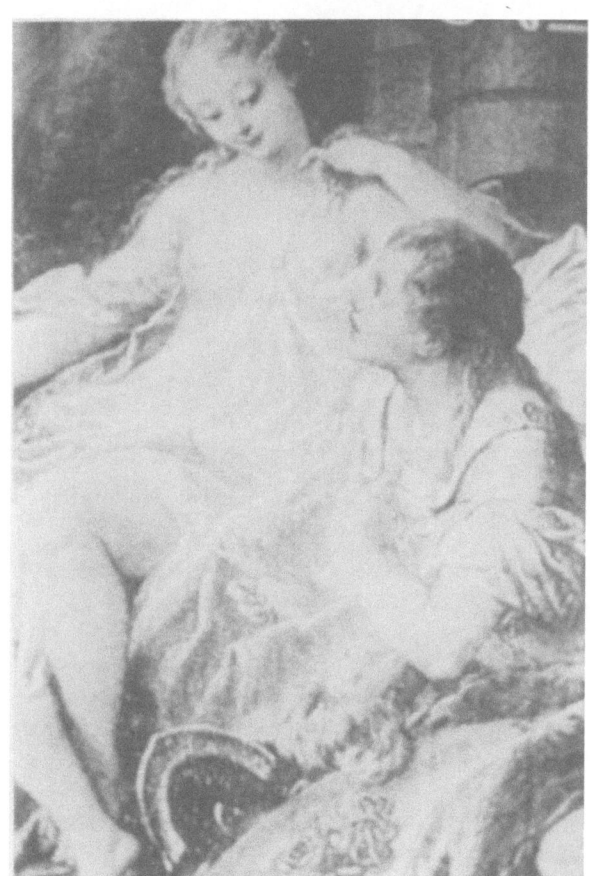

Figure 2.

[1] *Editorial Note.* On the evening of Monday, 19th June, the Western Regional Hospital Board in association with the University of Strathclyde, gave a Reception for participants and invited guests. The Chairman of the Western Regional Hospital Board, Mr Simpson Stevenson, and Mrs Stevenson, received the guests. Strathclyde was represented by the Deputy Principal of the University, Professor A. S. T. Thomson and Mrs Thomson. Mr Wolff delivered his Symposium Address following the Reception to an audience of some 400 people. The Address, a witty and entertaining discourse, was most enthusiastically received.
[2] Bioengineering Division, Medical Research Council, Clinical Research Centre, Northwick Park Hospital, London, England.

obvious pleasure to both (figure 3). This he compared with figure 4, in which an operator, totally impersonal, looks at a patient disembodiedly transformed into readings on a monitoring machine.

One of the reasons why technology was not making as much headway in clinical practice as had been hoped, was a series of misunderstandings and underestimations of human capability. Figure 5 illustrated the '1600' effect, which intended to

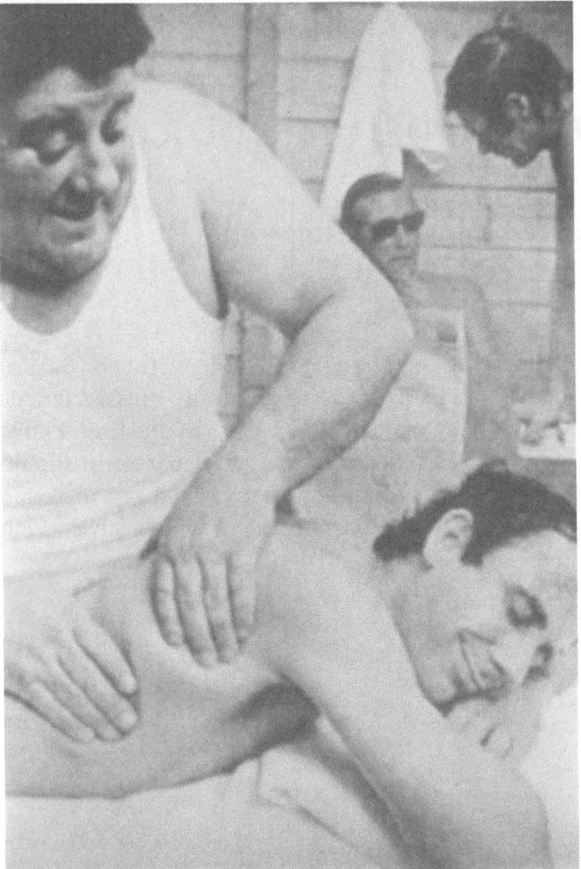

Figure 3.

Figure 4.

M   C   C   C   C   M
I   H   A   H   L   I
N   A   P   of   A   a   I   N
I   N   A   N   N   D
M   G   B   G   I
U   E   L   I   C
M   E   N   I
                                   G   A
                                                  N
                                                  S
                                                  .

The 1600 effect

Figure 5.

convey that the accuracy and resolution of common measuring instruments should be matched to the level required by the actual treatment. It was a common pitfall to provide instruments with capabilities much beyond this, producing, in fact, an embarrassing richness of meaningless information.

He then discussed the case of two doctors, good and less good (naturally there are no 'bad' doctors!) who as 'input' received the same amount of measured data of the same accuracy. The 'less good' made far less of this as he did not have the sensitivity to pick up the spectrum of impressions through his senses as the 'good' doctor who possesses superior central processors. Figure 6 showed the amount of

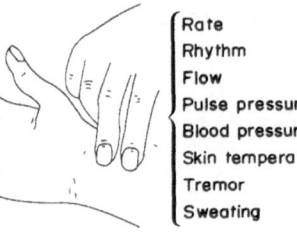

Rate   Cardiotachometer
Rhythm   Arhythmia computer
Flow   Cardiac output
Pulse pressure }   Systolic/diastolic
Blood pressure }   monitor
Skin temperature   Surface temp. meter
Tremor   Tremor monitor
Sweating   Skin resistance meter

Figure 6.

qualitative information that a physician might obtain via his educated finger on the patient's pulse and demonstrated how wrong the naive engineer was in thinking that a rate meter was a complete substitute. The instruments listed on the right hand side of figure 6, which represented the instrumental equivalents necessary to provide the information obtained by the physician's finger, almost exhausted the range available to patient monitoring at the present time.

Figures 7 to 11 presented evocative expressions or situations to illustrate how, even in normal usage, we had learned to pick up very quickly quite complex impressions. Figure 7 conveyed immediately to any onlooker a state of puzzlement. There was, however, neither puzzlement nor doubt regarding the message conveyed by the seductive young lady shown in a state of semi-'retirement' in figure 8. It was noteworthy that such interpretations were learnt and were culture-dependent: the facial expression of the Geisha girl shown in figure 9 appeared 'inscrutable' to occidental eyes yet on unimpeachable Japanese authority Mr Wolff assured his audience that the Geisha girl conveyed precisely the same message in content, implication and unequivocality as the young lady of figure 8. Illustrating a more complex inference, he showed figure 10 and left the audience to draw their own conclusion. Even in stylised features our learnt processing ability discovered personality, as demonstrated by a study of the picture of his favourite teddy bear (figure 11).

Following this, Mr Wolff commenced to ride

Figure 7.

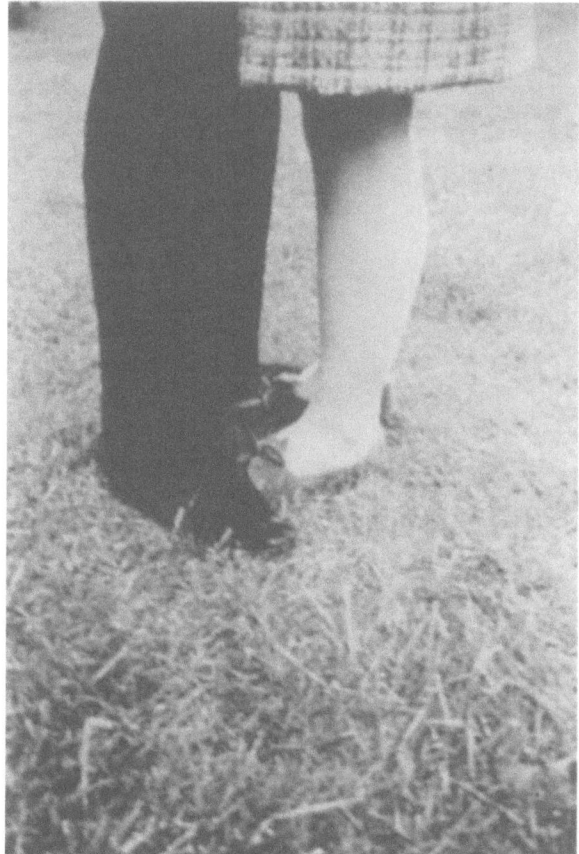

Figure 8.

Figure 9.

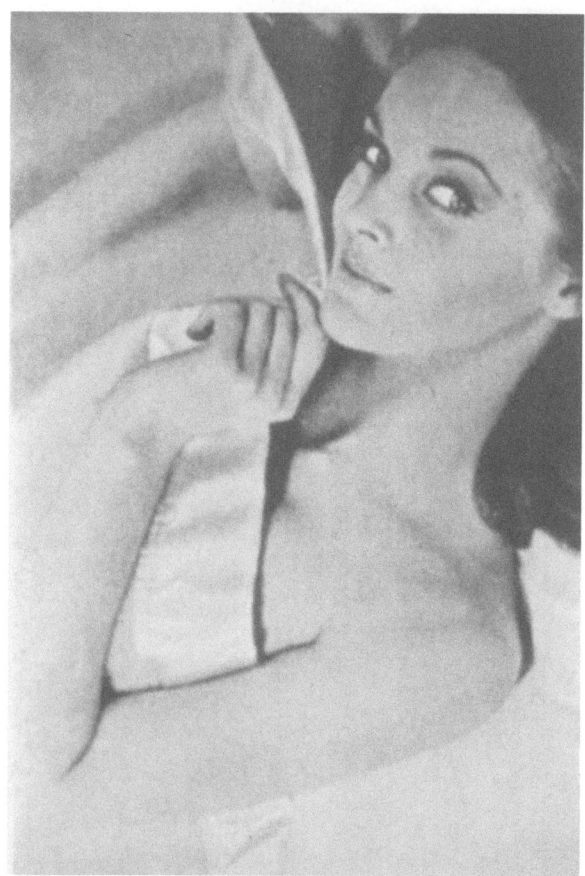

Figure 10.

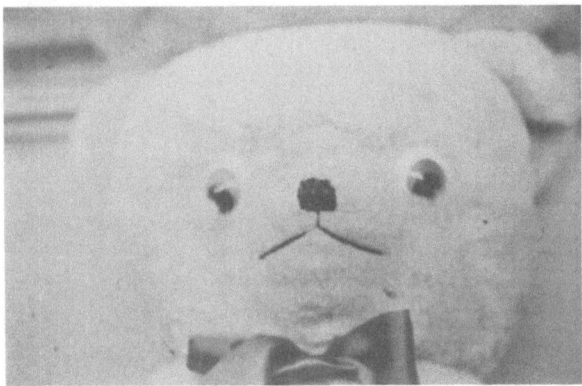

Figure 11.

Figure 13.

two of his hobby horses. The first was the comparative neglect of the study of normal man in modern environment. The drastic environmental transition that *homo sapiens* had to undergo in a very short time span not permitting of evolutionary adjustment was illustrated by figures 12 to 15 showing respectively: a primitive man chipping flint; the close juxtaposition of office workers; primitive man, this time kindling a fire, thereby initiating a process which culminated, as shown, in New York becoming smog-bound. In addition to such environmental stresses, modern man inflicted

Figure 14.

Figure 12.

Figure 15.

upon himself stresses arising out of his vices (figure 16). The vicissitudes of modern life during work and play were being investigated by instruments developed in Mr Wolff's division, and known as SAMI's (socially acceptable monitoring instruments). These could meter and record a variety of parameters such as blood pressure, heart

Figure 16.

rate, respiration rate, electrocardiograms, electromyograms etc., over extended periods. The latest of these, easily carried in a pocket, was a multichannel, sub-miniature tape recorder, shown in figure 17 compared with a pair of ordinary spectacles.

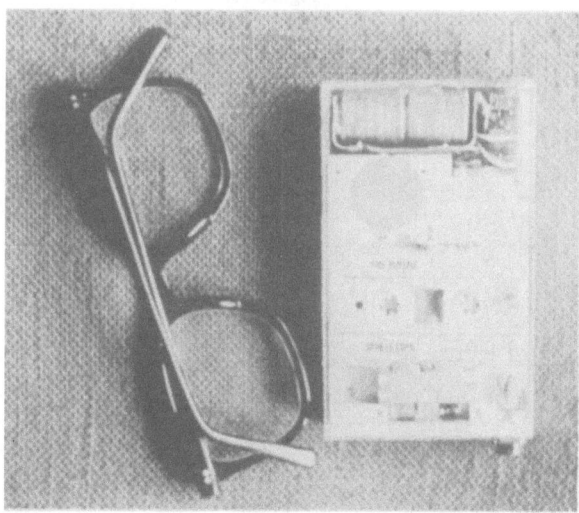

Figure 17.

Mr Wolff's second hobby horse was that of utilising minor technology for the care of the old and slightly disabled. He recommended particularly for the attention of the bioengineering members of his audience, project URINE, in other words Unexciting Research Into Necessary Equipment and quoted the following data on the pensionable age distribution in England and Wales in 1969: total of pensionable age 7·75 million, 15·8 per cent of the population. Of these, some 1 million were over 80 years of age.

The data available indicated that some 30 per

cent lived with their children, 15 per cent near their children but separately from them, 9 per cent with relatives, 10 per cent near relatives but separately from them, 6 per cent in residential homes and 30 per cent of the aged population totally by themselves. The problem of looking after the aged (who, because of smaller families and greater population mobility were increasingly being stripped of the support previously supplied by the 'extended family') was going to be one of the major social problems of the future. Technology could obviously not be the *complete* answer but it could be used in the following ways:

(i) to extend the independence of the people concerned, particularly if living by themselves or living in couples of comparable age;

(ii) to lessen the social load which aged dependents imposed on young families with whom they lived; for example an incontinent grandmother could be destructive of family life yet her social value could be changed from minus to plus by providing her with some anti-incontinence device which could then convert her into a 'built-in' baby-sitter;

(iii) to fashion tools which would make it possible to harness a greater number of people to the care industry who might otherwise be unsuitable for it because of lack of motivation or general personal qualities. He outlined as an example how unpleasant jobs, which few people were prepared to undertake, might be converted into ones where the smell, dirt etc. was mitigated, thus making people available to operate the devices that did the 'dirty' jobs and, at the same time, supply human contact to the aged.

This led to a general consideration of the care industry as being possibly one of the only two industries (the other being that of leisure) which would remain labour-intensive in the future. It seemed to him that labour intensivity would require to be carefully husbanded as a national resource because it had been shown that the socially erosive effects of unemployment were much more serious than those of low average productivity. This should reflect itself both in *political attitude*—because it really made no financial difference whether an individual was paid unemployment benefit or paid a salary as a functionary in the care industry—and in *education*, which should be orientated towards giving people the kind of training which the care industry required.

Many bioengineers tended to pick their projects because of the intellectual challenge which their complexity presented rather than because of their utility in medical care. Yet simple-appearing devices might be far more complex than they appeared at first sight. The pair of tongs shown in figure 18, which

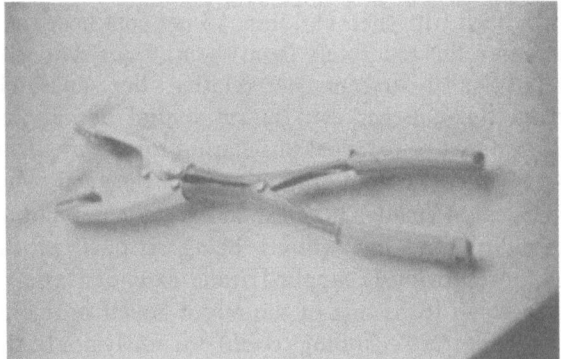

Figure 18.

could be used by wheelchair or bed-bound patients
to make their arms a foot longer, was a case in
point. The mechanical advantage had to be right so
as to transmit feel, the jaws had to be made of the
right shape and material so as to pick up a wide
variety of objects without slipping. The material had
to be light and stiff. Additionally, a really universally
applicable device might have an optional ratchet
so that patients with limited movement in their
fingers could add up small movements to produce
a big one and the handles might be made out of an
uncured thermo-setting plastic, which could then
be moulded to the hand of the patient and cured
in a domestic oven. Thus a *complete comprehensive
design* was as intellectually worthy as a project
involving complex electronics, for example. He
felt that Universities deserved a degree of chastise-
ment, as they tended to take a rather restrictive
view of the subjects suitable for Ph.D. theses.
Their spectrum, he felt, could be usefully extended
to the kind of project that he had just described.

There were further very simple devices which
could make a very considerable contribution. For
example, figure 19 showed a teapot on a tiltable plate
which avoided having to support the full weight of
the pot while pouring a cup of tea. This, in itself,
obviously made no very large contribution to independence, but the creation of the correct psychological climate to allow, for example, a disabled
mother to continue to perform a duty symbolic of
'mother', was an important consideration. Figure
20 showed an inflatable bag placed on the seat
of a chair which could raise grandad from his
favourite armchair by the application of low-
pressure air from the domestic vacuum cleaner,
activated by a self-cancelling timeswitch on the arm
of the chair. There was merit in considering using
such low-pressure air for other forms of person-
'handling', because of its safety and gentleness.
Incidentally, this particular application might also
make a larger contribution to the welfare of the
patient than at first appeared, because incontinence
could be due to the mere physical inability to get up
in time to reach the loo. Another very important

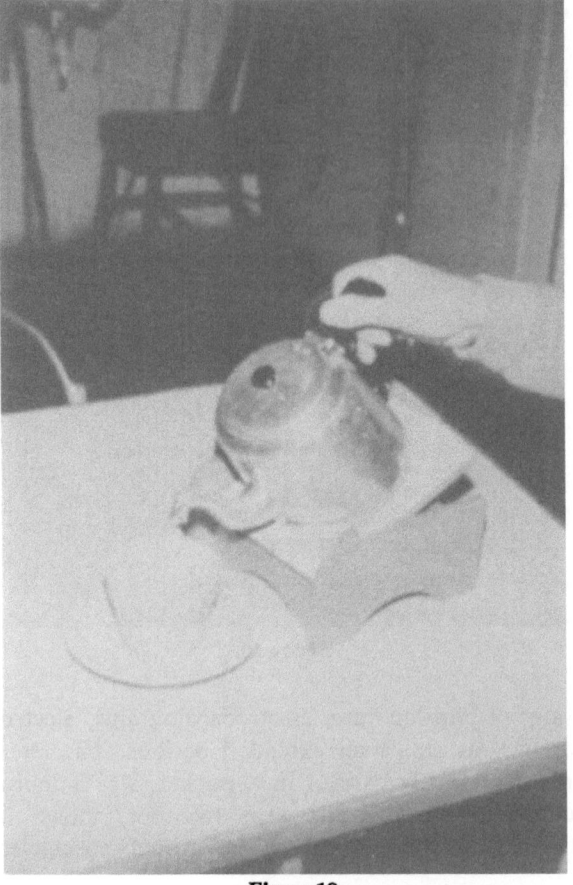

Figure 19.

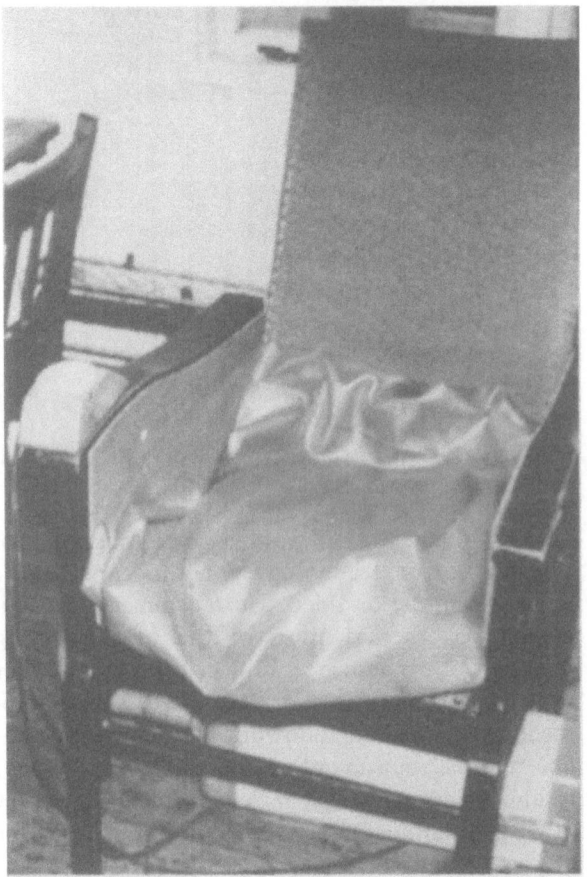

Figure 20.

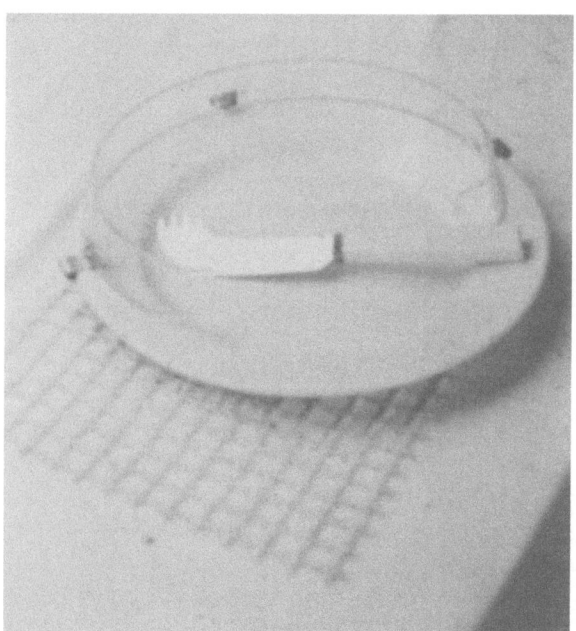

Figure 21.

range of aids were those designed for one-handed eating. Figure 21 showed a plate with a rim, on a non-slip mat, and a knife which cut by rocking on its curved edge rather than sawing. The tooth end of the knife was used as a fork. Aids of this kind played a very significant part in the maintenance of the disabled adult's self-respect.

Mr Wolff terminated his address by referring again to impression conveyance and receptance. As a parting shot, he wanted to convey clearly and beyond doubt his determination to get something done in the area of both of his hobby horses. He could think of no better way of doing this than by associating himself with the unspoken resolution expressed by the individual shown in figure 22.

Figure 22.

INDEX